Lecture Notes in Mathematics

Edited by A. Dold and B. Eckmann

836

Differential Geometrical Methods in Mathematical Physics

Proceedings of the Conferences Held at Aix-en-Provence, September 3 – 7, 1979 and Salamanca, September 10 – 14, 1979

Edited by
P. L. García, A. Pérez-Rendón, and J. M. Souriau

Springer-Verlag
Berlin Heidelberg New York 1980

Editors

P. L. García
A. Pérez-Rendón
Seccion de Matematicas
Facultad de Ciencias
Universidad de Salamanca
Salamanca
Spain

J. M. Souriau
Centre de Physique Théorique
Luminy
Case 907
13288 Marseille, Cédex 2
France

AMS Subject Classifications (1980): 53-06, 55Rxx, 57Rxx, 81–XX, 83Cxx

ISBN 3-540-10275-2 Springer-Verlag Berlin Heidelberg New York
ISBN 0-387-10275-2 Springer-Verlag New York Heidelberg Berlin

Printing and binding: Beltz Offsetdruck, Hemsbach/Bergstr.
2141/3140-543210

TABLE OF CONTENTS

PART I Proceedings of the International Colloquium of the C.N.R.S. Held at Aix-en-Provence, September 3 - 7, 1979 Edited by J.M. Souriau

*) References: 1) M. Casartelli, G. Marchesini and E. Onofri, J. Phys. A; Math. Gen. 13 (1980) 1217-1225

2) G. Marchesini and E. Onofri, J. Math. Phys. 21 (1980) 1103-1110

3) E. Onofri, "A Modified Bars-Durgut equation with polynomial eigenfunctions", J. Math. Phys. (in press).

Chapter III. Deformations of Lie Algebras

Chapter IV. Classical Field Theory

Chapter V. Supersymmetry and Supergravity

**) References: D. Arnal et J.C. Cortet - J. Math. Phys. 20, 556-563 (1979)
D. Arnal et J.C. Cortet - "La Notion de *-produit et ses applications aux représentations de groupe", Journées Relativistes 79, p. 27-45, Université d'Angers (France).

PART II Proceedings of the Conference Held at Salamanca, September 10 - 14, 1979

Edited by P.L. García and A. Pérez-Rendón

Chapter IV. Classical Field Theory and Analytical Mechanics

PART I

Proceedings of the International Colloquiu

of the C.N.R.S.

Held at Aix-en-Provence

September 3 - 7, 1979

Edited by J.M. Souriau

INTRODUCTION

Differential geometry plays, in contemporary physics, a very special role; thus an abstract theory such as that of connections is important for the analysis of cosmological models as well as the fundamental interactions of elementary particles; whereas the scales of these phenomena differ by 60 orders of magnitude. The hopes of unification which gave birth to gauge theories give to the idea of symmetry - that is : the theory of groups - a new role : symmetry is no longer just a particular property of objects, it becomes progressively the object itself - or at least one of the terms of a duality which will perhaps be our most useful way of viewing matter.

The intense activity in research in this area has been the motivation for many meetings of which we may be permitted to recall those held in Aix-en -Provence in June 1974 (Coll. Intern. C.N.R.S. N°237, éditions C.N.R.S.), in Bonn in 1975 (Springer Lecture Notes in Mathematics, 570) , in Warsaw in 1976 (Reports in Math. Physics), in Bonn in 1977 (Springer Lecture Notes in Mathematics, 676).

In this volume are published articles from the International Colloquium of the C.N.R.S. (Centre National de la Recherche Scientifique), held in Aix-en-Provence from 3rd to 7th of September 1979; this colloquium was organized with the assistance of the University of Provence and the Centre for Theoretical Physics in Marseille. The organizing committee consisted of K. Bleuler, A. Lichnerowicz and myself. A meeting on the same subject took place in Salamanca (10-14 September) immediately afterwards, and will be published conjointly with that of Aix.

I wish to thank all those who have helped me, both the members of the organizing committee and in the preparation of this work, especially C. Duval, J. Elhadad and N. Jean.

J.M. Souriau

Editor

List of participants

C. ABBATI (Milano - Italie)
J.P. ANTOINE (Louvain la Neuve
D. ARNAL (Dijon - France)
H. BACRY (Marseille - France)
A. BANYAGA (Harvard - USA)
M. BAUHAIN (Paris - France)
S. BENENTI (Torino - France)
L. BIBCO (Belgique)
K. BLEULER (Bonn - RFA)
F.J. BLOORE (Liverpool - Grande Bretagne)
J. BREUNVAL (Marseille - France)
G. BURDET (Dijon - France)
M. CAHEN (Bruxelles - Belgique)
M. CHEVALIER (Caen - France)
G. CICOGNA (Pisa - Italie)
A. COHEN (Villetaneuse - France)
J.C. CORTET (Dijon -France)
S. DESER (Waltham - USA)
D. DUBORGEL (CEA - France)
C. DUVAL (Marseille - France)
J. ELHADAD (Marseille - France)
S. FERRARA (Rome - Italie)
M. FLATO (Dijon - France)
H.H. FLICHE (Marseille - France)
M. FRANCAVIGLIA (Torino - Italie)
Y. GEORGELIN (Orsay - France)
A. GROSSMANN (Marseille - France)
D. GUTKIN (Lille -France)
S. GUTT (Bruxelles - Belgique)
H. HESS (Berlin - RFA)
P. HORVATHY (Marseille - France)
P. IGLESIAS (Marseille - France)
Y. ILAMED (Yavne - Israel)
A. JADCZYK (Wroclaw - Pologne)

./.

M. JASPERS (Liège -Belgique)
G. KAISER (Lowell - USA)
H. KERBRAT (Lyon - France)
J. KIJOWSKI (Varsovie - Pologne)
B. KOSTANT (M.I.T. -USA)
H.P. KUNZLE (Edmonton - Canada)
A. LICHNEROWICZ (Paris - France)
J. LUKIERSKI (Wroclaw - Pologne)
L. MARQUEZ (Bordeaux - France)
C.M. MARLE (Paris - France)
F. ONGAY (Lyon - France)
PHAN MAN QUAN (Villetaneuse - France)
E. ONOFRI (Parma - Italie)
M. PERRIN (Dijon - France)
G. PETIAU (Paris - France)
J.F. POMMARET (Boulogne - France)
J. RAWNSLEY (Coventry - Grande Bretagne)
J.L. RICHARD (Marseille - France)
D. SIMMS (Bonn - RFA)
M. SIRUGUE (Marseille - France)
M. SIRUGUE-COLLIN (Marseille - France)
F. SOLER (Paris - France)
J.M. SOURIAU (Marseille - France)
A. SPARZANI (Milano - Italie)
D. STERNHEIMER (Dijon - France)
S. SUMMERS (Marseille - France)
D. TESTARD (Marseille - France)
L. TISZA (M.I.T. USA)
R. TRIAY (Marseille - France)
M. TULCZYJEW (Calgary - Canada)
J.A. WOLF (Berkeley - USA)

Configuration spaces of identical particles

F.J. Bloore

D.A.M.T.P., The University, Liverpool L69 3BX

Abstract

We define the configuration space $C_m(M)$ of m identical particles moving on a manifold M and give several examples. We indicate how the cohomology groups $H^q(C_m(M), Z)$ may be calculated, and compute $H^2(C_3(R^n), Z)$.

Résumé

Nous définissons l'espace de configuration $C_m(M)$ de m particules identiques chacune avec son espace M et nous donnons plusieurs exemples. Nous indiquons dans quelle façon les groupes cohomologiques $H^q(C_m(M), Z)$ peuvent être calculés, et nous trouvons $H^2(C_3(R^n), Z)$.

1. Definition and motivation

Consider m particles, each with configuration space M. If the particles are distinguishable, then the configuration space of the system is the Cartesian product M^m, whose elements are ordered m-tuples of points of M.

If they are indistinguishable then the configurations $(p_1,\ldots,p_m)$ and $(p_{\pi(1)},\ldots,p_{\pi(m)})$, $\pi \in S_m$, S_m = symmetric group, are the same and the configuration space of the system is the quotient $C_m(M) = M^m/S_m$, sometimes called the symmetrised product space. We shall discuss later whether to include configurations in which the p_i are not all distinct.

In the Schrödinger representation, the wave functions are sections of a Hermitian line bundle with connection over $C_m(M)$ (or of a Hermitian vector bundle with connection in Yang-Mills theory).

In Souriau's book [1] , he shows that if one has m copies of a line bundle Y over a simply connected space M, there are two ways to take the quotient of Y^m to get a line bundle over $C_m(M)$, corresponding to the two characters $\chi_{\pm}$ of S_m. One obtains bosons for χ_+ and fermions for χ_-. We investigate here whether any other Hermitian line bundles exist over $C_m(M)$. If so, a quantum system could have configuration space $C_m(M)$ and so look like a system of m identical particles but not actually be composed of m identical particles. We find this is not possible - there are no other bundles.

A line bundle L over $C_m(M)$ is classified up to isomorphism of line bundles by its Chern class $c_2(L)$ which is an element of $H^2(C_m(M), Z)$. Higher dimensional vector bundles are partially classified by their higher Chern classes - elements of $H^{2q}(C_m(M), Z)$ - but a full classification requires the classifying space BS_m which has not yet been completely described. In fact the mathematicians use C_m as a tool to get information about BS_m - studying C_m is the easier problem! [2, 3].

2. Diagonals

An important question is whether to include those elements $\{p_1,\ldots,p_m\}$ in $C_m(M)$ in which the p_i are not all distinct. Intuitively, classical point particles should not be forbidden to collide. However, if we include these "diagonal" elements,

- if dim $M = 1$, then $C_m(M)$ is a manifold with boundary, the boundary being the diagonals,
- if dim $M = 2$, then $C_m(M)$ is a manifold
- if dim $M = n > 2$, then $C_m(M)$ is not a manifold; neighbourhoods of diagonal elements are not diffeomorphic to subsets of R^{nm}. Hence

in this case we must exclude diagonal elements in the definition of $C_m(M)$ to ensure we have a manifold. To see this last fact take $m = 2$ and consider a point O in the diagonal Δ of $M \times M$. A neighbourhood of O in $M \times M$

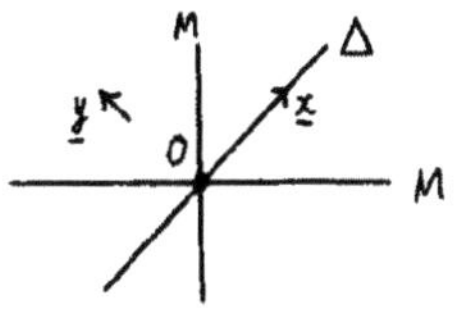

is diffe omorphic to a neighbourhood of the origin in $R^n \times R^n$.

Take local coordinates $x^1 \ldots x^n$ "along" Δ and $y^1 \ldots y^n$ "perpendicular" to Δ, i.e. so that O is the origin, Δ is the surface $\underline{y} = 0$ and the permutation $(p, p') \to (p', p)$ sends $(\underline{x}, \underline{y})$ to $(\underline{x}, -\underline{y})$. In the quotient space $C_2(M)$, we must identify $(\underline{x}, \underline{y})$ with $(\underline{x}, -\underline{y})$, so the neighbourhood of O in $C_2(M)$ is diffeomorphic to a neighbourhood of the origin in $R^n \times (R^n/S_2)$ where the first factor is the $\underline{x}$-part, and the second is the $\underline{y}$-part. A neighbourhood of the origin in R^n/S_2 is a cone whose vertex is the origin and whose base is $S^{n-1}/S_2 = P^{n-1}R$. To be coordinatisable it must be a cone with base S^{n-1}. However,

$$H_1(P^{n-1}R) = Z \text{ if } n = 2, \quad = Z_2 \text{ if } n > 2 \qquad \text{and} \qquad H_1(S^{n-1}) = Z \text{ if } n = 2, \quad = 0 \text{ if } n > 2.$$

Thus for $n > 2$ these cones are not even homeomorphic, let alone diffeomorphic. If $n = 2$, regard $\underline{y} = y^1 + iy^2 \in \mathbb{C}$, and then a diffeomorphism between the cones is given by the map $\underline{y} \to \underline{y}^2$. Here are some examples of $C_m(M)$ with dim $M = 1$ or 2 and the diagonals included; the groups $H^2(C_m(M), Z)$ are listed alongside.

3. Examples (i) $C_2(R)$ = half-plane with edge 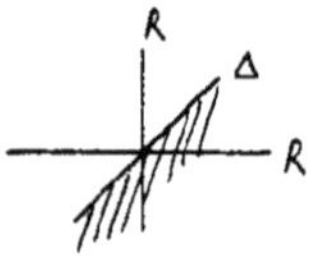

$H^2 = 0$.

(ii) $C_2(S^1)$ = Möbius band with edge 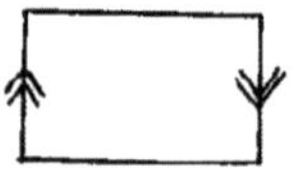 $H^2 = 0$

[Fold over the torus $S^1 \times S^1$, 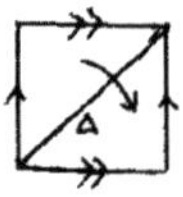

to get 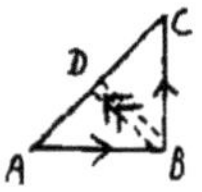

a triangle ABC with AB identified with BC. This space may be reassembled to produce the usual picture of the Möbius band by cutting along DB and attaching triangle ABD to triangle BCD using the identification of AB with BC.

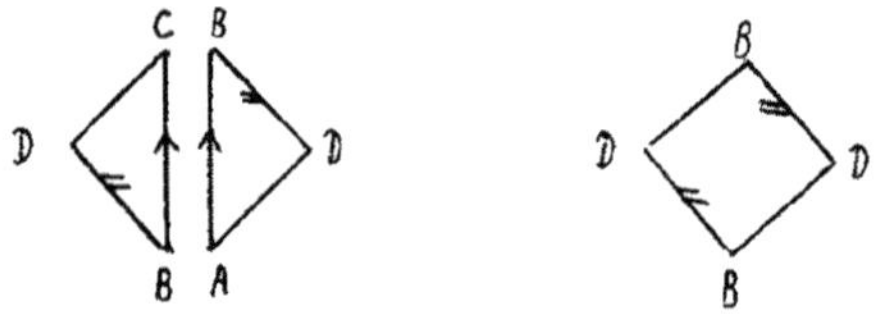

(A. Sudbery showed me this.)]

(iii) $C_m(S^1) = S^1 \times D^{m-1}$ if m is odd; $= S^1 \tilde{\times} D^{m-1}$ if m is even $\Big\}$ = tubular neighbourhood of a projective line in P^mR. $H^2 = 0$

Here D^{m-1} is the closed $(m - 1)$ - disk and $\tilde{\times}$ stands for the non-orientable bundle over S^1. This is a result of H. Morton [4].

(iv) $C_m(S^2) = P^m(C)$, see refs. [4] and [5] $H^2 = Z$

(v) $C_m(P^2R) = P^{2m}(R)$, see article of H. Bacry, these proceedings. $H^2 = Z_2$.

(vi) $C_2(R^2) = R^2 \times (R^2/S_2) = R^4$. $H^2 = 0$

(vii) For $n > 2$, $C_2(R^n) = R^n \times [(R^n \setminus \{0\})/S_2]$ which is retractable to $R^n \times [S^{n-1}/S_2] = R^n \times P^{n-1}R$. Here dim > 2 and we have excluded the diagonals.

4. Cohomology

We now show how to compute $H^2(C_m(R^n), Z)$ for $n \geq 3$, which classifies the line bundles over $C_m(R^n)$. We exploit the fact that $C_m(R^n) = \tilde{C}_m(R^n)/S_m$ where $\tilde{C}_m(R^n) = R^{nm} \setminus \Delta$ and Δ is the set of all diagonal points. Thus $\tilde{C}_m$ covers C_m with covering transformation group S_m. In such a situation the $H^q(C_m)$ are related to the $H^r(\tilde{C}_m)$ and $H^s(S_m, H^r(\tilde{C}_m))$ (which are more easily computable) by a spectral sequence [6]. I know of no other way to calculate $H^q(C_m)$. The $H^r(\tilde{C}_m)$ are calculable in two ways.

(1) $\tilde{C}_m = R^{nm} \setminus \Delta = S^{nm} \setminus \Delta_c$ where we add the point at ∞ to R^{nm} and Δ to get the compact spaces S^{nm} and Δ_c. $H^q(\tilde{C}_m)$ is related to $H^q(S^{nm}) = Z\delta_{q,nm} + Z\delta_{q,0}$ and $H^q(\Delta_c)$ by Alexander duality. Δ_c is a union of spheres; its cohomology is given by the Mayer-Vietoris sequence.

(2) As we must use spectral sequences to get $H^q(C_m)$ we may as well use them to get $H^q(\tilde{C}_m)$. $\tilde{C}_m$ is a fibred manifold over $\tilde{C}_{m-1}$, so we may compute $H^q(\tilde{C}_m)$ from $H^r(\tilde{C}_{m-1})$ and H^s (fibre) by a spectral sequence. We shall do it when $m = 3$. $\tilde{C}_3$ is the space of ordered triples of distinct points of R^n. The first point is any point of R^n. The second is then in $R^n \setminus \{$ first point $\}$, a space which is retractable to S^{n-1}. Hence $\tilde{C}_2$ has the same cohomology as $R^n \times S^{n-1}$, $H^q(\tilde{C}_2) = Z\delta_{q,n-1} + Z\delta_{q,0}$.

We now regard $\tilde{C}_2$ as the base of a fibration of $\tilde{C}_3$; when the first two points of the triple are fixed, the third point can move in $R^n \setminus \{$ two points $\}$ which retracts to the one-point union of two (n-1)-spheres, and $H^q(S^{n-1} \vee S^{n-1}) = 2Z\delta_{q,n-1} + Z\delta_{q,0}$.

The starting term of the spectral sequence for the cohomology of C_3 is

$$E^2_{pq} = H^p(C_2) \otimes H^q(S^{n-1} \vee S^{n-1})$$

drawn below.

q	H^q(fibre)			E^2_{pq}		
$n-1$	$2Z$	$2Z$	0	$---$	0	$2Z$
$\vdots$	0	0	0	$---$	0	0
$\vdots$	$\vdots$	$\vdots$				
1	0	0	0	$---$	0	0
0	Z	Z	0	$---$	0	Z
	H^p(base)	Z	0	$---$	0	Z
	p	0	1	$---$		$n-1$

Since in the spectral sequence (E^r, ∂^r), $\partial^r : E^r_{pq} \to E^r_{p-r,q+r-1}$, see [6] or [7], the ∂^r never connect two non-zero elements, so that $E^2_{pq} = E^\infty_{pq}$. The cohomology groups $H^a(\tilde{C}_3)$ are related to the cohomology groups D_{pq} of $(a = p + q)$ -cochains of weight $\leq p$, which satisfy the inclusions

$$0 = D_{-1,a+1} \subset D_{0,a} \subset D_{1,a-1} \subset \cdots \subset D_{a,0} = H^a(\tilde{C}_3)$$

and are given by $E^\infty_{pq} = D_{p,q}/D_{p-1,q+1}$. It follows that $H^a(\tilde{C}_3) = Z\delta_{a,0} + 3Z\delta_{a,n-1} + 2Z\delta_{a,2n-2}$. The book-keeping is neatly expressed in terms of the Poincaré polynomial $p_m(t)$ of a manifold M, [8]

$$p_M(t) = \sum_{q=0}^{\infty} a_q t^q \quad \text{where } H^q(M) = a_q \; Z.$$

Thus $p_{fibre}(t) = 1 + 2t^{n-1}$, $p_{base}(t) = 1 + t^{n-1}$ and the polynomial of the bundle is the product, $p_{\tilde{C}_3}(t) = (1 + 2t^{n-1})(1 + t^{n-1}) = 1 + 3t^{n-1} + 2t^{2n-2}$.

To obtain the cohomology of $C_3(R^n)$ we use the fact [6] that there is a spectral sequence (E^r, ∂^r) in which

$$E^2_{p,q} = H^p(S_3, H^q(\tilde{C}_3, Z))$$

and $E^\infty_{p,a-p}$ gives the filtration of $H^a(C_3, Z)$. Part of E^2 is drawn below.

q	$H^q(\tilde{C}_3, Z)$							
n-1	3Z	Z	?	?	?			$H^p(S_3, 3Z)$
n-2	0	0	0	0	0	0	0	
⋮	⋮	-	-	-	-	-	-	
1	0	0	0	0	0	0	0	
0	Z	Z	0	Z_2	0	Z_6	0	$H^p(S_3, Z)$
		0	1	2	3	4	5	p

The cohomology groups of S_3 may be deduced by methods given in [9] . We have $H^0(S_3, A) = \{\alpha \in A \; ; \; g\alpha = \alpha \;\forall\; g \in S_3\}$ so that $H^0(S_3, Z) = Z, H^0(S_3, 3Z) = 0$, since S_3 acts on Z trivially and on 3Z freely . The higher cohomology groups of S_3 are obtained from those of its Sylow subgroups Z_2 and Z_3.

Thus for $p + q \geq 2$, E^{∞}_{pq} is

q			
2	$Z\,\delta_{n,3}$		
1	0	0	
0	Z	0	Z_2
	0	1	2 p

The inclusion string of D-groups gives

$$H^0(C_3, Z) = Z, \; H^1(C_3, Z) = 0, \; H^2(C_3, Z) = Z_2$$

Thus , for $n > 3$, there are, up to bundle isomorphism, just two line bundles over $C_3(R^n)$, Bose and Fermi.

I am grateful to many colleagues for explaining things to me, particularly E. Ihrig, H. Morton, P. Newstead and C.T.C. Wall. Also to the Department of Physics, Dalhousie University for their good hospitality while most of this study was accomplished.

References

[1] J.M. Souriau - Structure des Systemes Dynamiques, Dunod, Paris 1970.

[2] G.B. Segal - Configuration Spaces and Iterated Loop-Spaces, Inventiones Math. 21 (1973) 213-221.

[3] D. McDuff - Configuration Spaces of Positive and Negative Particles, Topology, 14 (1975) 91-107.

- Configuration Spaces, Lecture Notes in Mathematics 575, Springer, Berlin, 88-95.

[4] H. Morton - Symmetric Products of the Circle, Proc. Camb. Phil. Soc. 63 (1967) 349-352.

[5] H. Bacry - Orbits of the Rotation Group on Spin States, J. Math. Phys. 15 (1974) 1686-1688.

[6] P.J. Hilton and S. Wylie - Homology Theory, Cambridge 1962, Chap. 10.

[7] R.C. Hwa and V.L. Teplitz - Homology and Feynman Integrals, Benjamin 1966, Chap. 5.

[8] M.F. Atiyah and J.D.S. Jones - Topological Aspects of Yang-Mills Theories, Comm. Math. Phys. 61 (1978) 97-118, § 5.

[9] H. Cartan and S. Eilenberg - Homological Algebra, Oxford 1956, Chap. 12.

THE GEOMETRICAL MEANING AND GLOBALIZATION OF THE HAMILTON-JACOBI METHOD

S. Benenti & W.M. Tulczyjew

Istituto di Meccanica Razionale
Istituto di Fisica Matematica
Università di Torino

This lecture gives an incomplete short account of a research on geometric foundations of analytical mechanics conducted at the Institute of Mathematical Physics and Institute of Rational Mechanics in Turin.

1.- Homogeneous systems.

A homogeneous system is a triple $(M,\mu;C)$ where (M,μ) is a symplectic manifold and C is an embedded connected coisotropic submanifold of (M,μ). Dynamics of many mechanical systems can be represented by homogeneous systems. Although dynamics can usually be represented in a different form, the description of the Hamilton-Jacobi theory is most natural within the homogeneous framework. We recall that a submanifold $C \subseteq M$ is coisotropic if at each $x \in C$, $(T_x C)^{\S} \subseteq T_x C$, where $(T_x C)^{\S}$ is the symplectic polar of $T_x C$ defined by

$$(T_x C)^{\S} = \{ v \in T_x M \; ; \; \langle v \wedge u, \mu \rangle = 0 \; , \; \forall u \in T_x C \} \quad .$$

We note that for a homogeneous system $(M,\mu;C)$ the spaces $D'_x = (T_x C)^{\S}$ are of constant dimensions equal to the codimension of C. It follows that

$$D' = \bigcup_{x \in C} D'_x$$

is a distribution on C. The distribution D' is the characteristic distribution of $\mu|C$:

$$D' = \{ v \in TC \ ; \ v \lrcorner (\mu|C) = 0 \} \quad ,$$

where $\mu|C$ is the pullback of μ to C. Since $\mu|C$ is closed it follows that D' is involutive. We call D' the <u>characteristic distribution</u> of $(M,\mu;C)$. A maximal connected integral manifold of D' is called a <u>characteristic</u> of $(M,\mu;C)$. Characteristics form the <u>characteristic foliation</u> of $(M,\mu;C)$.

We introduce the subset

(1.1) $$D = \{ (x,y) \in C \times C \ ; \ x \text{ and } y \text{ belong to the same characteristic of } (M,\mu;C) \} \quad .$$

It will be pointed out in Section 2 that D, considered as a subset of $M \times M$, is a symplectic relation.

For a homogeneous system $(M,\mu;C)$ associated with the dynamics of a mechanical system the distribution D' represents the differential dynamical equations, the characteristics are the phase-space trajectories of the mechanical system and the relation D represents dynamics in the integrated form. The Hamilton-Jacobi method is an indirect method of constructing the relation D by obtaining solutions of the so called Hamilton-Jacobi equation instead of integrating the characteristic distribution D'.

<u>Example</u> **1.**- <u>Homogeneous formulation of time-dependent Hamiltonian dynamics</u>. Let (P,ω) be a symplectic manifold and $H: P \times \mathbb{R} \longrightarrow \mathbb{R}$ a differentiable function. We construct a homogeneous system $(M,\mu;C)$ by setting $M = P \times \mathbb{R} \times \mathbb{R}$, $\mu = \pi^* \omega - de \wedge dt$ and $C = \{ (p,t,e) \in M \ ; \ e = H(p,t) \}$, where $\pi: M \longrightarrow P$, $t: M \longrightarrow \mathbb{R}$, $e: M \longrightarrow \mathbb{R}$ are the cartesian projections. If (q^i, p_j) are canonical coordinates of (P,ω), then $(q^i, t; p_j, -e)$ are canonical coordinates of (M,μ). The submanifold C is locally represented by the equation

(1.2) $$e = H(q^i, p_j, t) \quad .$$

If $(\dot{q}^i, \dot{p}_j; \dot{t}, \dot{e})$ are the components of a generic vector $v \in TM$ with respect to the coordinates $(q^i, p_j; t, e)$ of M, then TC is represented by (1.2) together with the equation

(1.3) $$\dot{e} - \frac{\partial H}{\partial q^i} \dot{q}^i - \frac{\partial H}{\partial p_j} \dot{p}_j - \frac{\partial H}{\partial t} \dot{t} = 0$$

Since

$$\mu|C = dp_i \wedge dq^i - dH \wedge dt$$

$$= dp_i \wedge dq^i - (\frac{\partial H}{\partial q^i} dq^i + \frac{\partial H}{\partial p_j} dp_j) \wedge dt \quad ,$$

we have

$$v \lrcorner \mu|C = (\dot{p}_i + \dot{t}\frac{\partial H}{\partial q^i})\, dq^i - (\dot{q}^i - \dot{t}\frac{\partial H}{\partial p_i})\, dp_i - (\frac{\partial H}{\partial q^i}\dot{q}^i + \frac{\partial H}{\partial p_i}\dot{p}_i)\, dt \quad .$$

Considering equations (1.3) and $v \lrcorner \mu|C = 0$, we see that the characteristic distribution D' is represented by equations

(1.4) $$\dot{q}^i = \dot{t}\frac{\partial H}{\partial p_i} \quad , \quad \dot{p}_i = -\dot{t}\frac{\partial H}{\partial q^i} \quad , \quad \dot{e} = \dot{t}\frac{\partial H}{\partial t}$$

together with (1.2). It is usual to represent D' by a vectorfield on C obtained from D' by imposing the condition $\dot{t} = 1$. This field is described by the Hamilton equations:

$$\dot{q}^i = \frac{\partial H}{\partial p_i} \quad , \quad \dot{p}_i = -\frac{\partial H}{\partial q^i}$$

and the energy balance equation

$$\dot{e} = \frac{\partial H}{\partial t} \quad .$$

<u>Example</u> 2.- <u>Energy surfaces of a time-independent Hamiltonian system</u>. Let (P,ω) be a symplectic manifold, $H:P \to \mathbb{R}$ a differentiable function, $e \in \mathbb{R}$ a regular value of H and $C_e = \{p \in P \; ; \; e = H(p)\}$ the corresponding 'energy surface'. Since C_e is of codimension 1, it is a coisotropic submanifold of (P,ω). Hence $(P,\omega;C_e)$ is a homogeneous system (C_e is assumed to be connected). If (q^i,p_j) are canonical coordinates of (P,ω), and $(q^i,p_j;\dot{q}^i,\dot{p}_j)$ are the corresponding coordinates of TP, then it can be easily seen that the characteristic distribution is represented by equations:

$$H(q^i,p_j) = e \quad , \quad \dot{q}^i = \lambda\frac{\partial H}{\partial p_i} \quad , \quad \dot{p}_i = -\lambda\frac{\partial H}{\partial q^i} \quad ,$$

where λ is an arbitrary real parameter.

<u>Example</u> 3.- <u>Homogeneous formulation of constrained Hamiltonian dynamics</u> (<u>Dirac</u> [3]). Let (P,ω) be a symplectic manifold, $K \subseteq P$ an embedded submanifold and $H:K \to \mathbb{R}$ a differentiable function. We construct a triple $(M,\mu;C)$ by setting $M = P \times \mathbb{R} \times \mathbb{R}$, $\mu = \pi^*\omega - de \wedge dt$ and $C = \{ x \in M \; ; \; \pi(x) \in K \; , \; e(x) = H(\pi(x)) \}$. Here $\pi:M \to P$, $e:M \to \mathbb{R}$ and $t:M \to \mathbb{R}$ are the canonical projections. It is easily seen that (M,μ) is a symplectic manifold and C is an embedded submanifold of M.

<u>Proposition 1.</u>- $(M,\mu;C)$ <u>is a homogeneous system if and only if</u> $(P,\omega;K)$ <u>is a homogeneous system and</u> H <u>is constant on the characteristics</u>.

<u>Proof.</u>- A vector $v\in T_yM$ is a triple (u,a,b) where $u\in T_{\pi(y)}P$ and $a,b\in\mathbb{R}$. From the definition of C we have

$$\text{(A)}\qquad T_yC=\{(u,a,b)\in T_yM;\ u\in T_{\pi(y)}K,\ \langle u,dH\rangle=a\}$$

and also $(T_yC)^{\S}=\{(u',a',b')\in T_yM;\ \langle u\wedge u',\omega\rangle-ab'+ba'=0,\ \forall(u,a,b)\in T_yC\}$.
Since $(0,0,b)\in T_yC$, it follows that $a'=0$. Hence

$$\text{(B)}\qquad (T_yC)^{\S}=\{(u',a',b')\in T_yM;\ a'=0,\ u'\lrcorner\,\omega-b'\,d\bar{H}\in(T_{\pi(y)}K)^{o}\}\ ,$$

where $\bar{H}:P\longrightarrow\mathbb{R}$ is any differentiable extension of H $(\bar{H}\,|\,K=H)$ and ^{o} denotes the polar operator (if V is a subspace of a vector space S, V^{o} is the subspace of the dual S^{*} whose elements are zero on all of V). We have to prove the equivalence of the following two conditions:

$$\text{i)}\qquad (T_xK)^{\S}\subseteq T_xK \quad\text{and}\quad dH\in((T_xK)^{\S})^{o}\ ,\ \forall x\in K\ ,$$

$$\text{ii)}\qquad (T_yC)^{\S}\subseteq T_yC\ ,\ \forall y\in C\ .$$

Let us assume that condition i) is satisfied. If $(u',0,b')\in(T_yC)^{\S}$, from $(T_{\pi(y)}K)^{o}\subseteq((T_{\pi(y)}K)^{\S})^{o}$ and (B) it follows that $u'\lrcorner\,\omega-b'\,d\bar{H}\in((T_{\pi(y)}K)^{\S})^{o}$. Since any extension $\bar{H}$ of H is also constant on the characteristics, i.e. $d\bar{H}\in((T_{\pi(y)}K)^{\S})^{o}$, we have $u'\lrcorner\,\omega\in((T_{\pi(y)}K)^{\S})^{o}$, hence: $u'\in((T_{\pi(y)}K)^{\S})^{\S}=T_{\pi(y)}K$. Consequently, by (B) we have $\langle u',dH\rangle=\langle u',d\bar{H}\rangle$. Then from (A) we can see that $(u',0,b')\in T_yC$. Hence condition ii) holds. Conversely, let us assume that ii) is satisfied. From (B) we see that $\bar{u}\lrcorner\,\omega\in(T_xK)^{o}$ implies $(\bar{u},0,0)\in(T_yC)^{\S}\subseteq T_yC$ (for each y such that $\pi(y)=x$). Then from (A) it follows that $\bar{u}\in T_xK$ and $\langle\bar{u},dH\rangle=0$, so that condition i) holds.

Q.E.D.

The quadruple $(P,\omega;K,H)$ is called a <u>constrained Hamiltonian system</u>. Proposition 1 shows that $(M,\mu;C)$ is a homogeneous system if and only if the constrained Hamiltonian system $(P,\omega;K,H)$ is <u>integrable</u> [5].

2.- Symplectic relations associated with homogeneous systems.

Let (P_1, ω_1) and (P_2, ω_2) be two symplectic manifolds. We define the product $(P_1, \omega_1) \times (P_2, \omega_2)$ as the symplectic manifold $(P_1 \times P_2, \pi_1^* \omega_1 + \pi_2^* \omega_2)$, where $\pi_i : P_1 \times P_2 \longrightarrow P_i$ $(i = 1,2)$ are the cartesian projections. A symplectic relation from (P_1, ω_1) to (P_2, ω_2) is an immersed Lagrangian submanifold of $(P_1, -\omega_1) \times (P_2, \omega_2)$. A symplectic relation from a symplectic manifold (P, ω) to (P, ω) will be called a symplectic relation in (P, ω).

Let B be the set of characteristics of the homogeneous system $(M, \mu; C)$. We consider the following relation from M to B:

(2.1) $$R = \{ (x,b) \in M \times B \; ; \; x \in C \; , \; b = \rho(x) \} \quad ,$$

where $\rho : C \longrightarrow B$ is the canonical projection. We observe that

(2.2) $$D = R^t \circ R \quad ,$$

i.e. that the integral relation D of D' is the composition of R with its transpose relation $R^t = \{ (b,x) \in B \times M \; ; \; (x,b) \in R \}$. It is known ([7],[4]) that if B admits a differentiable structure and the map ρ is a submersion, then there is a unique symplectic form β on B such that

(2.3) $$\rho^* \beta = \mu \,|\, C \quad .$$

In this case the homogeneous system $(M, \mu; C)$ is said to be globally reducible and (B, β) is called the reduced symplectic manifold.

Proposition 2.- If $(M, \mu; C)$ is globally reducible then R is an embedded symplectic relation from (M, μ) to (B, β).

Proof.- We say that a chart (U, φ) of C is adapted to the submersion ρ if there exists a chart (V, ψ) of B such that: i) $V = \rho(U)$, ii) $\forall b \in V$, $\varphi(U \cap \rho^{-1}(b)) = \varphi(U) \cap (\mathbb{R}^k \times \psi(b))$, where k is the dimension of the fibers of ρ (in our case k = codim C). Then for the map $\varphi : U \longrightarrow \mathbb{R}^c$ (c = dim C) we can consider a matrix representation $\varphi = (\varphi^{\parallel}, \varphi^{\perp})$, where $\varphi^{\parallel} : U \longrightarrow \mathbb{R}^k$, $\varphi^{\perp} : U \longrightarrow \mathbb{R}^{c-k}$ and $\varphi^{\perp} = \psi \circ \rho$. Frobenius theorem provides the existence of an adapted chart at each point of C. More precisely, for each $(x_o, b_o) \in R$ there is an adapted chart, as above, such that $(x_o, b_o) \in U \times V$. Let us consider the map $\chi : U \times V \longrightarrow \mathbb{R}^c \times \mathbb{R}^{c-k}$: $(x,b) \longmapsto (\varphi(x), \psi(b) - \varphi^{\perp}(x))$. The pair $(U \times V, \chi)$ is a chart of $C \times B$. Moreover, if $(x,b) \in (U \times V) \cap R$, then $\chi(x,b) =$

$(\varphi(x),0)$, that is to say: $\chi((U\times V)\cap R) = \chi(U\times V)\cap(\mathbb{R}^c\times\{0\})$. This proves that R is an embedded submanifold of $C\times B$, hence an embedded submanifold of $M\times B$, and that

(*) $$\dim R = \dim C = \tfrac{1}{2}\dim(M\times B) \quad .$$

The set R is the image of the differential map $(\iota_C,\rho):C\longrightarrow M\times B$, where $\iota_C:C\longrightarrow M$ is the embedding of C into M. If $pr_M:M\times B\longrightarrow M$, $pr_B:M\times B\longrightarrow B$ are the Cartesian projections, then $(\iota_C,\rho)^*(-pr_M^*\mu + pr_B^*\beta) = -(pr_M\circ(\iota_C,\rho))^*\mu + (pr_B\circ(\iota_C,\rho))^*\beta =$ $= -\iota_C^*\mu + \rho^*\beta = 0$. Hence R is an isotropic submanifold of $(M,-\mu)\times(B,\beta)$. From (*) it follows that R is Lagrangian.

Q.E.D.

The symplectic relation R is called the reduction relation of the symplectic manifold (M,μ) with respect to the coisotropic submanifold C [6]. The transpose relation R^t, which is a symplectic relation from (B,β) to (M,μ), is called the counter-reduction relation of (M,μ) with respect to C. We recall a useful property of reductions ([7,1]): if L is a Lagrangian submanifold of (M,μ) which has clean intersection with C (i.e.: $L\cap C$ is a submanifold and $TL\cap TC = T(L\cap C)$) then the reduction of L, i.e. the set

$$R(L) = \{\, b\in B \; ; \; \exists\, l\in L : (l,b)\in R \,\} \quad ,$$

is an immersed Lagrangian submanifold of (B,β).

Proposition 3.- Let $(M,\mu;C)$ be a homogeneous system, then the set D defined as in (1.1) is a symplectic relation in (M,μ). If $(M,\mu;C)$ is globally reducible, then D is an embedded symplectic relation.

Proof.- First we consider the case of B admitting a differentiable structure such that ρ is a submersion (i.e. $(M,\mu;C)$ is globally reducible). For each $(x_1,x_2)\in D$ it is always possible to find two charts (U_1,φ_1) and (U_2,φ_2) of C, at x_1 and x_2 respectively, both adapted to the submersion ρ (see the proof of Prop. 2), such that $\rho(U_1) = \rho(U_2)$ and $\Psi_1 = \Psi_2$. The map $\chi:U_1\times U_2\longrightarrow \mathbb{R}^k\times\mathbb{R}^k\times\mathbb{R}^{c-k}\times\mathbb{R}^{c-k} : (x_1,x_2)\longmapsto(\varphi_1^{\parallel}(x_1),$ $\varphi_2^{\parallel}(x_2), \varphi_1^{\perp}(x_1)+\varphi_2^{\perp}(x_2), \varphi_1^{\perp}(x_1)-\varphi_2^{\perp}(x_2))$ defines a chart of $C\times C$ such that $\chi((U_1\times U_2)\cap D) = \chi(U_1\times U_2)\cap(\mathbb{R}^k\times\mathbb{R}^k\times\mathbb{R}^{c-k}\times\{0\})$. It follows that D is an embedded submanifold of $C\times C$, hence of $M\times M$, and $\dim D = c+k$, i.e.:

(*) $$\dim D = \dim M = \tfrac{1}{2}\dim(M\times M).$$

We notice that a vector tangent to D, as a class of curves on D, is a class of pairs of curves on C which project point by point (through the map ρ) onto the same curve on

B. This implies that a vector tangent to D is a pair of vectors tangent to C such that their projections (through the tangent map $T\rho$) coincide: $(v_1,v_2)\in TD \Longleftrightarrow T\rho(v_1) = T\rho(v_2)$. Then, if (v_1,v_2) and $(v_1',v_2') \in T_{(x_1,x_2)}D$ and if $(-\mu,\mu)$ denotes the symplectic form of $(M,-\mu)\times(M,\mu)$, we have $\langle (v_1,v_2)\wedge(v_1',v_2'),(-\mu,\mu)\rangle = \langle v_1\wedge v_1',-\mu\rangle + \langle v_2\wedge v_2',\mu\rangle = -\langle T\rho(v_1)\wedge T\rho(v_1'),\beta\rangle + \langle T\rho(v_2)\wedge T\rho(v_2'),\beta\rangle = 0$. This means that D is an isotropic submanifold of $(M,-\mu)\times(M,\mu)$. From (*) it follows that D is Lagrangian. Hence D is an embedded symplectic relation on (M,μ), and the second part of the proposition is proved. Now, let us abandon the assumption of the existence of a differentiable structure on B. We can anyway consider an atlas $\{C_\alpha,\varphi_\alpha\}_{\alpha\in I}$ of C satisfying the following conditions for each index $\alpha\in I$:

i) C_α is connected;

ii) the set B_α of the maximal connected integral manifolds of the distribution $D'_\alpha = D'\cap TC_\alpha$ admits a differentiable structure such that the natural projection $\rho_\alpha : C_\alpha \longrightarrow B_\alpha$ is a submersion;

iii) the chart $(C_\alpha,\varphi_\alpha)$ is adapted to the distribution D'_α .

The existence of such an atlas is provided by Frobenius theorem. As in the proof of Prop. 2, we can decompose each φ_α into the matrix $(\varphi_\alpha^{\parallel},\varphi_\alpha^{\perp}):C_\alpha \longrightarrow \mathbb{R}^k\times\mathbb{R}^{c-k}$ in such a way that $\varphi_\alpha^{\perp}(x) = \varphi_\alpha^{\perp}(x')$ if and only if x and x' belong to the same element of B_α. Each C_α is the domain of a relation $D_\alpha \subseteq M\times M$ defined as follows: $D_\alpha = \{(x,y)\in C_\alpha\times C_\alpha ; \exists b\in B : x,y\in b\}$. Let (x_o,y_o) be an element of D. Among the open sets $\{C_\alpha\}_{\alpha\in I}$ we can consider a finite sequence $C_1,C_2,\dots,C_r$ covering an integral curve of D' joining x_o and y_o and such that: $x_o\in C_1$, $y_o\in C_r$, $C_i\cap C_{i+1} \neq \emptyset$ for $i = 1,\dots,r-1$. Let $\bar{C}$ be the domain of the composition $D_1\circ D_2\circ\dots\circ D_r$. It is a non-empty open set made of integral manifolds of D', passing through all the intersections C_i C_{i+1}. Let us consider the two open sets $\bar{C}_1 = C_1\cap\bar{C}$ and $\bar{C}_r = C_r\cap\bar{C}$. By suitable successive adaptations of the maps $\varphi_2,\dots,\varphi_r$ we can always adjust φ_r, hence $\varphi_r^{\perp}$, in such a manner that $\varphi_1^{\perp}|\bar{C}_1(x) = \varphi_r^{\perp}|\bar{C}_r(y)$ implies that x and y belong to the same element of B. Then the map $\chi : \bar{C}_1\times\bar{C}_r \longrightarrow \mathbb{R}^k\times\mathbb{R}^k\times\mathbb{R}^{c-k}$: $(x,y)\longmapsto(\varphi_1^{\parallel}(x),\varphi_r^{\parallel}(y),\varphi_1^{\perp}|\bar{C}_1(x))$ defines a chart of D at the point (x_o,y_o). It can be verified that charts of this kind define a differentiable structure on D such that the injection map $\iota_D : D\longrightarrow M\times M$ is an immersion. Moreover, for any chart constructed in this manner the triple $(M,\mu;\bar{C})$ forms a homogeneous system whose characteristic foliation $\bar{B}$ has a differentiable structure such that the natural projection $\bar{\rho}:\bar{C}\longrightarrow\bar{B}$ is a submersion (in fact $\bar{B}$ can be identified with an open submanifold of each space $B_1,\dots,B_r$). Hence, as we have seen in the first part of this proof, the corresponding relation $\bar{D}$ is a Lagrangian embedded submanifold of $(M,-\mu)\times(M,\mu)$. On the other hand $\bar{D}$ is an open submanifold of D. Since D can be covered by open submanifolds of this kind, we conclude that D is an immersed Lagrangian submanifold.

Q.E.D.

3.- Generating functions of Lagrangian submanifolds.

Let $\pi:\Lambda \longrightarrow U$ be a differentiable surjective submersion with connected fibers and let $K \subseteq T^*\Lambda$ be the set of vertical covectors:

$$K = \{ p \in T^*\Lambda ;\ \langle v,p\rangle = 0,\ \forall\, v \in T\Lambda \ :\ T\pi(v) = 0 \} \quad . \tag{3.1}$$

K is a coisotropic submanifold of $(T^*\Lambda,\ \omega_\Lambda)$ where ω_Λ is the canonical symplectic form. It can be shown that the homogeneous system $(T^*\Lambda, \omega_\Lambda ;K)$ is globally reducible and that the reduced symplectic manifold can be identified with (T^*U, ω_U), where ω_U is the canonical symplectic form in T^*U. Let us denote by $\tilde{R}$ the reduction relation from $(T^*\Lambda, \omega_\Lambda)$ to (T^*U, ω_U). Let $F:\Lambda \longrightarrow \mathbb{R}$ be a differentiable function and $dF: \Lambda \longrightarrow T^*\Lambda$ its differential. We call F a Morse family [7] if the Lagrangian submanifold $\tilde{L} = dF(\Lambda) \subseteq T^*\Lambda$ is transversal to K, i.e.

$$T_{\tilde{L}\cap K}\tilde{L} + T_{\tilde{L}\cap K}K = T_{\tilde{L}\cap K}T^*\Lambda \quad . \tag{3.2}$$

If F is a Morse family then the reduction of $\tilde{L}$:

$$\tilde{R}(\tilde{L}) = \{ p \in T^*U \ ;\ \exists\ l \in \tilde{L} \ :\ (p,l) \in \tilde{R} \} \quad , \tag{3.3}$$

is an immersed Lagrangian submanifold of (T^*U, ω_U).

A Lagrangian submanifold of a cotangent bundle can always be generated, at least locally, by Morse families [7]. More precisely, let N be a differentiable manifold, $\pi_N: T^*N \longrightarrow N$ the cotangent bundle projection, ω_N the canonical symplectic form of T^*N. If L is a Lagrangian submanifold of (T^*N, ω_N) then for each $p_o \in L$ there is a neighbourhood U of $\pi_N(p_o)$, a surjective submersion $\pi:\Lambda \longrightarrow U$ and a Morse family $F:\Lambda \longrightarrow \mathbb{R}$ such that $\tilde{R}(\tilde{L})$ (as defined in (3.3)) is an open submanifold of L containing p_o.

A function $F:U \longrightarrow \mathbb{R}$ is a particular case of a Morse family, with $\Lambda = U$ and π the identity mapping. The Lagrangian submanifold generated by F is simply the image of $dF : U \longrightarrow T^*U$.

If (x^i, λ^a) are coordinates of Λ adapted to the submersion π ($i = 1,\dots,n$; $a = 1,\dots,l$; $n = \dim N$; $n+l = \dim \Lambda$), then the transversality conditions (3.2) for a function $F(x^i, \lambda^a)$ is stated as the maximality of the rank of the matrix

$$(3.4) \qquad \left\| \frac{\partial^2 F}{\partial x^i \partial \lambda^a} \quad , \quad \frac{\partial^2 F}{\partial \lambda^b \partial \lambda^a} \right\|$$

In terms of natural canonical coordinates $(x^i; y_j)$ of T^*N the Lagrangian submanifold $\tilde{R}(\tilde{L})$ is described by equations

$$(3.5) \qquad \begin{cases} y_i = \dfrac{\partial F}{\partial x^i} & (i = 1,\dots,n) \\ 0 = \dfrac{\partial F}{\partial \lambda^a} & (a = 1,\dots,l) \end{cases}$$

4.- The Hamilton-Jacobi equation.

The Hamilton-Jacobi equation, in the classical sense, or in the generalized sense given below, is an object which arises in connection with a homogeneous system $(M,\mu;C)$. It specifies generating functions of Lagrangian submanifolds of (M,μ) contained in C. We assume that (M,μ) is the cotangent bundle $M = T^*N$ of a manifold N with the canonical 2-form $\mu = \omega_N$.

A function $F : U \longrightarrow \mathbb{R}$, where U is an open subset of N, is called *a solution of the Hamilton-Jacobi equation associated with the homogeneous system* $(T^*N, \omega_N; C)$ if it generates a Lagrangian submanifold contained in C:

$$(4.1) \qquad dF(U) \subseteq C \quad .$$

Equation (4.1) is the classical Hamilton-Jacobi equation associated with $(T^*N, \omega_N; C)$.

The Lagrangian submanifolds generated in this way are sections of the projection π_N, so that such solutions exist locally only when the projection of C by π_N is sufficiently regular.

If $(x^i; y_j)$ are natural canonical coordinates of T^*N the Lagrangian submanifold generated by F is defined by the equations

$$(4.2) \qquad y_i = \frac{\partial F}{\partial x^i} \qquad (i = 1,\dots,n).$$

Then, if C is represented by the equations

$$(4.3) \qquad C^A(x^i, y_j) = 0 \qquad (A = 1,\ldots,k \; ; \; k = \text{codim } C),$$

where C^A are independent functions, the Hamilton-Jacobi equation gives rise to the following system:

$$(4.4) \qquad C^A(x^i, \frac{\partial F}{\partial x^j}) = 0 \qquad (A = 1,\ldots,k).$$

The condition of coisotropy of C is equivalent to the involutivity of the system (see [2],[4],[5]):

$$(4.5) \qquad \{C^A, C^B\}|C = 0 \qquad .$$

If we want to consider Lagrangian submanifolds in C which are not sections of π_N we must consider Morse families as solutions of the Hamilton-Jacobi equation. This procedure allows us to deal with any Lagrangian submanifold (at least locally) including those with singular projections by π_N.

We say that a Morse family $F: \Lambda \longrightarrow \mathbb{R}$ is <u>a generalized solution of the Hamilton-Jacobi equation associated with the homogeneous system</u> $(T^*N, \omega_N; C)$ if it generates a Lagrangian submanifold contained in C:

$$(4.6) \qquad \tilde{R}(dF(\Lambda)) \subseteq C \qquad ,$$

where $\tilde{R}$ is the reduction relation defined as in Section 3.

If (x^i, λ^a) are coordinates of Λ adapted to the submersion $\pi : \Lambda \longrightarrow N$, then $F(x^i, \lambda^a)$ is a generalized solution of the Hamilton-Jacobi equation if matrix (3.4) has maximal rank and

$$(4.7) \qquad \begin{cases} C^A(x^i, \dfrac{\partial F}{\partial x^j}) = 0 & (A = 1,\ldots,k) , \\ \dfrac{\partial F}{\partial \lambda^a} = 0 & (a = 1,\ldots,l) . \end{cases}$$

5.- <u>The Hamilton-Jacobi method</u>.

We consider families of solutions of the Hamilton-Jacobi equation associated with a homogeneous system $(T^*N, \omega_N; C)$. Let A be a differentiable manifold and $S: V \longrightarrow \mathbb{R}$ a differentiable function defined on an open submanifold $V \subseteq A \times N$. Let $pr_A : A \times N \longrightarrow A$ be the cartesian projection. For each $a \in pr_A(V)$ we can consider the open submanifold

$$V_a = \{ x \in N \; ; \; (a,x) \in V \}$$

of N and the function

$$S_a : V_a \longrightarrow \mathbb{R} : x \longmapsto S(a,x) \quad .$$

Definition 1.- In terms of the notations introduced above, the function $S:V \longrightarrow \mathbb{R}$ is called a complete solution of the Hamilton-Jacobi equation associated with $(T^*N, \omega_N; C)$ if each function S_a is a solution and the Lagrangian submanifolds $L_a = dS_a(V_a)$ generated by the functions S_a form a foliation of an open connected submanifold C_S of C.

If S is a complete solution then $\dim A = \dim N + \operatorname{codim} C$. Moreover, if (a^κ) are coordinates on A $(\kappa = 1,\ldots,n-k)$, (x^i, y_j) are natural canonical coordinates of T^*N and C is represented as in (4.3), then the function $S(a^\kappa, x^i)$ satisfies the involutive system

$$C^A(x^i, \frac{\partial S}{\partial x^i}) = 0 \qquad (A = 1,\ldots,k) \quad , \tag{5.1}$$

and the matrix

$$\left\| \frac{\partial^2 S}{\partial a^\kappa \partial x^i} \right\| \tag{5.2}$$

is of maximal rank.

A complete solution $S:V \longrightarrow \mathbb{R}$ is the generating function of a symplectic relation $\overset{t}{R}_S$ from (T^*A, ω_A) (where ω_A is the canonical symplectic form on T^*A) to (T^*N, ω_N), in the sense that the set

$$\overset{t}{R}_S = \{ (b,y) \in T^*A \times T^*N \; ; \; (-b,y) = dS(a,x) \; , \; a = \pi_A(b) \; , \; x = \pi_N(y) \; , \; (a,x) \in V \} \quad , \tag{5.3}$$

where $\pi_A : T^*A \longrightarrow A$ and $\pi_N : T^*N \longrightarrow N$ are the cotangent bundle projections, is a Lagrangian submanifold of $(T^*A, -\omega_A)$ (T^*N, ω_N). In terms of natural canonical coordinates (a^κ, b_λ) and (x^i, y_j) of T^*A and T^*N respectively, $\overset{t}{R}_S$ is locally described by the equations:

$$b_\kappa = - \frac{\partial S}{\partial a^\kappa} \quad , \quad y_i = \frac{\partial S}{\partial x^i} \quad . \tag{5.4}$$

Let us denote by R_S the transpose relation of $\overset{t}{R}_S$.

For each $b \in T^*A$ we can consider the image set

$$\overset{t}{R}_S(b) = \{ y \in T^*N \; ; \; (b,y) \in \overset{t}{R}_S \} \quad . \tag{5.5}$$

If $R_S^t(b)$ is not empty, then it is a submanifold of the Lagrangian submanifold L_a, where $a = \pi_A(b)$. In this case, for each $y \in R_S^t(b)$ (i.e. for each $y \in T^*N$ such that $(b,y) \in R_S^t$) the relation R_S^t induces a linear symplectic relation from the symplectic vector space $(T_b(T^*A), \omega_A|_b)$ to the symplectic vector space $(T_y(T^*N), \omega_N|_y)$. It can be shown that this relation is isomorphic to the linear counter-reduction of $(T_y(T^*N), \omega_N|_y)$ with respect to the coisotropic subspace T_yC. This implies that $T_y(R_S^t(b)) = (T_yC)^{\S} = D'_y$, hence that $R_S^t(b)$ is the union of open submanifolds of characteristics of the homogeneous system $(T^*N, \omega_N; C_S)$. Then we can prove the following proposition.

Proposition 4.- Let $S: V \longrightarrow \mathbb{R}$ be a complete solution of the Hamilton-Jacobi equation associated with the homogeneous system $(T^*N, \omega_N; C)$ (Def.1) such that the submanifolds $R_S^t(b)$ ((5.5)) are connected. Then: i) the homogeneous system $(T^*N, \omega_N; C_S)$ is globally reducible; ii) the reduced symplectic manifold can be identified with an open submanifold of (T^*A, ω_A) and the relation R_S^t with the corresponding counter-reduction; iii) the composed relation $D_S = R_S^t \circ R_S$ is an open subrelation of the symplectic relation D ((1.1)) associated with $(T^*N, \omega_N; C)$.

It follows that with a complete solution S we can construct only a part of the relation D, unless S defines a Lagrangian foliation of all of C. In this case, if the manifolds $R_S^t(b)$ are connected, the homogeneous system $(T^*N, \omega_N; C)$ is globally reducible and the reduced symplectic space can be identified with an open submanifold of $T^*A, \omega_A)$ in such a way that the corresponding counter-reduction relation R^t can be identified with the symplectic relation R_S^t generated by S. Hence (cf.(2.2)) $D_S = D$. For the construction of D we can take into account the following proposition.

Proposition 5.- Let $(T^*N, \omega_N; C)$ be a homogeneous system. If C can be covered by open connected submanifolds $\{C_\alpha\}_{\alpha \in I}$ associated with a family of complete solutions $\{S_\alpha\}_{\alpha \in I}$ satisfying the condition of Prop.4 and if $\{D_\alpha\}_{\alpha \in I}$ are the corresponding subrelations of D, then

$$D = \bigcup D_\alpha \circ D_\beta \circ \dots \circ D_\gamma$$

where the union is taken over all the finite sequences $(\alpha, \beta, \dots, \gamma)$ of elements of the index set I.

Proof.- Let $\hat{D}$ be the union considered in the proposition. If $(x,y) \in \hat{D}$, we can take a sequence $(\alpha, \beta, \dots, \gamma)$ such that $(x,y) \in D_\alpha \circ D_\beta \circ \dots \circ D_\gamma$. Then x and y can be joined through a chain of integral manifolds of $D' \cap TC_\alpha$, $D' \cap TC_\beta, \ldots$, $D' \cap TC_\gamma$, whose union is an integral manifold of D'. Hence $(x,y) \in D$ and $\hat{D} \subseteq D$. Conversely, if $(x,y) \in D$, we take an integral curve of D' with end points x and y and cover it by a finite sequence $C_1, C_2, \ldots, C_r$ of elements of the set $\{C_\alpha\}_{\alpha \in I}$ such that $C_i \cap C_{i+1} \neq \emptyset$ for $i = 1, \ldots, r-1$ (compare with the proof of Proposition 3). Furthermore, we can choose a sequence of points $x_1, x_2, \ldots, x_r$ on the integral curve such that $x_1 = x$, $x_r = y$ and $x_i \in C_i \cap C_{i+1}$ for $i = 2, \ldots, r-1$. This means that $(x_i, x_{i+1}) \in D_i$ for $i = 1, \ldots, r-1$. Then $(x,y) = (x_1, x_r) \in D_1 \circ D_2 \circ \ldots \circ D_r \subseteq \hat{D}$. Hence $D \subseteq \hat{D}$.

Q.E.D.

In some cases it may be necessary to consider complete solutions in the generalized sense (Morse families); for example when C does not project onto an open submanifold of N. We leave this matter out of the present lecture.

References

[1] .- R.Abraham & J.Marsden, Foundations of Mechanics, Benjamin-Cummings (1978).

[2] .- C.Carathéodory, Calculus of variations and partial differential equations of first order, Part I, Holden-Day (1965).

[3] .- P.A.M.Dirac, Canad. J. Math., 2 (1950), 129 - 148.

[4] .- A.Lichnerowicz, C. R. Acad. Sci. Paris, A- 280 (1975), 523 - 527.

[5] .- M.R.Menzio & W.M.Tulczyjew, Ann. Inst. H. Poincaré, 27 (1978), 349 - 367.

[6] .- J.Sniatycki & W.M.Tulczyjew, Indiana Univ. Math. J., 22 (1972), 267 - 275.

[7] .- A.Weinstein, Lectures on Symplectic Manifolds, CBMS Regional Conferences, 29 (1976), A.M.S..

The present research has been sponsored by Consiglio Nazionale delle Ricerche - Gruppo Nazionale per la Fisica Matematica (CNR - GNFM).

THE EULER-LAGRANGE RESOLUTION

W. M. Tulczyjew

Department of Mathematics and Statistics
The University of Calgary
Calgary, Alberta T2N 1N4, Canada

Istituto di Fisica Matematica
Universitá di Torino
Via Carlo Alberto, 10
I-10123 Torino

Institut de Mathématique
Université Catholique de Louvain
Chemin du Cyclotron 2
B-1348 Louvain-la-Neuve

1. Introduction.

In this lecture I will present a differential geometric approach to the problem of characterizing the kernel and the image of the Euler-Lagrange operator of the calculus of variations. Several original elements are contained in this approach. The Euler-Lagrange operator is made a part of a cochain complex for which the Poincaré lemma is proved. The proof of the Poincaré lemma is derived from the exactness of a double complex. the construction of the Euler-Lagrange complex is interpreted in terms of a spectral sequence. I began the work on this problem in 1973. Results were published in a series of papers: [10],[11],[12]. Work on the same problem was started independently at about the same time at Moscow University, and remarkably similar results were obtained. The idea of constructing a complex was used by Kupershmidt [6], although his construction is different from mine. More recently a complex equivalent to mine was introduced by Vinogradov [14],[15], who also relates the complex to a spectral sequence of a double complex. I owe the interpretation of my construction in terms fo a spectral sequence to Dedecker [2]. Other approaches to the problem both geometric and based on functional analysis can be found in [1],[4],[8], [9],[13].

The content of the present lecture is essentially the same as that of reference [12]. In [12] the calculus of variations of parametrized submanifolds of a manifold was considered. Here we reformulate the problem in the context of the calculus of variations of sections of a fibration. In this setting it is easier to compare results with those of other researchers who used similar methods. Transition to the new setting required a proof of coordinate independence of some of the constructions. This proof is given in Proposition 5.3, which is the only significant addition to the material contained in [12].

Extensive use of derivations is made in the present lecture. The theory of derivations of the exterior algebra of differential forms was formulated by Frölicher and Nijenhuis [3]. Derivations of forms defined on jet bundles were used by Kumpera [5].

2. Jet bundles, the holonomic lift.

Let M denote a C^∞-manifold of dimension p. We consider a differential fibration (E,M,π) of rank n. For each $k \in \mathrm{N}$ we denote by J_kE the bundle of jets of sections of π. The bundle J_0E is identified with E. If $\sigma:M \to E$ is a section of π then $j_k\sigma:M \to J_kE$ denotes the *prolongation* of σ to J_kE. We have the *canonical jet-source projections*

$$\pi^k:J_kE \to M \tag{2.1}$$

and the *canonical jet-jet projections*

$$\rho_{k'}{}^k:J_kE \to J_{k'}E \tag{2.2}$$

for $k' \leq k$. The projections

$$\rho_0{}^k:J_kE \to J_0E = E \tag{2.3}$$

are the *canonical jet-target projections*. Relations

$$\rho_{k''}{}^{k'} \circ \rho_{k'}{}^k = \rho_{k''}{}^k \tag{2.4}$$

hold for $k'' \leq k' \leq k$.

For each k we introduce the *holonomic lift* [5]

$$\lambda_k:TM \times_M J_{k+1}E \to TJ_kE. \tag{2.5}$$

If $(u,a) \in TM \times_M J_{k+1}E$, $\gamma:\mathrm{R} \to M$ is an integral curve of u and $\sigma:M \to E$ is a representative of a then $j_k\sigma \circ \gamma$ is an integral curve of $\lambda_k(u,a)$. For each k a mapping

$$H_k:TJ_{k+1}E \to TJ_kE \tag{2.6}$$

is defined by

$$H_k(w) = \lambda_k(T\pi^{k+1}(w), \tau_{J_{k+1}E}(w)), \tag{2.7}$$

where

$$\tau_{J_{k+1}E}: TJ_{k+1}E \to J_{k+1}E \tag{2.8}$$

denotes the tangent bundle projection. Relations

$$H_{k'} \cdot T\rho_{k'+1}{}^{k+1} = T\rho_{k'}{}^{k} \cdot H_k \tag{2.9}$$

and

$$H_{k-1} \cdot H_k = H_{k-1} \cdot T\rho_k{}^{k+1} = T\rho_{k-1}{}^{k} \cdot H_k \tag{2.10}$$

are easily verified.

Local coordinates of M will be denoted by (t^α); $\alpha = 1,\ldots,p$. Adapted local coordinates (t^α, x^i); $\alpha = 1,\ldots,p$; $i = 1,\ldots,n$ will be used for E. These coordinates induce coordinates $(t^\alpha, x^i{}_m)$; $\alpha = 1,\ldots,p$; $i = 1, \ldots,n$; $|m| \leq k$ of J_kE. Here $m = (m^1,\ldots,m^p) \in \mathrm{N}^p$ is a multi-index and $|m| = m^1 + \ldots + m^p$. The coordinate $x^i{}_m$ of the jet of a section $\sigma: M \to E$ represented locally by functions $\sigma^i(t^\alpha)$ is the partial derivative

$$D^m\sigma^i(t^\alpha) = D_1{}^{m^1} \ldots D_p{}^{m^p}\sigma^i(t^\alpha). \tag{2.11}$$

If $u \in TM$ and $a \in J_{k+1}E$ have coordinates (t^α, t'^β) and $(t^\alpha, x^i{}_m)$ respectively then the coordinates of $\lambda_k(u,a)$ are $(t^\alpha, x^i{}_m, t'^\beta, x'^j{}_n)$, where $x'^j{}_n = \Sigma_\gamma x^j{}_{n+e_\gamma} t'^\gamma$ and the multi-index $e_\gamma = (e_\gamma{}^1,\ldots,e_\gamma{}^p)$ is defined by $e_\gamma{}^\alpha = \delta_\gamma{}^\alpha$.

3. Differential forms on jet bundles.

For each $k \in \mathrm{N}$ we denote by

$$\Phi_k = \Sigma_q \Phi_k^q \tag{3.1}$$

the exterior algebra of differential forms on J_kE. For each $q = 0,1,\ldots$ the space Φ_k^q consists of forms of degree q. The *exterior product* is denoted by $\wedge$ and the *exterior differential* is denoted by d.

It is easily seen from (2.4) that algebras Φ_k together with the pullback mappings

$$(\rho_{k'}{}^{k})^*:\Phi_{k'} \to \Phi_k \tag{3.2}$$

form a *direct system* over N [7]. We denote by

$$\Phi = \Sigma_q \, \Phi^q \tag{3.3}$$

the *direct limit* of this system. Spaces Φ^q are direct limits of systems formed by spaces Φ_k^q and mappings (3.2). The exterior product $\wedge$ and the exterior differential d are defined in Φ. We use the canonical construction of the direct limit. In this construction the space Φ is the quotient set of $\cup_{k\in N} \Phi_k$ by the equivalence relation according to which forms $\mu \in \Phi_k$ and $\mu' \in \Phi_{k'}$ are equivalent if $k' \leq k$ and $\mu = (\rho_{k'}{}^{k})^*\mu'$ or $k \leq k'$ and $\mu' = (\rho_k{}^{k'})^*\mu$. We identify the space Φ_k with its image by the canonical injection in Φ. Consequently if $\mu \in \Phi$ has a representative in Φ_k we say that μ belongs to Φ_k and write $\mu \in \Phi_k$.

We define a linear operator

$$P^{0,1}:\Phi^1 \to \Phi^1:\mu \mapsto P^{0,1}\mu. \tag{3.4}$$

If $\mu \in \Phi_k^1 \subset \Phi^1$ and $w \in TJ_{k+1}E$ then

$$\langle w, P^{0,1}\mu\rangle = \langle H_k(w), \mu\rangle. \tag{3.5}$$

Compatibility of this definition with the structure of the the direct limit is guaranteed by (2.9). Let

$$I^1:\Phi^1 \to \Phi^1:\mu \mapsto I^1\mu = \mu \tag{3.6}$$

be the identity operator. We denote by $P^{1,0}$ the operator $I^1 - P^{0,1}$. It follows from (2.10) that

$$P^{0,1}\,P^{0,1} = P^{0,1}. \tag{3.7}$$

Hence, also

$$P^{1,0}\,P^{1,0} = P^{1,0} \tag{3.8}$$

and

$$P^{0,1}\,P^{1,0} = P^{1,0}\,P^{0,1} = 0. \tag{3.9}$$

For two linear operators $A:\Phi^r \to \Phi^r$ and $B:\Phi^s \to \Phi^s$ we define the exterior product $A\wedge B:\Phi^{r+s} \to \Phi^{r+s}$ by

$$(A\wedge B)(\mu_1\wedge\ldots\wedge\mu_{r+s}) = \frac{1}{(r+s)!}\Sigma_{\sigma\in S_{r+s}} sgn(\sigma)A(\mu_{\sigma(1)}\wedge\ldots\wedge\mu_{\sigma(r)}) \wedge B(\mu_{\sigma(r+1)}\wedge\ldots\wedge\mu_{\sigma(r+s)}). \tag{3.10}$$

In terms of this definition and the obvious definition of the exterior power of an operator we introduce operators

$$P^{r,s} = \frac{(r+s)!}{r!s!}(\wedge^r P^{1,0})\wedge(\wedge^s P^{0,1}):\Phi^{r+s} \to \Phi^{r+s} \tag{3.11}$$

satisfying relations

$$P^{r,s}\,P^{r',s'} = \delta^{rr'}\,P^{r,s}, \tag{3.12}$$

where $r' + s' = r + s$, and

$$\Sigma_{r+s=q}\,P^{r,s} = I^q, \tag{3.13}$$

where I^q is the identity operator in Φ^q. With spaces $\Phi^{r,s}$, $\Phi^{r,\cdot}$ and $\Phi^{\cdot,s}$ defined by

$$\Phi^{r,s} = P^{r,s}(\Phi), \tag{3.14}$$

$$\Phi^{r,\cdot} = \Sigma_s\,\Phi^{r,s} \tag{3.15}$$

and

$$\Phi^{\cdot,s} = \Sigma_r\,\Phi^{r,s} \tag{3.16}$$

the algebra Φ becomes a *bigraded algebra*:

$$\Phi = \Sigma_r\,\Phi^{r,\cdot} = \Sigma_s\,\Phi^{\cdot,s} = \Sigma_{r,s}\,\Phi^{r,s} \tag{3.17}$$

and

$$\Phi^{r,s}\,\Phi^{r',s'} = \Phi^{r+r',s+s'}. \tag{3.18}$$

In addition to the *total degree* q we have the *bidegree* (r,s).

A linear operator $a:\Phi \to \Phi$ is called a *graded operator of total degree* q' if $a(\Phi^q) \subset \Phi^{q+q'}$. A graded operator a of degree q' is called a *derivation of total degree* q' if

$$a(\mu\wedge\nu) = a\mu\wedge\nu + (-1)^{qq'}\mu\wedge a\nu, \tag{3.19}$$

where $\mu \in \Phi^q$ and $\nu \in \Phi$. The *commutator*

$$[a,b] = ab - (-1)^{q'q''}ba \tag{3.20}$$

of derivations a and b of total degrees q' and q'' respectively is a derivation of total degree $q'+q''$. A derivation is completely characterized by its action on Φ^0 and Φ^1. Following Frölicher and Nijenhuis [3] we single out derivations of type i_*, which act trivially on Φ^0 and derivations of type d_*, which commute with d. To each derivation i_A of type i_* there corresponds a derivation $d_A = [i_A, d]$ of type d_*. An operator or a derivation a is said to be of bidegree (r',s') if $a(\Phi^{r,s}) \subset \Phi^{r+r',s+s'}$.

Derivations i_H and i_V of type i_*, total degree 0 and bidegree (0,0) are defined by

$$i_H\mu = P^{0,1}\mu, \quad i_V\mu = P^{1,0}\mu \tag{3.21}$$

if $\mu \in \Phi^1$. The corresponding derivations

$$d_H = [i_H, d] \tag{3.22}$$

and

$$d_V = [i_V, d] \tag{3.23}$$

of type d_* and total degree 1 satisfy

$$d_H + d_V = d. \tag{3.24}$$

Locally a derivation is characterized by its action on the coordinates t^α, $x^i{}_m$ and their differentials dt^α, $dx^i{}_m$. From the coordinate description of the holonomic lift λ_k we derive

$$i_H dt^\alpha = d_H t^\alpha = dt^\alpha, \tag{3.25}$$

$$i_H dx^i{}_m = d_H x^i{}_m = \Sigma_\gamma\, x^i{}_{m+e_\gamma} dt^\gamma, \tag{3.26}$$

$$i_V dt^\alpha = d_V t^\alpha = 0 \tag{3.27}$$

and

$$i_V dx^i{}_m = d_V x^i{}_m = dx^i{}_m - \Sigma_\gamma\, x^i{}_{m+e_\gamma} dt^\gamma. \tag{3.28}$$

When giving local characterizations of derivations or writing local expressions of elements of Φ it is convenient to replace differentials $dx^i{}_m$ by $d_V x^i{}_m$. We have

$$i_H d_V x^i{}_m = 0 \tag{3.29}$$

and

$$i_V d_V x^i{}_m = d_V x^i{}_m. \tag{3.30}$$

From (3.25) and (3.27) it follows that dt^α is of bidegree (0,1). Equations (3.26) and (3.28) show that $dx^i{}_m$ is of mixed bidegree and from (3.29) and (3.30) we see that $d_V x^i{}_m$ is of bidegree (1,0). It follows that d_H and d_V are derivations of bidegrees (0,1) and (1,0) respectively.

Commutation relations

$$[d_H, d_H] = 2d_H d_H = 0, \tag{3.31}$$

$$[d_V, d_V] = 2d_V d_V = 0, \tag{3.32}$$

and

$$[d_H, d_V] = d_H d_V + d_V d_H = 0 \tag{3.33}$$

are easily established. Consequently, we have the double complex

$$\begin{array}{ccccccccccc}
\Phi^{0,0} & \xrightarrow{d_V} & \Phi^{1,0} & \xrightarrow{d_V} & \Phi^{2,0} & \xrightarrow{d_V} & \cdots & \xrightarrow{d_V} & \Phi^{r,0} & \xrightarrow{d_V} & \cdots \\
\downarrow d_H & & \downarrow d_H & & \downarrow d_H & & & & \downarrow d_H & & \\
\Phi^{0,1} & \xrightarrow{d_V} & \Phi^{1,1} & \xrightarrow{d_V} & \Phi^{2,1} & \xrightarrow{d_V} & \cdots & \xrightarrow{d_V} & \Phi^{r,1} & \xrightarrow{d_V} & \cdots \\
\downarrow d_H & & \downarrow d_H & & \downarrow d_H & & & & \downarrow d_H & & \\
\vdots & & \vdots & & \vdots & & & & \vdots & & \\
\downarrow d_H & & \downarrow d_H & & \downarrow d_H & & & & \downarrow d_H & & \\
\Phi^{0,s} & \xrightarrow{d_V} & \Phi^{1,s} & \xrightarrow{d_V} & \Phi^{2,s} & \xrightarrow{d_V} & \cdots & \xrightarrow{d_V} & \Phi^{r,s} & \xrightarrow{d_V} & \cdots \\
\downarrow d_H & & \downarrow d_H & & \downarrow d_H & & & & \downarrow d_H & & \\
\vdots & & \vdots & & \vdots & & & & \vdots & &
\end{array} \tag{3.34}$$

Let $\Omega = \Sigma_s \Omega^s$ be the algebra of differential forms on M with the exterior product $\wedge$ and the exterior differential d. If $\mu \in \Omega$ then $\pi^*\mu \in \Phi_0 \subset \Phi$ belongs to $\Phi^{0,\bullet}$. The mapping

$$\eta : \Omega \to \Phi^{0,\bullet} : \mu \to \pi^*\mu \tag{3.35}$$

is a monomorphism and satisfies relations

$$d_V \eta = 0 \tag{3.36}$$

and

$$\eta d = d_H \eta. \tag{3.37}$$

The diagram

$$\begin{array}{ccccccccccccc}
0 & \longrightarrow & \Omega^0 & \xrightarrow{\eta} & \Phi^{0,0} & \xrightarrow{d_V} & \Phi^{1,0} & \xrightarrow{d_V} & \cdots & \xrightarrow{d_V} & \Phi^{r,0} & \xrightarrow{d_V} & \cdots \\
 & & \downarrow d & & \downarrow d_H & & \downarrow d_H & & & & \downarrow d_H & & \\
0 & \longrightarrow & \Omega^1 & \xrightarrow{\eta} & \Phi^{0,1} & \xrightarrow{d_V} & \Phi^{1,1} & \xrightarrow{d_V} & \cdots & \xrightarrow{d_V} & \Phi^{r,1} & \xrightarrow{d_V} & \cdots \\
 & & \downarrow d & & \downarrow d_H & & \downarrow d_H & & & & \downarrow d_H & & \\
 & & \vdots & & \vdots & & \vdots & & & & \vdots & & \\
 & & \downarrow d & & \downarrow d_H & & \downarrow d_H & & & & \downarrow d_H & & \\
0 & \longrightarrow & \Omega^s & \xrightarrow{\eta} & \Phi^{0,s} & \xrightarrow{d_V} & \Phi^{1,s} & \xrightarrow{d_V} & \cdots & \xrightarrow{d_V} & \Phi^{r,s} & \xrightarrow{d_V} & \cdots \\
 & & \downarrow d & & \downarrow d_H & & \downarrow d_H & & & & \downarrow d_H & & \\
 & & \vdots & & \vdots & & \vdots & & & & \vdots & &
\end{array} \tag{3.38}$$

can be described as an *augmented double complex* since $\eta : \Omega \to \Phi^{0,\bullet}$ is a cochain mapping and horizontal sequences are augmented cochain complexes.

For $s > p$ we have $\Omega^s = 0$ and $\Phi^{\cdot,s} = 0$. We introduce the *spectral sequence*

$$0 \longrightarrow \Xi^0 \xrightarrow{\gamma^0} \Xi^1 \xrightarrow{\gamma^1} \ldots \xrightarrow{\gamma^{r-1}} \Xi^r \xrightarrow{\gamma^r} \ldots , \qquad (3.39)$$

where for each r

$$\Xi^r = \Phi^{r,p} \Big/ d_H(\Phi^{r,p-1}) \qquad (3.40)$$

and the operators γ^r are defined by the commutativity of the bottom row of squares in the diagram

$$\begin{array}{ccccccccccccc}
0 & \longrightarrow & \Omega^0 & \xrightarrow{\eta} & \Phi^{0,0} & \xrightarrow{d_V} & \Phi^{1,0} & \xrightarrow{d_V} & \ldots & \xrightarrow{d_V} & \Phi^{r,0} & \xrightarrow{d_V} & \ldots \\
 & & \downarrow d & & \downarrow d_H & & \downarrow d_H & & & & \downarrow d_H & & \\
0 & \longrightarrow & \Omega^1 & \xrightarrow{\eta} & \Phi^{0,1} & \xrightarrow{d_V} & \Phi^{1,1} & \xrightarrow{d_V} & \ldots & \xrightarrow{d_V} & \Phi^{r,1} & \xrightarrow{d_V} & \ldots \\
 & & \downarrow d & & \downarrow d_H & & \downarrow d_H & & & & \downarrow d_H & & \\
 & & \vdots & & \vdots & & \vdots & & & & \vdots & & \\
 & & \downarrow d & & \downarrow d_H & & \downarrow d_H & & & & \downarrow d_H & & \\
0 & \longrightarrow & \Omega & \xrightarrow{\eta} & \Phi^{0,p} & \xrightarrow{d_V} & \Phi^{1,p} & \xrightarrow{d_V} & \ldots & \xrightarrow{d_V} & \Phi^{r,p} & \xrightarrow{d_V} & \ldots \\
 & & \downarrow & & \downarrow \xi^0 & & \downarrow \xi^1 & & & & \downarrow \xi^r & & \\
 & & 0 & \longrightarrow & \Xi^0 & \xrightarrow{\gamma^0} & \Xi^1 & \xrightarrow{\gamma^1} & \ldots & \xrightarrow{\gamma^{r-1}} & \Xi^r & \xrightarrow{\gamma^r} & \ldots
\end{array} \qquad (3.41)$$

Mappings

$$\xi^r : \Phi^{r,p} \to \Xi^r \qquad (3.42)$$

are the canonical projections.

We denote by $\iota : \mathbf{R} \to \Omega^0$ and $\varepsilon : \mathbf{R} \to \Phi^{0,0}$ mappings which map numbers into constant functions.

PROPOSITION 3.1. *All horizontal and vertical sequences in the diagram*

$$
\begin{array}{ccccccccccccc}
 & & 0 & & 0 & & & & & & & & \\
 & & \downarrow & & \downarrow & & & & & & & & \\
 & & \mathbb{R} & = & \mathbb{R} & & 0 & & & & 0 & & \\
 & & \downarrow\iota & & \downarrow\varepsilon & & \downarrow & & & & \downarrow & & \\
0 & \longrightarrow & \Omega^0 & \xrightarrow{\eta} & \Phi^{0,0} & \xrightarrow{d_V} & \Phi^{1,0} & \xrightarrow{d_V} & \cdots & \xrightarrow{d_V} & \Phi^{r,0} & \xrightarrow{d_V} & \cdots \\
 & & \downarrow d & & \downarrow d_H & & \downarrow d_H & & & & \downarrow d_H & & \\
0 & \longrightarrow & \Omega^1 & \xrightarrow{\eta} & \Phi^{0,1} & \xrightarrow{d_V} & \Phi^{1,1} & \xrightarrow{d_V} & \cdots & \xrightarrow{d_V} & \Phi^{r,1} & \xrightarrow{d_V} & \cdots \\
 & & \downarrow d & & \downarrow d_H & & \downarrow d_H & & & & \downarrow d_H & & \\
 & & \vdots & & \vdots & & \vdots & & & & \vdots & & \\
 & & \downarrow d & & \downarrow d_H & & \downarrow d_H & & & & \downarrow d_H & & \\
0 & \longrightarrow & \Omega^p & \xrightarrow{\eta} & \Phi^{0,p} & \xrightarrow{d_V} & \Phi^{1,p} & \xrightarrow{d_V} & \cdots & \xrightarrow{d_V} & \Phi^{r,p} & \xrightarrow{d_V} & \cdots \\
 & & \downarrow & & \downarrow\xi^0 & & \downarrow\xi^1 & & & & \downarrow\xi^r & & \\
 & & 0 & \longrightarrow & \Xi^0 & \xrightarrow{\gamma^0} & \Xi^1 & \xrightarrow{\gamma^1} & \cdots & \xrightarrow{\gamma^{r-1}} & \Xi^r & \xrightarrow{\gamma^r} & \cdots \\
 & & & & \downarrow & & \downarrow & & & & \downarrow & & \\
 & & & & 0 & & 0 & & & & 0 & &
\end{array}
\tag{3.43}
$$

are exact.

A complete proof of Proposition 3.1 is given in the Appendix. Here we define certain auxiliary objects and prove the least obvious part of the proposition. In Proposition 3.1 and everywhere else in this lecture exactness is understood as local exactness.

For every coordinate system (t^α) of M we introduce derivations ι_α of type i_*, total degree -1 and bidegree $(0,-1)$, and the corresponding derivations $\partial_\alpha = [\iota_\alpha, d]$ of type d_*, total degree 0 and bidegree $(0,0)$ characterized by

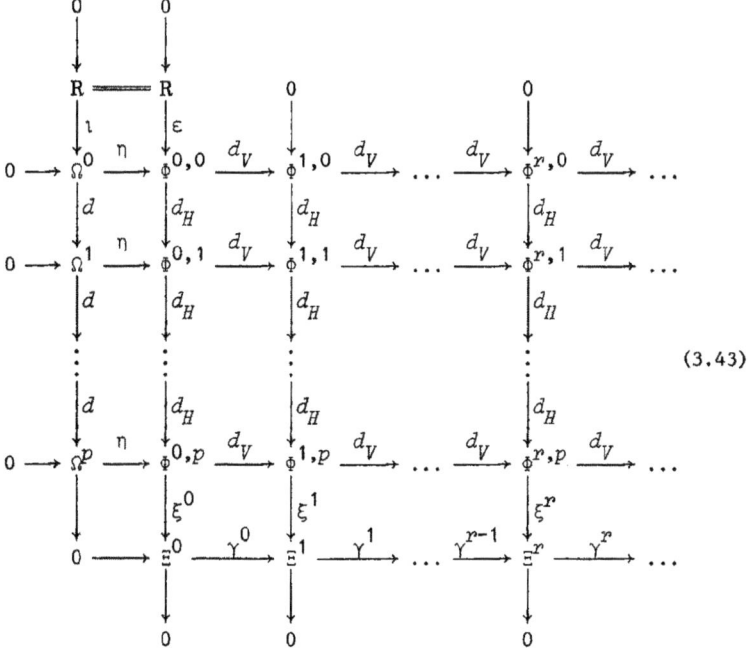

$$\iota_\alpha dt^\beta = \partial_\alpha t^\beta = \delta_\alpha{}^\beta, \tag{3.44}$$

$$\iota_\alpha d_V x^i{}_m = 0 \tag{3.45}$$

and

$$\partial_\alpha x^i{}_m = x^i{}_{m+e_\alpha}. \tag{3.46}$$

Derivations d_H and ∂_α are related by the formula

$$d_H\mu = \Sigma_\alpha dt^\alpha \wedge \partial_\alpha \mu\ . \tag{3.47}$$

To prove this formula we observe that both d_H and the operator

$$\mu \mapsto \Sigma_\alpha\ dt^\alpha \wedge \partial_\alpha \mu$$

are derivations of type d_* and total degree 1, and their action on coordinate functions t^α, $x^i{}_m$ is the same. An element $\mu \in \Phi^{r,s}$ can be represented locally by

$$\mu = \Sigma_{\alpha_1<\alpha_2\ldots<\alpha_{p-s}}\ (d^s t)_{\alpha_1\alpha_2\ldots\alpha_{p-s}} \wedge \mu^{\alpha_1\alpha_2\ldots\alpha_{p-s}}\ , \tag{3.48}$$

where $\mu^{\alpha_1\alpha_2\ldots\alpha_{p-s}} \in \Phi^{r,0}$, $d^p t = dt^1 \wedge dt^2 \wedge \ldots \wedge dt^p$ and $(d^s t)_{\alpha_1\alpha_2\ldots\alpha_{p-s}}$ $= \iota_{\alpha_1}\iota_{\alpha_2}\ldots\iota_{\alpha_{p-s}} d^p t$. In terms of this representation we have

$$d_H\mu = \Sigma_{\alpha_1<\ldots<\alpha_{p-s}} \sum_{p=1}^{p-s} (-1)^{i+1} (d^s t)_{\alpha_1\ldots\hat{\alpha}_i\ldots\alpha_{p-s}} \wedge \partial_{\alpha_i} \mu^{\alpha_1\ldots\alpha_{p-s}}. \tag{3.49}$$

Forms $d_V x^i{}_m$ can be obtained from $d_V x^i$ by the action of derivations ∂_α. It follows that a type i_* derivation will be defined if its action on dt^α and $d_V x^i$ is specified and its commutators with derivations ∂_α are given. For each $m \in \mathbb{N}^p$ we define a derivation θ_m of type i_*, total degree 0 and bidegree (0,0) by

$$\theta_0 = i_V, \tag{3.50}$$

if $m > 0$ then

$$\theta_m dt^\alpha = 0, \quad \theta_m d_V x^i = 0 \tag{3.51}$$

and

$$[\theta_m, \partial_\alpha] = \begin{cases} \theta_{m-e_\alpha} & \text{if } m \geq e_\alpha, \\ 0 & \text{otherwise.} \end{cases} \tag{3.52}$$

Formula (3.52) holds also for $m = 0$. From (3.51) and (3.52) we derive

$$\theta_m d_V x^i{}_n = \begin{cases} \binom{n}{m} d_V x^i{}_{n-m} & \text{if } n \geq m, \\ 0 & \text{otherwise,} \end{cases} \tag{3.53}$$

where $\binom{n}{m} = \binom{n^1}{m^1}\binom{n^2}{m^2}\cdots\binom{n^p}{m^p}$.

For each $\alpha = 1,\ldots,p$ we define a set

$$I_\alpha = \{m \in \mathbb{N}^p;\ m^\alpha > 0,\ m^\beta = 0 \text{ for } \beta > \alpha\} \tag{3.54}$$

and an operator

$$\sigma^\alpha = -\Sigma_{m\in I_\alpha} (-1)^{|m|} \partial^{m-e_\alpha}\theta_m, \tag{3.55}$$

where $\partial^n = \partial_1^{n^1}\partial_2^{n^2}\ldots\partial_n^{n^p}$. The operators σ^α are of total degree 0 and bidegree (0.0). For each $\mu \in \Phi$, $\theta_m\mu = 0$ for sufficiently high m. Hence, only a finite number of terms in (3.55) give non-zero contributions when σ^α is applied to μ. In terms of the representation (3.49) we define an operator

$$D:\Phi \to \Phi:\mu \to \Sigma_{\alpha<\alpha_1\ldots<\alpha_{p-s}} (d^{s-1}t)_{\alpha\alpha_1\ldots\alpha_{p-s}} \wedge\sigma^\alpha\mu^{\alpha_1\ldots\alpha_{p-s}} \tag{3.56}$$

of total degree -1 and bidegree $(0,-1)$.

LEMMA 3.1. *For $r > 0$ the sequence*

$$0 \longrightarrow \Phi^{r,0} \xrightarrow{d_H} \Phi^{r,1} \xrightarrow{d_H} \cdots \xrightarrow{d_H} \Phi^{r,p} \xrightarrow{\xi^r} \Xi^r \longrightarrow 0 \tag{3.57}$$

is exact.

Proof. From (3.52) we derive the relations

$$\sigma^\alpha\partial_\beta = \begin{cases} \partial_\beta\sigma^\alpha & \text{if } \alpha < \beta, \\ \theta_0 - \Sigma_{\gamma<\alpha}\partial_\gamma\sigma^\gamma & \text{if } \alpha = \beta, \\ 0 & \text{if } \alpha > \beta, \end{cases} \tag{3.58}$$

A simple calculation based on these relations shows that operators

$$D^{r,s}:\Phi^{r,s} \to \Phi^{r,s-1}:\mu \mapsto \frac{1}{r} D\mu;\ r > 0,\ s > 0 \tag{3.59}$$

satisfy

$$D^{r,s+1}d_H + d_H D^{r,s} = 1 \tag{3.60}$$

for $s > 0$ and

$$D^{r,1}d_H = 1. \tag{3.61}$$

It follows that the sequence (3.57) is exact.∎

4. The Euler-Lagrange operator and the calculus of variations.

In the present section we construct a complement $\Lambda^1 \subset \Phi^{1,p}$ of the subspace $d_H(\Phi^{1,p-1}) \subset \Phi^{1,p}$. Subsequently we define the *Euler-Lagrange operator* $\delta:\Phi^{0,p} \to \Lambda^1 : L \mapsto \delta L$, where δL is the component belonging to Λ^1 of $dL = d_V L \in \Phi^{1,p}$.

LEMMA 4.1. *Let* τ *denote the operator*

$$\tau = \Sigma_m (-1)^{|m|} \partial^m \theta_m \tag{4.1}$$

associated with a local chart (t^α) *of* M. *Then*

$$\tau \partial_\alpha = 0 \tag{4.2}$$

for each α,

$$\tau\tau = \tau i_V, \tag{4.3}$$

$$\tau d_H = 0 \tag{4.4}$$

and

$$i_V \mu - \tau\mu = d_H D\mu \tag{4.5}$$

for each $\mu \in \Phi^{\bullet,p}$.

Proof. Using formula (3.52) we obtain

$$\begin{aligned}\tau\partial_\alpha &= \Sigma_m (-1)^{|m|} \partial^m \theta_m \partial_\alpha \\ &= \Sigma_m (-1)^{|m|} \partial^{m+e_\alpha} \theta_m + \Sigma_m (-1)^{|m|} \partial^m [\theta_m, \partial_\alpha] \\ &= \Sigma_m (-1)^{|m|} \partial^{m+e_\alpha} \theta_m + \Sigma_{m \geq e_\alpha} (-1)^{|m|} \partial^m \theta_{m-e_\alpha} \\ &= \Sigma_m (-1)^{|m|} \partial^{m+e_\alpha} \theta_m + \Sigma_m (-1)^{|m|} \partial^{m+e_\alpha} \theta_m = 0.\end{aligned} \tag{4.6}$$

Consequently

$$\begin{aligned}\tau\tau &= \tau \Sigma_m (-1)^{|m|} \partial^m \theta_m \\ &= \tau\theta_0 = \tau i_V.\end{aligned} \tag{4.7}$$

For each μ we have

$$\begin{aligned}\tau d_H \mu &= \tau(\Sigma_\alpha dt^\alpha \wedge \partial_\alpha \mu) \\ &= \Sigma_\alpha dt^\alpha \wedge \tau \partial_\alpha \mu = 0,\end{aligned} \tag{4.8}$$

hence

$$\tau d_H = 0. \tag{4.9}$$

An element $\mu \in \Phi^{\bullet,p}$ can be represented as a product $d^p t \wedge \nu$, where $\nu \in \Phi^{\bullet,0}$. It follows that

$$\begin{aligned} d_H D\mu &= d_H D(d^p t \wedge \nu) \\ &= d_H \Sigma_\alpha \, (d^{p-1}t)_\alpha \wedge \sigma^\alpha \nu \\ &= \Sigma_\alpha \Sigma_\beta \; dt^\beta \wedge \partial_\beta [(d^{p-1}t)_\alpha \wedge \sigma^\alpha \nu] \\ &= \Sigma_\alpha \Sigma_\beta \; dt^\beta \wedge (d^{p-1}t)_\alpha \wedge \partial_\beta \sigma^\alpha \nu \\ &= \Sigma_\alpha \; d^p t \wedge \partial_\alpha \sigma^\alpha \nu \\ &= \Sigma_\alpha \; \partial_\alpha \sigma^\alpha \mu. \end{aligned} \tag{4.10}$$

Also

$$\begin{aligned} \tau &= \Sigma_m \; (-1)^{|m|} \partial^m \theta_m \\ &= \theta_0 + \Sigma_\alpha \; \partial_\alpha \Sigma_{m \in I_\alpha} (-1)^{|m|} \partial^{m-e_\alpha} \theta_m \\ &= i_V - \Sigma_\alpha \; \partial_\alpha \sigma^\alpha. \end{aligned} \tag{4.11}$$

Hence

$$i_V \mu - \tau\mu = d_H D\mu. \; \blacksquare \tag{4.12}$$

PROPOSITION 4.1. *The subspace*

$$\Lambda^1 = \tau(\Phi^{1,p}) \subset \Phi^{1,p} \tag{4.13}$$

is a complement of the subspace $d_H(\Phi^{1,p-1}) \subset \Phi^{1,p}$:

$$\Phi^{1,p} = \Lambda^1 + d_H(\Phi^{1,p-1}) \tag{4.14}$$

and

$$\Lambda^1 \cap d_H(\Phi^{1,p-1}) = 0. \tag{4.15}$$

Proof. If $\mu \in \Phi^{1,p}$ then $\tau\mu \in \Lambda^1$, from (4.5) it follows that $(1 - \tau)\mu = d_H D\mu \in d_H(\Phi^{1,p-1})$ and also $\mu = \tau\mu + (1 - \tau)\mu$. Hence $\Phi^{1,p} = \Lambda^1 + d_H(\Phi^{1,p-1})$. If $\nu \in \Lambda^1$ and $\nu \in d_H(\Phi^{1,p-1})$ then $\nu = \tau\nu$ due to (4.3) and $\nu = d_H\kappa$ for some $\kappa \in \Phi^{1,p-1}$. It follows from (4.4) that $\nu = \tau\nu = \tau d_H \kappa = 0$. Hence $\Lambda^1 \cap d_H(\Phi^{1,p-1}) = 0$. $\blacksquare$

An arbitrary element $\mu \in \Phi^{1,p}$ is locally represented by

$$\mu = \Sigma_i \Sigma_n \ \mu_i{}^n d_V x^i{}_n \wedge d^p t = \Sigma_i \Sigma_n \ \mu_i{}^n dx^i{}_n \wedge d^p t, \tag{4.16}$$

where $\mu_i{}^n \in \Phi^0$. We apply τ separately to each term in this representation:

$$\begin{aligned}
\tau(\mu_i{}^n d_V x^i{}_n \wedge d^p t) &= \Sigma_m \ (-1)^{|m|} \partial^m \theta_m (\mu_i{}^n d_V x^i{}_n \wedge d^p t) \\
&= \Sigma_m \ (-1)^{|m|} \partial^m \mu_i{}^n \theta_m d_V x^i{}_n \wedge d^p t \\
&= \Sigma_{m=0}^n \ (-1)^{|m|} \binom{n}{m} \partial^m \mu_i{}^n d_V x^i{}_{n-m} \wedge d^p t \\
&= \Sigma_{m=0}^n \Sigma_{k=0}^m \ (-1)^{|m|} \binom{n}{m} \binom{m}{k} (\partial^k \mu_i{}^n)(\partial^{m-k} d_V x^i{}_{n-m}) \wedge d^p t \\
&= \Sigma_{m=0}^n \Sigma_{k=0}^m \ (-1)^{|m|} \binom{n}{m} \binom{m}{k} (\partial^k \mu_i{}^n) d_V x^i{}_{n-k} \wedge d^p t \\
&= \Sigma_{k=0}^n \Sigma_{m=k}^n \ (-1)^{|m|} \binom{n}{m} \binom{m}{k} (\partial^k \mu_i{}^n) d_V x^i{}_{n-k} \wedge d^p t \\
&= \Sigma_{k=0}^n \Sigma_{m=k}^n \ (-1)^{|m|} \binom{n}{k} \binom{n-k}{m-k} (\partial^k \mu_i{}^n) d_V x^i{}_{n-k} \wedge d^p t \\
&= \Sigma_{k=0}^n \Sigma_{l=0}^{n-k} \ (-1)^{|l|} \binom{m-k}{l} (-1)^{|k|} \binom{n}{k} (\partial^k \mu_i{}^n) d_V x^i{}_{n-k} \wedge d^p t.
\end{aligned} \tag{4.17}$$

Since

$$\Sigma_{l=0}^{n-k} \ (-1)^{|l|} \binom{n-k}{l} = \begin{cases} 1 & \text{if } k = n \\ 0 & \text{if } k \quad n \ , \end{cases} \tag{4.18}$$

we obtain

$$\begin{aligned}
\tau(\mu_i{}^n d_V x^i{}_n \wedge d^p t) &= (-1)^{|n|} (\partial^n \mu_i{}^n) d_V x^i \wedge d^p t \\
&= (-1)^{|n|} (\partial^n \mu_i{}^n) dx^i \wedge d^p t.
\end{aligned} \tag{4.19}$$

Hence

$$\tau\mu = \Sigma_i \Sigma_n \ (-1)^{|n|} (\partial^n \mu_i{}^n) dx^i \wedge d^p t. \tag{4.20}$$

Formula (4.20) leads to the following proposition.

PROPOSITION 4.2. *Let $\Lambda' \subset \Phi$ be the ideal generated by Φ_0^{p+1}. Then*

$$\Lambda^1 = \Phi^{1,p} \cap \Lambda'. \tag{4.21}$$

Proof. Formula (4.20) shows that $\tau\mu \in \Lambda'$ for each $\mu \in \Phi^{1,p}$. Conversely if $\mu \in \Phi^{1,p}$ belongs to Λ' then the local expression of μ is $\mu = \Sigma_i \ \mu_i dx^i \wedge d^p t$ and from (4.20) we have $\tau\mu = \mu$. Hence $\mu \in \Lambda^1$. ∎

Originally we defined the space Λ^1 in terms of a local chart. Proposition 4.2 globalizes the construction of Λ^1 giving it an intrinsic, coordinate independent meaning.

If $L = Ld^pt$ is an element of $\Phi^{0,p}$ then

$$dL = \Sigma_i\Sigma_n \frac{\partial L}{\partial x^i_n} dx^i_n \wedge d^pt \tag{4.22}$$

and

$$dL = \Sigma_i\Sigma_n (-1)^{|n|}\left(\partial^n \frac{\partial L}{\partial x^i_n}\right)dx^i \wedge d^pt. \tag{4.23}$$

Equations

$$\Sigma_n (-1)^{|n|}\partial^n \frac{\partial L}{\partial x^i_n} = 0 \tag{4.24}$$

are the Euler-Lagrange equations well known from the traditional calculus of variations. This suggests the following definitions.

DEFINITION 4.1. The mapping

$$\delta:\Phi^{0,p} \to \Lambda^1 : L \mapsto \tau dL \tag{4.25}$$

is called the *Euler-Lagrange operator*.

DEFINITION 4.2. An element $\lambda \in \Lambda^1$ is called an *Euler-Lagrange form* if $\lambda = \delta L$ for some $L \in \Phi^{0,p}$.

5. The Euler-Lagrange complex.

We incorporate the Euler-Lagrange operator in an exact sequence. Exactness of the sequence provides a criterion for systems of differential equations to be Euler-Lagrange equations. Two versions are presented.

Proposition 4.1 implies that spaces Λ^1 and Ξ^1 are isomorphic. Using this isomorphism and introducing the mapping $\tau^1:\Phi^{1,p} \to \Lambda^1 : \mu \mapsto \tau\mu$ we replace the diagram (3.43) by the following diagram with exact vertical and horizontal sequences:

$$
\begin{array}{ccccccccccccc}
 & & 0 & & 0 & & & & & & & & \\
 & & \downarrow & & \downarrow & & & & & & & & \\
 & & \mathbf{R} & = & \mathbf{R} & & 0 & & & & 0 & & \\
 & & \downarrow \iota & & \downarrow \varepsilon & & \downarrow & & & & \downarrow & & \\
0 & \longrightarrow & \Omega^0 & \xrightarrow{\eta} & \Phi^{0,0} & \xrightarrow{d_V} & \Phi^{1,0} & \xrightarrow{d_V} & \cdots & \xrightarrow{d_V} & \Phi^{r,0} & \xrightarrow{d_V} & \cdots \\
 & & \downarrow d & & \downarrow d_H & & \downarrow d_H & & & & \downarrow d_H & & \\
0 & \longrightarrow & \Omega^1 & \xrightarrow{\eta} & \Phi^{0,1} & \xrightarrow{d_V} & \Phi^{1,1} & \xrightarrow{d_V} & \cdots & \xrightarrow{d_V} & \Phi^{r,1} & \xrightarrow{d_V} & \cdots \\
 & & \downarrow d & & \downarrow d_H & & \downarrow d_H & & & & \downarrow d_H & & \\
 & & \vdots & & \vdots & & \vdots & & & & \vdots & & \\
 & & \downarrow d & & \downarrow d_H & & \downarrow d_H & & & & \downarrow d_H & & \\
0 & \longrightarrow & \Omega^p & \xrightarrow{\eta} & \Phi^{0,p} & \xrightarrow{d_V} & \Phi^{1,p} & \xrightarrow{d_V} & \cdots & \xrightarrow{d_V} & \Phi^{r,p} & \xrightarrow{d_V} & \cdots \\
 & & \downarrow & & \downarrow \xi^0 & & \downarrow \tau^1 & & & & \downarrow \xi^r & & \\
 & & 0 & \longrightarrow & \Xi^0 & \xrightarrow{\chi^0} & \Lambda^1 & \xrightarrow{\chi^1} & \cdots & \xrightarrow{\gamma^{r-1}} & \Xi^r & \xrightarrow{\gamma^r} & \cdots \\
 & & & & \downarrow & & \downarrow & & & & \downarrow & & \\
 & & & & 0 & & 0 & & & & 0 & &
\end{array}
\qquad (5.1)
$$

THEOREM 5.1. *The sequence*

$$
\begin{array}{l}
0 \longrightarrow \mathbf{R} \xrightarrow{\varepsilon} \Phi^{0,0} \xrightarrow{d_H} \Phi^{0,1} \xrightarrow{d_H} \cdots \\
\cdots \xrightarrow{d_H} \Phi^{0,p} \xrightarrow{\delta} \Lambda^1 \xrightarrow{\chi^1} \Xi^2 \xrightarrow{\gamma^2} \cdots
\end{array}
\qquad (5.2)
$$

is exact.

Proof. If $L \in \Phi^{0,p}$ and $\delta L = 0$ then $\chi^0\xi^0 L = 0$ and $\xi^0 L = 0$. Hence $L \in d_H(\Phi^{0,p-1})$. If $\lambda \in \Lambda^1$ and $\chi^1\lambda = 0$ then $\lambda = \chi^0\kappa$ for some $\kappa \in \Xi^0$. Since $\kappa = \xi^0 L$ for some $L \in \Phi^{0,p}$, we have $\lambda = \chi^0\xi^0 L = \delta L$. The rest of the proof follows from Proposition 3.1. ∎

Conclusions applicable to the calculus of variations are formulated in the following corollaries.

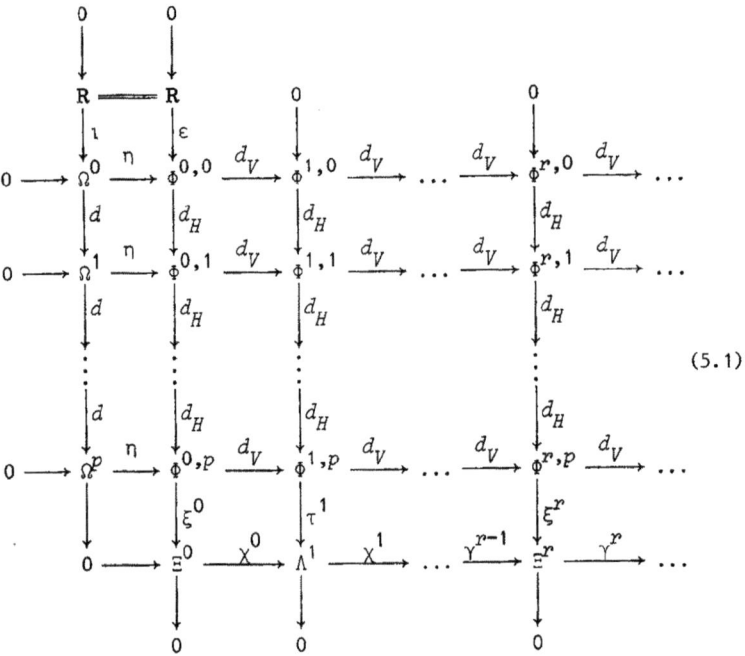

COROLLARY 5.1. *Let L be an element of $\Phi^{0,p}$. The Euler–Lagrange form δL vanishes identically if and only if $L \in d_H(\Phi^{0,p-1})$.*

COROLLARY 5.2. *An element λ of $\Lambda^1 \subset \Phi^{1,p}$ is an Euler–Lagrange form if and only if $d\lambda \in d_H(\Phi^{2,p-1})$.*

The criterion contained in Corollary 5.2 is not effective. To turn it into an effective criterion we need an operator whose kernel is the space $d_H(\Phi^{2,p-1})$.

PROPOSITION 5.1. *For each $r > 0$ the subspace*

$$\Lambda^r = \tau(\Phi^{r,p}) \subset \Phi^{r,p} \tag{5.3}$$

is a complement of the subspace $d_H(\Phi^{r,p-1}) \subset \Phi^{r,p}$.

The proof of Proposition 5.1 is analogous to that of Proposition 4.'

Let

$$\mu = \Sigma_{i,j}\Sigma_{n,l}\ \mu_{ij}{}^{nl} dx^i{}_n \wedge dx^j{}_l \wedge d^p t \tag{5.4}$$

be an arbitrary element of $\Phi^{2,p}$. Repeating the calculations used to derive formula (4.20) we find

$$\begin{aligned}\tau\mu &= \Sigma_{i,j}\Sigma_{m,n,l}\ (-1)^{|m|}\partial^m(\mu_{ij}{}^{nl} dx^i{}_n \wedge \theta_m dx^j{}_l \wedge d^p t)\\ &\quad - \Sigma_{i,j}\Sigma_{m,n,l}\ (-1)^{|m|}\partial^m(\mu_{ij}{}^{nl} dx^j{}_l \wedge \theta_m dx^i{}_n \wedge d^p t)\\ &= \Sigma_{i,j}\Sigma_{n,l}\ (-1)^{|l|}(\partial^l \mu_{ij}{}^{nl} dx^i{}_n) \wedge dx^j \wedge d^p t\\ &\quad - \Sigma_{i,j}\Sigma_{n,l}\ (-1)^{|n|}(\partial^n \mu_{ij}{}^{nl} dx^j{}_l) \wedge dx^i \wedge d^p t.\end{aligned} \tag{5.5}$$

PROPOSITION 5.2. *Let $\Lambda'' \subset \Phi$ denote the subalgebra generated by Φ_0 and Φ^0. Then*

$$\Phi^{r,p} \cap \Lambda'' \subset \Lambda^r \subset \Phi^{r,p} \cap \Lambda' \tag{5.6}$$

for each $r > 0$.

Proof. For $r = 1$ the equality

$$\Phi^{1,p} \cap \Lambda'' = \Lambda^1 = \Phi^{1,p} \cap \Lambda' \tag{5.7}$$

is proved in Proposition 4.2. For $r = 2$ the inclusion

$$\Lambda^2 \subset \Phi^{2,p} \cap \Lambda' \tag{5.8}$$

follows from (5.4). An element μ of $\Phi^{2,p}\cap\Lambda''$ is represented locally by $\mu = \Sigma_{i,j}\, \mu_{ij} dx^i \wedge dx^j \wedge d^p t$ and $\tau\mu = \theta_0\mu = 2\mu$. Hence

$$\Phi^{2,p} \cap \Lambda'' \subset \Lambda^2. \tag{5.9}$$

For $r > 2$ the proof is analogous. ∎

Examples show that neither of the inclusions in (5.6) is an equality for $r > 1$. Hence Proposition 5.2 does not imply a globalization of the construction of spaces Λ^r for $r > 1$. Not having found a coplete intrinsic characterization of these spaces we resort to a technique of globalization based on coordinate invariance.

PROPOSITION 5.3. *Let* (t^α) *and* (t'^α) *be two local charts of* M *and let* τ *and* τ' *be the two associated operators defined by* (4.1). *Then*

$$\tau\mu = \tau'\mu \tag{5.10}$$

for each $\mu \in \Phi^{\bullet,p}$.

Proof. Let ∂'_α and θ'_m be derivations associated with the chart (t'^α). The action of ∂'_α on $\Phi^{\bullet,0}$ is determined by its action on elements $f \in \Phi^{0,0}$ and $d_V f \in \Phi^{1,0}$. From

$$\partial'_\alpha f = \Sigma_\beta \frac{\partial t^\beta}{\partial t'^\alpha} \partial_\beta f \tag{5.11}$$

and

$$\begin{aligned} \partial'_\alpha d_V f = d_V \partial'_\alpha f &= d_V \Sigma_\beta \frac{\partial t^\beta}{\partial t'^\alpha} \partial_\beta f \\ &= \Sigma_\beta \frac{\partial t^\beta}{\partial t'^\alpha} \partial_\beta d_V f \end{aligned} \tag{5.12}$$

we conclude that

$$\partial'_\alpha \nu = \Sigma_\beta \frac{\partial t^\beta}{\partial t'^\alpha} \partial_\beta \nu \tag{5.13}$$

for each $\nu \in \Phi^{\bullet,0}$. Each $\mu \in \Phi^{\bullet,p}$ can be represented by $\nu \wedge d^p t$ where $\nu \in \Phi^{\bullet,0}$. Since

$$\begin{aligned}\partial'_\alpha d^p t &= \partial'_\alpha Det\left(\frac{\partial t^\beta}{\partial t'^\gamma}\right) d^p t' \\ &= Det\left(\frac{\partial t^\beta}{\partial t'^\gamma}\right)\Sigma_{\mu,\nu}\frac{\partial t'^\mu}{\partial t^\nu}\frac{\partial^2 t^\nu}{\partial t'^\mu \partial t'^\alpha}\, d^p t' \\ &= \Sigma_{\gamma,\beta}\frac{\partial t'^\gamma}{\partial t^\beta}\frac{\partial^2 t^\beta}{\partial t'^\gamma \partial t'^\alpha}\, d^p t \\ &= \Sigma_\beta\, \partial_\beta \frac{\partial t^\beta}{\partial t'^\alpha}\, d^p t,\end{aligned} \tag{5.14}$$

we have

$$\begin{aligned}\partial'_\alpha \mu &= \partial'_\alpha \nu \wedge d^p t \\ &= (\partial'_\alpha \nu)\wedge d^p t + \nu\wedge\partial'_\alpha d^p t \\ &= \Sigma_\beta \frac{\partial t^\beta}{\partial t'^\alpha}(\partial_\beta \nu)\wedge d^p t + \nu\wedge\Sigma_\beta\, \partial_\beta\left(\frac{\partial t^\beta}{\partial t'^\alpha}\, d^p t\right) \\ &= \Sigma_\beta\, \partial_\beta \frac{\partial t^\beta}{\partial t'^\alpha}\, \nu\wedge d^p t \\ &= \Sigma_\beta\, \partial_\beta \frac{\partial t^\beta}{\partial t'^\alpha}\, \mu\end{aligned} \tag{5.15}$$

Let $a_m^{\;n}$ be numbers defined for $m,n \in \mathrm{N}^p$ by

$$a_m^{\;0} = \begin{cases} 1 & \text{for } m = 0 \\ 0 & \text{for } m \neq 0 \end{cases} \tag{5.16}$$

and

$$a_m^{\;n} = \frac{\partial a_m^{\;n-e_\alpha}}{\partial t^\alpha} + \Sigma_{\substack{\beta \\ e_\beta \geqq m}} \frac{\partial t'^\beta}{\partial t^\alpha}\, a_{m-e_\beta}^{\;n-e_\alpha} \tag{5.17}$$

for $n \geq e_\alpha$. The equality

$$[\theta_m, \partial'_\alpha] = \Sigma_\beta \frac{\partial t^\beta}{\partial t'^\alpha}\,[\theta_m, \partial_\beta] \tag{5.18}$$

is proved by applying both sides to elements $f \in \Phi^0$, $d_V f$ and dt^γ. With the help of this equality it is easily shown that

$$\theta'_m = \Sigma_n\, a_m^{\;n}\theta_n. \tag{5.19}$$

Relations derived below are valid in $\Phi^{\bullet,p}$. If $n > 0$ then $n \geq e_\alpha$ for some α and

$$\Sigma_m (-1)^{|m|} \partial'^m a_m{}^n = \Sigma_m (-1)^{|m|} \partial'^m \frac{\partial a_m{}^{n-e_\alpha}}{\partial t^\alpha}$$

$$+ \Sigma_\beta \Sigma_{m \geq e_\beta} (-1)^{|m|} \partial'^m \frac{\partial t'^\beta}{\partial t^\alpha} a_{m-e_\beta}{}^{n-e_\alpha}$$

$$= \Sigma_m (-1)^{|m|} \partial'^m \frac{\partial a_m{}^{n-e_\alpha}}{\partial t^\alpha}$$

$$+ \Sigma_\beta \Sigma_{m \geq e_\beta} (-1)^{|m|} \partial'^{m-e_\beta} \partial'_\beta \frac{\partial t'^\beta}{\partial t^\alpha} a_{m-e_\beta}{}^{n-e_\alpha}$$

$$= \Sigma_m (-1)^{|m|} \partial'^m \frac{\partial a_m{}^{n-e_\alpha}}{\partial t^\alpha}$$

$$- \Sigma_\beta \Sigma_m (-1)^{|m|} \partial'^m \partial'_\beta \frac{\partial t'^\beta}{\partial t^\alpha} a_m{}^{n-e_\alpha} \tag{5.20}$$

$$= \Sigma_m (-1)^{|m|} \partial'^m \frac{\partial a_m{}^{n-e_\alpha}}{\partial t^\alpha}$$

$$- \Sigma_m (-1)^{|m|} \partial'^m \partial_\alpha a_m{}^{n-e_\alpha}$$

$$= - \Sigma_m (-1)^{|m|} \partial'^m a_m{}^{n-e_\alpha} \partial_\alpha .$$

Repeating this procedure as many times as necessary we prove that

$$\Sigma_m (-1)^{|m|} \partial'^m a_m{}^n = (-1)^{|n|} \Sigma_m (-1)^{|m|} \partial'^m a_m{}^0 \partial^n$$
$$= (-1)^{|n|} \partial^n . \tag{5.21}$$

Hence

$$\tau' = \Sigma_m (-1)^{|m|} \partial'^m \theta'_m$$
$$= \Sigma_m \Sigma_n (-1)^{|m|} \partial'^m a_m{}^n \theta_n \tag{5.22}$$
$$= \Sigma_n (-1)^{|n|} \partial^n \theta_n = \tau . \quad ∎$$

We introduce operators

$$\tau^r : \Phi^{r,p} \to \Lambda^r : \mu \mapsto \frac{1}{r} \tau\mu \tag{5.23}$$

for each $r > 0$. From Lemma 4.1 we have

$$\tau^r \tau^r \mu = \tau^r \mu \tag{5.24}$$

for each $\mu \in \Phi^{r,p}$. Hence $\mu \in \Lambda^r$ if and only if $\tau^r \mu = \mu$.

It follows from Proposition 5.1 that spaces Λ^r and Ξ^r are isomorphic. Consequently we can replace (3.43) by the following diagram

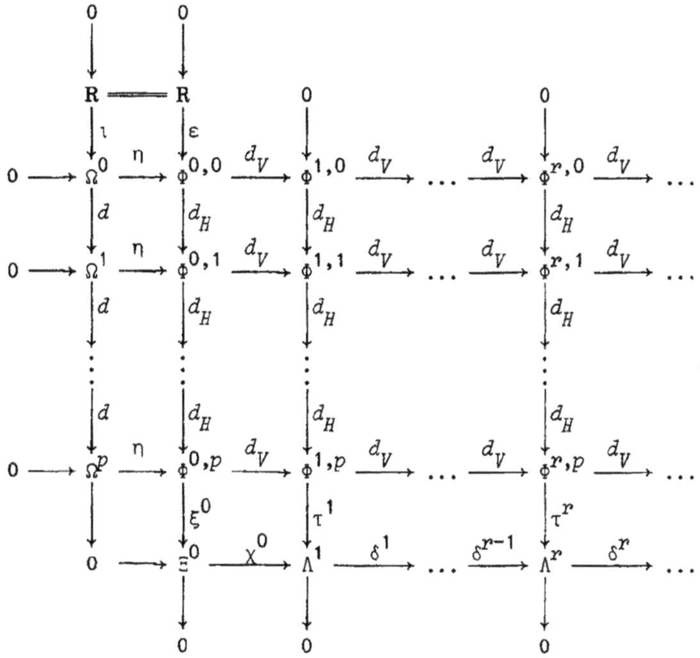

(5.25)

with exact rows and columns. It is easily seen that operators δ^r are defined by

$$\delta^r : \Lambda^r \to \Lambda^{r+1} : \mu \mapsto \tau^{r+1} d\mu. \tag{5.26}$$

For $r = -p, -p+1, \ldots, 0$ we write

$$\Lambda^r = \Phi^{0,r+p}. \tag{5.27}$$

Also

$$\delta^r = d_H \tag{5.28}$$

for $r = -p, -p+1, \ldots, -1$, and

$$\delta^0 = \delta. \tag{5.29}$$

DEFINITION 5.1. The sequence

$$0 \longrightarrow \mathbf{R} \xrightarrow{\varepsilon} \Lambda^{-p} \xrightarrow{\delta^{-p}} \dots$$

$$\dots \xrightarrow{\delta^{-1}} \Lambda^{0} \xrightarrow{\delta^{0}} \Lambda^{1} \xrightarrow{\delta^{1}} \dots \xrightarrow{\delta^{r-1}} \Lambda^{r} \xrightarrow{\delta^{r}} \dots \tag{5.30}$$

is called the *Euler-Lagrange sequence* and $\{\Lambda^r, \delta^r\}$ is called the *Euler-Lagrange complex.*

THEOREM 5.2. *The Euler-Lagrange sequence is exact.*

The proof of this theorem is the same as that of Theorem 5.1.

COROLLARY 5.3. *An element λ of $\Lambda^1 \subset \Phi^{1,p}$ is an Euler-Lagrange form if and only if $\tau d\lambda = 0$.*

The criterion expressed in Corollary 5.3 is effective. Corollaries 5.1 and 5.3 express that part of the content of Thoerem 5.2 which is applicable to the calculus of variations.

Appendix. Proof of Proposition 3.1.

LEMMA A.1. *The sequence*

$$0 \longrightarrow \mathbf{R} \xrightarrow{\iota} \Omega^{0} \xrightarrow{d} \Omega^{1} \xrightarrow{d} \dots \xrightarrow{d} \Omega^{p} \longrightarrow 0 \tag{A.1}$$

is exact.

This lemma is the standart Poincaré lemma for differential forms on M.

LEMMA A.2. *The sequence*

$$0 \longrightarrow \Omega^{0} \xrightarrow{\eta} \Phi^{0,0} \xrightarrow{d_V} \Phi^{1,0} \xrightarrow{d_V} \dots \xrightarrow{d_V} \Phi^{r,0} \xrightarrow{d_V} \dots \tag{A.2}$$

is exact.

Proof. An element $\mu \in \Phi^{\cdot,0} \cap \Phi_k$ can be viewed as a family of differential forms on fibres of π^k parametrized over M. An element $\mu \in \Omega^0$ is a function on M. Exactness of (A.2) follows from a version of the Poincaré lemma with parameters. ∎

LEMMA A.3. *For each $s = 1,\ldots,p$ the sequence*

$$0 \longrightarrow \Omega^s \xrightarrow{\eta} \Phi^{0,s} \xrightarrow{d_V} \Phi^{1,s} \xrightarrow{d_V} \cdots \xrightarrow{d_V} \Phi^{r,s} \xrightarrow{d_V} \cdots \tag{A.3}$$

is exact.

Proof. If $\mu \in \Phi^{r,s}$ is represented as in formula (3.48) and $d_V\mu = 0$ then $d_V\mu^{\alpha_1\ldots\alpha_{p-s}} = 0$ and by Lemma A.3, $\mu^{\alpha_1\ldots\alpha_{p-s}} = d_V\nu^{\alpha_1\ldots\alpha_{p-s}}$ for some $\nu^{\alpha_1\ldots\alpha_{p-s}} \in \Phi^{r-1,0}$ or $\mu^{\alpha_1\ldots\alpha_{p-s}} = \eta\nu^{\alpha_1\ldots\alpha_{p-s}}$ for some $\nu^{\alpha_1\ldots\alpha_{p-s}} \in \Omega^0$. Hence $\mu = d_V\nu$ or $\mu = \eta\nu$, where

$$\nu = \Sigma_{\alpha_1<\ldots<\alpha_{p-s}} (d^s t)_{\alpha_1\ldots\alpha_{p-s}} \wedge \mu^{\alpha_1\ldots\alpha_{p-s}}. \quad \blacksquare \tag{A.4}$$

LEMMA A.4. *The sequence*

$$\begin{aligned} 0 \longrightarrow \mathbf{R} \xrightarrow{\varepsilon} \Phi^{0,0} \xrightarrow{d_H} \Phi^{0,1} \xrightarrow{d_H} \cdots \\ \cdots \xrightarrow{d_H} \Phi^{0,p} \xrightarrow{\xi^0} \Xi^0 \longrightarrow 0 \end{aligned} \tag{A.5}$$

is exact.

Proof. For each $s = 1,\ldots,p-1$ the top statement in the sequence of statements

$$\begin{aligned} &ker(d_Vd_H:\Phi^{s,0} \to \Phi^{s+1,1}) = d_V(\Phi^{s-1,0}) \\ &ker(d_Vd_H:\Phi^{s-1,1} \to \Phi^{s,2}) = d_V(\Phi^{s-2,1}) + d_H(\Phi^{s-1,0}) \\ &\cdots\cdots\cdots\cdots\cdots\cdots\cdots\cdots\cdots\cdots\cdots\cdots \\ &ker(d_Vd_H:\Phi^{1,s-1} \to \Phi^{2,s}) = d_V(\Phi^{0,s-1}) + d_H(\Phi^{1,s-2}) \\ &ker(d_Vd_H:\Phi^{0,s} \to \Phi^{1,s+1}) = \eta(\Omega^s) + d_H(\Phi^{0,s-1}) \end{aligned} \tag{A.6}$$

follows from Lemmas A.3 and 3.1 and each of the remaining statements is a consequence of the statement immediately above and Lemmas A.3 and 3.1. Separately we prove

$$ker(d_H\eta:\Omega^s \to \Phi^{0,s+1}) = d(\Omega^{s-1}) \tag{A.7}$$

from Lemmas A.3 and 3.1. For $s = 0$ the bottom statement of (A.6) and the statement (A.7) must be replaced by

$$\begin{aligned} ker(d_V d_H : \Phi^{0,0} \to \Phi^{1,1}) &= \eta(\Omega^0) + \varepsilon(\mathbf{R}) \\ &= \eta(\Omega^0) \end{aligned} \tag{A.8}$$

and

$$ker(d_H \eta : \Omega^0 \to \Phi^{0,1}) = \iota(\mathbf{R}). \tag{A.9}$$

If $s = 1,..,p-1$, $\mu \in \Phi^{0,s}$ and $d_H\mu = 0$ then $d_V d_H \mu = 0$ and there exist elements $\kappa \in \Omega^s$ and $\lambda \in \Phi^{0,s-1}$ such that $\mu = \eta\kappa + d_H\lambda$. The element κ satisfies $d_H\eta\kappa = d_H(\mu - d_H\lambda) = 0$. Hence $\kappa = d\nu$ for some $\nu \in \Omega^{s-1}$. Consequently $\mu = \eta d\nu + d_H\lambda = d_H(\eta\nu + \lambda) \in d_H(\Phi^{0,s-1})$.

If $\mu \in \Phi^{0,0}$ and $d_H\mu = 0$ then $d_V d_H \mu = 0$ and there exists a function $\kappa \in \Omega^0$ such that $\mu = \eta\kappa$. Since $d_H\eta\kappa = d_H\mu = 0$ we have $\kappa = \iota\nu$ for some number $\nu \in \mathbf{R}$. Consequently $\mu = \eta\iota\nu = \varepsilon\nu \in \varepsilon(\mathbf{R})$.

We have shown that

$$ker(d_H : \Phi^{0,s} \to \Phi^{0,s+1}) \subset d_H(\Phi^{0,s-1}) \tag{A.10}$$

for each $s = 1,...,p-1$ and

$$ker(d_H : \Phi^{0,0} \to \Phi^{0,1}) \subset \varepsilon(\mathbf{R}). \tag{A.11}$$

The rest of the proof is obvious. ∎

LEMMA A.5. *The sequence*

$$0 \longrightarrow \Xi^0 \xrightarrow{\gamma^0} \Xi^1 \xrightarrow{\gamma^1} \cdots \xrightarrow{\gamma^{r-1}} \Xi^r \xrightarrow{\gamma^r} \cdots \tag{A.12}$$

is exact.

Proof. For each $r > 0$ we have the sequence of statements

$$\begin{aligned} &ker(d_V d_H : \Phi^{r+p,0} \to \Phi^{r+p+1,1}) = d_V(\Phi^{r+p-1,0}) \\ &ker(d_V d_H : \Phi^{r+p-1,1} \to \Phi^{r+p,2}) = d_V(\Phi^{r+p-2,1}) + d_H(\Phi^{r+p-1,0}) \\ &\cdots\cdots\cdots\cdots\cdots\cdots\cdots\cdots\cdots\cdots \\ &ker(d_V d_H : \Phi^{r+1,p-1} \to \Phi^{r+2,p}) = d_V(\Phi^{r,p-1}) + d_H(\Phi^{r+1,p-2}) \\ &ker(\gamma^r \xi^r) = d_V(\Phi^{r-1,p}) + d_H(\Phi^{r,p-1}) \end{aligned} \tag{A.13}$$

proved as in Lemma A.4. The bottom statement of (A.13) is replaced by

$$ker(\gamma^0 \xi^0) = \eta(\Omega^p) + d_H(\Phi^{0,p-1}) \tag{A.14}$$

for $r = 0$.

If $r > 0$ and $\mu \in \Xi^r$ then there is an element $\nu \in \Phi^{r,p}$ such that $\mu = \xi^r\nu$. If $\gamma^r\mu = 0$ then $\gamma^r\xi^r\nu = 0$ and $\nu = d_V\kappa + d_H\lambda$ for some $\kappa \in \Phi^{r-1,p}$ and some $\lambda \in \Phi^{r,p-1}$. Consequently $\mu = \xi^r\nu = \xi^r d_V\kappa = \gamma^{r-1}\xi^{r-1}\kappa \subset \gamma^{r-1}(\Xi^{r-1})$.

If $\mu \in \Xi^0$ and $\gamma^0\mu = 0$ then there is an element $\nu \in \Phi^{0,p}$ such that $\mu = \xi^0\nu$. Since $\gamma^0\xi^0\nu = 0$ we have $\nu = \eta\kappa + d_H\lambda$ for some $\kappa \in \Omega^p$ and some $\lambda \in \Phi^{0,p-1}$. Consequently $\mu = \xi^0\nu = \xi^0\eta\kappa = 0$.

We have shown that $ker(\gamma^r) \subset \gamma^{r-1}(\Xi^{r-1})$ for $r > 0$ and $ker(\gamma^0) = 0$. Hence the sequence (A.12) is exact. ∎

Lemmas A.1 through A.5 and 3.1 express the content of Proposition 3.1.

References.

[1] Dedecker, P., *Sur un problème inverse du calcul des variations*, Bull. Acad. Roy. Belgique, 36, 1950, p. 63-70.

[2] Dedecker, P., *Applications of homological algebra to calculus of variations and mathematical physics*, to appear.

[3] Frölicher, A. and Nijenhuis, A., *Theory of vector valued differential forms*, Nederl. Akad. Wetensch., Proc., A 59, 1956, p. 338-359.

[4] Hughes, T.J.R. and Marsden, J.E., *Some applications of geometry in continuum mechanics*, Reports on Math. Phys., 12, 1977, p. 35-44.

[5] Kumpera, A., *Invariants différentiels d'un pseudogroupe de Lie, I.*, J. Differential Geometry, 10, 1975, p. 289-345.

[6] Kupershmidt, B.A., *The Lagrangian formalism in the calculus of variations*, Func. Anal. Appl., 10, 1976, p. 147-149.

[7] Mitchell, B., *Theory of categories*, Academic Press, 1965.

[8] Takens, F., *Symmetries, conservation laws and variational principles*, Lecture Notes in Mathematics, 597, Springer-Verlag, 1977.

[9] Tonti, E., *Variational formulations of nonlinear differential equations*, Acad. Roy. Belg. Bull., 55, 1969, p. 137-165.

[10] Tulczyjew, W.M., *Sur la différentielle de Lagrange*, C. R. Acad. Sc. Paris, 280, 1975, p. 1295-1298.

[11] Tulczyjew, W.M., *The Lagrange differential*, Bull. Acad. Polon. Sc., 24, 1976, p. 1089-1096.

[12] Tulczyjew, W.M., *The Lagrange complex*, Bull. Soc. math. France, 105, 1977, p. 419-431.

[13] Veinberg, M.M., *Variational methods in the theory of nonlinear operators*, Holden-Day, 1964.

[14] Vinogradov, A.M., *On the algebro-geometric foundations of Lagrangian field theory*, Soviet Math. Dokl., 18, 1977, p. 1200-1204.

[15] Vinogradov, A.M., *A spectral sequence associated with a nonlinear differential equation, and algebro-geometric foundations of Lagrangian field theory with constraints*, Soviet Math. Dokl., 19, 1978, p. 144-148.

ON THE PREQUANTUM DESCRIPTION OF SPINNING PARTICLES IN AN EXTERNAL GAUGE FIELD

C. DUVAL

Faculté des Sciences de Luminy
et Centre de Physique Théorique CNRS Marseille (*)

Introduction.

In the last few years there has been a renewed interest in the so called problem of the equations of motion. The reason for this must be ascribed to the recent achievements in gauge theories which prompted physicists and geometers to tackle the question of the dynamics of test particles in an external Yang-Mills field.

The case of gravitation and electromagnetism has been investigated for a long time. Since the geodesic motion in a gravitational field and the Laplace law had to be modified to take into account the spin of test particles, new points of view on the formulation of classical mechanics itself had to be adopted. We refer at this stage to the decisive break-through of symplectic mechanics which is associated with the names of Kirillov, Kostant and Souriau. Although quantization is the main programme that has been pursued from the beginning, new insights into the very nature of "classical" systems helped for a better understanding of the principle of correspondence. There is a wide unanimity on the specific contribution of Souriau [17] to the physical interpretation of that theory.

Let us quote the basic references [16], [18] which have influenced this work. On the other hand, techniques initiated by Papapetrou [12] have been thoroughfully exploited by Dixon [2]. They are known as the pole-dipole approximation procedures. Souriau has proposed a synthetic treatment of both approaches in [18].

(*) Postal address : CNRS-Luminy - Case 907
Centre de Physique Théorique - F 13288 Marseille Cedex 2

As far as abelian gauge theories (except gravitation wich plays sort of a priviledged role as will soon be discussed) are concerned, the situation is quite clear and every model is due to lead to the Bargmann-Michel-Telegdi equations (the Papapetrou equations) which have received neat experimental confirmations.

The case of non abelian gauge theories is somewhat different because no experimental evidence of what should be the behaviour of e.g. isospin in a Yang-Mills field is up to now available. It would be helpless to hide the highly speculative character of all attempts to describe internal structure by means of classical objects. Nevertheless, inasmuch as classical spin has successfully been described on purely geometrical grounds, it seems reasonable to think that geometry would again allow for a description of internal structure which would not be devoid of physical meaning. This point of view has been espoused, for the first time to our knowledge, by Sternberg in a series of papers listed in the reference [15]. Developments along these lines can be found in [5],[14],[21]. The striking point is that symplectic mechanics turns out to be well adapted to handle such notions as phase spaces, internal degrees of freedom, minimal coupling, symmetry breaking, etc... in a geometrical, and hence intrinsic manner. Also prequantization (Kostant-Souriau) helps to elucidate (iso)spin, isospin-hypercharge quantization as a first stage towards full quantization.

There is, however, another standpoint (the pseudomechanics of Grassmann variables) which has been adopted by Balachandran et al. [1],[13] to derive the equations of motion for a spinning particle in a gauge field. This work constitutes an interesting alternative. However, no clearcut physical justification of anticommuting variables at the classical level has been proposed by the authors, and the relevance of these structures for classical mechanics is still under dispute. As for us, we think that Kirillov-Kostant-Souriau theory avoids these drawbacks and yields a perfectly well behaved description of phase spaces for spin, isospin, etc... in terms of homogeneous Kähler manifolds.

The purpose of this paper is to show how the derivation of the equations of motion of a spinning particle moving under the influence of an external gauge field can be carried out starting from two complementary points of view : the principle of general covariance and the symplectic geometry. The dialectic relationship between both approaches is analyzed in section 5.

Section 1 is devoted to the introduction of minimal coupling of a spinless particle to a Yang-Mills field. The basic geometrical structures show up there. The Wong equations of motion are given as an outcome of the model. Let us recall that parallel transport of the isospin was first established by Wong [22] by quasi-classical arguments.

The principle of general covariance is presented in section 2. In order to deal with test particles in an external gauge field, we have found it necessary to come back to the basic principle of gauge independance of any physical theory. The point of view we develop here stresses the role of the automorphisms of the principal G-fibre bundle of the theory we start with. The striking feature of this approach is that gravitation is necessarily taken into account through the spacetime diffeomorphisms. Einstein-Yang-Mills identities of conservation for a continuum are interpreted as the dual counterpart of the gauge freedom for the choice of the E.Y.M. potentials (g , A). As for the case of concentrated matter (a particle travelling on a curve of spacetime), the proposed procedure leads to the definition of the momentum P, the spin S, the internal pole Q and dipole M structure of a test particle. One gets in this manner a set of universal equations of motion which generalize the Papapetrou-Dixon-Souriau - ones [12],[2],[18]. See also [4]. By standard arguments, we propose to introduce the dipole coupling to the Yang-Mills field by letting the "mass" depend in an unspecified manner on the term Q.F(S) - F is the field strength. In the case of a weak and slowly varying electromagnetic field (G = U(1)), these equations reduce to the B.M.T. equations [17].

One of the most appealing results of this theory is the consistency of the previous equations of motion with a symplectic model.

An introductory presentation of free spinning particle symplectic models is performed in section 3,ii. It is shown that the Poincaré group, as an autonomous entity, plays a central role in the construction of the phase space as a coadjoind orbit of the Poincaré group. This fact has first been emphasized by Künzle [11]. Also prequantization of the physically relevant orbits is performed with the help of the Maurer-Cartan 1 - form of the Poincaré Group. Spin quantization appears then as part of a constructive description of the prequantum bundles. Section 3,iii deals with internal phase spaces which are introduced as coadjoint orbits of a given internal symmetry Lie group G. Prequantization of these orbits is carried out in the physically relevant case of a compact semi-simple Lie group G. The special case G = SU(n) is investigated in full details. We have also proposed a geometrical interpretation of the classical multiplets such as the "quark orbit", the "meson orbit", etc... associated with the representations $\{3\}$, $\{8\}$, etc... of SU(3). The phase space of massive spinning particles with internal degrees of freedom is naturally introduced as a coadjoint orbit of the direct product of the Poincaré group with the internal symmetry group G.

The symplectic models for spinning test particles in a Yang-Mills field are presented in section 4. They are labelled by a real positive function f, $P^2 = f(Q \cdot F(S))$. As long as one insists on the permanence of a symplectic structure for dynamical systems in interaction, one must take this mass coupling as a reasonable one. The point is that this coupling allows for much flexibility especially in the case of strong external fields. Several remarks concerning the notion of minimal coupling end this section. It would be interesting to find other coupling procedures which still admit a symplectic model.

We would like to list, by way of conclusion, several open questions which would deserve further investigations. First of all, the procedure of symmetry breaking (Marsden-Weinstein) should lead, to some extent, to mass formulas. Also statistical mechanics of hadrons could be formulated according to the idea that the "temperature" vectorfield might be taken as an infinitesimal automorphism (of the G-fibre bundle of the theory) which projects onto a Killing vector field of spacetime. In case G is compact, the notion of critical temperature of hadrons should presumably be an outcome of the theory.

§1. The Wong equations of motion.

The geometrical description of the minimal coupling of a relativistic spinless particle to an external gauge field is originally due to Sternberg [15] and is presented here with minor modifications leading to more simplicity.

Let $\pi : P \to M$ be a principal G-fibre bundle over Minkowski spacetime (M,g) and ω be a given connection form on P [10]. Consider now the "evolution" space (V, ϖ) of a free spinless particle of mass m ;
$V := \{(P,X) \in TM,\ g(P,P) = m^2 = \text{const.},\ P \text{ future pointing}\}$;
$\varpi := g(P, dX)$. $\tilde{\pi} : V \to M$ denotes the projection $\tilde{\pi}(P,X) = X$.
Introduce then the pull back $\tilde{P} := \tilde{\pi}^* P$. Let us further choose a fixed point q in $\mathfrak{g}^*$, the dual of the Lie algebra $\mathfrak{g}$ of G in order to define

$$\alpha := q \cdot \omega \tag{1}$$

The 1-forms ϖ and α can be pulled back to $\tilde{P}$ and we can thus define with a slight abuse of notations the "minimal coupling" 1-form θ on $\tilde{P}$ as follows

$$\theta := \varpi + \alpha \tag{2}$$

Using the definition of the curvature Ω, we have $d\theta = d\varpi + q\cdot(\Omega - [\omega,\omega])$. But $d\theta$ is degenerate and in particular $\hat{Z} \in \ker(d\theta)$, $\hat{Z}$ denotes the fundamental vector field on $\tilde{P}$ associated with $Z \in \mathfrak{g}$ [10], if $q \cdot Ad(Z) = 0$ i.e. if $Z \in \mathfrak{g}_q$ ($\mathfrak{g}_q$ is the Lie algebra of the isotropy subgroup G_q of q for the coadjoint action ad^*). Clearly $d\theta$ passes to the quotient $\tilde{P}/G_q = \tilde{P} \times_G \mathcal{O}_q$ [10] which is interpreted as the new evolution space of dimension 2n + 7 (2n is the dimension of the coadjoint orbit $\mathcal{O}_q := \overset{*}{ad}(G)q)$. See [15] for an equivalent point of view. If we define

$$Q := \overset{*}{ad}(\tilde{u})\, q \qquad (3)$$

$\tilde{u} \in \tilde{P}$, then $Q \in \tilde{P}/G_q$. Introduce the field strength F by

$$\Omega =: ad(\tilde{u}^{-1})\, \pi^* F$$

in order to compute $\ker(d\theta)$ which yields the following equations of motion

$$\begin{cases} V = \lambda P & \lambda \in \mathbb{R} \\ \nabla_V \bar{P} = -Q \cdot F(V) \\ D_V Q = 0 \end{cases} \qquad (4)$$

where V denotes the projection on M of $\zeta \in \ker(d\theta)$. D is the covariant derivative .

We have used the notation $\bar{P} := g(P,\cdot) =: g(P)$. These equations are known as the Wong equations of motion. The generalized isospin Q ($\mathcal{O}_q$ is the internal phase space) is parallel transported along the worldline of the particle. ∇ denotes here the (flat) Lorentzian connection of (M,g) ; in the case of a curved spacetime, these equation retain the same expression.

§2. The principle of general covariance.

Let us start with a principal G-fibre bundle $\pi : P \to M$ over spacetime (M,g). We will call Einstein-Yang-Mills potential the triple $\varphi := (g, s, \omega)$ where ω denotes a connection form on P and s a local cross section of P. We also define $A := s^*\omega$, the Yang-Mills potential in the gauge s. Let us recall that the group $\mathrm{Aut}(P) = \{\alpha \in \mathrm{Diff}(P),\ \alpha \circ r_a = r_a \circ \alpha\ \forall a \in G\}$ acts on the set of connections of P according to $\alpha(\omega) = (\alpha^{-1})^*\omega$ [20]. Furthermore $\alpha \in \mathrm{Aut}(P)$ uniquely defines $\hat{\alpha} \in \mathrm{Diff}(M)$ by $\pi \circ \alpha = \hat{\alpha} \circ \pi$. Trivially $\widehat{\alpha' \circ \alpha''} = \hat{\alpha}' \circ \hat{\alpha}''$. It is natural to define the following action of Aut(P), the full gauge group of the theory, on the set of Einstein-Yang-Mills potentials

$$\alpha(g,s,\omega) := \left((\hat{\alpha}^{-1})^{*} g \ , s, (\alpha^{-1})^{*}\omega \right) \qquad (5)$$

$\forall \alpha \in$ Aut(P) (= Diff(M) $\times$ $\Gamma(P \times_G G)$, $\times :=$ extension). The "Lie algebra" aut(P) has the structure $\mathfrak{X}(M) \times \Gamma(P \times_G \mathfrak{g})$ where $\mathfrak{X}(M) = \Gamma(TM)$ is the Lie algebra of vector fields of M and $P \times_G \mathfrak{g}$ the associated bundle with fibre ($\mathfrak{g}$, ad). Let $Z \in$ aut(P), put $X := \pi_* Z$ and $\xi := s^*(\omega(Z - (s\circ\pi)_* Z))$. The infinitesimal version of (5) is easily found

$$(X, \xi)(g, A) = (\mathcal{L}_X g \ , \mathcal{L}_X A + D\xi) \qquad (6)$$

$X \in \mathfrak{X}(M)$, $\xi \in \Gamma(P \times_G \mathfrak{g})$ (locally $C^\infty(M, \mathfrak{g})$). Again D denotes the covariant derivative (locally $D\xi = d\xi + Ad(A)(\xi)$).

i) <u>The Einstein-Yang-Mills continuum</u>. A straightforward generalization of Souriau's approach [17] leads us to define the E.Y.M. continuum as a covector of the set of E.Y.M. potentials given by the completely continuous functional μ

$$\langle \mu, \delta\varphi \rangle := \int \left(\tfrac{1}{2}\ T(\delta g) + J \lrcorner\, \delta A\right) \text{vol}$$

where $T \in \Gamma(V^2 TM)$ and $J \in \Gamma((P \times_G \mathfrak{g}^*) \otimes TM)$; vol denotes the pseudoriemannian volume element of (M,g). The principle of general covariance which expresses that physical laws should be gauge independant can be stated as follows : let $\text{Aut}_0(P)$ denote the group of all automorphisms of P which project onto compactly supported diffeomorphisms of M, its orbits for the action (5) define a foliation with respect to which μ is required to be semi-basic, i.e.

$$\langle \mu, (\mathcal{L}_X g, \mathcal{L}_X A + D\xi)\rangle = 0 \quad \forall X \in \mathfrak{X}_0(M), \forall \xi \in C_0^\infty(M, \mathfrak{g}) \qquad (7)$$

We readily obtain that necessarily Div(J) = 0 and div(T) + $J \lrcorner F = 0$ (the Einstein-Yang-Mills identities of conservation) (1). Whence the physical interpretation : T is the energy momentum tensor and J the

(1) $\text{div}(J.\xi) = \text{Div}(J).\xi + J \lrcorner D\xi$, div denotes the Lorentzian divergence. $F = dA + [A, A]$.

Yang-Mills current. Conservation laws can be derived if there exist $X \in \mathfrak{X}(M)$ and $\xi \in C^{\infty}(M, \mathfrak{g})$ such that $\mathcal{L}_X g = 0$ and $\mathcal{L}_X A + D\xi = 0$ (see (6)). Under these circumstances

$$div\,(\,T(X) + J.\,(A(X) + \xi\,)) = 0$$

ii) <u>The pole-dipole approximation</u>. <u>Test particles endowed with spin and internal structure</u>. We consider now a first order distribution [(1)]

$$\langle \mu, \delta\varphi \rangle := \int_c \{ \tfrac{1}{2} T^{\alpha\beta} \delta g_{\alpha\beta} + J^{\alpha}_{a} \delta A^{a}_{\alpha} + \Theta^{\alpha\beta}_{\gamma} \delta \Gamma^{\gamma}_{\alpha\beta} + \tfrac{1}{2} M^{\alpha\beta}_{a} \delta F^{a}_{\alpha\beta} \}\, dt$$

supported on a curve $c : t \mapsto c(t)$ of M. The $\Gamma^{\gamma}_{\alpha\beta}$ are the Christoffel symbols of the Lorentzian connection ∇ of (M,g). The geometrical objects T, J, Θ, M are defined on c and turn out to satisfy some algebraic properties, by virtue of the principle of general covariance (7). For example, if X = 0, we must have $\int J^{\alpha}_{a} D_{\alpha} \xi^{a} - \tfrac{1}{2} M^{\alpha\beta}_{a} \epsilon^{a}_{\cdot bc} F^{c}_{\alpha\beta}\, dt = 0 \quad \forall\, \xi^{a}$ with compact support. In particular, if we replace ξ by $f\xi$ ere $f \in C^{\infty}(M)$, f = 0 on c, we get $\int J^{\alpha}_{a} \xi^{a} \partial_{\alpha} f\, dt = 0 \quad \forall\, \xi^{a}$ with compact support, hence

$$J = Q \otimes V$$

where $V := c_*(\partial/\partial t)$ is the velocity. $Q \in \Gamma(P \times_G \mathfrak{g}^*)$ is actually defined on c and is interpreted as the internal structure of the particle (e.g. isospin). Furthermore

$$\int (Q_a \xi^a)^{\cdot} - \xi^a (\dot{Q}_a + \tfrac{1}{2} \epsilon^{c}_{\cdot ab} M^{\alpha\beta}_{c} F^{b}_{\alpha\beta})\, dt = 0 \quad \forall\, \xi^a$$

and thus $\dot{Q}_a + \tfrac{1}{2} \epsilon^{c}_{\cdot ab} M^{\alpha\beta}_{c} F^{b}_{\alpha\beta} = 0$ where $\dot{Q} := D_V Q$ [(2)].

Taking into account the diffeomorphisms of M we end up with the final expression [(3)]

$$\langle \mu, \delta\varphi \rangle = \int_c \{ \tfrac{1}{2} (P^{\alpha} V^{\beta} - M^{\gamma\alpha}_{a} F^{\beta a}_{\gamma})\, \delta g_{\alpha\beta} + \tfrac{1}{2} S^{\alpha\beta} V^{\gamma} \nabla_{\alpha} \delta g_{\beta\gamma} + Q_a V^{\alpha} \delta A^{a}_{\alpha} + M^{\alpha\beta}_{a} \overset{D}{\nabla}_{\alpha} \delta A^{a}_{\beta} \}\, dt \tag{8}$$

The uniquely defined quantities P, S, Q, M are respectively interpreted as the linear momentum, the spin tensor ($S^{(\alpha\beta)} = 0$), the

(1) α, β, γ = 1, 2, 3, 4 and a, b, c = 1, 2, ..., dim(G) ; the $\epsilon^{a}_{\cdot bc}$'s are the structure constans of $\mathfrak{g}$.

(2) locally DQ = dQ − Q.Ad(A).

(3) $\overset{D}{\nabla} \delta A = \nabla \delta A + [A, \delta A]$. The Bianchi identities $dF + [A \wedge F] = 0$ [20] have been used to find (8).

internal structure, the internal dipole structure of the test particle. Theses quantities obey furthermore the following universal equations of motion [(1)]

$$\begin{cases} \dot{\bar{P}} = 1/2\ \bar{V} R(S) - Q.F(V) + 1/2\ M.(\overset{D}{\nabla} F) \\ \dot{S} = P\bar{V} - V\bar{P} + [M ; F] \\ \dot{Q} = 1/2\ M\ (Ad(F)) \end{cases} \qquad (9)$$

which generalize the Balachandran et al equations [1],[13].

If the potential (g,A) is non trivially stabilized in aut(P) just as in the above mentionned case of continua, the Noether theorem takes the form

$$(\bar{P} + Q.A)(x) + 1/2\ S(\nabla \bar{x}) + Q.\xi = \text{const.}$$

regardless to the non deterministic character of the latter set of equations. Some subsidiary conditions are necessary : monolocality ($P \in \ker(S)$) and the following generalization of the Maxwell case

$$M = \chi\ Q \otimes S \qquad (10)$$

where χ is any density of c (the generalized magnetic moment). A tedious calculation gives the expression of the velocity V (see (16) below), also $\bar{P}V(\bar{P}P)^{\cdot} = \bar{P}P(Q.F(S))^{\cdot}$ which forces the mass to depend in an uspecified manner on the quantity Q.F(S) - see [15],[4].

§3. The symplectic model of a spinning particle with internal structure. Prequantization.

The phase space of a free relativistic spinning particles will be introduced as a coadjoint orbit (of mass m and spin s) of the universal covering $\hat{P}_o$ of the restricted Poincaré group P_o [17]. It looks

(1) $\dot{P} := \nabla_V P$. R denotes the Riemann-Christoffel curvature of (M,g). (   ) means Lorentz contraction and [,] Lorentz commutator.

reasonable to let the internal degrees of freedom come into play by considering, from the outset, coadjoint orbits of $\hat{\mathcal{P}}_0 \times G$, where G is the internal symmetry (Lie) group of the theory.

i) <u>The canonical symplectic structure of Kirillov orbits</u> [9],[7]. Suppose we are given a point q in $\mathfrak{g}^*$, the dual of the Lie algebra of a Lie group G, then the real 1-form of G

$$\varpi := q \cdot \omega \tag{11}$$

(ω denotes the Maurer-Cartan 1-form of G) has the following property : $d\varpi = \pi^* \sigma$ where σ denotes the canonical symplectic structure of the coadjoint orbit $\mathcal{O}_q = ad^*(G)q = \pi(G)$. G is thus endowed in this manner with a left-invariant presymplectic structure $d\varpi$ whose kernel is generated by $\mathfrak{g}_q$ (the Lie algebra of the isotropy subgroup G_q with respect to ad^*). Note that these considerations will prove useful for prequantization [17]. We claim that $\hat{\mathcal{P}}_0 \times G$ plays the role of generalized "evolution" space since ker($d\varpi$) will appear to project onto a 1-dimensional distribution of Minkowski spacetime (1), whose integral curves are the possible worldlines of the particle. Ker($d\varpi$) yields thus the equations of motion.

ii) <u>The free massive spinning particle</u>. Let us illustrate the previous considerations in the case of $\hat{\mathcal{P}}_0$. Let us start with E $= (\mathbb{C}^{2,2}, H, J)$, the space of Dirac spinors, where we have chosen a pseudo-hermitian scalar product $H = 1/2 \begin{pmatrix} 0 & 1 \\ 1 & 0 \end{pmatrix}$ and where $J = B \begin{pmatrix} 0 & j \\ j & 0 \end{pmatrix} B^{-1}$ stands for the associated quaternionic (2) structure. B is a basis of E such that $\bar{B}\, B = H$ (the bar means adjoint with respect to H). If we are

(1) $\hat{\mathcal{P}}_0 \to M$ is a principal $\hat{L}_0$ fibre bundle (L_0 is the restricted Lorentz group). In the non-relativistic framework, the extended Galilei group should be considered instead of the Galilei group itself in order to overcome a well known cohomological obstruction [6].

(2) $j = \begin{pmatrix} 0 & -C \\ C & 0 \end{pmatrix}$ where C denotes the complex conjugation.

given a Lorentz basis of $(M, - - - +)$, we define the Dirac matrices γ_α ($\alpha = 1, \ldots, 4$) by [17]

$$\gamma(X) = \gamma_\alpha X^\alpha = B \begin{pmatrix} 0 & \sigma_A X^A + X^4 \\ -\sigma_A X^A + X^4 & 0 \end{pmatrix} B^{-1} \quad (X \in M)$$

Also $\gamma_5 := \gamma_1 \gamma_2 \gamma_3 \gamma_4$.

We have $\hat{L}_0 = \mathrm{Spin}(M) = \{ a \in U(E) ;\ aJ = Ja ;\ a\gamma_5 = \gamma_5 a \}$.
As a classical result Spin(M) is diffeomorphic with Sl(2,C) and moreover $\hat{L}_0 \cong \{ Z \in \mathbb{C}^{4,2} ;\ \bar{Z} Z = 1 , \bar{Z} \gamma_5 Z = 0 \}^{(1)}$.

We can now realize $\hat{\mathcal{P}}_0$ as the multiplicative group of the matrices

$$g := \begin{pmatrix} ad(a) & \gamma(X) \\ 0 & 1 \end{pmatrix} , \quad a \in \hat{L}_0 , \quad X \in M$$

with $ad(a)\gamma(X) := a\gamma(X)a^{-1} = \gamma(AX)$ $(A \in L_0)$. The Lie algebra $\mathcal{P}_0$ is the set of the matrices of the form $^{(2)}$

$$(\Lambda , \Gamma) := \begin{pmatrix} Ad(\gamma(\Lambda)) & \gamma(\Gamma) \\ 0 & 0 \end{pmatrix} , \quad \Lambda \in so(M), \ \Gamma \in M.$$

Let us denote by (M,P), an element of the dual $\mathcal{P}_0^*$. The pairing between $\mathcal{P}_0$ and $\mathcal{P}_0^*$ is defined by $(M,P).(\Lambda,\Gamma) = \frac{1}{2}\mathrm{Tr}(M\Lambda) + \bar{P}\Gamma$.
It is a simple matter to check that the Maurer-Cartan form of $\hat{\mathcal{P}}_0$ reads

$$\omega_{\mathcal{P}} = g^{-1} dg = \begin{pmatrix} Ad(\omega_L) & \gamma(A^{-1} dX) \\ 0 & 0 \end{pmatrix}$$

Let us choose in $\mathcal{P}_0^*$

$$q = \begin{pmatrix} Ad(2s\,\gamma_1\gamma_2) & m\,\gamma_4 \\ 0 & 0 \end{pmatrix}$$

whose orbit is labelled by the invariants s, m $\in \mathbb{R}^+ - \{0\}$ (spin and mass). After a short calculation, we find that $q.\omega_{\mathcal{P}}$ is the following

(1) Every $a \in \hat{L}_0$ is of the form $a = (IZ\ IJZ\ \bar{I}Z\ \bar{I}JZ)B^{-1}$, $I := (1 + i\gamma_5)/2$.
(2) $\gamma(\Lambda) := \frac{1}{4} \gamma_\alpha \gamma_\beta \Lambda^{\alpha\beta}$.

1-form of $\hat{\mathcal{P}}_0$

$$\varpi = 2s\, \bar{z}\, dz\, i^{-1} + m\, \bar{z}\, \gamma(dx)\, z \qquad (12)$$

This is precisely the expression introduced by Souriau [17] in order to prequantize the orbits $\mathcal{O}_{sm} \cong S^2 \times \mathbb{R}^6$ (1). It is well known that $\mathcal{O}_{sm}$ is prequantizable if $s \in \mathbb{N}/2$ (see below).

On the other hand, the evolution space of the model is $\tilde{E} = \{(X, P, S)$; $X, P \in M$; $S \in so(M), Tr(S^2) = -2s^2, SP = 0, \bar{P}P = m^2$; P future pointing$\}$. Putting

$$\bar{P}W := m\, \bar{z}\, \gamma(W)\, z \qquad \forall W \in M \qquad (13)$$

$$\bar{W} S W' := s\, Re\,(\bar{z}\, \gamma(W)\gamma(W')\, z\, i^{-1}) \quad \forall W, W' \in M \qquad (14)$$

we get $\tilde{E} = \hat{\mathcal{P}}_0 / \mathbb{T}$ $((z, x) \sim (z', x')$ iff $\exists\, z \in \mathbb{T}$ such that $z' = z z$ and $x' = x)$. $d\varpi$ passes to the quotient and the equations of motion ($\ker(d\varpi)$) are :

$$\begin{cases} V = \lambda P \quad , \quad \lambda \in \mathbb{R} \\ \nabla_V P = 0 \\ \nabla_V S = 0 \end{cases}$$

where V denotes here the projection on M of $\zeta \in \ker(d\varpi)$. P is the momentum of the particle, S its spin tensor, X its spacetime location.

iii) <u>The internal phase spaces</u>. We will call internal phase space a (prequantizable) (2) coadjoint orbit of a given internal symmetry group (e.g. SU(2) : isospin , SU(3) : isospin-hypercharge). From now

(1) The stabilizer of q is diffeomorphic to U(1) x $\mathbb{R}$.

(2) A symplectic manifold is prequantizable if there exists a prequantization for it. A prequantization of a symplectic manifold (U, σ) is defined as a principal T-fibre bundle $\mu : Y \to U$ with connection form $i\varpi$ such that $d\varpi = \mu^* \sigma$.

on we are considering the physically relevant case of compact semi-simple (and simply connected) gauge groups. Just as in (i), choose $q \in \mathfrak{g}^*$ which meets a Cartan subalgebra $\mathcal{H}$ ($\mathfrak{g}^*$ is identified to $\mathfrak{g}$ via the Killing form) whose underlying Cartan subgroup H is diffeomorphic to $\mathbb{T}^r$ (r = rank(G)). The 1-form ϖ (11) of G serves to define the prequantum bundle of the orbit $\mathcal{O}_q = G/G_q$ under certain conditions we are going to determine. Due to the connectedness of G_q (which contains H as a subgroup), $\exp : \mathfrak{g}_q \to G_q$ is onto and we can define $\chi : G_q \to \mathbb{T}$ by $\chi(\exp(A)) := e^{i\,\varpi(A)} \quad \forall A \in \mathfrak{g}_q$. This χ is a well defined character of G_q if $[\iota^*\varpi] \in H^1(G_q, \mathbb{Z})$ where $\iota : G_q \hookrightarrow G$ stands for inclusion. This condition implies that $[\sigma] \in H^2(G/G_q, \mathbb{Z})$ where σ is the canonical symplectic structure of G/G_q (the prequantization conditions).

It is easy to show at under these circumstances $(G/\tilde{G}_q, \varpi)$ - slight abuse of notation - $\tilde{G}_q := \ker(\chi)$ - prequantizes $(\mathcal{O}_q, \sigma)$ [7]. All prequantizations are equivalent to the latter since $\mathcal{O}_q$ is simply connected.

<u>Example</u> : SU(n). Put $g := (z_1, \ldots, z_n) \in SU(n)$; $z_j \in \mathbb{C}^n$, $\bar{z}_j z_k = \delta_{jk}$ $\forall j,k$ $\det(z_1 \cdots\cdots z_n) = 1$. Let us choose $q = \mathrm{diag}(q_1, \ldots\ldots, q_n)\, i^{-1}$ with $\sum_{j=1}^{n} q_j = 0$, $q_j \in \mathbb{R}$ $\forall j = 1, \ldots, n$. Suppose then without loss of generality that $q_1 \geq \cdots\cdots \geq q_n$. Since $\sum_{j=1}^{n} z_j \bar{z}_j = \mathbb{1}$, we have

$$Q := ad^*(g)q = \Big(\sum_{j=1}^{n-1} p_j z_j \bar{z}_j - \sum_{j=1}^{n-1} p_j / n \;\mathbb{1} \Big) i^{-1} \in \mathcal{O}_q$$

with $p_j := q_1 + \cdots + 2q_j + \cdots + q_{n-1}$; $p_1 \geq \cdots\cdots \geq p_{n-1} \geq 0$.
A short calculation gives

$$\varpi = \sum_{j=1}^{n-1} p_j \bar{z}_j dz_j \, i^{-1} \tag{15}$$

The p's are the Casimirs of the orbit $\mathcal{O}_q$ which turns out to be prequantizable if $p_1, \ldots\ldots, p_{n-1}$ are positive integers. By virtue of the Borel-Weil-Bott theorem, these prequantizable orbits are associated with the unitary irreducible representations $\mathcal{D}_{p_1 \ldots\ldots p_{n-1}}$ of SU(n) which are interpreted in the physical context as the multiplets of hadrons. As for us, we will naturally call the orbits

$\mathcal{O}_{p_1 p_2 \cdots\cdots p_{n-1}}$ $(p_1, p_2, \ldots, p_{n-1} \in \mathbb{N})$ "classical multiplets".

It is worth noticing that the SU(n) (co)adjoint orbits can alternatively be worked out the Stiefel manifolds $V_{k,n-k}(\mathbb{C})$ when $p_{k+1} = \cdots\cdots = p_{n-1} = 0$. The form (15) whose exterior derivative passes to $\mathcal{O}_{p_1 p_2 \cdots\cdots p_{n-1}}$ gives rise to the Kähler structure of these orbits.

We propose now the following physical interpretation [7], [8].

a) n = 2. If we put $p_1 =: 2s$, s is interpreted as the isospin. $(\mathbb{P}_1(\mathbb{C}), d\bar{z}_1 \wedge dz_1 i^{-1})$: the nucleon orbit ({2}) ; $(\mathbb{P}_1(\mathbb{C}), 2\, d\bar{z}_1 \wedge dz_1\, i^{-1})$: the meson orbit ({3}) ; etc...

b) n = 3. There is no direct physical interpretation of the invariants p_1, p_2. There are two non trivial strata, namely the minimal orbits $\mathcal{O}_{p_1 0}$, $\mathcal{O}_{p_1 p_1}$ $(= - \mathcal{O}_{p_1 0})$ which are diffeomorphic to $\mathbb{P}_2(\mathbb{C})$, and the maximal orbits $\mathcal{O}_{p_1 p_2}$ $(p_1 > p_2)$ with topology $\mathbb{P}T\,\mathbb{P}_2(\mathbb{C})$. The orbits $\mathcal{O}_{2p_2\, p_2}$ are self-conjugate. For example, $(\mathbb{P}_2(\mathbb{C}), d\bar{z}_1 \wedge dz_1\, i^{-1})$: the quark orbit ({3}) ; $(\mathbb{P}_2(\mathbb{C}), (d\bar{z}_1 \wedge dz_1 + d\bar{z}_2 \wedge dz_2)\, i^{-1})$: the antiquark orbit ({$\bar{3}$}) ; $(\mathbb{P}_2(\mathbb{C}), 3\, d\bar{z}_1 \wedge dz_1\, i^{-1})$: the "decimet" orbit ({10}) ; etc... ; $(\mathbb{P}T\,\mathbb{P}_2(\mathbb{C}), (2\, d\bar{z}_1 \wedge dz_1 + d\bar{z}_2 \wedge dz_2)\, i^{-1})$ the "octet" orbit ({8}) ; etc...

iv) <u>Remark</u>. Since the direct product of symplectic manifolds is prequantizable if each of them is separately prequantizable, the phase space $\mathcal{O}_{sm} \times \mathcal{O}_q$ (§3, ii, iii) of a spinning particle with internal degrees of freedom is prequantizable if s is half integral and the internal phase space $\mathcal{O}_q$ prequantizable.

§4. The spinning particle in an external Yang-Mills field.

Let $\hat{\pi} : \hat{P}_0 \to M$ denote the universal covering of the restricted Lorentz bundle over Minkowski spacetime M and $\pi : P \to M$ a given G-principal bundle over M with connection ω. Pick a point q in $\mathfrak{g}^*$ and define, just as in (1) $\alpha := q \cdot \omega$. We again define (see (3))

$Q := ad^*(u)q \ (u \in P)$ - $Q \in P/G_q$. The coupling of a massive spinning particle to the external Yang-Mills field ω is performed by considering the following 1 - form of $\hat{\pi}^* P$

$$\theta := \varpi + \alpha \qquad \text{(see (2))}$$

where ϖ is given by (12). There is now a slight subtelety concerning the "mass" m of the particle. As suggested in (§2,ii), the dipole interaction can be taken into account if we let m (20) depend upon the coupling term Q.F(S) where F is the field strength and S the spin tensor (22). We thus label models of spinning particle in a Yang-Mills field by a positive real function f such that

$$m^2 = f(Q.F(S))$$

Since the curvature itself enters the definition of the coupling in the 1-form θ , instead of the sole connection (see § 1), we may not consider that the particle is minimally coupled to the Yang-Mills field. This situation much looks like the Dirac-Pauli case where it has been shown that f has the specific form : $f(x) = m_0^2 + (gx)/2.m_0$ is the naked mass ; g = 2 defines the minimal coupling (the Dirac equation) [3],[5].

We will skip over the computation of ker($d\theta$). With the help of the definitions((3), (13), (14)), the resulting equations of motion read

$$\begin{cases} V = \lambda\{P + (\bar{P}P + Q.F(S)/2)^{-1} S((1-f')Q.F P + f'/2\ \overline{Q.DF(S)})\} \\ \nabla_V \bar{P} = -Q.F(V) + \lambda/2\ f'\ Q.DF(S) \\ \nabla_V S = P\bar{V} - V\bar{P} + \lambda f' Q.[S,F] \\ D_V Q = \lambda/2\ f' Q.Ad(F(S)) \qquad , \quad \lambda \in \mathbb{R} \end{cases} \qquad (16)$$

(Compare with (9)). The generalized magnetic moment (10) is $\chi = \frac{f'}{f} \bar{P}V$. Note that the generalized isospin is no longer parallel transported (see also [1],[13]). We would like to point out an intriguing fact : the minimal coupling might be introduced by requiring that the velocity V should be parallel to the momentum P in a constant Yang-Mills field (put $f' = 1$ and $DF = 0$ in (16)).

§5. Gravitational and Yang-Mills scattering.

Suppose that a particle, travelling in spacetime, encounters a compactly supported perturbation (δg , δA) of the background E.Y.M. field (g , A) (cf. § 2). It has been assumed in [19] that if x_{in} (resp. x_{out}) denotes the incoming (resp. outgoing) motion of a scattered particle, the correspondence : $x_{in} \mapsto x_{out}$ is realized by a symplectomorphism of the unperturbed space of motion (phase space) ($\mathcal{U}, \sigma$). The infinitesimal version of the preceding assumption can now be stated as follows : the variation (δg , δA) generates an infinitesimal symplectomorphism δx (1) of ($\mathcal{U}, \sigma$), i.e. σ is Lie propagated by δx . Things can however be best formulated at the prequantum level. Let (Y, ϖ) denote a prequantization of ($\mathcal{U}, \sigma$). We can say that (δg , δA) gives rise to an infinitesimal quantomorphisms $\delta \xi$ ($\xi \in Y$), so that there exists a hamiltonian h of U (the eikonal) such that $d\varpi(\delta\xi) = -dh, \varpi(\delta\xi) = h$.

We will put $\delta \xi = \underset{h}{\delta} \xi$. Let us compute h in the case of a spinless particle, for the sake of simplicity.

Since the bundle $\tilde{P}$ (§ 1) serves to construct the prequantum bundle (§ 3,ii), we may work directly on $\tilde{P}$ endowed with the 1-form (2)

$$\theta(\delta\tilde{u}) = g(P, \delta x) + q \cdot \omega(\delta\tilde{u})$$

Choosing a local chart, we write $\omega = a^{-1}da + a^{-1}Aa$ ($a \in G$), and then $\theta(\delta\tilde{u}) = g(P,\delta x) + Q \cdot A(\delta x) + q \cdot a^{-1}\delta a$ where $Q := ad^*(a)\, q$ (3). Let us compute now $d\theta(\delta\tilde{u}, \underset{h}{\delta}\tilde{u})$ with $\delta\tilde{u} \in \ker(d\theta)$. Putting $\delta X = V = dX/dt$ (t is the curve parameter of the worldline c defined by $dX/dt = P$ (4)), we find

$$\begin{aligned} dh/dt = {} & \delta(g(P,V)) - \delta g(P,V) - g(\hat{\delta}P, V) \\ & + Q\cdot[A(V), A(\delta x)] + Q\cdot d(A(\delta x))/dt - Q\cdot F(V,\delta x) \\ & + q\cdot[a^{-1}A(V)a, a^{-1}\delta a] - q\cdot \delta(a^{-1}A(V)a) \end{aligned}$$

(1) In the context of the calculus of variations, we use the notation δx for an element of $T_x\mathcal{U}$.

where $\hat{\delta} P$ denotes the Riemannian covariant derivative. Since $\delta(g(P,P)) = 0 = \delta g(P,P) + 2\, g(\hat{\delta} P, P)$, we are left, after a tedious calculation, with the formula $dh/dt = -\frac{1}{2}\delta g(P,V) - Q.\delta A(V)$. At last

$$h = \theta(\delta_h \tilde{u}) = -\int_c \left\{ \frac{1}{2}\delta g\left(P, \frac{dx}{dt}\right) + Q.\delta A\left(\frac{dx}{dt}\right) \right\} dt$$

which is precisely the expression (8) in the pole approximation. No scattering if the infinitesimal E. Yang-Mills field is generated by an E.Y.M. gauge (7) !

References.

[1] A.P. BALACHANDRAN, P. SALOMONSON, B.S. SKAGERSTAM, J.O. WINNBERG. Classical Description of a Particle Interacting with a Non Abelian Gauge Field. Phys. Rev. D, vol 15 n° 8, 2308-2317 (1977).

[2] W.G. DIXON - Dynamics of Extended Bodies in General Relativity I. Proc. Roy. Soc. A 314, 499-527 (1970).

[3] W.G. DIXON - On a Classical Theory of Charged Particles with Spin and the Classical Limit of the Dirac Equation. Nuov. Cim. vol. XXXVIII n° 4, 1616 (1965).

[4] C. DUVAL - Sur les mouvements classiques dans un champ de Yang-Mills. Preprint 78/P.1056 - C.P.T. Marseille (1978).

[5] C. DUVAL - The General Relativistic Dirac-Pauli Particle : an Underlying Classical Model. Ann. Inst. Henri Poincaré, vol. XXV n° 4, 345-362 (1976).

[6] C. DUVAL, H.P. KUNZLE - Dynamics of Continua and Particles from General Covariance of Newtonian Gravitation Theory. Rep. on Math. Phys. vol. 13 n° 3, 351-368 (1978).

[7] C. DUVAL - On the Polarizers of Compact Semi-Simple Lie Groups. Applications. Preprint 80/P.E. 1185 - C.P.T. Marseille (1980).

[8] B.T. FELD - Models of Elementary Particles. Blaisdell (1969).

[9] A. KIRILLOV - Eléments de la théorie des représentations. Ed. Mir, Moscou (1974).

[10] S. KOBAYASHI, K. NOMIZU - Foundations of Differential Geometry. Volume 1. Interscience, New-York (1962).

[11] H.P. KUNZLE - Canonical Dynamics of Spinning Particles in Gravitational and Electromagnetic Fields.
J. Math. Phys. 13, 739-744 (1972).

[12] A. PAPAPETROU - Spinning Test Particles in General Relativity.
Proc. Roy. Soc. A 209, 248-258 (1951).

[13] P. SALOMONSON, B.S. SKAGERSTAM, J.O. WINNBERG
Equations of Motion of a Yang-Mills particle.
Phys. Rev. D, vol. 16 n° 8, 2581-2585 (1977).

[14] J. SNIATYCKI - On Hamiltonian Dynamics of Particles with Gauge Degrees of Freedom.
Hadr. J. 2, 642-656 (1979).

[15] S. STERNBERG - On the Role of Field Theories in our Physical Conception of Geometry. Differential Methods in Mathematical Physics II.
Proceedings Bonn (1977) - Springer-Verlag Berlin (1978).

[16] S. STERNBERG, T. UNGAR - Classical and Prequantized Mechanics without Lagrangians or Hamiltonians.
Preprint Tel Aviv (1978).

[17] J.M. SOURIAU - Structure des systèmes dynamiques.
Dunod Paris (1970).

[18] J.M. SOURIAU - Modèle de Particule à Spin dans le Champ Electromagnétique et Gravitationnel.
Ann. Inst. Henri Poincaré, vol. XX n° 4, 315-364 (1974).

[19] J.M. SOURIAU - Thermodynamique et Géométrie.
in "Differential Geometrical Methods in Mathematical Physics" II.
Proceedings Bonn (1977) - K. Bleuler, A. Reetz ed. -
Springer-Verlag (1978).

[20] A. TRAUTMAN - Elementary Introduction to Fibre Bundles and Gauge Fields.
Preprint Warsaw (1978).

[21] A. WEINSTEIN - A Universal Phase Space for Particles in Yang-Mills Fields.
Lett. in Math. Phys. 2, 417-420 (1978).

[22] S.K. WONG - Field and Particle Equations for the Classical Yang-Mills Field and Particles with Isotopic Spin.
Nuov. Cim., A 65, 689 (1970).

CLASSICAL ACTION, THE WU-YANG PHASE FACTOR AND PREQUANTIZATION

Péter A. HORVÁTHY *
Université d'Aix-Marseille I
et
Centre de Physique Théorique, CNRS Marseille

ABSTRACT : For local variational systems (like a charged particle in the field of a Dirac monopole) a quantum mechanically well-defined action (Q.M.W.D.A.) can be introduced iff the system is prequantizable in the Kostant-Souriau sense. If the configuration space is multiply connected (as in the Bohm-Aharonov experiment), different expressions for the classical action may emerge; they are quantum mechanically equivalent (Q.M.E.) iff the corresponding prequantizations are equivalent. In both cases the situation depends on the behaviour of the non integrable phase factor of Wu and Yang.

* On leave from Veszprém University of Chemical Engineering Veszprém, (Hungary).

0. INTRODUCTION

The importance of classical action in quantum mechanics emerges the clearest way from Feynman's path integral approach [1]. To a path γ in spacetime between x and x' is associated the amplitude

$$\exp\left[\frac{i}{\hbar} S(\gamma)\right] \tag{1}$$

where $S(\gamma)$ is the classical action along γ ; the propagator is expressed as

$$K(x',x) = \int_{\mathcal{P}} \exp\left[\frac{i}{\hbar} S(\gamma)\right] \mathcal{D}\gamma \tag{2}$$

$\mathcal{P}$ being the "infinite dimensional manifold" of paths joining x to x'.

We are not concerned here with the tremendous problem of defining and computing this integral ; we shall accept its intuitive meaning and focus our attention to the amplitude (1).

The point is that in some interesting situations, as in the Bohm-Aharonov experiment [2] [3] the expression of classical action may be ambiguous [4]; in other cases, as for the motion of a charged particle in the field of a Dirac monopole [6], it may be even ill-defined [5].

Motivated by ordinary gauge transformation, we introduce the notion of quantummechanically well-defined action (Q.M.W.D.A.) and the idea of equivalent (Q.M.E.) actions.

The requirement of having a Q.M.W.D.A. leads to quantum conditions (like quantization of the monopole's strength) ; the equivalence of actions provides us with a classification scheme and with a simple proof of the C. DeWitt-Laidlaw theorem [7] [8] [9] on propagators.

These results can be reexpressed in a rather elegant geometric form : a Q.M.W.D.A. exists iff the system is prequantizable in the Kostant-Souriau (K-S) sense [10] [11] [13]. The classification scheme turns out to be just that of inequivalent prequantum bundles.

Our approach shows some similarities to that of Wu and Yang [15] who describe gauge fields in terms of a "non integrable phase factor". The relation is explained in the U(1) (electromagnetic) case.

1. LOCAL VARIATIONAL SYSTEMS [15][16]

Let Q be the manifold of all possible configurations of a classical system. If we are given a Lagrangian function $L: TQ \times R \longrightarrow R$, the variational problem can be translated to symplectic terms [11], [24], [25]: from L we can derive a 1-form Θ such, that the Euler-Lagrange equations have the geometric form

$$\dot{\gamma} \in \operatorname{Ker} d\Theta \tag{3}$$

The curves γ satisfying (3) - the lifts to TQxR of the classical motions- are the extremals of the variational problem. $\sigma = d\Theta$ is a presymplectic form on the manifold E = TQxR ("evolution space").

Souriau proposed [11] to enlarge classical mechanics by describing systems with such a pair (E, σ), without bothering about Lagrangians. The existence of a Lagrangian function is, however, a basic requirement in mechanics [23]. Also, as it will appear from the discussion which follows, (Sections 3, 4,5) in order to have a meaningful quantization procedure, we need some additional condition which rules out the velocity-dependence of potentials.

The exact relations between symplectic and variational description are the best established using the homogeneous formalism [17], [11], [15], [16] which we review here briefly.

Write X = QxR for (configuration) space-time, denote $\pi\; TX \to E$ (E = TQxR) the projection given locally as $\pi(x, \dot{x}) = (q, \frac{\dot{q}}{\dot{t}}, t)$, where $x = (q,t)$, $\dot{x} = (\dot{q}, \dot{t})$; suppose $\dot{t} > 0$. The homogenized Lagrangian reads $\mathcal{L}(x, \dot{x}) = \dot{t}\, L \circ \pi$. We have then a unique 1-form Λ on TX such that for any curve $\gamma \subset TX$

$$\int \mathcal{L} \circ \gamma(\tau)\, d\tau = \int_{\gamma} \Lambda \tag{4}$$

where $\tau \to \gamma(\tau) = (\gamma_q(\tau), \gamma_t(\tau), \gamma_{\dot{j}}(\tau), \gamma_{\dot{t}}(\tau))$ is any parametrization with $d\gamma_t/d\tau > 0$.

Explicitely, Λ is the fiber derivative of $\mathcal{L}$ [18],

$$\Lambda = \dot{d}\mathcal{L} \tag{5}$$

(recall the definition of $\dot{d}$:

For a function $f : TX \to R$ $\dot{d}f = (\partial f/\partial \dot{x}^\alpha) dx^\alpha$; the extension to forms is made by the requirements

$$\dot{d}(dx^\alpha) = \dot{d}(d\dot{x}^\alpha) = 0$$

$$\dot{d}(\omega \wedge \beta) = \dot{d}\omega \wedge \beta + (-1)^{\deg \omega} \omega \wedge \dot{d}\beta$$

this Λ is

- semibasic,

$$\Lambda_{(x,\dot{x})} = a_\alpha(x,\dot{x})\, dx^\alpha \tag{6a}$$

- homogeneous of order 0 in $\dot{x}$; for $0 \neq c \in \mathbb{R}$

$$a_\alpha(x,\dot{x}) = a_\alpha(x, c\dot{x}) \tag{6b}$$

- of the form

$$\Lambda = \pi^* \Theta \tag{6c}$$

with a 1-form Θ on E (this is just the usual Cartan form [11], used in (3)).

Conversely, if we are given a Λ with these properties (6), we can always reconstruct a Lagrangian function

$$L(q,v,t) = \sum_{\alpha=1}^{n} v^\alpha a_\alpha(q,v,t) + a_{n+1} \tag{7}$$

Thus it is justified to call 1-forms on TX satisfying (6) global variational 1-forms ; (TX, Λ) is a global variational system.

Denote $\Sigma = d\Lambda$; then L is regular (i.e. $\partial^2 L/\partial v^\alpha \partial v^\beta$ is a regular nxn matrix) iff

$$\dim \operatorname{Ker} \Sigma = 2 \tag{8}$$

If (8) holds then the smooth distribution $(x, \dot{x}) \to \operatorname{Ker} \Sigma_{(x, \dot{x})}$ is integrable : the characteristic leaves [13] , [18] (which are in 1-1 correspondence with the curves in E satisfying (3)) are 2-dimensional submanifolds in TX . They project to the world lines in X , and thus it is justified to consider these leaves as the generalized solutions of the variational problem.

Σ satisfies [14]

$$d\Sigma = 0 \tag{9a}$$

$$\Sigma = \pi^* \sigma \quad , \quad \sigma, \qquad \text{presymplectique form on E} \tag{9b}$$

$$\dot{d}\Sigma = 0 \tag{9c}$$

In our case $\sigma = d\Theta$.

This is just this condition (9c) which singles out variational system among (pre)symplectic ones.

Unfortunately, <u>global variational systems</u> do not exhaust all the physically interesting situations : for a charged particle moving in the field of a Dirac monopole (see example [1] below) for instance, no global Λ exists. Conditions (9) are however satisfied.

On the other hand, Klein has shown [17] that (9) assures the existence of a <u>local variational description</u> at least.

Theorem, Definition 1.1

Let Σ be a 2-form on TX satisfying (9). Then, in a neighbourhood of any point at least, the equations

$$\Sigma = d\Lambda \qquad \text{or} \qquad \sigma = d\Theta \tag{10}$$

admit solutions such that Λ (or Θ) satisfy (6). Such 1-forms will be called local variational or action forms, (TX, Σ) or (E, σ) being a local variational system.

It is well-known (e.g. [19]) that the possibility of extending a local solution depends on the topology : if $H^2(TX, \mathbb{R}) = 0$ every local solution of (10) extends to the entire TX (or E).

Proposition 1.2

Let Λ and Λ' (or Θ, Θ') be local variational solutions of (10), then in the intersection of their domain

$$\alpha = \Lambda' - \Lambda = \Theta' - \Theta = A(q,t)\, dq + V(q,t)\, dt \tag{11}$$

is a closed 1-form on X , $d\alpha = 0$.

If this intersection is simply connected then α is exact.

Proof : α is obviously closed ; a closed semibasic 1-form can not depend on $\dot{x}$.

Theorem 1.3 [15], [16]

If (E, σ) is a regular local variational system, $\mathrm{Ker}\, \pi^* \sigma$ defines a foliation of TX by 2-dimensional leaves. These leaves-considered as generalized solutions of the variational problem-project onto curves in X .

Thus, at a purely classical level, these systems admit a completely satisfactory variational description.

Remark 1.4

If we replace (8) by $\dim \mathrm{Ker}\, \Sigma = 2K$, $K > 1$, the whole formalism keeps on working ; this allows for including spin [15]. We study here, however, only spinless systems.

In what follows, we shall use the (E,σ) setting, (8) and (9) supposed being satisfied.

2. THE CLASSICAL ACTION

Consider first a global system with action form Θ. For $\gamma \subset E$ set

$$S(\gamma) = \int_\gamma \Theta \tag{12}$$

and call it classical action along γ. (If $\gamma \subset X$ is a curve, lift it to E : call the lift again γ to save characters) ; by (4), (12) reduces then to the usual expression).

Note however, that his definition is ambiguous : we are always allowed to change Θ to Θ' which also satisfies $d\Theta' = \sigma$; the requirements (6) imply (Prop. 1.2) that $\Theta' = \Theta + \alpha$ with a 1-form α on X . This has the effect of changing (12) by an additional term $\int_\gamma \alpha$.

If the configuration space is simply connected, then α is exact: $\alpha = df$ with $f : X \to \mathbb{R}$; thus the additional term is just a constant $\{f(x') - f(x)\}$, which changes the amplitude (1) and thus the propagator (2) only by an overall phase factor

$$c\ (x',x) = \exp \frac{i}{\hbar} \{ f(x') - f(x) \} \tag{13}$$

which is physically unobservable.

However, if the underlying space is multiply connected (as in the Bohm-Aharonov experiment, see example 2 below), this term will depend on γ , and will change essentially the physics at the quantum level.

For local systems the situation is even worse : an action from Θ_α exists only locally, over an open set $\mathcal{U}_\alpha$. Consequently, the corresponding classical action $S_\alpha(\gamma) := \int_\gamma \Theta_\alpha$ will be meaningful only for paths γ contained entirely in $\mathcal{U}_\alpha$.

But even for such paths, we have an essential ambiguity: if we change $(\mathcal{U}_\alpha, \Theta_\alpha)$ to $(\mathcal{U}_\beta, \Theta_\beta)$ with $\gamma \subset \mathcal{U}_\beta$, then the new $S_\beta(\gamma) = \int_\gamma \Theta_\beta$ will be, generally, completely different from $S_\alpha(\gamma)$ (see Example 1 below). This is due again to topology: $\mathcal{U}_\alpha \cap \mathcal{U}_\beta$ may be non-simply connected, and thus $\Theta_\alpha - \Theta_\beta$ may be not exact, and so

$$S_\alpha(\gamma) - S_\beta(\gamma) = \int_\gamma (\Theta_\alpha - \Theta_\beta) \tag{14}$$

will be path dependent. Consequently, for local systems, it is generally meaningless to speak of classical actions.

3. A QUANTUMMECHANICALLY WELL-DEFINED ACTION

Fortunately, as it is clear from (2) it is the amplitude (1) rather than the action itself, which is important for quantum mechanics.

Consider a local system (E, σ).

Definition 3.1

The classical action is quantummechanically well-defined (Q.M.W.D.) if to any choice $(\mathcal{U}_\alpha, \Theta_\alpha)$, and any path γ whose end points x, x' belong to $\mathcal{U}_\alpha$, we can associate an expression

$$\text{"} \exp\left[\frac{i}{\hbar} S_\alpha(\gamma)\right] \text{"} \tag{15}$$

such that

a) a change $(\mathcal{U}_\alpha, \Theta_\alpha) \to (\mathcal{U}_\beta, \Theta_\beta)$ introduces merely a phase factor

$$\text{"} \exp\left[\frac{i}{\hbar} S_\alpha(\gamma)\right] \text{"} = C_{\alpha\beta}(x', x) \cdot \text{"} \exp\left[\frac{i}{\hbar} S_\beta(\gamma)\right] \text{"} \tag{16}$$

where $|C_{\alpha\beta}(x', x)| = 1$, $C_{\alpha\beta}(x', x)$ depends only on x,x' and not the particular path γ between them.

b) for $\gamma \subset \mathcal{U}_\alpha$ (15) reduces to $\exp\left[\frac{i}{\hbar} S_\alpha(\gamma)\right]$ with $S_\alpha(\gamma) = \int_\gamma \Theta_\alpha$

If a Q.M.W.D.A. exists, then a change $(\mathcal{U}_\alpha, \Theta_\alpha) \to (\mathcal{U}_\beta, \Theta_\beta)$ will introduce only a phase factor in the propagator (2).

Study first paths in $\mathcal{U}_\alpha \cap \mathcal{U}_\beta$. It is easier to use loops:

Proposition 3.2

If a Q.M.W.D.A. exists, then for a loop $\gamma \subset \mathcal{U}_\alpha \cap \mathcal{U}_\beta$, we have

$$\exp\left[\frac{i}{\hbar} S_\alpha(\gamma)\right] = \exp\left[\frac{i}{\hbar} S_\beta(\gamma)\right] \tag{17}$$

Proof : In fact, split up γ to $\gamma_1 \circ \gamma_2$; apply (16) to γ_1 and γ_2^{-1} ; divide, noting that $\exp\left[\frac{i}{\hbar} S(\gamma)\right] = \left(\exp\left[\frac{i}{\hbar} S(\gamma^{-1})\right]\right)^{-1}$

In other form :

Proposition 3.3

A necessary condition for the existence of a Q.M.W.D.A. is that for a loop $\gamma \subset U_\alpha \cap U_\beta$ we have

$$\exp\left[\frac{i}{\hbar} \oint_\gamma (\Theta_\alpha - \Theta_\beta)\right] = 1 \tag{18}$$

It may happen, that it is possible to pull "caps" s_α and s_β over γ in U_α resp. U_β , each cap being diffeomorphic to $\mathbb{R}^2$; $s = s_\alpha \cup s_\beta$ is then diffeomorphic to S^2 ; let's apply Stokes' theorem to s_α, resp. s_β ; we get :

Proposition 3.4

(17) is equivalent to

$$\frac{1}{2\pi\hbar} \int_s \sigma \in \mathbb{Z} \tag{19}$$

In Section 5, we shall show that these conditions are in fact sufficient.

Remark 3.5

As $\Theta_\alpha - \Theta_\beta$ is in fact a 1-form over X ,(Prop. 1.2) (18) and (19) hold if they hold for loops, resp. 2-surfaces in X .

EXAMPLE 1 (Charged particle moving in the field of Dirac's monopole)

Suppose we have a magnetic monopole of strength g fixed in the origin ; an electron moving in its field has the symplectic description [12] $Q = \mathbb{R}^3 \setminus \{0\}$, $E = TQ \times \mathbb{R}$, $\sigma = \sigma_{free} + e\mathbb{B}$, i.e.

$$\sigma = d\left(m v\, dq - m\frac{v^2}{2}\, dt\right) + eg\left\langle \frac{q}{|q|^3}, dq \times dq \right\rangle \tag{20}$$

It is easy to see that no global Θ with $\sigma = d\Theta$ (and thus no global vector potential), exists : if σ was $d\Theta$, $\int \sigma$ would be 0 by Stokes' theorem ; however one computes at once that $\int_{S^2} \sigma = 4\pi e g$.

Nevertheless, local solutions of (10) can be found on any chart corresponding to $\mathcal{U}_{\underline{n}} = \mathbb{R}^3 \setminus \{$a "string" in the direction of $\underline{n}\}$ e.g.

$$\Theta_{\underline{n}} = \left\{ m v \, dq - m \frac{v^2}{2} dt \right\} + e A^{(\underline{n})}(q) \, dq \tag{21}$$

with the local vector potential [12]

$$A^{(\underline{n})} = g \frac{\underline{n} \times q}{q^2 + |q| \langle \underline{n}, q \rangle} \tag{22}$$

The ambiguity in the classical action can be tested on $\mathcal{U}_\alpha = \mathcal{U}_{(0,0,1)}$ $\mathcal{U}_\beta = \mathcal{U}_{(0,0,-1)}$, $\gamma(\varphi) = (\cos\varphi, \sin\varphi, 0)$ with $0 \leq \varphi < 2\pi$ (the equator)

$$S_\alpha(\gamma) - S_\beta(\gamma) = \oint_\gamma \left(A^{(\underline{n})} - A^{(-\underline{n})}\right) dq = 4\pi e g \tag{23}$$

Thus a Q.M.W.D.A. exists iff the <u>monopole is quantized</u> as

$$2eg = \hbar k \qquad k \in \mathbb{Z} \tag{24}$$

More generally, one shows that a Q.M.W.D.A. exists iff

$$\exp\left[\frac{ie}{\hbar} \oint A_\alpha \right] \tag{25}$$

has the same value for all α (with $\sigma|_{\mathcal{U}_\alpha} = d\Theta_\alpha$). (25) is just the phase factor of <u>Wu and Yang</u> [14].

4. A CLASSIFICATION SCHEME. THE PROPAGATOR IN MULTIPLY CONNECTED SPACES [4]

Let's consider a global system with multiply connected configuration space. The general solution of (10) among variational 1-forms is by Prop. 1.2

$$\Theta = \Theta_0 + \alpha \tag{26}$$

with Θ_0 a particular solution ; as a consequence of (6a), (6b), α is a 1-form on X

$$\alpha = A(q,t)\,dq + V(q,t)\,dt \tag{27}$$

Definition 4.1

Let $\Theta_1 = \Theta_0 + \alpha_1$ and $\Theta_2 = \Theta_0 + \alpha_2$ two actions forms for a global system. Two expressions $S_1(\gamma) = \int_\gamma \Theta_1$ and $S_2(\gamma) = \int_\gamma \Theta_2$ are told to be quantummechanically equivalent (Q.M.E.), (denoted also $\Theta_1 \sim \Theta_2$) iff

$$\exp\left[\frac{i}{\hbar} S_1(\gamma)\right] = C(x',x)\cdot \exp\left[\frac{i}{\hbar} S_2(\gamma)\right] \tag{28}$$

with a phase factor $C(x',x)$ depending only on (the projection onto X of the) end points of γ; $|C(x',x)| = 1$

Proposition 4.2

$\Theta_1 \sim \Theta_2$ iff for any loop γ

$$\exp\left[\frac{i}{\hbar}\oint_\gamma (\Theta_1 - \Theta_2)\right] = \exp\left[\frac{i}{\hbar}\oint_\gamma (\alpha_1 - \alpha_2)\right] = 1 \tag{29}$$

or

$$\frac{1}{2\pi\hbar}\oint_\gamma (\Theta_1 - \Theta_2) = \frac{1}{2\pi\hbar}\oint_\gamma (\alpha_1 - \alpha_2) \in \mathbb{Z} \tag{30}$$

(As the space is not simply connected, Stokes' theorem does not apply, and thus we cannot transform this to integrals over 2-cycles). Again, by Prop. 1.2, we can limit ourselves to path in X .

$\exp\left[\frac{i}{\hbar}\oint_\gamma \alpha\right]$ is studied the easiest way if we climb to the universal covering $(\tilde{X}, \Pi, P)$ of X : $\tilde{X} = \tilde{Q}\times R$, where $\tilde{Q}$ is the universal covering of Q ; Π is the (first) homotopy group of Q (and X) ; $P : \tilde{X} \ni (\tilde{q},t) = (q,t) \in X$ projection.

Set $\tilde{\alpha} = P^*\alpha$. As $\tilde{X}$ is already simply connected, $\tilde{\alpha} = d\tilde{f}$, with $\tilde{f} : \tilde{X} \longrightarrow \mathbb{R}$.
Let $\gamma \subset X$ be any path, $x \in \gamma$, $\tilde{x} \in P^{-1}(x)$; γ has a unique lift $\tilde{\gamma}$ to $\tilde{X}$ through $\tilde{x}$. Evidently, $\int_\gamma \alpha = \int_{\tilde{\gamma}} \tilde{\alpha}$

In particular, if γ is a closed loop $\tilde{\gamma}$ will end at $g\tilde{x}$, where $g = [\gamma]$ is the homotopy class of γ. Consequently

$$\oint_\gamma \alpha = \tilde{f}(g\tilde{x}) - \tilde{f}(\tilde{x}) \tag{31}$$

Note that (31) depends only on g . Thus

Proposition 4.3

$$\chi(g) = \exp\left[\frac{i}{\hbar}\oint_\gamma \alpha\right] \tag{32}$$

is well-defined, and is in fact, a character of the homotopy group Π . In this way we get the following classification theorem:

Theorem 4.4

$\Theta_1 \sim \Theta_2$ iff for any loop γ

$$\chi_1(g) = \exp\left[\frac{i}{\hbar}\oint_\gamma \alpha_1\right] = \exp\left[\frac{i}{\hbar}\oint_\gamma \alpha_2\right]$$

where $g = [\gamma]$.

<u>The different situations are thus labelled by the characters of the homotopy group.</u>

Now we can prove an interesting theorem,[4]first stated explicitely by C. de Witt and Laidlaw [7] (see also [8],[9]). Consider $x, x' \in X$, let $\mathcal{P}$ be the set of paths between them ; choose any $\varrho \in \mathcal{P}$; any $\gamma \in \mathcal{P}$ can be written -up to homotopy- as $\gamma = \varrho \circ \beta$, where β is a loop through x . γ and γ' are homotopic if β and β' are. The classical action is

$$S(\gamma) = S_0(\gamma) + \int_\varrho \alpha + \oint_\beta \alpha \tag{34}$$

where $S_0(\gamma) = \int_\gamma \Theta_0$.

Note that

- $\int_\varrho \alpha$ is independent of γ; denote

$$\exp\left[\frac{i}{\hbar}\int_\varrho \alpha\right] =: c \tag{35}$$

- $\exp\left[\frac{i}{\hbar}\oint_\beta \alpha\right] = \chi(g)$, with $g = [\beta]$, is constant on a homotopy class,

- define the partial amplitude

$$K_g(x',x) := \int_{\mathcal{P}_g} \exp\left[\frac{i}{\hbar} S_0(\gamma)\right] \mathcal{D}\gamma \tag{36}$$

where $\mathcal{P}_g \subset \mathcal{P}$ is the class of paths in $\mathcal{P}$ labelled by the same g ; as $\mathcal{P} = \bigcup_{g \in \pi} \mathcal{P}_g$, the additivity of the path integral gives

<u>**Theorem 4.5**</u>

$$K(x',x) = c \sum_{g \in \pi} \chi(g)\, K_g(x',x) \tag{37}$$

(a different choice in ϱ , the map $\gamma \leftrightarrow \beta$, or in Θ_0 introduces only an unobservable phase factor.)

EXAMPLE 2 (Bohm–Aharonov Experiment) [2][3][4][14]

As the electron is classically excluded from the interior of the solenoid, the configuration space is $\mathbb{R}^2 \setminus \{\text{a disk}\}$; the presymplectic form is just that of a free particle restricted to TQxR : $\sigma = \sigma_{free}\big|_{TQ\times\mathbb{R}}$. It is of course exact, $\sigma = d\Theta_0$ with

$$\Theta_0 = \Theta_{free} = mv\,dq - m\frac{v^2}{2}\,dt$$

But, as we have pointed out, we can add any 1-form

$$\alpha = e\left(A(q,t)\,dq + V(q,t)\,dt\right)$$

with $d\alpha = 0$. However, as Q is not simply connected $\alpha \neq df$.

Now, as far as we take seriously geometry and do not look into the solenoid, there is no reason to call A, resp. V, vector, resp. scalar, potential ; in order to identify them, we have to consider our system to be the part of a larger one, consisting of the electron and the solenoid and the magnetic field. [3], [8].

The homotopy group is here $\mathbb{Z}$, thus the characters (32) are written as

$$\chi(n) = \left(\exp\left[\frac{i}{\hbar}\oint_{\gamma_0} A\right]\right)^n = \left(\exp\left[\frac{ie\Phi}{\hbar}\right]\right)^n$$

γ_0 being a loop going once around the solenoid ; Φ is the enclosed magnetic flux.

Here we recognize again the n-th power of the <u>non-integrable phase factor of Wu and Yang</u> [14].

Theorem 4.6

Two expressions of the classical action are Q.M.E. iff the corresponding Wu-Yang factors are the same : $\Theta_1 \sim \Theta_2$ iff $\chi_1(1) = \chi_2(1)$ iff

$$\Phi_1 - \Phi_2 = \frac{h}{e} k \quad , \quad k \in \mathbb{Z} \tag{38}$$

confirming the conclusions of Bohm and Aharonov. $(h = 2\pi\hbar)$.

5. PREQUANTIZATION

The fact that conditions (18), (19) are sufficient to the existence of a Q.M.W.D.A., will follow from noting the relation to prequantization.

Theorem 5.1 [21]

A Q.M.W.D.A. exists iff the system is prequantizable in the K-S (Kostant-Souriau) sense [10],[11],[13].

Proof : By Weil's theorem, the system is prequantizable iff (18) or (19) holds. They ensure the possibility of constructing a U(1) principal bundle Y over E with connection form ω whose curvature form is $i\pi^*\sigma$, $\pi : Y \to E$ being the projection.

On the other hand, if the system is prequantizable, then, for any path γ in E , and $Y \ni \xi \in \pi^{-1}(y)$ with $y \in \gamma$, we have a unique horizontal lift $\hat{\gamma}$ through ξ .

If $\gamma \subset U_\alpha$, where Y has the local trivialization $\pi^{-1}(U_\alpha) \simeq U_\alpha \times U($
$\hat{\gamma}$ is written here as $\hat{\gamma} = (\gamma , z^\alpha)$; furthermore, [12] [13]

$$\frac{z^\alpha(0)}{z^\alpha(1)} = \exp\left[\frac{i}{\hbar} S_\alpha(\gamma)\right] \tag{39}$$

(γ being parametrized by $t \in [0,1]$; $\xi = (y, z^\alpha(0))$).

Thus, the classical action can be recovered by dividing the "heights" above a U_α of the horizontal lift of γ .

Now if we change our local trivialization, the new expression will be related to the old one as

$$\frac{z^\beta(0)}{z^\beta(1)} = \frac{z_{\alpha\beta}(y)}{z_{\alpha\beta}(y')} \cdot \frac{z^\alpha(0)}{z^\alpha(1)} \tag{40}$$

with $y' = \gamma(1)$; the $z_{\alpha\beta}$'s are here the transition functions of the U(1) bundle Y .

For local variational systems the transition functions depend only of x , the projection of y onto X ; this is the consequence of fact, that in (18) we could restrict ourselves to paths in X by (11). Thus

$$C_{\alpha\beta}(x',x) = \frac{Z_{\alpha\beta}(y)}{Z_{\alpha\beta}(y')} \tag{41}$$

will be a phase factor required in Definition 3.1.

On the other hand, as the horizontal lift of curve is well-defined independently of any local trivialization, $Z^{\alpha}(0)$ and $Z^{\alpha}(1)$ will have a meaning as soon as y and y' are contained in U_{α} , even if γ zigzags out and back from U_{α}. Let's define " $\exp[\frac{i}{\hbar}S_{\alpha}(\gamma)]$ " then just by (39) ; this will give a Q.M.W.D.A., as required (cf. [5]).

Q.E.D.

Consider a global system (E,σ) . It is always prequantizable: Define $Y = E \times U(1)$, $\pi : Y \ni (y,z) \to y \in E$; if Θ is a 1-form on E with $d\Theta = \sigma$, then

$$\omega = \pi^{*}\Theta + \hbar \cdot \frac{dz}{iz} \tag{42}$$

is a connection form, and any solution is written like this.

Two such constructions are told to be equivalent, if there exists a diffeomorphic map $\hat{F} : Y \to Y$ projecting onto E as identity, intertwinning the actions of U(1) and carrying one connection form to the others. ($\hat{F}$ is then necessarily of the form $\hat{F}(y,z) = (y, F(y)\cdot z)$ with $|F(y)| = 1$).

As it is well-known, if the underlying space is not simply connected, we may have different inequivalent prequantizations for the same classical system. We propose to rederive this theorem ([10], [11], [13]) by establishing

<u>Theorem 5.2</u> [4]

Let Θ_1, Θ_2 be action forms for a global variational system (E,σ). Then Θ_1 and Θ_2 are Q.M.E. (Def. 4.1) iff the corresponding prequantizations are equivalent in the prequantum bundle sense.

<u>Proof</u> : If $\hat{F} : (Y, \omega_1, \pi) \to (Y, \omega_2, \pi)$ is a map establishing the equivalence between the prequantizations, then $\hat{F}^* \omega_2 = \omega_1$ implies that $\Theta_1 = \Theta_2 + \hbar \frac{dF}{iF}$ and thus, for any loop γ, we have

$$\exp\left[\frac{i}{\hbar} \oint_\gamma \Theta_1\right] = \exp\left[\frac{i}{\hbar} \oint_\gamma \Theta_2\right], \text{ i.e. } \Theta_1 \sim \Theta_2$$

On the other hand, write $d\tilde{f} = \tilde{\Theta}_1 - \tilde{\Theta}_2$ (notation of Section 4) ; then $\Theta_1 \sim \Theta_2$ implies that $F(y) = \exp\left[\frac{i}{\hbar} \tilde{f}(\tilde{y})\right]$ is well-defined $(\tilde{y} \in p^{-1}(y))$, $\hat{F}(y,z) := (y, F(y) \cdot z)$ establishes the equivalence.

6. RELATION TO THE WU-YANG APPROACH TO GAUGE THEORY

Primarily interested in describing the quantized motion of a system, we investigated the conditions under which the amplitude (1) is well-defined and is unique. In geometric terms these properties could be expressed using the prequantum bundle (Y, ω, π).

On the other hand, in their approach to gauge theory, Wu and Yang [14] proposed to study an expression slightly similar to the amplitude (1), namely (in U(1) case)

$$\exp\left[\frac{ie}{\hbar}\int_{\gamma} A_{\alpha}\, dx^{\alpha}\right] \tag{43}$$

In our examples (monopole and Bohm-Aharonov experiment) also our conditions turned out to depend on this expression. It is not difficult to understand that this happens quite generally, at least for electromagnetic interactions.

Electromagnetism can, in fact, be conceived as constructing a U(1) principal bundle P with connection form $\mathcal{E}$ above space-time [21], [22]. The curvature form of this bundle is just then the electromagnetic 2-form F . $\mathcal{E}$ has the local expression

$$\mathcal{E} = \alpha + \hbar\,\frac{dz}{iz} \tag{44}$$

where the local 1-form α is written as

$$\alpha = e A_{\alpha}\, dx^{\alpha} \quad \left(= e A_j\, dq^j + eV\, dt\right) \tag{45}$$

The existence and uniqueness of this construction depends on the "non integrable phase factor" (43).

Now, to see the connection between these two theories, imagine that a particle with charge e and mass m moves in the electromagnetic field. In describing the interacting quantum system electron + field the latter can be studied by the W.K.B. approximation. Then, the electron wave equation factors out from that of the composite system ; the effect of the field is retained by an interaction term ("minimal coupling") in the Schrödinger equation [3].

Minimal coupling has the following geometric expression [21], [22]: Let σ_0 denote the symplectic form of a free particle. Then the prequantum bundle for the electron + (passive) field is constructed as

$$Y = pr_X^{\ast} P \tag{46}$$

where $pr_X : TQ \times R \ni (q, v, t) \longrightarrow (q, t) = x \in X$; the connection form itself is written (with a slight ambiguity) as

$$\omega = \Theta_0 + \mathcal{E} = \text{locally} = (\Theta_0 + \alpha) + \hbar \frac{dz}{iz} \tag{47}$$

where Θ_0 is the global action form $(mv\,dq - m\frac{v^2}{2}\,dt)$ for σ_0

Now, as we have shown, the properties of the prequantum bundle depend on the amplitude

$$\exp\left[\frac{i}{\hbar}\int_\gamma (\Theta_0 + \alpha)\right] \tag{48}$$

$(\gamma \subset TQ \times \mathbb{R})$

but $\int_\gamma \Theta_0$ exists always and is unique. Thus all the problems of existence and uniqueness come from the factor $\exp[\frac{i}{\hbar}\int_\gamma \alpha]$. Finally, as α projects to X by construction, we can always use curves γ lying in X . Thus, we can explain the role of the Wu-Yang factor ; it is just the factor which determines whether a test particle moving in the exterior classical field has a meaningful quantum description.

ACKNOWLEDGEMENTS

I am indebted to Jean-Marie SOURIAU for hospitality, help and encouragement at Marseille. Discussions with Christian DUVAL and David SIMMS are also gratefully acknowledged.

REFERENCES

[1] R.P. FEYNMAN and A.R. HIBBS, Quantum Mechanics and Path Integrals, Mc. Graw-Hill Book Co., (1965).

[2] Y. AHARONOV and D. BOHM, Phys. Rev. 115, 3, 145 (1959).

[3] Y. AHARONOV and D. BOHM, Phys. Rev. 123, 4, 1511 (1961).

[4] P.A. HORVÁTHY, Phys. Letters 76A, 11 (1980), (to appear).

[5] T.T. WU and C.N. Yang, Phys. Rev. D14, 2, 437 (1976).

[6] P.A.M. DIRAC, Phys. Rev. 74, 817 (1948).

[7] C. DEWITT and M.G.G. LAIDLAW, Phys. Rev. D3, 6, 1375 (1971).

[8] L. SCHULMAN, J. Math. Phys. 12, 2, 304 (1971).

L. SCHULMAN, in "Functional Integration and its Applications", Proc. Intern. Conf. London 1974, Clarendon Press (1975) A.M. Arthurs Ed.

[9] J.S. DOWKER, J. Phys. A (Gen. Phys.) 5, 936 (1972).

[10] B. KOSTANT, in Lecture Notes in Math. 170, Springer (1970) Taam Ed.

[11] J.M. SOURIAU, Structure des systèmes dynamiques, Dunod (1970).

[12] J.M. SOURIAU, Structure of Dynamical Systems, to appear at North-Holland.

[13] N.M.J. WOODHOUSE and D.J. SIMMS, Lecture Notes in Physics 53, Springer (1976).

[14] T.T. WU and C.N. YANG, Phys. Rev. D12, 3845 (1975).

[15] P.A. HORVÁTHY, J. Math. Phys. 20, 1,49 (1979).

[16] P.A. HORVÁTHY, Ph.D. Thesis, (in Hungarian) (1978).

[17] J. KLEIN, Ann. Inst. Fourier (Grenoble) 12, 1-124 (1962).

J. KLEIN, Ann. Inst. Fourier (Grenoble) 13, 191 (1963).

[18] C. GODBILLON, "Géométrie différentielle et mécanique analytique" Hermann, Paris (1969).

[19] SULANKE and WINTGEN, Differentialgeometrie und Faserbündel, VEB Deutscher Verlag der Wissenschaften, Berlin (1972).

[20] P.A. HORVÁTHY, Feynman Integral for Spin, Preprint CPT Marseille 79/P.1099 (1979) (unpublished).

[21] S. STERNBERG, in Lecture Notes in Math. 676, 1-80, Bleuler et al. Eds., Springer (1978).

[22] Ch. DUVAL, Sur les mouvements classiques dans un champ de Yang-Mills, CPT Preprint Marseille 78/P.1056 (1978) (unpublished).

[23] L. LANDAU and E. LIFCHIFTZ, Mécanique, Mir (1965).

[24] R. ABRAHAM and J. MARSDEN, Foundations of Mechanics, Benjamin, (1978).

[25] P.A. HORVÁTHY and L. ÚRY, Acta Physica Hungarica 42, 3 (1977).

G R O U P E S D I F F E R E N T I E L S

J . M . S O U R I A U

Université de Provence
Centre de Physique théorique, CNRS, Marseille

Adresse postale

Centre de Physique théorique
CNRS - LUMINY - CASE 907
F - 13288 MARSEILLE CEDEX 2 (France)

Ce travail fait partie d'une tentative de formalisation de la mécanique quantique; mais il est rédigé sous forme mathématique autonome.

L'axiomatique des "difféologies" (§1) définit la catégorie des groupes différentiels ; catégorie suffisamment vaste pour contenir, par exemple, tous les groupes de difféomorphismes; cependant les groupes différentiels conservent la plupart des propriétés élémentaires des groupes de Lie, que nous passons en revue:

■ Topologie (§ 4): un groupe différentiel est canoniquement un groupe topologique; la structure topologique est construite par des méthodes d'analyse harmonique, exposées au § 3 , et qui seront utilisées systématiquement aux §§ 5,6.

■ Homotopie (§ 2) : on définit directement les revêtements des groupes différentiels; tout groupe différentiel connexe possède un revêtement universel, donc un groupe d'homotopie, qui sont construits explicitement.

■ Analyse infinitésimale (§ 5) : tout groupe différentiel G possède un espace tangent $\mathcal{G}$, qui est canoniquement un espace vectoriel topologique localement convexe, et sur lequel agit la représentation adjointe; on peut définir le crochet de Lie de deux vecteurs tangents, et extraire de $\mathcal{G}$ une algèbre de Lie $\mathcal{G}_o$. On définit corrélativement un espace cotangent $\hat{\mathcal{G}}$ (partie séparante du dual $\mathcal{G}'$), et la représentation coadjointe.

■ Enfin certains vecteurs de $\mathcal{G}$ (les vecteurs complets) possèdent une exponentielle dans le groupe; cette opération conserve quelques propriétés caractéristiques (§ 6). ■

Le présent exposé est destiné à montrer l'existence d'une telle théorie; nous n'abordons ici aucune application.

-o-o-o-o-o-o-o-o-o-o-o-o-

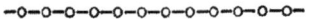

§ 0 : NOTATIONS

(0.1) F,G étant des applications quelconques, l'ensemble de définition de F sera noté def(F) ; son ensemble de valeurs im(F) ; nous noterons $<$ la relation d'ordre du <u>prolongement</u>: $\{$ def(F) $\subset$ def(G) ; $x \in$ def(F) $\Rightarrow$ G(x) = F(x)$\}$. Toute famille F_j qui est <u>compatible</u> $\{\forall x \in \text{def}(F_j) \cap \text{def}(F_k) \quad F_j(x) = F_k(x)\}$ admet une borne supérieure pour la relation $<$ (le plus petit prolongement commun des F_j) qui sera noté sup(F_j).

(0.2) Les <u>variétés</u> que nous rencontrerons seront de dimension finie, de classe C^∞, séparées. X et Y étant deux variétés, $C^\infty(X,Y)$ désigne la classe des applications (infiniment) différentiables de X dans Y ; nous noterons D(X,Y) l'ensemble des applications infiniment différentiables <u>d'un ouvert</u> de X dans Y . Dans cette notation X ou Y pourront en particulier être des espaces numériques $\mathbb{R}^n$.

(0.3) En général , nous noterons $\times$ une loi de groupe , e l'élément neutre. Nous appellerons <u>morphisme de groupe</u> $\Phi : G \to G'$ toute application d'un groupe G dans un groupe G' vérifiant
$\Phi(g_1 \times g_2) = \Phi(g_1) \times \Phi(g_2) \qquad \forall g_1, g_2 \in G$.

§ 1 : DIFFEOLOGIES ET GROUPES DIFFERENTIELS

<u>Définitions</u>

(1.1) Soit $G, \times$ un groupe.

Nous appellerons <u>difféologie</u> de G le choix, $\forall\, n \in \mathbb{N}^*$, d'un ensemble $D(\mathbb{R}^n, G)$, dont les éléments sont des applications d'un <u>ouvert</u> de $\mathbb{R}^n$ dans G, si les 5 axiomes suivants sont vérifiés:

a) Toute application <u>constante</u> $\mathbb{R}^n \to G$ appartient à $D(\mathbb{R}^n, G)$.

b) Si $\mathcal{F}_j \in D(\mathbb{R}^n, G)$, $\mathcal{F}_j$ compatibles (voir (0.1)), sup($\mathcal{F}_j$)$\in$ $D(\mathbb{R}^n, G)$.

c) $\forall \mathcal{F} \in D(\mathbb{R}^n, G)$, $\forall\, A \in D(\mathbb{R}^n , \mathbb{R}^m)$, $\mathcal{F} \circ A \in D(\mathbb{R}^m, G)$.

d) $\forall \mathcal{F} \in D(\mathbb{R}^n, G) \qquad [r \mapsto \mathcal{F}(r)^{-1}] \in D(\mathbb{R}^n, G)$.

e) $\forall \mathcal{F} \in D(\mathbb{R}^n, G), \forall \mathcal{F}' \in D(\mathbb{R}^{n'}, G), \left[(r,r') \mapsto \mathcal{F}(r) \times \mathcal{F}'(r')\right] \in D(\mathbb{R}^{n+n'}, G)$

(1.2) Une difféologie D' de G sera dite plus fine que D si elle a moins d'éléments : $\forall n$, $D'(\mathbb{R}^n, G) \subset D(\mathbb{R}^n, G)$ (nous verrons ci-dessous que la topologie associée est aussi plus fine).

(1.3) Si des $\mathcal{F}_j$ sont des applications quelconques d'ouverts de $\mathbb{R}^{n_j}$ dans un groupe G, la difféologie la plus fine contenant les $\mathcal{F}_j$ existe, et sera dite engendrée par les $\mathcal{F}_j$; en particulier, la difféologie la plus fine de G est la difféologie discrète : les éléments $\mathcal{F}$ de $D(\mathbb{R}^n, G)$ sont alors les applications localement constantes ($\mathcal{F}$ ne prend qu'une valeur sur chaque composante connexe de l'ouvert def($\mathcal{F}$)).

(1.4) Nous appellerons groupe différentiel un groupe G muni d'une difféologie. Les applications appartenant à l'un des $D(\mathbb{R}^n, G)$ seront alors dites différentiables.

Exemples de groupes différentiels.

(1.5) Soit G un groupe de Lie réel; $\mathbb{R}^n$ et G étant considérés comme variétés, nous pouvons définir $D(\mathbb{R}^n, G)$ par la règle (0.2); les axiomes (1.1) des groupes différentiels sont satisfaits : tout groupe de Lie est un groupe différentiel.

(1.6) Soit X une variété ; l'ensemble des difféomorphismes de X, sous-groupe du groupe X! des permutations de X :

$$\mathrm{diff}(X) = \left\{ g \in X! \;/\; g \text{ et } g^{-1} \in D(X,X) \right\}$$

devient un groupe différentiel si on note $D(\mathbb{R}^n, \mathrm{diff}(X))$ l'ensemble des applications $\mathcal{F}$ ($\mathrm{im}(\mathcal{F}) \subset \mathrm{diff}(X)$) telles que

$$\left[(r,x) \mapsto \mathcal{F}(r)(x)\right] \in D(\mathbb{R}^n \times X, X)$$

[la vérification de l'axiome (1.1 d) utilise le théorème de différentiabilité des solutions d'équations implicites dépendant de paramètres] .

Morphismes de groupes différentiels.

(1.7) Soient G et G' deux groupes différentiels; Φ un morphisme de groupe $G \to G'$ (0.3).

Nous dirons que Φ est un <u>morphisme de groupe différentiel</u> (en abrégé <u>D-morphisme</u>) si

$$\mathcal{F} \in D(\mathbb{R}^n, G) \Rightarrow \Phi \circ \mathcal{F} \in D(\mathbb{R}^n, G')$$

(1.8) Exemple: si G est un groupe de Lie, X une variété, un D-morphisme $G \to \mathrm{diff}(X)$ est ce qu'on appelle une <u>action différentiable</u> de G sur X.

(1.9) Il est clair que les D-morphismes se composent, constituant les flèches de la <u>catégorie des groupes différentiels</u>.

En particulier, un <u>D-isomorphisme</u> $\Phi: G \to G'$ est un isomorphisme de groupe tel que Φ et Φ^{-1} soient des D-morphismes.

(1.10) Exemple: Tout <u>automorphisme intérieur</u>

$$g \mapsto a \times g \times a^{-1} \qquad \text{(a choisi dans } G\text{)}$$

d'un groupe différentiel est un <u>D-automorphisme</u> (conséquence facile des axiomes (1.1)).

<u>Produit direct</u>

(1.11) Si G_1 et G_2 sont des groupes différentiels, on munit le produit direct $G = G_1 \times G_2$ de la difféologie la moins fine pour laquelles les projections canoniques $i_1 : G \to G_1$ et i_2 $G \to G_2$ soient des D-morphismes. Explicitement, $\mathcal{F} \in D(\mathbb{R}^n, G)$ ssi $i_1 \circ \mathcal{F}$ et $i_2 \circ \mathcal{F}$ appartiennent à $D(\mathbb{R}^n, G_1)$ et $D(\mathbb{R}^n, G_2)$.

<u>Sous-groupes</u>

(1.12) Soit G un sous-groupe d'un groupe différentiel G'.
On définit sur G une difféologie (<u>difféologie induite</u>) en posant

$$\left[\mathcal{F} \in D(\mathbb{R}^n, G)\right] \Longleftrightarrow \left[\mathcal{F} \in D(\mathbb{R}^n, G') \text{ et } \mathrm{im}(\mathcal{F}) \subset G\right]$$

c'est la difféologie la moins fine pour laquelle l'injection canonique $G \to G'$ soit un D-morphisme.

(1.13) Exemple : <u>tout groupe de difféomorphismes</u> est canoniquement un groupe différentiel , comme sous-groupe d'un groupe différentiel diff(X) (voir (1.6)).

(1.14) Si G est un sous-groupe d'un groupe différentiel G', et si k appartient au normalisateur de G dans G', l'application

$$g \mapsto k \times g \times k^{-1} \qquad [g \in G]$$

est un <u>D-automorphisme</u> de G (muni de sa difféologie induite).

<u>Quotients</u>

(1.15) Soit G un groupe différentiel, H un sous-groupe invariant; notons G' le <u>groupe quotient</u> G/H , Φ le morphisme canonique $G \to G'$.

Nous munirons G' de la difféologie la plus fine qui fait de Φ un D-morphisme; explicitement

$$[\mathcal{F}' \in D(\mathbb{R}^n, G')] \Leftrightarrow [\exists \mathcal{F}_j \in D(\mathbb{R}^n, G) \qquad \mathcal{F}' = \sup \Phi \circ \mathcal{F}_j]$$

(notation (0.1)).

<u>Morphismes stricts</u>

(1.16) Soit $\Phi : G \to G'$ un D-morphisme. On peut factoriser canoniquement Φ sous la forme

$$G \to G/\ker(\Phi) \xrightarrow{\Psi} \mathrm{im}(\Phi) \to G' \quad ;$$

on vérifie que Ψ est un isomorphisme de groupe et un D-morphisme ($G/\ker(\Phi)$ et $\mathrm{im}(\Phi)$ étant munis de leurs D-structures (1.15) et (1.12)). Nous dirons que Φ est un <u>D-morphisme strict</u> si Ψ est un D-isomorphisme, c'est-à-dire si Ψ^{-1} est aussi un D-morphisme.

<u>Exemples</u>:

(1.17) ■ Soit X une variété, G un groupe de difféomorphismes de X (voir (1.13)). Soit Y l'espace fibré des <u>vecteurs tangents</u> à X (resp. des <u>covecteurs tangents</u> , des <u>repères</u> , des <u>tenseurs</u> de variance donnée, des <u>α-densités</u> , des <u>connexions linéaires</u> , etc). Nous savons <u>relever</u> chaque élément g de G par un difféomorphisme de Y (technique des "objets géométriques") ; on définit ainsi un D-morphisme injectif $G \rightarrow \mathrm{diff}(Y)$. C'est un D-morphisme <u>strict</u> : on le constate en utilisant la différentiabilité de la projection $Y \rightarrow X$.

Ainsi, si un groupe différentiel est D-isomorphe à un groupe de difféomorphismes, il l'est d'une infinité de façons; il est en particulier isomorphe à un groupe de <u>symplectomorphismes</u> (il suffit de choisir pour Y le fibré cotangent).

(1.18) ■ Soit G un groupe différentiel quelconque.
Nous appellerons <u>arc</u> de G toute application γ de $\mathbb{R}$ dans G qui est différentiable (Cf.(1.4)) et qui vérifie la condition

$$\spadesuit \qquad \gamma(0) = e$$

On constate que l'ensemble Γ des arcs de G devient un <u>groupe différentiel</u> si on convient que :

$$\heartsuit \qquad [\gamma \times \gamma'](t) = \gamma(t) \times \gamma'(t) \qquad \forall \gamma, \gamma' \in \Gamma, \forall t \in \mathbb{R}$$

et que

$$\diamondsuit \left[\mathcal{F} \in D(\mathbb{R}^n, \Gamma) \right] \Leftrightarrow \left[(r,t) \mapsto \mathcal{F}(r)(t) \right] \in D(\mathbb{R}^{n+1}, G) \; ;$$

on vérifie alors que l'application P :

$$\clubsuit \qquad P(\gamma) = \gamma(1) \qquad \forall \gamma \in \Gamma$$

est un <u>D-morphisme strict</u> de Γ dans G. ■

§ 2 HOMOTOPIE DES GROUPES DIFFERENTIELS

(2.1) Soit G un groupe différentiel; considérons l'ensemble

$$G_0 = \left\{ g_0 \in G \; / \text{ il existe un } \underline{\text{arc}} \; \gamma, \gamma(0)=e, \gamma(1)=g_0 \right\}$$

G_0 est l'image du <u>groupe des arcs</u> de G par le D-morphisme $\gamma \mapsto \gamma(1)$ (voir ci-dessus (1.18)); c'est donc un sous-groupe de G , que nous appellerons <u>composante neutre</u> de G (voir pourquoi en (4.13)).

(2.2) Soient G et G' deux groupes différentiels, Φ un D-morphisme $G \rightarrow G'$. Si nous désignons par Γ et Γ' les groupes des arcs correspondants, il est immédiat qu'il existe un relèvement

$\bar{\Phi} : \Gamma \to \Gamma'$ de Φ qui est un D-morphisme :

$$\begin{array}{ccc} \Gamma & \xrightarrow{\bar{\Phi}} & \Gamma' \\ P\downarrow & & \downarrow P' \\ G & \xrightarrow{\Phi} & G' \end{array}$$

il suffit de poser $\bar{\Phi}(\gamma) = \Phi_\circ\gamma \quad \forall \gamma \in \Gamma$. On a évidemment $\Phi(P(\Gamma)) \subset P'(\Gamma')$; par conséquent :

[L'image par un D-morphisme $G \to G'$ de la composante neutre de G est incluse dans la composante neutre de G'.

(2.3) Il en résulte en particulier que tout D-isomorphisme $G \to G'$ envoie la composante neutre de G sur celle de G'; en considérant le cas des automorphismes intérieurs (voir (1.10)), on voit que la composante neutre G_0 est un sous-groupe invariant de G. Les classes selon G_0 s'appelleront composantes de G (ce seront les composantes connexes de la topologie qui sera définie au § 4); onvérifie la proposition:

[Soit G un groupe différentiel, G_0 sa composante neutre.
a) La difféologie quotient du groupe des composantes G/G_0 est discrète (voir (1.15) et (1.3)) ;
b) $[\ G$ discret $] \Leftrightarrow [\ G_0 = \{e\}\ ]$

(2.4) Nous dirons qu'un groupe différentiel est connexe s'il est égal à sa composante neutre (définition compatible avec la topologie du § 4); le diagramme (2.2) montre que l'image par un D-morphisme d'un groupe connexe est connexe et contenue dans la composante neutre du groupe d'arrivée. Par conséquent tout sous-groupe connexe d'un groupe différentiel G est contenu dans la composante neutre de G.

(2.5) Exemple : pour tout groupe différentiel G, le groupe des arcs Γ est connexe.

[Si $\gamma \in \Gamma$, $t \in \mathbb{R}$, la fonction $\gamma_\circ t$ définie par $[\gamma_\circ t](u) = \gamma(tu)\ \forall u \in \mathbb{R}$ est un arc ; il est immédiat que $\tilde{\gamma} : t \mapsto \gamma_\circ t$ est un arc de Γ, et que $P(\tilde{\gamma}) = \gamma$ (notation (1.18))]

Revêtements

(2.6) Soit G un groupe différentiel, H un sous-groupe invariant de G.

Puisque les $h \mapsto g \times h \times g^{-1}$ associés aux éléments g de G sont des D-automorphismes de H (voir (1.14)), ils laissent fixe la composante neutre H_0 de H (voir (2.3)); H_0 est donc sous-groupe invariant, non seulement de H (2.3) , mais aussi de G .

On connait dans ce cas l'isomorphisme de groupes

$$\Diamond \qquad G/H \sim [G/H_0] \,/\, [H/H_0]$$

qui correspond à la factorisation suivante du morphisme canonique $P : G \to G/H$:

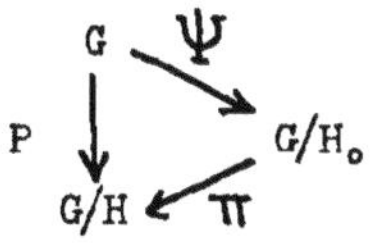

où $\ker(\pi) = \Psi(H)$.

On vérifie facilement que la relation ◊ est en fait un isomorphisme de groupes différentiels ; plus précisément

- π est un D-morphisme strict (1.16) ;
- $\ker(\pi)$ est discret ((1.12), (1.3)).

(2.7) Soient G et $\tilde{G}$ deux groupes différentiels; nous dirons que $\tilde{G}$ est un revêtement de G si G est isomorphe au quotient de $\tilde{G}$ par un sous-groupe discret (pour sa difféologie induite); en d'autres termes, s'il existe un D-morphisme $\pi : \tilde{G} \to G$, surjectif, strict, à noyau discret . Avec cette terminologie, le résultat (2.6) exprime que G/H_0 est un revêtement de G/H.

On vérifie aisément les deux propositions suivantes :

(2.8) Soient G , G_1 , G_2 des groupes différentiels; Φ_1 et Φ_2 des D-morphismes sur G :

On considère le produit croisé $\overline{G}$:

$$\overline{G} = \left\{ (g_1,g_2) \in G_1 \times G_2 \ / \ \Phi_1(g_1) = \Phi_2(g_2) \right\}$$

sous-groupe du produit direct $G_1 \times G_2$ (voir (1.11)), muni des D-morphismes $\theta_1 : (g_1,g_2) \mapsto g_1$ et $\theta_2 : (g_1,g_2) \mapsto g_2$.

Si (G_2,Φ_2) est un revêtement de G, $(\overline{G},\theta_1)$ est un revêtement de G_1 ; de même, si (G_1,Φ_1) est un revêtement de G, $(\overline{G}, \theta_2)$ est un revêtement de G_2.

(2.9) Tout revêtement de revêtement est un revêtement: si (G,π) est un revêtement de G', (G',π') un revêtement de G'' ,$(G, \pi' \circ \pi)$ est un revêtement de G''.

(2.10) Exemple de revêtement : Soit G le groupe des difféomorphismes d'une variété connexe X ; soit $\hat{X}$ la variété revêtement universel de X, P la projection de $\hat{X}$ sur X. Le groupe d'homotopie H de X est un sous-groupe de diff$(\hat{X})$; soit $\hat{G}$ son normalisateur dans diff$(\hat{X})$. $\forall \hat{g} \in \hat{G}$, il existe un difféomorphisme $\pi(\hat{g})$ de X défini par $\pi(\hat{g})(P(\hat{x})) = P(\hat{g}(\hat{x}))$ $\forall \hat{x} \in \hat{X}$.

On peut montrer que $(\hat{G}, \pi)$ est un revêtement de diff(X) , au sens (2. 7) ; de plus le noyau de π est discret dans $\hat{G}$, non seulement pour sa difféologie induite, mais aussi comme partie de l'espace topologique $\hat{G}$ (voir le §4).

(2.11) Lemme : Soit G un groupe différentiel, $(\tilde{G}, \pi)$ un revêtement de G ; soit $\mathcal{F} \in D(\mathbb{R}^n,G)$.
Si def$(\mathcal{F}) = \mathbb{R}^n$, il existe un relèvement global $\tilde{\mathcal{F}}$ de $\mathcal{F}$: $\tilde{\mathcal{F}} \in D(\mathbb{R}^n,\tilde{G})$, $\mathcal{F} = \pi \circ \tilde{\mathcal{F}}$

[désignons par E l'ensemble non vide des nombres $r > 0$ tels que $\mathcal{F}$ soit relevable dans la boule $B(0,r)$ (Cf.(1.15)); on peut fixer le relèvement $\tilde{\mathcal{F}}_r$ en choisissant $\tilde{\mathcal{F}}_r(0)$, ce qui rend compatibles les $\tilde{\mathcal{F}}_r$. La borne supérieure de toute partie majorée de E appartient à E ; en recouvrant la sphère $S(0,r)$ par des boules relevables, on montre que $[r \in E] \Rightarrow \exists\, r' > r, \ ]0, r'] \subset E$; d'où $E = \mathbb{R}^+$; il suffit de prendre $\tilde{\mathcal{F}} = \sup \tilde{\mathcal{F}}_r$]

(2.12) Soit G' un groupe différentiel, (G, Π) un revêtement de G'. Posons $H = \ker(\Pi)$ et définissons τ par

$$\tau(g)(h) = g \times h \times g^{-1} \qquad \forall g \in G, \forall h \in H$$

τ est un morphisme de groupe $G \to \mathrm{auto}(H)$, dont le noyau G_1 est le commutant de H dans G.

Soit γ un arc de G, $h \in H$. L'application $\mathbb{R} \to H \quad t \mapsto \tau(\gamma(t))(h)$ est différentiable à valeurs dans le groupe discret H, donc localement constante, donc constante; en faisant $t=0$ et $t=1$, on voit que la composante neutre G_0 de G est incluse dans le commutant G_1 de H.

Le morphisme Π envoie G_0 dans la composante neutre G'_0 de G'; le lemme (2.11), avec $n=1$, montre que $\Pi(G_0)$ est <u>égal</u> à G'_0 ; on vérifie ensuite que (G_0, Π) est un <u>revêtement</u> de G'_0 .

Quelques raisonnements standard permettent alors de construire un diagramme commutatif de D-morphismes:

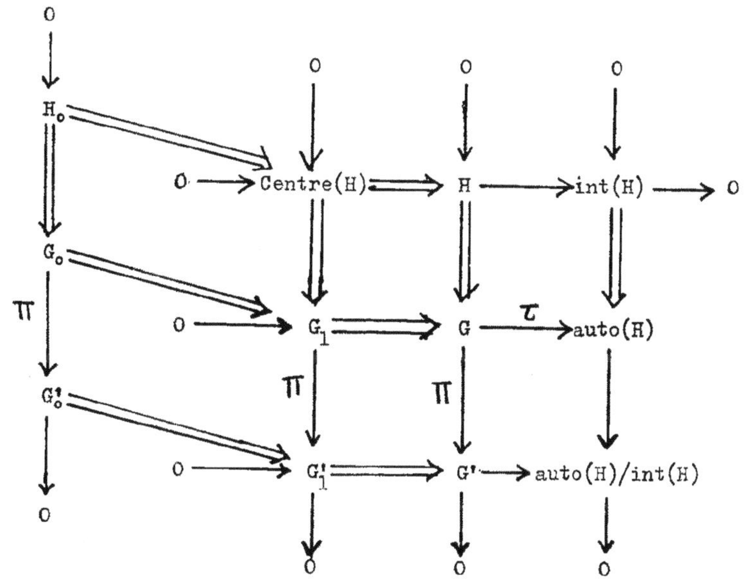

où les suites horizontales et verticales sont exactes; les flèches doubles indiquent une injection canonique sous-groupe $\Longrightarrow$ groupe.

Revêtement universel

(2.13) Nous nous occupons ici des revêtements connexes : si G est un revêtement connexe de G', G' est lui-même connexe; avec les notations (2.12), on a alors $G = G_0 = G_1 = \text{commutant}(H)$; H est central, donc abélien : tout revêtement connexe est donc une extension centrale .

(2.14) Soit G un groupe différentiel connexe quelconque.
Notons Γ le groupe des arcs de G (1.18); P le morphisme $\Gamma \to G$ défini par $P(\gamma) = \gamma(1)$; K son noyau.

Le fait que G soit connexe signifie que P est surjectif ; puisque P est strict (1.18) , G est isomorphe au quotient G/K . En considérant la composante neutre K_0 de K, on peut donc effectuer la construction (2.6) :

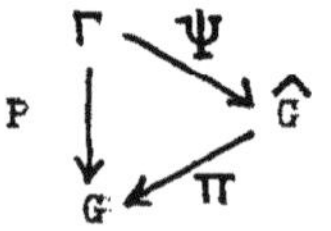

on note $\hat{G}$ le quotient Γ / K_0 , Ψ le morphisme canonique $\Gamma \to \hat{G}$, Π le morphisme $\hat{G} \to G$ défini par $\Pi \circ \Psi = P$. Nous savons alors que $(\hat{G}, \Pi)$ est un revêtement de G : Π est surjectif et strict, son noyau $H = \Psi(K)$ est discret.

Nous savons d'autre part que Γ est connexe (2.5); puisque Ψ est surjectif, $\hat{G}$ est connexe ; il en résulte que H est central (2.13), donc commutatif.

(2.15) Définition, théorème:
Un groupe différentiel G sera dit simplement connexe s'il est isomorphe à tous ses revêtements connexes.

Pour que G soit simplement connexe, il faut et il suffit que le groupe H construit en (2.14) soit égal à $\{e\}$.

a) L'implication est triviale dans un sens.

b) Supposons H réduit à e; soit $(\tilde{G}, \pi)$ un revêtement connexe de G; $\tilde{g}$ un élément du noyau de π ; il s'agit de montrer que $\tilde{g} = \tilde{e}$.

Puisque $\tilde{G}$ est connexe, il existe un arc $\tilde{\gamma}$ de $\tilde{G}$ tel que

$\tilde{\gamma}(1) = \tilde{g}$; posons $\gamma = \pi \circ \tilde{\gamma}$. Alors $\gamma \in \Gamma$, et $P(\gamma) = \gamma(1) = \pi(\tilde{g}) = e$. Donc $\gamma \in \ker(P) = K$. Puisque H est réduit à $\{e\}$, et que $H \sim K/K_0$, $K = K_0$; donc K est connexe; il existe un arc $\mathfrak{F}$ de K tel que $\mathfrak{F}(1) = \gamma$. Par construction, l'application $(u,t) \mapsto \mathfrak{F}(u)(t)$ appartient à $D(\mathbb{R}^2, G)$, est définie dans tout le plan et prend la valeur e sur les trois droites dessinées.

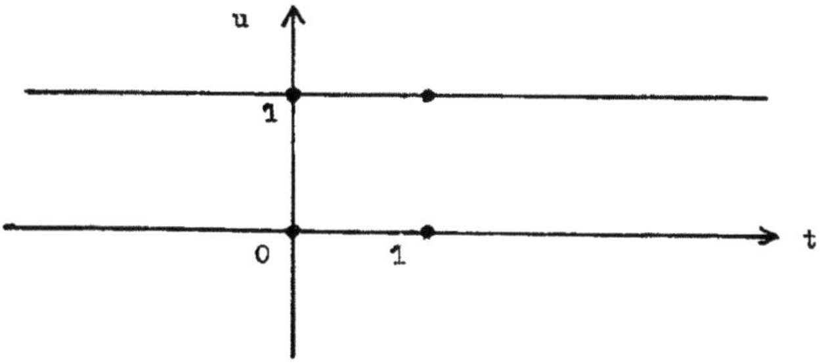

Le lemme (2.11), avec $n=2$, montre qu'il existe une application différentiable $\tilde{\mathfrak{F}}$ de $\mathbb{R}^2$ dans $\tilde{G}$ telle que $\pi(\tilde{\mathfrak{F}}(t,u)) = \mathfrak{F}(t)(u)$; on peut la choisir telle que $\tilde{\mathfrak{F}}(0,0) = \tilde{e}$.
Puisque $\ker(\pi)$ est discret, $\tilde{\mathfrak{F}}(t,u)$ est localement constante sur chacune de ces trois droites; puisque elle vaut $\tilde{e}$ à l'origine, elle vaut $\tilde{e}$ sur chaque droite; donc la fonction différentiable $\tilde{\gamma}'$: $u \mapsto \tilde{\mathfrak{F}}(1,u)$ vérifie $\tilde{\gamma}'(0) = \tilde{\gamma}'(1) = \tilde{e}$ et $\pi(\tilde{\gamma}'(u)) = \mathfrak{F}(1)(u) = \gamma(u) = \pi(\tilde{\gamma}(u))$; la fonction différentiable $u \mapsto \tilde{\gamma}(u) \times \tilde{\gamma}'(u)^{-1}$ prend ses valeurs dans $\ker(\pi)$ et donc est localement constante, donc constante; $u = 0$ et $u = 1$ donnent $\tilde{e} = \tilde{\gamma}(1) = \tilde{g}$

C.Q.F.D.

Théorème :

(2.16) Pour tout groupe différentiel connexe G , le revêtement $\hat{G}$ construit en (2.14) est simplement connexe.

Soit $\hat{\Gamma}$ le groupe des arcs de $\hat{G}$; $\hat{P} : \hat{\gamma} \mapsto \hat{\gamma}(1)$ sa projection sur $\hat{G}$; $\hat{K}$ le noyau de $\hat{P}$; d'après (2.15), il suffit de montrer que $\hat{K}$ est connexe.

On définit deux morphismes $\overline{\Pi} : \hat{\Gamma} \to \Gamma$ et $\overline{\Psi} : \Gamma \to \hat{\Gamma}$ (notations (2.14)) par

$$\overline{\Pi}(\hat{\gamma})(t) = \Pi(\hat{\gamma}(t))$$

$$\overline{\Psi}(\gamma)(t) = \Psi(\gamma_0 t) \qquad \text{(Cf.(2.5))};$$

il est élémentaire que $\overline{\Pi} \circ \overline{\Psi} \circ \overline{\Pi} = \overline{\Pi}$; puisque $H = \ker(\Pi)$ est discret, on en déduit que $\overline{\Psi}\circ\overline{\Pi}$ est le morphisme identique $\hat{\Gamma}\to\hat{\Gamma}$; il en résulte élémentairement que le noyau $\hat{K}$ de $\hat{P}$ est égal à $\overline{\Psi}(K_o)$, K_o étant le noyau de Ψ , connexe par hypothèse (2.14); par conséquent $\hat{K}=\overline{\Psi}(K_o)$ est connexe (2.4).

C.Q.F.D.

Théorème:

(2.17) Soit $\Phi : G_1 \to G$ un D-morphisme; $(\widetilde{G}, \pi)$ un revêtement de G. Si G_1 est simplement connexe, il existe un seul D-morphisme $\widetilde{\Phi} : G_1 \to \widetilde{G}$ qui relève Φ :

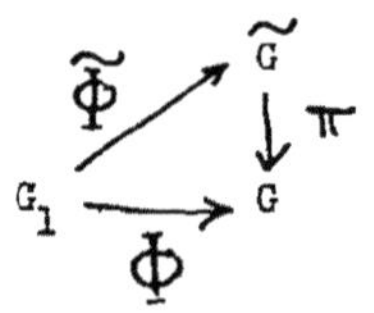

D'après (2.8) , le produit croisé

$$\overline{G} = \left\{ (g_1,\widetilde{g}) \in G_1 \times \widetilde{G} / \Phi(g_1) = \pi(\widetilde{g}) \right\}$$

et le morphisme $P : (g_1,\widetilde{g}) \mapsto g_1$ définissent un revêtement de G_1; $Q : (g_1,\widetilde{g}) \mapsto \widetilde{g}$ est un morphisme de $\overline{G}$ sur $\widetilde{G}$, qui vérifie $\pi\circ Q = \Phi\circ P$.

Puisque G_1 est connexe, la composante neutre $\overline{G}_o$ et la restriction P_o de P à $\overline{G}_o$ constituent un revêtement connexe de G_1 (2.12) ; puisque G_1 est simplement connexe, P_o est un isomorphisme (définition (2.15)).

Alors $\widetilde{\Phi} = Q\circ P_o^{-1}$ est un D-morphisme de G_1 dans G, qui vérifie $\pi\circ\widetilde{\Phi} = \Phi\circ P\circ P_o^{-1} = \Phi$. - L'unicité est immédiate.

C.Q.F.D.

Plaçons nous dans le cas particulier où (G_1,Φ) est un revêtement de G. Le théorème (2.8) montre que $(\overline{G},Q)$ est un revêtement de $\widetilde{G}$; si $\widetilde{G}$ est lui-même connexe, $(\overline{G}_o , Q)$ sera encore un revêtement de $\widetilde{G}$ (2.12) ; en composant avec l'isomorphisme P_o^{-1} , on voit que $(G_1,\widetilde{\Phi})$ est un revêtement de $\widetilde{G}$. D'où, après un changement de notations, le théorème :

(2.18) Soient G_1 et G_2 deux revêtements connexes d'un groupe différentiel G.

Si G_1 est <u>simplement</u> connexe, il existe un seul D-morphisme ψ qui factorise π_1 :

$$G_1 \xrightarrow{\psi} G_2$$
$$\pi_1 \searrow \swarrow \pi_2$$
$$G$$

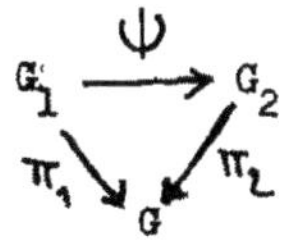

(G_1, ψ) est un revêtement de G_2.

(2.19) Pour tout groupe différentiel connexe G , tout revêtement simplement connexe G_1 est donc <u>universel</u> , en ce sens que l'on obtient tous les revêtements connexes de G en faisant le quotient de G_1 par un sous-groupe de $H_1 = \ker(\pi_1)$; en particulier tout autre revêtement simplement connexe lui est D-isomorphe; le groupe abstrait H_1 ainsi défini à un isomorphisme près s'appellera <u>groupe d'homotopie</u> de G.

Nous savons qu'un tel revêtement universel <u>existe</u>, parce que nous l'avons rencontré (construction (2.14), théorème (2.16)) ; le groupe H construit en (2.14) est donc un exemplaire du <u>groupe d'homotopie</u> de G ; (2.15) exprime donc que les groupes simplement connexes sont ceux dont l'homotopie est triviale.

§ 3 HARMONIES

(3.1) Soit G un groupe, m une fonction $G \to \mathbb{C}$.

On dit que m est <u>de type positif</u>, et on notera $m \in P(G)$, si $\forall n \in \mathbb{N}^*$, $\forall c_1, \ldots c_n \in \mathbb{C}$, $\forall g_1, \ldots g_n \in G$, on a

$\diamond \quad \sum_{k,\ell} \overline{c_k} c_\ell \, m\left(g_k^{-1} \times g_\ell\right) \geqslant 0$

Exemples : tout <u>caractère</u> de G (: morphisme dans U(1)) est de type positif; la <u>fonction caractéristique</u> de tout sous-groupe de G aussi.

L'inégalité $\diamond$ signifie que la matrice d'éléments $m\left(g_k^{-1} \times g_\ell\right)$ est hermitienne positive; elle a donc un déterminant $\geqslant 0$. On en déduit les formules suivantes, valables $\forall g, g' \in G$:

(3.2) $$m(g^{-1}) = \overline{m(g)}$$

(3.3) $$|m(g)| \leq m(e)$$ (n = 2)

(3.4) $$|m(g \times g')\, m(e) - m(g)\, m(g')| \leq \sqrt{m(e)^2 - |m(g)|^2} \times \sqrt{m(e)^2 - |m(g')|^2}$$

(n=3)

cette dernière formule entraîne

(3.5) $$|m(g) - m(g')| \leq \sqrt{2\, m(e) \left[m(e) - \mathcal{R}e\left(m(g^{-1} \times g')\right)\right]}$$

(3.6) - Il est évident que $P(G)$ est un <u>cône convexe</u> (dans l'espace vectoriel des fonctions bornées sur G) ; nous noterons $P_o(G)$ l'ensemble des fonctions m de $P(G)$ normalisées par la condition

$$m(e) = 1$$

il est clair (grâce à (3.3)) que

$$P(G) = \mathbb{R}^+ \times P_o(G)$$

$P_o(G)$ est évidemment un <u>convexe</u>.

- Quelques propriétés élémentaires des matrices positives montrent que :

(3.7) $m \in P(G) \Rightarrow \overline{m} \in P(G)$ $\quad [\overline{m} = g \mapsto \overline{m(g)}]$

(3.8) $m, m' \in P(G) \Rightarrow m\,m' \in P(G)$ $\quad [m\,m' = g \mapsto m(g)\, m'(g)]$

(3.9) - Soit $m \in P(G)$; choisissons $n \in \mathbb{N}^*$, $c_1, \ldots c_n \in \mathbb{C}$, $g_1, \ldots g_n \in G$. Il est immédiat que la fonction m' :

$$m'(g) = \sum_{k,\ell} \overline{c_k}\, c_\ell\, m(g_k^{-1} \times g \times g_\ell)$$

est elle-même <u>de type positif</u> ; nous dirons qu'elle est <u>subordonnée</u> à m ; cette relation est transitive.

Théorème:

(3.10) Soit G un groupe, m une fonction de type positif sur G. a et b des éléments de G .

Il existe trois fonctions m_o, m_1, m_2, de type positif sur G, subordonnées à m, telles que, $\forall g \in G$:

$$m(a \times g \times b) = \frac{1}{3}\left[m_o(g) + j\ m_1(g) + j^2\ m_2(g)\right]$$

$$\left[j = e^{2i\pi/3} \right]$$

[Il suffit de poser, pour $p = 0,1,2$: $c_{p,1} = 1$, $c_{p,2} = j^p$

$g_1 = b$, $g_2 = a^{-1}$ et

$$m_p(g) = \sum_{k,\ell \in \{1,2\}} \overline{c_{p,k}}\ c_{p,\ell}\ m(g_k^{-1} \times g \times g_\ell) \quad]$$

Définition :

(3.11) Soit G un groupe.

Nous appellerons harmonie de G tout ensemble M, non-vide, de fonctions de type positif sur G, vérifiant les 2 conditions

a) $[m_1, m_2 \in M] \Rightarrow [m_1 + m_2 \in M]$

b) $[m \in M$, m' subordonnée à m (3.9)$] \Rightarrow [m' \in M]$

En faisant $n=1$, $g_1 = e$ dans (3.9), on constate qu'une harmonie est un cône convexe.

Théorème :

(3.12) a) Soit p un morphisme de groupe $G \to G'$, M' une harmonie de G'. L'image réciproque

$$M = M' \circ p = \{ m' \circ p \ / \ m' \in M' \}$$

est une harmonie de G.

b) Soit M une harmonie d'un groupe G . Le "noyau" de M :

$$\ker(M) = \{ g \in G \ / \ m(g) = m(e)\ \forall m \in M \}$$

est un sous-groupe invariant de G; il existe une harmonie M' du groupe quotient $G' = G/\ker(M)$ caractérisée par la relation $M = M' \circ p$ (p : morphisme canonique). Nous dirons que M' est l'harmonie réduite de M.

c) Une harmonie sera dite irréductible si son noyau est égal à $\{e\}$; toute harmonie réduite (voir b) est irréductible.

[simple conséquence de (3.10) et (3.4)]

Théorème :

(3.13) Soit G un groupe; M une harmonie de G.

a) La topologie la moins fine de G qui rende continus les éléments de M est une topologie de groupe : l'application $G \times G \to G$ $(g, g') \mapsto g^{-1} \times g'$ est continue.

b) Une partie V de G est un voisinage de e si et seulement si il existe $m \in M$ et $\varepsilon > 0$ tels que

$$V \supset U_{m,\varepsilon} \; ; \qquad U_{m,\varepsilon} = \{ g \in G \; / \; \mathcal{R}e(m(g)) > m(e) - \varepsilon \}$$

c) Pour que cette topologie soit séparée (Haussdorff), il faut et il suffit que l'harmonie M soit irréductible (3.12 c).

[La démonstration est basée sur (3.2), (3.10) et sur les remarques suivantes : les ensembles $U_{m,\varepsilon}$ définis en b), contiennent e , sont symétriques : $U_{m,\varepsilon}^{-1} = U_{m,\varepsilon}$ et vérifient les propriétés suivantes : $\forall\, m_1, \varepsilon_1, m_2, \varepsilon_2, \; m, \varepsilon$

$$U_{m_1,\varepsilon_1} \cap U_{m_2,\varepsilon_2} \supset U_{m',\varepsilon'} \quad \text{si} \quad m' = m_1 + m_2 \, , \; \varepsilon' = \inf(\varepsilon_1, \varepsilon_2)$$

$$U_{m,\varepsilon} \supset U_{m',\varepsilon'} \times U_{m',\varepsilon'} \quad \text{si} \quad m' = m, \; \varepsilon' = \frac{1}{2}\left[\sqrt{m(e)+2\varepsilon} - \sqrt{m(e)}\right]^2$$

cette dernière propriété est une conséquence de (3.5)]

§ 4 TOPOLOGIE DES GROUPES DIFFERENTIELS

Etats.

(4.1) Soit G un groupe différentiel.
Nous noterons DP(G) l'ensemble des fonctions $m : G \to \mathbb{C}$ qui sont de type positif (3.1) et qui vérifient

$$[\varphi \in D(\mathbb{R}^n, G)] \Rightarrow [m \circ \varphi \in D(\mathbb{R}^n, \mathbb{C})]$$

(notations (1.4),(0.2)).

Nous appellerons états de G les éléments m de DP(G) qui sont normalisés : $m(e) = 1$ (Cf.(3.6)); leur ensemble sera noté $DP_0(G)$.

(4.2) Il est clair que $DP(G) = \mathbb{R}^+ \times DP_0(G)$; que $DP(G)$ est une <u>harmonie</u> (3.11), donc un cône convexe ; que le <u>conjugué</u> d'un état (3.7), le <u>produit de deux états</u> (3.8) sont encore des états; ainsi que la <u>partie réelle</u> $\mathcal{R}e_0\, m$ d'un état.

(4.3) Si Φ est un <u>D-morphisme</u> $G \to G'$, et si m' est un <u>état</u> de G', alors $m = m' \circ \Phi$ est un <u>état de</u> G.

<u>Topologie canonique</u>

Puisque $M = DP(G)$ est une harmonie, il suffit d'appliquer le théorème (3. 13) pour obtenir :

(4.4) <u>Tout groupe différentiel</u> G peut être muni canoniquement d'une <u>topologie de groupe</u>, définie comme la moins fine qui rende continus les <u>états</u> de G.

Une partie V de G est un <u>voisinage de e</u> si et seulement si il existe un état m et un nombre $\varepsilon > 0$ tels que

$$V \supset U_{m,\varepsilon} \quad ; \quad U_{m,\varepsilon} = \left\{ g \in G \,/\, \mathcal{R}e\,(m(g)) > 1 - \varepsilon \right\}$$

On vérifie immédiatement les résultats suivants, dans lesquels nous adoptons systématiquement la topologie (4.4) :

(4.5) Si $\mathcal{F} \in D(\mathbb{R}^n, G)$, $\mathcal{F}$ est <u>continue</u>.

(4.6) Tout D-morphisme est <u>continu</u> ; les D-isomorphismes sont des homéomorphismes.

(4.7) Une difféologie plus fine définit une topologie plus fine.

(4.8) Si G est un sous-groupe du groupe différentiel G', muni de sa difféologie induite (1.12), la topologie de G est <u>plus fine</u> que la topologie induite de celle de G'.

(4.9) Tout sous-groupe d'un groupe différentiel séparé est séparé.

(4.10) Tout sous-groupe topologiquement discret est muni de la difféologie discrète.

(4.11) La topologie d'un groupe quotient (1.15) est moins fine que la topologie quotient.

(4.12) Pour qu'un quotient G/H soit séparé, il est nécessaire que H soit fermé dans G ; la condition suivante est nécessaire et suffisante :

$$\left[g \in G,\ g \notin H \right] \Rightarrow \left[\text{il existe un état } m \,/\, m(g) \neq 1,\ m(H) = \{1\} \right]$$

–Le vocabulaire topologique que nous avons adopté au § 2 (en particulier dans (2.1), (2.3), (2.4)) sera justifié si on établit la proposition suivante :

(4.13) Soit G un groupe différentiel, G_o le sous-groupe :

$$\left\{ g_o \in G \ / \text{ il existe un arc } \gamma \text{ tel que } \gamma(0) = e , \gamma(1) = g_o \right\}$$

Alors G_o est la composante neutre de G pour la topologie (4.4)

Puisque les arcs sont continus (4.5) , G_o est connexe par arcs, donc connexe; nous allons montrer que G_o est le plus grand sous-groupe connexe (pour la topologie) en établissant que G_o est <u>ouvert et fermé</u> .

Soit m_o la fonction caractéristique de G_o ; on sait que m_o est de type positif (3.1); $\forall \mathcal{F} \in D(\mathbb{R}^n, G)$ on vérifie que $m_o \circ \mathcal{F}$ (qui ne prend que les valeurs 0 ou 1) est localement constante , donc différentiable; la condition (4.1) est vérifiée , m_o est un état, et par conséquent $m_o^{-1}(\,]1/2, 3/2[\,) = G_o$ est ouvert;de même $G - G_o$ est ouvert.

C.Q.F.D.

<u>Exemples</u>

(4.14) ■ Soit X une variété; désignons par φ un champ infiniment différentiable, à support compact, de demi-densités complexes de X. Si g est un difféomorphisme de X , nous savons définir l'image $g(\varphi)$ de φ par g (Cf.(1.17)). Le produit $\overline{\varphi}\ g(\varphi)$ est une 1-densité différentiable à support compact,et possède donc une intégrale intrinsèque sur X. Nous pouvons donc poser

$$m_\varphi(g) = \int_X \overline{\varphi}\ g(\varphi) \qquad \forall\, g \in \text{diff}(X);$$

on vérifie que $m \in DP(\text{diff}(X))$ (Cf.(1.6), (4.1)).

Si g n'est pas l'élément neutre e de diff(X), il est facile de choisir φ pour que $m_\varphi(g) \neq m_\varphi(e)$; d'où le théorème:

Tout groupe de difféomorphismes est un groupe différentiel <u>séparé</u>

(appliquer (3.13 c) et (4.9)).

■ La même technique des demi-densités (ou, si l'on préfère,des mesures de Haar) conduit au résultat suivant :

(4.15) Si G est un <u>groupe de Lie</u>, sa topologie de variété et sa topologie de groupe différentiel (1.5),(4.4) <u>coïncident</u> , et sont donc séparées.

(4.16) ■ Nous pourrions munir le même groupe de Lie G d'une autre difféologie D' (moins fine) en désignant par $D'(\mathbb{R}^n, G)$ l'ensemble des applications continues d'un ouvert de $\mathbb{R}^n$ dans G; on peut vérifier dans ce cas que G n'est plus séparé - et plus précisément que le noyau de l'harmonie D'P(G) est égal à la composante neutre de G.

(4.17) ■ Le quotient $\mathbb{R}/\mathbb{Q}$ ($\mathbb{R}$ et $\mathbb{Q}$ étant considérés comme groupes additifs) est un groupe différentiel connexe; son revêtement universel est $\mathbb{R}$; $\mathbb{R}/\mathbb{Q}$ n'est pas séparé en vertu de (4.12) ■

Axiome de séparation.

(4.18) Les exemples précédents nous montrent que certaines circonstances qui peuvent sembler pathologiques sont évitées si on se restreint aux groupes différentiels qui sont séparés: l'axiome de Haussdorff apparait donc comme un 6ème axiome facultatif que l'on peut adopter pour les groupes différentiels.

Il existe d'ailleurs une méthode systématique pour se ramener à ce cas :

(4.19) [Soit G un groupe différentiel quelconque; soit K le noyau de l'harmonie DP(G) (3.12 b). Alors le groupe différentiel G/K est séparé.

[On vérifie que DP(G/K) coïncide avec l'harmonie réduite de DP(G), au sens (3.12 b); elle est donc irréductible; par conséquent G/K est séparé (3.13c)]

Homotopie séparée

(4.20) Nous allons donner un exemple de cette réduction (4.19) : on se donne un groupe différentiel connexe séparé G; soit $(\hat{G}, \pi)$ son revêtement universel (2.14),(2.19).

Si $\hat{G}$ n'est pas séparé, on vérifie en appliquant à $\hat{G}$ la condition (4.12) que le noyau K de l'harmonie $DP(\hat{G})$ est un sous-groupe du groupe d'homotopie $H = \ker(\pi)$; que $\tilde{G} = \hat{G}/K$ est un revêtement connexe de G , séparé grâce à (4.19) ; que tout revêtement connexe séparé $\tilde{G}'$ est de la forme $\hat{G}/K'$, $K' \supset K$; par conséquent le D-morphisme $\hat{G} \to \tilde{G}'$ (voir (2.18)) se factorise par l'intermédiaire de $\tilde{G}$: $\tilde{G}$ est un revêtement séparé universel ; le groupe d'homotopie séparée , noyau de $\tilde{G} \to G$, est isomorphe à H/K ; il est encore central dans $\tilde{G}$:

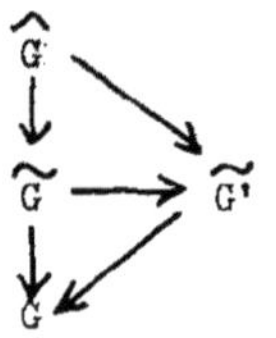

§ 5 TANGENT ET COTANGENT D'UN GROUPE DIFFERENTIEL

(5.1) Soit G un groupe différentiel quelconque.
Si γ et γ' sont deux arcs de G (Cf.(1.18)) et m un état de G (4.1), les fonctions $m \circ \gamma$ et $m \circ \gamma'$ sont deux applications différentiables de $\mathbb{R}$ dans $\mathbb{C}$, qui prennent la valeur 1 à l'origine; nous dirons que les arcs γ et γ' sont tangents si, $\forall m$, $m \circ \gamma$ et $m \circ \gamma'$ ont même dérivée à l'origine.

Cette relation est évidemment une équivalence; les classes correspondantes - les "jets" des arcs - s'appelleront vecteurs tangents à G (au point e).

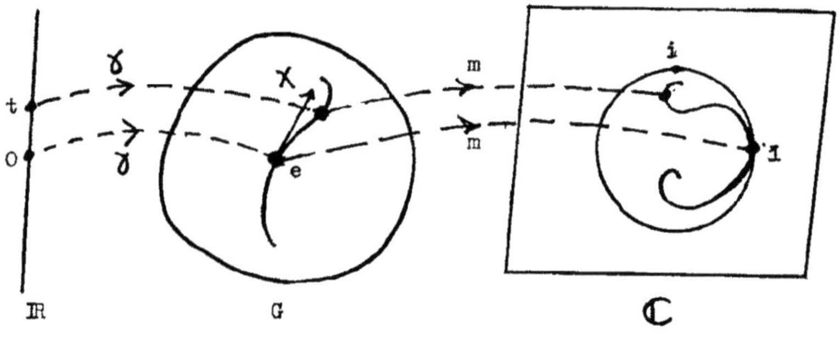

- Si γ est un arc et m un état, on sait que $[m \circ \gamma]\,(0) = 1$ et que m prend ses valeurs dans le disque unité $|z| \leq 1$ (3.3); ceci suffit à montrer que le développement de MacLaurin de $m \circ \gamma$ à l'ordre 2 s'écrit :

(5.2) $$\boxed{m(\gamma(t)) = 1 + it\rho - \frac{t^2}{2}\left[\rho^2 + \sigma^2 + i\tau\right] + o(t^3)}$$

avec $\rho, \sigma, \tau \in \mathbb{R}$, $\sigma \geq 0$. On en tire d'ailleurs:

(5.3) $$\boxed{|m(\gamma(t))|^2 = 1 - t^2\sigma^2 + o(t^3)}$$

Le jet de γ sera caractérisé par l'application $m \mapsto \rho$; nous pourrons donc poser

(5.4) $$\boxed{\text{jet}(\gamma)(m) = \frac{1}{i}\frac{d}{dt}\Big[m(\gamma(t))\Big]_{t=0}}$$

identifiant ainsi les vecteurs tangents à G avec des fonctions réelles définies sur le convexe des états.

- Soit γ' un autre arc quelconque; posons $\gamma'' = \gamma\times\gamma'$ (le produit des arcs étant défini comme en (1.18 ▽)), et effectuons les développements homologues de (5.2) pour γ' et γ''. Nous savons, grâce à (3.4), que

(5.5) $$\left|m(\gamma(t))\,m(\gamma'(t)) - m(\gamma''(t))\right|^2 \leqslant \left[1 - |m(\gamma(t))|^2\right]\left[1 - |m(\gamma'(t))|^2\right]$$

compte tenu de (5.3) on en déduit entre les trois développements les relations suivantes :

(5.6) $$\boxed{\rho + \rho' = \rho''}$$

(5.7) $$\boxed{\left[\tau+\tau'-\tau''\right]^2 \leqslant (\sigma+\sigma'+\sigma'')(\sigma+\sigma'-\sigma'')(\sigma-\sigma'+\sigma'')(-\sigma+\sigma'+\sigma'')}$$

que nous allons interpréter.

(5.6) peut évidemment s'écrire

(5.8) $$\boxed{\text{jet}(\gamma\times\gamma') = \text{jet}(\gamma) + \text{jet}(\gamma')}$$

en remarquant d'autre part (notation (2.5)) que

(5.9) $$\boxed{\text{jet}(\gamma_\circ r) = r\ \text{jet}(\gamma)} \qquad \forall r\in\mathbb{R}$$

on voit que <u>l'ensemble $\mathcal{G}$ des "vecteurs tangents" est un espace vectoriel</u>.

En dualité, nous dirons que deux <u>états</u> m et m' sont <u>tangents</u> si , pour tout arc γ, $m\circ\gamma$ et $m'\circ\gamma$ ont même dérivée à l'origine; nous pourrons poser

(5.10) $$\boxed{\text{jet}(m)(\gamma) = \frac{1}{i}\frac{d}{dt}\Big[m(\gamma(t))\Big]_{t=0}}$$

Soit $\hat{\mathcal{G}}$ l'ensemble des jets des états; ceux-ci constituant un convexe, on remarque que

(5.11) $$\text{jet}(r\,m + (1-r)\,m') = r\ \text{jet}(m) + (1-r)\ \text{jet}(m') \qquad \forall r\in[0,1]$$

donc que $\hat{\mathcal{G}}$ est un convexe; en utilisant (4.2) , on établit

(5.12) $$\text{jet}(\bar{m}) = -\ \text{jet}(m)$$

qui montre que ce convexe est symétrique, et

(5.13) $$\text{jet}(m\ m') = \text{jet}(m) + \text{jet}(m')$$

et en particulier $\text{jet}(m^2) = 2\ \text{jet}(m)$, qui montre que $\hat{\mathcal{G}}$ est en fait un <u>espace vectoriel</u> ; nous l'appellerons <u>cotangent</u> de G.

La comparaison de (5.4) et (5.10) montre que le réel $\frac{1}{i}\frac{d}{dt} m(\gamma(t))_{t=0}$ ne dépend de m et de γ que par l'intermédiaire de leurs <u>jets</u> respectifs μ et χ ; nous le noterons

(5.14) $$\{\mu, \chi\}$$

mettant ainsi $\hat{\mathcal{G}}$ et $\mathcal{G}$ <u>en dualité</u> : chacun de ces espaces vectoriels s'identifie à une partie séparante du dual algébrique de l'autre.

<u>Topologie de $\mathcal{G}$</u>

(5.15) La formule (5.8) exprime que $\gamma \mapsto \text{jet}(\gamma)$ est un morphisme de groupe $(G, \times) \to (\mathcal{G}, +)$; ce qui permet de munir $\mathcal{G}$ d'une structure de groupe différentiel quotient et, partant, d'une topologie; on peut vérifier qu'elle est séparée.

Nous allons procéder autrement, en utilisant l'inégalité (5.7)

(5.16) $$[\tau + \tau' - \tau'']^2 \leq (\sigma+\sigma'+\sigma'')(\sigma+\sigma'-\sigma'')(\sigma-\sigma'+\sigma'')(-\sigma+\sigma'+\sigma'')$$

Remarquons d'abord que trois nombres $\sigma, \sigma', \sigma''$ positifs (5.2) ne peuvent vérifier (5.16) que s'ils forment les <u>côtés d'un triangle</u> ; l'aire de ce triangle, on le sait, est

$$A = \frac{1}{4}\sqrt{(\sigma+\sigma'+\sigma'')(\sigma+\sigma'-\sigma'')(\sigma-\sigma'+\sigma'')(-\sigma+\sigma'+\sigma'')}$$

et il est clair que $A \leq \frac{1}{2}\sigma\sigma'$; en portant dans (5.16) on obtient donc

(5.17) $$|\tau + \tau' - \tau''| \leq 2\sigma\sigma'$$

(5.18) Soit χ un vecteur tangent à G; choisissons un état m. <u>A tout arc</u> γ <u>tel que</u> $\text{jet}(\gamma) = \chi$, les formules (5.2),(5.3) associent un nombre positif σ ; nous noterons

$$\|\chi\|_m$$

la <u>borne inférieure</u> de ces nombres.

le fait que $\sigma, \sigma', \sigma''$ soient les côtés d'un triangle permet d'établir l'inégalité triangulaire

(5.19) $$\|\chi + \chi'\|_m \leq \|\chi\|_m + \|\chi'\|_m \,;$$

l'égalité

(5.20) $$\|r\chi\|_m = |r| \times \|\chi\|_m \qquad \forall r \in \mathbb{R}, \forall \chi \in \mathcal{G}$$

résulte de (5.9) ; on voit donc que $\chi \mapsto \|\chi\|_m$ est une semi-norme sur l'espace vectoriel $\mathcal{G}$.

En utilisant la convexité de l'ensemble des états, on trouve

(5.21) $$\|\chi\|_{(m+m')/2} \geq \frac{1}{4}\{\mu-\mu',\chi\}^2 + \frac{1}{2}\|\chi\|_m^2 + \frac{1}{2}\|\chi\|_{m'}^2$$

formule qui devient, dans le cas $m' = \bar{m}$, l'égalité

(5.22) $$\boxed{\|\chi\|^2_{\mathrm{Re}\circ m} = \{\mu,\chi\}^2 + \|\chi\|_m^2}$$

ce qui montre que

$$\left[\|\chi\|_m = 0 \;\forall m\right] \Rightarrow \left[\{\mu,\chi\} = 0 \;\forall \mu\right] \Rightarrow \chi = 0 \quad :$$

les semi-normes associées aux divers états forment donc un système complet, et donnent à $\mathcal{G}$ une structure d'espace vectoriel topologique localement convexe ; on constate même, grâce à (5.21) que les convexes symétriques

(5.23) $$U_{m,\varepsilon} = \{\chi \in \mathcal{G} / \|\chi\|_m < \varepsilon\}$$

constituent un système fondamental de voisinages de 0; ainsi d'ailleurs que les $U_{m,1}$. Enfin, en utilisant la notion de fonction conditionnellement de type positif, on peut établir la proposition suivante :

(5.24) Le morphisme de groupe jet : $\Gamma \to \mathcal{G}$ (5.8) est continu (pour la topologie du groupe différentiel des arcs Γ et la topologie (5.23) de l'espace tangent $\mathcal{G}$).

-Il résulte de (5.22) que les éléments μ de $\hat{\mathcal{G}}$ s'identifient à des formes linéaires continues sur $\mathcal{G}$, donc que

(5.25) $$\hat{\mathcal{G}} \subset \mathcal{G}'$$

$\mathcal{G}'$ désignant le dual topologique de $\mathcal{G}$.

Représentations adjointe et co-adjointe

Soit $\Phi : G_1 \to G_2$ un D-morphisme.

Si γ est un arc de G_1 et m un état de G_2, $\Phi \circ \gamma$ est un arc de G_2, $m \circ \Phi$ un état de G_1 (4.3), et on a clairement

(5.26) $$\{ \text{jet}(m \circ \Phi), \text{jet}(\gamma) \} = \{ \text{jet}(m), \text{jet}(\Phi \circ \gamma) \}$$

il en résulte l'existence de deux applications linéaires, $T\Phi$ (tangente à Φ) et $T^*\Phi$ (cotangente à Φ) définies par

(5.27) $$T\Phi(\text{jet}(\gamma)) = \text{jet}(\Phi \circ \gamma)$$

(5.28) $$T^*\Phi(\text{jet}(m)) = \text{jet}(m \circ \Phi)$$

$T\Phi$ applique $\mathcal{G}_1$ dans $\mathcal{G}_2$, $T^*\Phi$ $\hat{\mathcal{G}}_2$ dans $\hat{\mathcal{G}}_1$, et elles sont transposées:

(5.29) $$\{ T^*\Phi(\mu), \chi \} = \{ \mu, T\Phi(\chi) \} \quad \forall \mu \in \hat{\mathcal{G}}_2, \forall \chi \in \mathcal{G}_1$$

ce qui permet de prolonger $T^*\Phi$ du dual $\mathcal{G}_2^*$ dans le dual $\mathcal{G}_1^*$, $T\Phi$ du dual $\hat{\mathcal{G}}_1^*$ au dual $\hat{\mathcal{G}}_2^*$.

La formule immédiate :

(5.30) $$\| T\Phi(\chi) \|_m \leq \| \chi \|_{m \circ \Phi}$$

montre que l'application linéaire $T\Phi$ est continue. Il en résulte que le prolongement de $T^*\Phi$ défini par (5.29) envoie le dual topologique $\mathcal{G}_2'$ dans $\mathcal{G}_1'$.

La composition de deux D-morphismes Φ et Ψ conduit aux formules

(5.31) $$\begin{cases} T[\Phi \circ \Psi] = T\Phi \circ T\Psi \\ T^*[\Phi \circ \Psi] = T^*\Psi \circ T^*\Phi \end{cases}$$

(5.32) Considérons un D-automorphisme intérieur (1.10) d'un groupe différentiel G $\Phi_g : g' \mapsto g \times g' \times g^{-1}$ et posons $Ad(g) = T\Phi_g$. Il résulte de (5.31) que Ad, ainsi défini, est une représentation linéaire de G sur son espace tangent $\mathcal{G}$, que nous appellerons représentation adjointe ; $Ad(g)$ est continu, et peut se définir explicitement par

(5.33) $$\boxed{Ad(g)(jet(\gamma)) = jet(\, t \mapsto g \times \gamma(t) \times g^{-1})}$$

Parallèlement, on définit la <u>représentation coadjointe</u> Ad^* de G sur $\hat{\mathcal{G}}$ par $Ad^*(g) = T\Phi_{g^{-1}}$ ou

(5.34) $$\boxed{Ad^*(g)(jet(m)) = jet\left(g' \mapsto m(g^{-1} \times g' \times g)\right)}$$

Ad et Ad^* sont liées par la relation d'équivariance

(5.35) $$\boxed{\left\{Ad^*(g)(\mu),\ Ad(g)(\chi)\right\} = \left\{\mu, \chi\right\}}$$

qui permet de prolonger $Ad^*(g)$ sur le dual topologique $\mathcal{G}'$ ou sur le dual algébrique $\mathcal{G}^*$.

Enfin, si Φ est un D-morphisme $G \to G'$, on trouve grâce à (5.31)

(5.36) $$\left\{\begin{array}{l} T\Phi \circ Ad(g) = Ad(\Phi(g)) \circ T\Phi \\ T^*\Phi \circ Ad^*(\Phi(g)) = Ad^*(g) \circ T^*\Phi \end{array}\right. \qquad \forall g \in G$$

<u>Crochet de Lie</u>

<u>Théorème</u>:

(5.37)

Soit G un groupe différentiel, m un état (4.1), γ et γ' deux arcs (1.18).

Le nombre

$$\diamondsuit \qquad \frac{1}{i}\left[\frac{\partial^2}{\partial t \partial u}\ m\left(\gamma(t) \times \gamma'(u) \times \gamma(t)^{-1} \times \gamma'(u)^{-1}\right)\right]_{(t,u)=0}$$

est réel et <u>ne dépend que des jets</u> μ, χ, χ' de m, γ, γ' ; nous le noterons

$$\{\mu, \chi, \chi'\}$$

$\{\mu, \chi, \chi'\}$ est une forme <u>trilinéaire</u> sur $\hat{\mathcal{G}} \times \mathcal{G} \times \mathcal{G}$, <u>antisymétrique</u> pour ses deux arguments vectoriels:

$$\heartsuit \qquad \{\mu, \chi', \chi\} = - \{\mu, \chi, \chi'\}$$

On a aussi

$$\clubsuit \qquad \{\mu, \chi, \chi'\} = \frac{d}{dt}\{\mu, Ad(\gamma(t))(\chi')\}_{t=0}$$

Soit $\varphi(t,u) = m(\gamma(t) \times \gamma'(u) \times \gamma(t)^{-1} \times \gamma'(u)^{-1})$

φ est une application de $\mathbb{R}^2$ dans $\mathbb{C}$; les axiomes (1.1) et la définition (4.1) montrent que φ est C^∞.

Posons, pour t fixé, $\gamma_t^{!}(u) = \gamma(t) \times \gamma'(u) \times \gamma(t)^{-1}$; on a alors

$$\frac{1}{i} \frac{d}{du} \varphi(t,u)\Big|_{u=0} = \frac{1}{i} \frac{d}{du} m([\gamma_t^{!} \times \gamma'^{-1}](u))_{u=0} =$$

$\{\mu, \text{jet}(\gamma_t^{!}) - \chi'\}$ (grâce à (5.8)); d'où, en utilisant la définition (5.33) de Ad :

$$♠ \quad \frac{1}{i} \frac{d}{du} \varphi(t,u)\Big|_{u=0} = \Big\{ \mu, \left[\mathrm{Ad}(\gamma(t)) - I\right] (\chi') \Big\}$$

en dérivant par rapport à t pour $t=0$, on trouve que ◇ est égal au second membre de ♣ , et par conséquent ne dépend de m et γ' que par leurs jets μ et χ', et bi-linéairement.
Si on échange les arcs γ et γ', la formule (3.2) montre que $\varphi(t,u)$ est remplacée par $\overline{\varphi(u,t)}$; la commutation des dérivées partielles montre que ◇ change de signe ; ◇ ne dépend donc de γ que par l'intermédiaire de son jet χ , et est antisymétrique en χ et χ' ; la tri-linéarité en découle immédiatement.

C.Q.F.D.

La fonction φ que nous venons d'utiliser vérifie évidemment $\varphi(t,0) = \varphi(0,u) = 1 \quad \forall t,u \in \mathbb{R}$; son développement de Taylor à l'origine se déduit alors immédiatement de (5.37 ◇) :

(5.38) $$m(\gamma(t) \times \gamma'(u) \times \gamma(t)^{-1} \times \gamma'(u)^{-1}) = 1 + itu\{\mu, \chi, \chi'\} + O((t^2+u^2)^{3/2})$$

Posons maintenant $\gamma_I'' = \gamma \times \gamma'$, $\gamma_{II}'' = \gamma' \times \gamma$,et considérons les développements limités analogues à (5.2) :

$$m(\gamma_I''(t)) = 1 + it\rho'' - \frac{t^2}{2}\left[\rho''^2 + \sigma_I''^2 + i\tau_I''\right] + O(t^3) \qquad ◇$$

$$m(\gamma_{II}''(t)) = 1 + it\rho'' - \frac{t^2}{2}\left[\rho''^2 + \sigma_{II}''^2 + i\tau_{II}''\right] + O(t^3) \qquad ♡$$

avec $\rho'' = \rho + \rho'$ (Cf (5.6)) ; l'inégalité (5.17) nous donne

$$|\tau + \tau' - \tau_I''| \leq 2\sigma\sigma' \qquad ⊙$$

$$|\tau + \tau' - \tau_{II}''| \leq 2\sigma\sigma' \qquad ☾$$

en utilisant (5.5), (5.38) et les développements ◇ et ♡ ci-dessus, il vient tous calculs faits

$$t^4\left[\left(\sigma_{I}''^{2} - \sigma_{II}''^{2}\right)^2 + \left(\tau_{I}'' - \tau_{II}'' + 2\{\mu, \chi, \chi'\}\right)^2\right] = o(t^5)$$

d'où

$$\sigma_{II}'' = \sigma_{I}'' \qquad \text{et} \qquad \{\mu, \chi, \chi'\} = \frac{1}{2}(\tau_{II}'' - \tau_{I}'')$$

la soustraction de ◇ et ♡ donne alors

(5.39) $$m(\gamma(t) \times \gamma'(t)) - m(\gamma'(t) \times \gamma(t)) = it^2 \{\mu, \chi, \chi'\} + o(t^3)$$

et le collationnement avec ⊙ et ☾ :

$$|\{\mu, \chi, \chi'\}| \leqslant 2\sigma\sigma'$$

d'où, en choisissant judicieusement γ et γ' dont les jets valent χ et χ':

(5.40) $$\boxed{|\{\mu, \chi, \chi'\}| \leqslant 2\|\chi\|_m \times \|\chi'\|_m}$$

dans cette inégalité, μ est arbitraire dans $\hat{\mathcal{G}}$, χ et χ' dans $\mathcal{G}$, et m désigne n'importe quel état dont le jet est égal à μ ; elle implique que $\{\mu, \chi, \chi'\}$ est fonction <u>continue</u> de χ et de χ'.

–Soit Φ un D-morphisme $G_1 \to G_2$; soient $\mu \in \hat{\mathcal{G}}_2$, $\chi, \chi' \in \mathcal{G}_1$. En utilisant la définition (5.37 ◇), on vérifie la formule

(5.41) $$\{\mu, T\Phi(\chi), T\Phi(\chi')\} = \{T^*\Phi(\mu), \chi, \chi'\}$$

avec comme cas particulier la formule d'équivariance

(5.42) $$\{Ad^*(g)(\mu), Ad(g)(\chi), Ad(g)(\chi')\} = \{\mu, \chi, \chi'\}$$

–Soient χ et χ' deux vecteurs tangents à G. Nous définirons leur <u>crochet de Lie</u>

(5.43) $$[\chi, \chi']$$

comme l'application linéaire

$$\mu \mapsto \{\mu, \chi, \chi'\} \qquad \mu \in \hat{\mathcal{G}}$$

$[\chi, \chi']$ est donc un élément du dual $\hat{\mathcal{G}}^*$; il dépend linéairement de χ et de χ', et il est antisymétrique :

(5.44) $$[\chi', \chi] = -[\chi, \chi']$$

Théorème:

(5.45) Soit G un groupe différentiel, $\mathcal{G}$ son tangent et $\hat{\mathcal{G}}$ son cotangent.

Désignons par $\mathcal{G}_\circ$ l'ensemble des $\chi_\circ \in \mathcal{G}$ ayant les deux propriétés suivantes :

a) $\forall \chi \in \mathcal{G}$, le crochet $[\chi_\circ, \chi]$: $\mu \mapsto \{\mu, \chi_\circ, \chi\}$ appartient à $\mathcal{G}$;

b) $\forall \mu \in \hat{\mathcal{G}}$, l'application linéaire $\chi \mapsto \{\mu, \chi_\circ, \chi\}$ appartient à $\hat{\mathcal{G}}$.

Alors $\mathcal{G}_\circ$, muni du crochet (5.43), est une algèbre de Lie.

Cet énoncé utilise évidemment l'identification de chacun des espaces $\mathcal{G}, \hat{\mathcal{G}}$ avec une partie du dual de l'autre (5.14).

Si $\chi_\circ \in \mathcal{G}_\circ$, nous pouvons utiliser cette identification pour définir $\alpha(\chi_\circ) \in L(\mathcal{G}, \mathcal{G})$ et $\beta(\chi_\circ) \in L(\hat{\mathcal{G}}, \hat{\mathcal{G}})$ par

(5.46) $$\{\mu, \chi_\circ, \chi\} = \{\mu, \alpha(\chi_\circ)(\chi)\} = \{\beta(\chi_\circ)(\mu), \chi\} \quad \forall \mu \in \hat{\mathcal{G}}, \forall \chi \in \mathcal{G}$$

$\alpha(\chi_\circ)(\chi)$ et $\beta(\chi_\circ)(\mu)$ sont précisément les applications linéaires dont l'existence est exigée en a) et b).

–Soient χ_1 et χ_2 deux éléments de $\mathcal{G}_\circ$; désignons par γ_1 et γ_2 deux arcs dont ils soient les jets.

Soit γ un arc quelconque, dont le jet sera noté χ, et m un état, de jet μ.

Posons, $\forall\, t,t',s,s' \in \mathbb{R}$:

$$\omega(t,t',s,s') = m\big(\gamma(t)\times\gamma_1(s)\times\gamma(t)^{-1}\times\gamma(t')\times\gamma_2(s')\times\gamma(t')^{-1}\times\gamma(t)\times\gamma_1(s)^{-1}\times\gamma(t)^{-1}\big)$$

il résulte des axiomes (1.1) et de (4.1) que la fonction ω est de classe C^∞ ; donc aussi la fonction ψ :

$$\psi(t,t',s) = \frac{1}{i}\,\frac{\partial}{\partial s'}\,\omega(t,t',s,s')\Big|_{s'=0}$$

, qui est égale (grâce à (5.14) et (5.33)) à

$$\{\mu, \operatorname{Ad}\big(\gamma(t)\times\gamma_1(s)\times\gamma(t)^{-1}\times\gamma(t')\big)(\chi_2)\};$$

ou encore, grâce à (5.35)

$$\left\{ Ad^*(\gamma(t)^{-1})(\mu),\ Ad(\gamma_1(s))\left(Ad(\gamma(t)^{-1}\times\gamma(t'))(\chi_2)\right)\right\}$$

donc enfin la fonction φ :

$$\varphi(t,t') = \frac{\partial}{\partial s}\psi(,t,t',s)\Big|_{s=0}$$ qui vaut, grâce à (5.37 ♣)

$$\left\{ Ad^*(\gamma(t)^{-1})(\mu),\ \chi_1,\ Ad(\gamma(t)^{-1}\times\gamma(t'))(\chi_2)\right\};$$

d'où finalement, par usage de (5.42) :

$$\varphi(t,t') = \left\{\mu,\ Ad(\gamma(t))(\chi_1),\ Ad(\gamma(t'))(\chi_2)\right\} \qquad \diamondsuit$$

Il est immédiat (5.46) que

$$\varphi(0,t') = \left\{\beta(\chi_1)(\mu),\ ad(\gamma(t'))(\chi_2)\right\};$$

en dérivant par rapport à t' on trouve

$$\frac{\partial}{\partial t'}\varphi(t,t')\Big|_{(t,t')=0} = -\left\{[\beta(\chi_2)\circ\beta(\chi_1)](\mu),\ \chi\right\}$$
$$= -\left\{\mu,\ [\alpha(\chi_1)\circ\alpha(\chi_2)]\ (\chi)\right\}$$

On trouve de même en dérivant $\varphi(t,0)$:

$$\frac{\partial}{\partial t}\varphi(t,t')\Big|_{(t,t')=0} = \left\{[\beta(\chi_1)\circ\beta(\chi_2)](\mu),\ \chi\right\}$$
$$= \left\{\mu,\ [\alpha(\chi_2)\circ\alpha(\chi_1)]\ (\chi)\right\}$$

d'où finalement

$$\frac{d}{du}\varphi(u,u)\Big|_{u=0} = \left\{[\beta(\chi_1)\circ\beta(\chi_2) - \beta(\chi_2)\circ\beta(\chi_1)](\mu),\ \chi\right\}$$
$$\heartsuit \qquad = \left\{\mu,\ [\alpha(\chi_2)\circ\alpha(\chi_1) - \alpha(\chi_1)\circ\alpha(\chi_2)]\ (\chi)\right\}$$

D'autre part la relation d'équivariance (5.42) appliquée à $\diamondsuit$ donne

$$\varphi(u,u) = \left\{ Ad^*(\gamma(u)^{-1})(\mu),\ \chi_1,\ \chi_2\right\}$$
$$= \left\{ Ad^*(\gamma(u)^{-1})(\mu),\ \alpha(\chi_1)(\chi_2)\right\} \qquad (5.46)$$
$$= \left\{\mu,\ Ad(\gamma(u))\ (\alpha(\chi_1)(\chi_2))\right\} \qquad (5.35)$$

d'où

$$\frac{d}{du}\varphi(u,u)\Big|_{u=0} = \left\{\mu , \chi , \alpha(\chi_1)(\chi_2)\right\} \qquad (5.37♣)$$

$$= -\left\{\mu , \alpha(\chi_1)(\chi_2) , \chi\right\} \qquad (5.37♡)$$

La comparaison avec ♡ montre que le vecteur $\chi_0 = \alpha(\chi_1)(\chi_2)$ vérifie les conditions (5.45 a et b) , donc que c'est un élément de $\mathcal{G}_0$; en appliquant (5.46) , on trouve

(5.47)
$$\begin{cases} \alpha(\alpha(\chi_1)(\chi_2)) = \alpha(\chi_1)\circ\alpha(\chi_2) - \alpha(\chi_2)\circ\alpha(\chi_1) \\ \beta(\alpha(\chi_1)(\chi_2)) = \beta(\chi_2)\circ\beta(\chi_1) - \beta(\chi_1)\circ\beta(\chi_2) \end{cases}$$

La démonstration du théorème (5.45) s'achève par les remarques suivantes :

$\mathcal{G}_0$ est un sous-espace vectoriel de $\mathcal{G}$;

le vecteur $\chi_0 = \alpha(\chi_1)(\chi_2)$ dont nous avons montré qu'il est un élément de $\mathcal{G}_0$ est égal à $[\chi_1, \chi_2]$;

en appliquant la première des égalités (5.47) à un troisième vecteur χ_3 <u>pris dans</u> $\mathcal{G}_0$, on trouve

$$\left[[\chi_1 , \chi_2] , \chi_3\right] = \left[\chi_1, [\chi_2 , \chi_3]\right] - \left[\chi_2, [\chi_1, \chi_3]\right]$$

soit l'identité de Jacobi.

C.Q.F.D.

On vérifie immédiatement, si $g \in G$ et $\chi_0 \in \mathcal{G}_0$ que $Ad(g)(\chi_0)$ est encore un élément de $\mathcal{G}_0$; et plus précisément que

(5.48)
$$\begin{cases} \alpha(Ad(g)(\chi_0)) = Ad(g)\circ\alpha(\chi_0)\circ Ad(g^{-1}) \\ \beta(Ad(g)(\chi_0)) = Ad^*(g)\circ\beta(\chi_0)\circ Ad^*(g^{-1}) \end{cases}$$

en appliquant la première de ces égalités à $Ad(g)(\chi_1)$ $[\chi_1 \in \mathcal{G}_0]$, on trouve

(5.49)
$$\left[Ad(g)(\chi_0), Ad(g)(\chi_1)\right] = Ad(g)\left([\chi_0 , \chi_1]\right)$$

ce qui s'interprète de la façon suivante :

(5.50) [La représentation adjointe de G sur $\mathcal{G}$ induit sur l'algèbre de Lie $\mathcal{G}_0$ une représentation de G <u>par automorphismes</u>.

(5.51) Il est clair que $\mathcal{G}_0$ coïncidera avec $\mathcal{G}$ si G est commutatif, ou si $\mathcal{G}$ est de dimension finie; en particulier, bien entendu, si G est un groupe de Lie.

§ 6 EXPONENTIELLE

Rayons

(6.1)

Soit G un groupe différentiel. Nous appellerons rayon de G tout D-morphisme $\mathbb{R} \to G$.

La composition des D-morphismes (1.9) nous montre aussitôt que:

(6.2) [Si f est un rayon de G, si $t \in \mathbb{R}$, $f_o t$ est un rayon.

(6.3) [Si Φ est un D-morphisme $G \to G'$, si f est un rayon de G, $\Phi_o f$ est un rayon de G',

et dans le cas où Φ est un automorphisme intérieur :

(6.4) [Si f est un rayon de G, si $g \in G$

$$t \mapsto g \times f(t) \times g^{-1}$$

est un rayon.

En revenant à la définition (1.7) des D-morphismes, on transforme la définition (6.1) en :

(6.5) [Soit G un groupe différentiel. Un rayon f de G est un arc de G (1.18) vérifiant

$$f(t+t') = f(t) \times f(t') \qquad \forall\ t,t' \in \mathbb{R}$$

Soit f un rayon de G, m un état. Alors $m_o f$ est un état de $\mathbb{R}$ (4.3), c'est-à-dire une fonction de type positif sur $\mathbb{R}$ qui est différentiable. Le théorème de Bochner nous apprend qu'il existe une loi de probabilité ν de $\mathbb{R}$ (: une mesure positive de masse 1) dont $m_o f$ est la transformée de Fourier:

(6.6)
$$m(f(t)) = \int_{\mathbb{R}} e^{it\omega}\ d\nu(\omega)$$

la différentiabilité de $m_o f$ signifie que ν possède des moments de tous les ordres. On peut obtenir sur l'expression (6.6) le développement de MacLaurin (voir (5.2)) :

(6.7) $$m(f(t)) = 1 + it\rho - \frac{t^2}{2}\left[\rho^2 + \sigma^2 + i\tau\right] + o(t^3)$$

on trouve immédiatement :

(6.8) $$\rho = \{\mu, \varphi\} = \int_{\mathbb{R}} \omega \, d\gamma(\omega)$$ = valeur moyenne de la variable aléatoire γ

(μ et φ sont les jets de m et f)

(6.9) $$\sigma^2 = \int_{\mathbb{R}} (\omega - \rho)^2 \, d\gamma(\omega)$$ = variance de γ

σ est donc l'écart-type (standard error) de γ .

(6.10) $$\tau = 0.$$

Si f' est un autre rayon de G, auquel m associe un écart-type σ' , l'inégalité (5.40) nous montre que

(6.11) $$\sigma\sigma' \geqslant \frac{1}{2}\left|\{\mu, \varphi, \varphi'\}\right|$$

[dans le cas particulier où G est le groupe de Weyl-Heisenberg associé au plan symplectique $\mathbb{R}^2$, et où les rayons f et f' sont associés aux hamiltoniens-coordonnées p et q, (6.11) nous fournit la relation d'incertitude de Heisenberg , sous sa forme la plus précise (le facteur $\frac{1}{2}$ ne peut pas être majoré)] .

Théorème:

(6.12) Soit G un groupe différentiel ; ($\widetilde{G}$, Π) un revêtement de G. La correspondance (voir (6.3))

$$\diamond \qquad \tilde{f} \mapsto \Pi \circ \tilde{f}$$

entre rayons de $\widetilde{G}$ et rayons de G est bijective.

-L'injectivité est immédiate(du fait que ker(Π) est discret).

-Soit f un rayon de G. Le lemme (2.11) montre qu'il existe $\tilde{f} \in D(\mathbb{R}, \widetilde{G})$ tel que $\Pi \circ \tilde{f} = f$; $\tilde{f}$ est déterminé M° ker(Π), et peut être choisi pour que $\tilde{f}(0) = \tilde{e}$. L'application $\mathbb{R}^2 \to \ker(\Pi)$: $(t,t') \mapsto \tilde{f}(t)^{-1} \times \tilde{f}(t+t') \times \tilde{f}(t')^{-1}$ est localement constante donc constante; il en résulte que $\tilde{f}$ est un rayon de $\widetilde{G}$ (Cf.(6.5)), envoyé sur f par (6.12 $\diamond$).

C.Q.F.D.

Difféologie forte

(6.13) Soit G un groupe différentiel.
Nous appellerons difféologie forte de G la difféologie engendrée par les rayons (voir (1.3),(6.5)).

Désignons par D la difféologie donnée de G, par D' la difféologie forte. Si on choisit $g \in G$, des rayons $f_1 \ldots . f_p$ de G, et une application différentiable $(r_1, \ldots r_n) \mapsto (u_1, \ldots . u_p)$ il est clair que la fonction $\mathfrak{F}$:

(6.14) $$\mathfrak{F}(r_1, \ldots \ldots . r_n) = g \times f_1(u_1) \times \ldots . \times f_p(u_p)$$

appartient à $D'(\mathbb{R}^n, G)$. Réciproquement on vérifie (en utilisant (6.4)) que tout élément de $D'(\mathbb{R}^n, G)$ est borne supérieure (0.1) de fonctions du type (6.14) .

(6.15) Par construction même de la difféologie D', les D-rayons et les D'- rayons sont les mêmes; il en résulte que le passage de D à D' est idempotent.

(6.16) La difféologie forte D' est plus fine que la difféologie initiale D (1.3); elle implique donc une topologie plus fine (4.7); si la topologie D était séparée, la topologie D' le sera.

La composante neutre forte est donc un sous-groupe (invariant) de la composante neutre initiale; à l'aide de (6.14), on voit que cette composante neutre forte est l'ensemble des produits finis

(6.17) $$f_1(t_1) \times \ \ldots . \times f_p(t_p) \qquad (f_j = \text{rayons})$$

ou , si on préfère, des produits finis

(6.18) $$f_1(1) \times \ \ldots . \ \times f_p(1)$$

(utiliser (6.2)).

(6.19) Soit G un groupe différentiel, $(\widetilde{G}, \pi)$ un revêtement de G. A l'aide de (6.14) et (6.12) , on constate que $(\widetilde{G}, \pi)$ reste un revêtement de G si on munit $\widetilde{G}$ et G de la difféologie forte .S'ils étaient connexes initialement, ils ne sont pas nécessairement fortement connexes; la situation s'analyse alors par les considérations du § 2 (en particulier (2.12)) qui sont valables pour les difféologies fortes comme pour les autres.

(6.20) Si G est un groupe différentiel <u>fort</u>, il résulte de (6.14) que l'espace vectoriel tangent $\mathcal{G}$ est <u>engendré par les jets des rayons</u>.

Exemples:

(6.21) ■ La difféologie standard des groupes de Lie (1.5) est <u>forte</u>.

(6.22) ■ Nous avons construit pour les groupes de Lie une "mauvaise" difféologie non séparée (4.16); la difféologie forte correspondante est "la bonne" (1.5). ■

<u>Exponentielle.</u>

Tout rayon d'un groupe différentiel G est <u>caractérisé par son germe à l'origine</u> ; ceci résulte du théorème suivant :

(6.23) Soit G un groupe différentiel, $\mathfrak{F}$ une fonction $\in D(\mathbb{R}, G)$ définie sur un intervalle I contenant 0, et vérifiant

$\diamondsuit \qquad \mathfrak{F}(t+t') = \mathfrak{F}(t) \times \mathfrak{F}(t') \qquad$ si t, t' et $t+t' \in I$

alors il existe <u>un seul rayon</u> f <u>prolongeant</u> $\mathfrak{F}$.

[On peut construire f en remarquant qu'il existe un seul prolongement $A(\mathfrak{F})$ de $\mathfrak{F}$, défini sur l'intervalle $2 \times I$ et vérifiant la relation $\diamondsuit$; à savoir $A(\mathfrak{F})(t) = \mathfrak{F}(t/2)^2$; alors $f = \sup A^n(\mathfrak{F})$.]

Si le groupe G est <u>séparé</u> , <u>un rayon est même défini par son jet</u> :

(6.24) Soit G un groupe différentiel séparé, φ un vecteur tangent à G. Nous dirons que φ est <u>complet</u> s'il existe un rayon f tel que $\mathrm{jet}(f) = \varphi$. Alors f est <u>unique</u>.

Supposons qu'il existe deux rayons f_1 et f_2 de même jet φ; soit m un état. On considère la fonction :

$$\psi(t,u) = m(\ f_1(t) \times f_2(u)\ )$$

Quels que soient t_0 et $u_0 \in \mathbb{R}$, on a

$$\psi(t_0+t,\ u_0+u\ ) = m(\ \ f_1(t_0) \times \left[\ f_1(t) \times f_2(u)\right] \times f_2(u_0)\ \ ),$$

et le théorème (3.10) nous montre qu'il existe trois états m_0, m_1, m_2 et trois complexes c_0, c_1, c_2 tels que

$$\psi\ (t_0+t,\ u_0+u\ ) = \sum_{p=0}^{2} c_p\ m_p(\ f_1(t) \times f_2(u)\ )$$

On en déduit la valeur des dérivées $\frac{\partial \psi}{\partial t}$ et $\frac{\partial \psi}{\partial u}$ pour $t = t_0$, $u = u_0$: $i \sum_p c_p \left\{ \mathrm{jet}(m_p),\ \varphi \right\}$ dans les deux cas; par conséquent $\frac{\partial\psi}{\partial t} - \frac{\partial\psi}{\partial u}$ est identiquement nul, $\psi(t, -t) = Cte = \psi(0,0) = 1$; on a donc $m(f_1(t)\times f_2(-t)) = 1$ pour tout état m; G étant séparé, ceci entraînc $f_1(t)\times f_2(-t) = e$, d'où $f_1 = f_2$.

C.Q.F.D.

(6.25) Si G est séparé et si φ est un vecteur complet, nous poserons

$$\exp(\varphi) = f(1)$$

f étant le rayon vérifiant

$$jet(f) = \varphi$$

$\forall t \in \mathbb{R}$, on sait que $jet(f_o t) = t\ jet(f)$ (5.9); par conséquent

(6.26) $$\boxed{f(t) = \exp(t\varphi)} \qquad t \in \mathbb{R}$$

-Si Ψ est un D-morphisme $G \to G'$, si G et G' sont séparés et si φ est un vecteur complet de G, il résulte de (6.3) que $T\Psi(\varphi)$ est un vecteur complet de G', et que

(6.27) $$\boxed{\exp(T\Psi(\varphi)) = \Psi(\exp(\varphi))}$$

En particulier, si G est séparé, φ complet et $g \in G$, $Ad(g)(\varphi)$ est complet et

(6.28) $$\boxed{\exp(Ad(g)(\varphi)) = g \times \exp(\varphi) \times g^{-1}}$$

Théorème :

(6.29) Si G est un groupe séparé, φ_1 et φ_2 des vecteurs complets de G, et si

$$[\varphi_1, \varphi_2] = 0$$

$\varphi_1 + \varphi_2$ est complet, et

$$\exp(\varphi_1 + \varphi_2) = \exp(\varphi_1) \times \exp(\varphi_2) = \exp(\varphi_2) \times \exp(\varphi_1)$$

Posons $f_1(t) = \exp(t\varphi_1)$, $f_2(t) = \exp(t\varphi_2)$; soit m un état, μ son jet, et considérons la fonction réelle ψ :

$$\psi(t) = \{\mu, Adj(f_1(t))(\varphi_2)\}$$

$\forall t_o \in \mathbb{R}$, on trouve $\psi(t_o + t) = \{\mu_o , Adj(f_1(t))(\varphi_2)\}$, avec $\mu_o = Adj^*(f_1(-t_o))(\mu)$; d'où grâce à (5.37 ♣) , $\psi'(t_o) =$

$\{\mu_0, \varphi_1, \varphi_2\} = [\varphi_1, \varphi_2](\mu_0) = 0$ (Cf.(5.43)); d'où $\psi(t) = \psi(0) = \{\mu, \varphi_2\}$. μ étant arbitraire, on a $Ad(f_1(t))(\varphi_2) = \varphi_2$; d'où, grâce à (6.28), $f_1(t) \times f_2(t') \times f_1(t)^{-1} = f_2(t')$ $\forall t, t' \in \mathbb{R}$.

$f_1(t)$ et $f_2(t')$ commutent, $f_1 \times f_2$ est donc un rayon; on sait que son jet est égal à $\varphi_1 + \varphi_2$; donc $\exp(\varphi_1 + \varphi_2) = [f_1 \times f_2](1) = \exp(\varphi_1) \times \exp(\varphi_2)$.

C.Q.F.D.

Ce résultat comprend comme cas particulier la formule

(6.30) $$\exp([t+t']\varphi) = \exp(t\varphi) \times \exp(t'\varphi)$$

évident sur la définition (6.26).

(6.31) On peut construire des exemples de vecteurs complets φ_1 et φ_2 <u>dont la somme</u> $\varphi_1 + \varphi_2$ <u>n'est pas un vecteur complet</u> [ainsi, dans le groupe des symplectomorphismes de $\mathbb{R}^2$, les vecteurs associés aux hamiltoniens $p^2/2$ et $q^3/3$] : l'ensemble des vecteurs complets est donc une <u>étoile</u> du plan tangent $\mathcal{G}$ (:une partie invariante par les homothéties) , <u>invariante par la représentation adjointe</u> (Cf.(6.28)) , <u>mais pas nécessairement un espace vectoriel</u>. Dans le cas de la difféologie forte, on sait que l'espace vectoriel engendré est égal à $\mathcal{G}$ (6.20).

REPRESENTATIONS THAT REMAIN IRREDUCIBLE ON PARABOLIC SUBGROUPS

Joseph A. Wolf
Department of Mathematics
University of California
Berkeley, California 94720

§1. Introduction

In 1969 Mack and Todorov [6] showed that ladder representations of the conformal group remain irreducible on the Poincaré group. Then in 1975 (see [7 , Theorem 4. 27]) Sternberg and I extended that irreducibility result to ladder representations of any $U(k, \ell)$ on a maximal parabolic subgroup or its maximal unimodular subgroup. We noted that this seemed to be a combination of near transitivity and totally complex polarizations, as implicit in the reproducing kernel arguments ([3], [4]) of Kobayashi.

"Differential-Geometrical Methods in Mathematical Physics," Aix-en-Provence, September 1979. Research partially supported by N. S. F. Grants MCS 76-01692 and MCS 79-02522.

Here I explicitly combine a notion of near transitivity with the uniqueness principle for reproducing kernels and obtain a very strong irreducibility theorem in §2. After a digression (§3) to explain dual reductive pairs and their classification over the real and complex numbers, I apply the result of §2 to

the group G of an irreducible dual reductive pair,

a summand π of the G-restriction of the metaplectic representation,

a subgroup $L \subset G$ constructed from parabolic subgroups,

in §4, and to certain other types of subgroups L in §5. The case $G = U(k, \ell)$ and L a maximal parabolic or the maximal unimodular subgroup of one, is a very special case. But this does seem to be the natural geometric setting for the Mack-Todorov irreducibility result.

I spoke on Theorem 2.1 and some of the examples in §4B at Oberwolfach in the summer of 1977. The example of §5A appeared in [10], in another setting. Now the general pattern for irreducible dual reductive pairs of type I seems clear, so publication is appropriate.

Finally let us note that there is a connection between representations that remain irreducible on parabolics and construction of L-functions in number theory. See ([3], §§3 and 4).

§2. An Irreducibility Criterion

We formalize an argument that I used in [7 , Theorem 4.27] to show that certain unitary representations remain irreducible on parabolic subgroups. The argument is a variation on Kobayashi's use of reproducing kernels to prove irreducibility theorems.

<u>2.1. Theorem.</u> <u>Let</u> $\mathcal{H}$ <u>be a Hilbert space of holomorphic functions on a complex manifold</u> M <u>such that point evaluations</u> $f \mapsto f(m)$ <u>are continuous functionals on</u> $\mathcal{H}$. <u>Let</u> G <u>be a Lie group acting on</u> M <u>such that the action lifts to a unitary representation</u> π <u>of</u> G <u>on</u> $\mathcal{H}$.

<u>Let</u> L <u>be a subgroup of</u> G <u>that satisfies the following near-transitivity condition on</u> G-<u>orbits. There is a</u> G-<u>invariant open set</u> $U \subset M$ <u>that meets every topological component of</u> M <u>such that: if</u> $\mathfrak{G} \subset U$ <u>is a</u> G-<u>orbit then there is a point</u> $m \in \mathfrak{G}$ <u>such that</u> $L(m)$ <u>is open in</u> $\mathfrak{G}$ <u>and meets every component of</u> $\mathfrak{G}$.

<u>Then every closed</u> $\pi(L)$-<u>invariant subspace of</u> $\mathcal{H}$ <u>is</u> $\pi(G)$-<u>invariant. In particular if</u> π <u>is irreducible then</u> $\pi|_L$ <u>is irreducible.</u>

Remarks. G need not be a Lie group so long as its orbits are real analytic submanifolds of M. Also, as in [5], the theorem holds for spaces of holomorphic sections of vector bundles.

<u>Proof.</u> As point evaluations are continuous $\mathcal{H}$ has a reproducing Bergman kernel $K(z,\zeta)$, holomorphic in z, antiholomorphic in ζ, and hermitian in the sense $K(\zeta,z) = \overline{K(z,\zeta)}$. If $\{\varphi_i\}$ is any complete orthonormal set in $\mathcal{H}$ then $K(z,\zeta) = \Sigma\,\varphi_i(z)\overline{\varphi_i(\zeta)}$, uniformly and absolutely convergent on compact subsets of M.

Let $\mathcal{H}_1$ be a closed subspace of $\mathcal{H}$. As above, it has a reproducing Bergman kernel $K_1(z,\zeta) = \Sigma\,\psi_\alpha(z)\overline{\psi_\alpha(\zeta)}$ where $\{\psi_\alpha\} \subset \mathcal{H}_1$ is any complete orthonormal set. If $f \in \mathcal{H}$ then $f_1(z) = \langle K_1(z,\zeta), f(\zeta)\rangle_\zeta$ is its orthogonal projection to $\mathcal{H}_1$.

Now suppose that $\mathcal{H}_1$ is $\pi(L)$-invariant. Then $K_1(z,\zeta)$ is L-invariant,

$$K_1(gz, g\zeta) = K_1(z,\zeta) \quad \text{for} \quad g \in L \quad \text{and} \quad z,\zeta \in M \;,$$

by uniqueness. In particular the function $K_1(z,z)$ is constant on L-orbits. Given $L(m) \subset \mathfrak{G} \subset U$ as in the statement of the theorem, $z \mapsto K_1(z,z)$ is a real analytic function on the real analytic manifold $\mathfrak{G}$ that is constant on the open subset $L(m)$ which meets every component; so $K_1(z,z)$ is constant on the G-orbit $\mathfrak{G}$. Now the function $K_1(z,z)$ is G-invariant on U, hence also on M. But $K_1(z,\zeta)$ is determined by its restriction to the diagonal of $M \times M$, so it must be G-invariant,

$$K_1(gz, g\zeta) = K_1(z,\zeta) \quad \text{for} \quad g \in G \quad \text{and} \quad z,\zeta \in M .$$

Now $\mathcal{H}_1 = \{\langle K_1(z,\zeta), f(\zeta)\rangle_\zeta = f \in \mathcal{H}\}$ is a $\pi(G)$-invariant subspace of $\mathcal{H}$.

q.e.d.

2.2. Corollary. Assume the conditions of Theorem 2.1 and decompose

$$\pi = \int \pi_\alpha \, d\nu(\alpha), \qquad \text{direct intergral of irreducibles}.$$

Then $\pi_\alpha|_L$ is irreducible for ν-almost-every α.

Proof (C. C. Moore). The result of Theorem 2.1 can be phrased: every projection in $\mathcal{H}$ that commutes with $\pi(L)$ also commutes with $\pi(G)$. As these projections generate the commuting rings $A(\pi)$ and $A(\pi|_L)$ now $A(\pi) = A(\pi|_L)$. A direct integral decomposition into irreducibles corresponds to a choice of maximal abelian subalgebra of the commuting ring. Let $B \subset A(\pi)$ be the maximal abelian subalgebra whose Boolean algebra of projections gives $\pi = \int \pi_\alpha \, d\nu(\alpha)$. It also gives $\pi|_L = \int (\pi_\alpha|_L) \, d\nu(\alpha)$, and ν-almost-every $\pi_\alpha|_L$ is irreducible because B is maximal abelian in $A(\pi|_L)$.

§3. Dual Reductive Pairs

W is a symplectic vector space over a field $\mathbb{F}$ of characteristic $\neq 2$. Thus W has finite dimension over $\mathbb{F}$ and is equipped with a nondegenerate antisymmetric bilinear form $\langle \, , \, \rangle_W$. The automorphism group of W is the symplectic group

$Sp(W) \cong Sp(\frac{1}{2} \dim W ; \mathbb{F})$.

A pair of reductive subgroups $G_1, G_2 \subset Sp(W)$ forms a dual reductive pair if each is the centralizer of the other inside $Sp(W)$. If W is direct sum $W' \oplus W''$ of symplectic subspaces (i. e. if $\langle\ ,\ \rangle_W$ is nondegenerate on each and if $\langle W', W'' \rangle_W = 0$) that are invariant under G_1 and G_2, then the dual reductive pair (G_1, G_2) is reducible. Otherwise it is irreducible.

In §§ 4 and 5 we will apply the result of §2 to the groups $G = G_1 G_2$ where (G_1, G_2) is an irreducible dual reductive pair of type I in a real or complex symplectic group. There, π will be a summand of the G-restriction of the metaplectic representation. Here we are going to describe those pairs explicitly.

First let us recall Howe's general classification [1] of irreducible dual reductive pairs in $Sp(W)$.

Type I. $G = G_1 G_2$ acts irreducibly on W. Then there exist

(3.1) $\left\{ \begin{array}{l} \text{a division algebra } \mathbb{D} \text{ over } \mathbb{F} \text{ with involution } x \mapsto \bar{x} \\ \text{a right vector space } W_1 \text{ over } \mathbb{D} \text{ and a left vector space } W_2 \\ \text{forms } (\ ,\ )_1 \text{ on } W_1 \text{ and } (\ ,\ )_2 \text{ on } W_2 \text{, one hermitian and the other} \\ \qquad \text{skew hermitian} \end{array} \right.$

such that

(3.2) $$W \cong W_1 \otimes_{\mathbb{D}} W_2 \quad \text{and} \quad \langle\ ,\ \rangle_W = \text{trace}_{\mathbb{D}/\mathbb{F}} (\ ,\ )_1 \otimes (\ ,\ )_2$$

in such a way that

(3.3) $$G_1, G_2 \text{ go to the isometry groups of } (W_1, (\ ,\ )_1) \text{ and } (W_2, (\ ,\ )_2) .$$

Type II. $G = G_1 G_2$ acts reducibly on W. Then W is the vector space direct sum of G-invariant maximal totally isotropic subspaces V, V' and there exist

(3.4) $\begin{cases} \text{a division algebra } \mathbb{D} \text{ over } \mathbb{F} \\ \text{a right vector space } V_1 \text{ over } \mathbb{D} \text{ and a left vector space } V_2 \end{cases}$

such that

(3.5) $$V \cong V_1 \otimes_{\mathbb{D}} V_2 \quad \text{with } G_i \text{ identified to } GL(V_i; \mathbb{D}) \quad .$$

Here V' is identified to the $\mathbb{F}$-linear dual space of V under $\langle\ ,\ \rangle_W$. The $\mathbb{D}$-linear dual spaces of V_1 and V_2 are

(3.6) a left vector space V_1' over $\mathbb{D}$ and a right vector space V_2' such that

(3.7) $$V' \cong V_2' \otimes V_1' \quad \text{with } G_i \text{ identified to } GL(V_i'; \mathbb{D})$$

where

(3.8) the action of G_i on V_i' is the $\mathbb{D}$-dual to its action on V_i.

Next we specialize to the case $\mathbb{F} = \mathbb{C}$. Then $\mathbb{D} = \mathbb{C}$, and $\bar{x} = x$ in (3.1), so up to interchange of G_1, G_2 the only irreducible dual reductive pairs are

(3.9) $$O(u;\mathbb{C}),\ Sp(v;\mathbb{C}) \quad \text{in} \quad Sp(uv;\mathbb{C}) \quad ,$$

(3.10) $$GL(u;\mathbb{C}),\ GL(v;\mathbb{C}) \quad \text{in} \quad Sp(uv;\mathbb{C}) \quad .$$

In (3.9), $W \cong \mathbb{C}^u \otimes \mathbb{C}^{2v} = \mathbb{C}^{2uv}$ where $\langle\ ,\ \rangle_W$ is the tensor product of the symmetric form on $\mathbb{C}^u$ with the antisymmetric form on $\mathbb{C}^{2v}$. In (3.10), $W \cong V \oplus V'$ where $V = \mathbb{C}^{u \times v}$ space of $u \times v$ complex matrices, $V' = \mathbb{C}^{v \times u}$ with pairing $x'(x) = \operatorname{trace}(xx')$, and

$$(a,b) \in GL(u;\mathbb{C}) \cdot GL(v;\mathbb{C})$$

acts on $(x,x') \in V \oplus V' = W$ by

(3.11) $$(a,b) : (x,x') \mapsto (axb^{-1}, bx'a^{-1}) \quad .$$

Next consider type I pairs with $\mathbb{F} = \mathbb{R}$. Then $\mathbb{D}$ is $\mathbb{R}$, $\mathbb{C}$ or the quaternion algebra $\mathbb{H}$, and $x \to \bar{x}$ has its usual meaning in (3.1). Suppose $(\ ,\ )_1$ hermitian and $(\ ,\ )_2$ skew hermitian. Then we can take W_1 to be

$$(3.12)\quad \begin{cases} \mathbb{D}^{k,\ell} : \text{right vector space of } (k+\ell)\text{-tuples over } \mathbb{D} \text{ with form} \\ \qquad (x,y)_1 = \sum_{1}^{k} x_j \bar{y}_j - \sum_{k+1}^{k+\ell} x_j \bar{y}_j \quad . \end{cases}$$

Its isometry group G_1 is the (indefinite)

$$(3.13)\quad \begin{cases} \text{orthogonal group} & O(k,\ell) & \mathbb{D} = \mathbb{R} \\ \text{unitary group} & U(k,\ell) & \mathbb{D} = \mathbb{C} \\ \text{unitary symplectic group} & Sp(k,\ell) & \mathbb{D} = \mathbb{H} \end{cases}$$

A skew-hermitian form over $\mathbb{R}$ is just an antisymmetric bilinear form. If $\mathbb{D} = \mathbb{R}$ now W_2 and G_2 are $\mathbb{R}^v$ and $Sp(\frac{1}{2}v;\mathbb{R})$. A skew hermitian form over $\mathbb{C}$ is just i times a hermitian one, and they have the same isometry group. If $\mathbb{D} = \mathbb{C}$ now W_2 and G_2 are $\mathbb{C}^v$ and $U(p,q)$, $p+q = v$. A skew hermitian form on a left vector space $\mathbb{H}^v$ over $\mathbb{H}$ is equivalent to $(x,y)_2 = \Sigma\, \bar{x}_j\, i\, y_j$. Its isometry group is the real form $SO^*(2v)$ of $SO(2v;\mathbb{C})$ whose maximal compact subgroup is $U(v)$. See [9] and [10] for details on this. Now, G_2 and W_2 are

$$(3.14)\quad \begin{cases} Sp(\frac{1}{2}v;\mathbb{R}) & \text{and } \mathbb{R}^v & \mathbb{D} = \mathbb{R} \\ U(p,q) & \text{and } \mathbb{C}^v,\ v = p+q & \mathbb{D} = \mathbb{C} \\ SO^*(2v) & \text{and } \mathbb{H}^v & \mathbb{D} = \mathbb{H} \end{cases}$$

Thus, the irreducible dual reductive pairs of type I with $\mathbb{F} = \mathbb{R}$ are

$$
(3.15)\qquad \begin{cases} O(k,\ell)\ ,\ Sp(\tfrac{1}{2}v\,;\mathbb{R}) & \text{in} \quad Sp(\tfrac{1}{2}(k+\ell)v\,;\mathbb{R}) \\ U(k,\ell)\ ,\ U(p,q)\ ,\ v=p+q & \text{in} \quad Sp((k+\ell)v\,;\mathbb{R}) \\ Sp(k,\ell)\ ,\ SO^*(2v) & \text{in} \quad Sp(2(k+\ell)v\,;\mathbb{R}) \end{cases}
$$

In each case, the action on $W \cong \mathbb{D}^{k,\ell} \otimes_{\mathbb{D}} \mathbb{D}^{v}$, viewed as the space $\mathbb{D}^{(k+\ell)\times v}$ of $(k+\ell)\times v$ matrices over $\mathbb{D}$, is $(a,b) : x \mapsto axb^{-1}$.

Finally consider type II pairs with $\mathbb{F} = \mathbb{R}$. As in the case $\mathbb{F} = \mathbb{C}$, the pairs are

$$
(3.16)\qquad \begin{cases} GL(u\,;\mathbb{R})\ ,\ GL(v\,;\mathbb{R}) & \text{in} \quad Sp(uv\,;\mathbb{R}) \\ GL(u\,;\mathbb{C})\ ,\ GL(v\,;\mathbb{C}) & \text{in} \quad Sp(2uv\,;\mathbb{R}) \\ GL(u\,;\mathbb{H})\ ,\ GL(v\,;\mathbb{H}) & \text{in} \quad Sp(4uv\,;\mathbb{R}) \end{cases}
$$

They act on $W = \mathbb{D}^{u\times v} \oplus \mathbb{D}^{v\times u}$ as in (3.11). Here $\mathbb{D}^{u\times v} = \mathbb{D}^{u} \otimes_{\mathbb{D}} \mathbb{D}^{v} = V_1 \otimes V_2 = V$ $V' = V_2' \otimes V_1' = \mathbb{D}^{v\times u}$ under $x'(x) = \mathrm{Re\,trace}\,(xx')$.

§4. Patterns of Near Transitivity in Type I Pairs

Let (G_1, G_2) be an irreducible dual reductive pair of type I in a complex or real symplectic group $Sp(W)$. It is given by (3.9) or (3.15). W is a space $W_1 \otimes W_2$ of matrices and $G = G_1 \cdot G_2$ acts on it by $(a,b) : x \mapsto axb^{-1}$. In the setting of the metaplectic representation, W will be viewed as a complex vector space.

4A. $G = Sp(W)$. This is (3.9) with $u = 1$ or (3.15) with $k + \ell = 1$. The parabolic subgroups of G are the normalizers of "flags" $0 \subsetneqq E_1 \subsetneqq E_2 \subsetneqq \cdots \subsetneqq E_t$ where the E_i are $\mathbb{F}$-subspaces of W that are totally isotropic, i.e. $\langle E_i, E_i\rangle_W = 0$.

The parabolic is

$$(4.1)\qquad P = P_{E_1,\ldots E_t} = \{g \in G : gE_i = E_i \quad \text{for} \quad 1 \leqq i \leqq t\} \quad .$$

The set $L(m)$ in Theorem 2.1 will be

$$(4.2)\qquad Q = \{x \in W : x \notin E_t \quad \text{and} \quad \langle x, E_1\rangle_W \neq 0\} \quad .$$

Note that Q is a single P-orbit on W. For if $x, y \in Q$ then we have bases

$$\{v_1, \ldots v_q, x\} \text{ of } E_t + x\mathbb{F} \text{ and } \{w_1, \ldots, w_q, y\} \text{ of } E_t + y\mathbb{F}$$

such that $\{v_1, \ldots, v_{\dim E_i}\}$ and $\{w_1, \ldots w_{\dim E_i}\}$ are bases of E_i $(1 \leqq i \leqq t)$ and

$$\langle v_1, x\rangle = 1 = \langle w_1, y\rangle, \text{ and } \langle v_j, x\rangle = 0 = \langle w_j, y\rangle \text{ for } j > 1 \quad .$$

Witt's Theorem provides $g \in G$ with $gv_i = w_i$ $(1 \leqq i \leqq q)$ and $gx = y$. So $g \in P$ sends x to y.

We just showed that the parabolic P is transitive on Q. By rescaling the w_i, $i > 1$, we may assume that the element $g \in G$ there, belongs to

$$(4.3)\qquad \begin{cases} \text{if} \quad \dim E_1 > 1 : L = \{g \in P : \det(g|_{E_i/E_{i-1}}) = 1 \quad \text{for} \quad 1 \leqq i \leqq t\} \\ \text{if} \quad \dim E_1 = 1 : L = \{g \in P : \det(g|_{E_i/E_{i-1}}) = 1 \quad \text{for} \quad 2 \leqq i \leqq t\} \end{cases}$$

This group L now satisfies the near transitivity condition of Theorem 2.1 with $M = W$, $U = W\setminus\{0\}$ a single G-orbit, and $L(m) = Q$ there.

<u>4B.</u> $G = O(k,\ell) \cdot Sp(\tfrac{1}{2}v;\mathbb{R})$ with $k + \ell > 1$, or $U(k,\ell) \cdot U(p,q)$ with $p + q = v$, or $Sp(k,\ell) \cdot SO^*(2v)$; and $\operatorname{rank}_{\mathbb{R}} G_1 = \min(k,\ell) \geqq v$. The parabolic subgroups of $G_1 = U(k,\ell;\mathbb{D})$ are the normalizers

$$(4.4)\qquad P = P_{E_1,\ldots E_t} = \{g \in G_1 : gE_i = E_i \quad \text{for} \quad 1 \leqq i \leqq t\}$$

of $(\, , \,)_1$-isotropic flags $0 \subsetneqq E_1 \subsetneqq \ldots \subsetneqq E_t$. Thus these E_i are $\mathbb{D}$-subspaces of $W_1 = \mathbb{D}^{k,\ell}$ with $(E_i, E_i)_1 = 0$. Similarly the parabolics in G_2 are the normalizers

(4.5) $$P' = P'_{F_1,\ldots F_s} = \{g \in G : gF_i = F_i \quad \text{for} \quad 1 \leqq i \leqq s\}$$

of $(\, , \,)_2$-isotropic flags $0 \neq F_1 \subsetneqq \ldots \subsetneqq F_s$ in $W_2 = \mathbb{D}^v$.

We suppose $\min(k, \ell) \geqq v$, i.e. that W_1 has a totally isotropic subspace of dimension $\geqq \dim W_2$. In [2] this is the defining condition for (G_1, G_2) to be "stable." It says that G_1 has a parabolic subgroup $P = P_{E_1,\ldots,E_t}$ as in (4.4) with $\dim E_1 \geqq \dim W_2$. Define

(4.6) $$\begin{cases} \text{if} \quad \dim E_1 > \dim W_2 : L_1 = \{g \in P : \det(g|_{E_i/E_{i-1}}) = 1 \quad \text{for} \quad 1 \leqq i \leqq t\} \\ \text{if} \quad \dim E_1 = \dim W_2 : L_1 = \{g \in P : \det(g|_{E_i/E_{i-1}}) = 1 \quad \text{for} \quad 2 \leqq i \leqq t\} \end{cases}$$

where we note that the condition $\det = r$, r real, is well defined over $\mathbb{H}$.

Since $\min(k, \ell) \geqq \dim E_1 \geqq v$, $W_1 = \mathbb{D}^{k,\ell}$ has subspaces $V \cong \mathbb{D}^{k',\ell'}$ of every signature (k', ℓ') with $k' + \ell' = v$, such that $V \cap E_1^{\perp} = 0$. Fix one such (k', ℓ'); that gives us

$$U(k', \ell'; \mathbb{D}) \subset GL'(v; \mathbb{D}) = GL'(W_2)$$

where $GL' = \{g \in GL : |\det g| = 1\} = \{g \in GL : g \text{ preserves Lebesgue measure}\}$. Let

(4.7) $$\begin{cases} L_2 : \text{any subgroup of } G_2 \text{ such that } L_2 \text{ and} \\ U(k', \ell'; \mathbb{D}) \text{ generate } GL'(W_2) \end{cases},$$

(4.8) $$L = L_1 \cdot L_2 \subset G_1 \cdot G_2 = G \quad .$$

There are many groups L_2 because $U(k', \ell'; \mathbb{D})$ is a maximal subgroup of $GL'(W_2)$.

Our G-invariant open set $U \subset W$ will consist of all $x \in \mathbb{D}^{(k+\ell)\times v} = W$ whose columns span a subspace $V \cong \mathbb{D}^{k',\ell'}$ with $V \cap E_1^{\perp} = 0$. We will verify that every G-orbit on V is an L-orbit.

Fix $y \in U$. We first check that every L-orbit on U contains a positive real multiple ry of y. For that, let y_i be the columns of y, $y = (y_1, \ldots, y_v)$, and let T be their span. Given $x \in U$ we have $x = (x_1, \ldots, x_v)$; let S be its column span. Since $S \cong \mathbb{D}^{k',\ell'} \cong T$ we have an isometry $\varphi : S \cong T$. As $S \cap E_1^{\perp} = 0$ we have $\{f_j\}$ with $\{f_1, \ldots, f_{\dim E_i}\}$ basis of E_i for $1 \leqq i \leqq t$, such that $(f_j, x_i)_1 = \delta_{ij}$. Similarly $T \cap E_1^{\perp} = 0$ gives $\{e_j\}$ such that $\{e_1, \ldots, e_{\dim E_i}\}$ is a basis of E_i and $(e_j, \varphi x_i)_1 = \delta_{ij}$. Now $f_j \mapsto e_j$, $x_i \mapsto \varphi x_i$ is an isometry of $E_t + S$ onto $E_t + T$ that sends each E_i to itself. Witt's Theorem extends this to an isometry g_1 of W_1. Thus we have $g_1 \in P_{E_1, \ldots, E_t}$ with $g_1 S = T$. We can freely replace the f_j, $j > v$, by nonzero elements of $f_j \mathbb{D}$, so we may assume $g_1 \in L_1$.

Let $L_1^T = \{g|_T : g \in L_1 \text{ and } gT = T\}$. If we repeat the above argument with $x = y$ and any isometry of T we see $L_1^T \cong U(k', \ell'; \mathbb{D})$. The action $z \mapsto z \cdot g_2^{-1}$ of G_2 does not change column span. Let L^T denote the group of all transformations of T generated by L_1^T and the $z \mapsto z g_2^{-1}$, $g_2 \in L_2$. By hypothesis on L_2 we have $L^T = GL'(T)$. Now, for a unique $r > 0$ some $\gamma \in L^T$ carries φx_i to $r y_i$ for $1 \leqq i \leqq v$. So $ry \in L(x)$.

We next check that the positive multiple of y in a G-orbit on U is unique. For let $a \cdot b \in G$ and $r, r' > 0$ with $a \cdot ry \cdot b^{-1} = r'y$. As $ry \cdot b^{-1}$ and ry

have the same column span, our old T, now a preserves T acting there as an element of $U(k',\ell';\mathbb{D})$, so $a_{|T} \in GL'(T)$. Also $b \in G_2 \subset GL'(W_2)$ acts on T as an element of $GL'(T)$. It follows that $r = r'$.

We have shown that the group L of (4.10) is transitive on every G-orbit in U, thus proving the near-transitivity condition of Theorem 2.1 with $M = W$ and U as above.

4C. $G = O(k,\ell) \cdot Sp(\frac{1}{2}v;\mathbb{R})$ with $k+\ell > 1$, or $U(k,\ell) \cdot U(p,q)$ with $p+q = v$, or $Sp(k,\ell) \cdot SO^*(2v)$; and $\operatorname{rank}_{\mathbb{R}} G_2$ ($= \frac{1}{2}v$, $\min(p,q)$ or $[v/2]$, resp.) $\geqq k+\ell$. Thus W_2 has a totally isotropic subspace of dimension $\geqq \dim W_1$, i.e. G_2 has a parabolic subgroup $P' = P'_{F_1,\dots,F_s}$ with $\dim F_1 \geqq \dim W_1$. Define

$$(4.9)\quad \begin{cases} \text{if } \dim F_1 > \dim W_1 : L_2 = \{g \in P' : \det(g_{|F_i/F_{i-1}}) = 1 \text{ for } 1 \leqq i \leqq s\} \\ \text{if } \dim F_1 = \dim W_1 : L_2 = \{g \in P' : \det(g_{|F_i/F_{i-1}}) = 1 \text{ for } 2 \leqq i \leqq s\} \end{cases}$$

As $\dim F_1 \geqq \dim W_1$, so $W_2 = \mathbb{D}^v$ has subspaces V of dimension $k+\ell$ with $V \cap F_1^\perp = 0$ and $(\ ,\ )_{2|V\times V}$ of maximal possible rank. If

$$(4.10)\qquad J = \{g_{2|V} : g_2 \in G_2 \text{ and } g_2 V = V\}$$

then the possibilities are

(i) $\mathbb{D} = \mathbb{R}$ and $k+\ell$ odd : $(\ ,\ )_{2|V\times V}$ has rank $k+\ell-1$ and

$$J \cong \left\{\begin{pmatrix} \alpha & 0 \\ \beta & \gamma \end{pmatrix} : 0 \neq \alpha \in \mathbb{R},\ \beta \in \mathbb{R}^{k+\ell-1},\ \gamma \in Sp(\tfrac{k+\ell-1}{2};\mathbb{R})\right\}$$

(ii) $\mathbb{D} = \mathbb{R}$ and $k+\ell$ even : $(\ ,\ )_{2|V\times V}$ has rank $k+\ell$ and

$$J \cong Sp(\tfrac{k+\ell}{2};\mathbb{R})$$

(iii) $\mathbb{D} = \mathbb{C} : (\ ,\ )|_{V \times V}$ has rank $k + \ell$, in fact $\sqrt{-1}\,(\ ,\ )|_{V \times V}$ has has any specified signature (p', q') with $p' + q' = k + \ell$, and $J \cong U(p', q')$

(iv) $\mathbb{D} = \mathbb{H} : (\ ,\ )_{V \times V}$ has rank $k + \ell$ and $J \cong SO^*(2(k + \ell))$.

Identify J to a subgroup of $GL(W_1)$ and let

$$\text{(4.11)} \qquad \begin{cases} L_1 : \text{any subgroup of } G_1 \text{ such that } L_1 \text{ and } J \text{ generate} \\ \qquad GL'(W_1) \text{ or } GL(W_1), \end{cases}$$

$$\text{(4.12)} \qquad L = L_1 \cdot L_2 \subset G_1 \cdot G_2 = G .$$

J is maximal in $GL'(W_1)$ or nearly maximal in $GL(W_1)$, so there are many groups L_1.

Our G-invariant open set $U \subset W$ will consist of all $x \in \mathbb{D}^{(k+\ell) \times v} = W$ whose rows span a subspace V of dimension $k + \ell$ in W_2 with $V \cap F_1^{\perp} = 0$ and $(\ ,\ )_2|_{V \times V}$ of maximal possible rank. If $\mathbb{D} = \mathbb{C}$ we specify the signature (p', q') of $\sqrt{-1}\,(\ ,\ )_2|_{V \times V}$.

Fix $y \in U$. As in the second and third paragraph following (4.10), if $x \in U$ then $ry \in L(x)$ for some $r > 0$. As in the paragraph after that, when $(\ ,\ )_2|_{V \times V}$ is nondegenerate and L_1, J generate $GL'(W_1)$, ry is the only positive real multiple of y in $G(x)$, and when $(\ ,\ )_2|_{V \times V}$ degenerates and L_1, J generate $GL(W_1)$, $L(x)$ contains every real multiple of y. Thus L is transitive on every G-orbit in U, and we have the near-transitivity condition of Theorem 2.1.

§5. Isolated Cases of Near Transitivity in Type I Pairs

As in §4, (G_1, G_2) is an irreducible dual reductive pair of type I in a symplectic group $Sp(W)$. Here we describe certain subgroups $L \subset G = G_1 G_2$ that satisfy the near transitivity condition of Theorem 2.1 but, in contrast to those of §4, are not modeled on parabolic subgroups.

5A. $G = U(2k, 2\ell)$. This is the second case of (3.15) with $v = 1$. The G-orbits on $W = \mathbb{C}^{2k, 2\ell}$ are $\{0\}$, the light cone and the mass shells. The group

(5.1) $$L = Sp(k, \ell) \subset U(2k, 2\ell) = G$$

is transitive on them. We mentioned this example in [10, (4.20)].

5B. $G = O(7) \cdot Sp(1; \mathbb{R})$. This is the first case of (3.15) with $(k, \ell) = (7, 0)$ and $v = 2$. Here $W = \mathbb{R}^{7 \times 2}$. Let U be the subspace of rank 2 matrices; the G-orbits are the sets

$$U_r = \{x = (x_1\, x_2) \in W : \|x_1 \wedge x_2\| = r\} \quad , \quad 0 < r < \infty \quad .$$

Every $x \in U_r$ is in the $Sp(1, \mathbb{R})$-orbit of some $x' = (x_1', x_2')$ where $\|x_1'\| = 1$, $\|x_2'\| = r$ and $(x_1', x_2')_1 = 0$. The subgroup $G_2 \subset O(7)$ is transitive on orthonormal 2-frames, hence for each r is transitive on pairs x_1', x_2' of vectors with $\|x_1'\| = 1$, $\|x_2'\| = r$, $(x_1', x_2') = 0$. See [8] for the transitivity. Now

(5.2) $$L = G_2 \cdot Sp(1; \mathbb{R}) \subset O(7) \cdot Sp(1; \mathbb{R}) = G$$

is transitive on every G-orbit in U.

5B'. $G = O(3, 4) \cdot Sp(1; \mathbb{R})$. This is the first case of (3.15) with $(k, \ell) = (7, 0)$ and $v = 2$. Let U be all $x \in W = \mathbb{R}^{7 \times 2}$ whose column span is a positive definite

2-plane in $W_1 = \mathbb{R}^{3,4}$. The subgroup $G_2^* \subset O(3,4)$, the noncompact real group of type G_2, is transitive on (positive) orthonormal 2-frames. So, as above,

$$L = G_2^* \cdot Sp(1;\mathbb{R}) \subset O(3,4) \cdot Sp(1;\mathbb{R}) = G \tag{5.3}$$

is transitive on every G-orbit in U.

5C. $G = O(8) \cdot Sp(1;\mathbb{R})$, the first case of (3.15) with $(k,\ell) = (8,0)$ and $v = 2$. Let U = all rank 2 matrices in $W = \mathbb{R}^{8\times 2}$. Just as in §5B,

$$L = Spin(7) \cdot Sp(1;\mathbb{R}) \subset O(8) \cdot Sp(1;\mathbb{R}) = G \tag{5.4}$$

is transitive on every G-orbit in U.

5C'. $G = O(4,4) \cdot Sp(1;\mathbb{R})$, the first case of (3.15) with $(k,\ell) = (4,4)$ and $v = 2$. Just as in §5B',

$$L = Spin(3,4) \cdot Sp(1;\mathbb{R}) \subset O(4,4) \cdot Sp(1;\mathbb{R}) = G \tag{5.5}$$

is transitive on every G-orbit in the space of x in $W = \mathbb{R}^{8\times 2}$ whose column span is a positive definite 2-plane.

References

[1] R. Howe, θ-series and invariant theory, Proc. Symp. Pure Math. 33 (Proceedings, Corvallis 1977), Part 1, pp. 275-285. Amer. Math. Soc., Providence, 1979.

[2] R. Howe, L^2-duality for stable dual reductive pairs, Yale University preprint, 1979.

[3] H. Jacquet, Principal L-functions of the linear group, Proc. Symp. Pure Math. 33 (Proceedings, Corvallis 1977) Part 2, pp. 63-86. Amer. Math. Soc., Providence, 1979.

[4] S. Kobayashi, On automorphism groups of homogeneous complex manifolds, Proc. Amer. Soc. 12 (1961), pp. 359-361.

[5] S. Kobayashi, Irreducibility of certain unitary representations, J. Math. Soc. Japan 20 (1968), pp. 638-642.

[6] G. Mack and I Todorov, Irreducibility of the ladder representations of U(2,2) when restricted to the Poincaré subgroup, J. Math. Phys. 10 (1969), pp. 2078-2085.

[7] S. Sternberg and J. A. Wolf, Hermitian Lie algebras and metaplectic representations, I, Trans. Amer. Math. Soc. 238 (1978), pp. 1-43.

[8] J. A. Wolf, Isotropic manifolds of indefinite metric, Comment. Math. Helv. 39 (1964), pp. 21-64.

[9] J. A. Wolf, "Unitary Representations of Maximal Parabolic Subgroups of the Classical Groups," Memoirs Amer. Math. Soc., No. 180. Providence, 1976.

[10] J. A. Wolf, Representations associated to minimal co-adjoint orbits, Differential Geometrical Methods in Physics, II (Proceedings, Bonn 1977), Springer Lecture Notes in Math. 676 (1978), pp. 329-349.

Comment added March 21, 1980

Along the lines of [2, Theorem 1] we note that, for the subgroups $L \subset G$ considered in this paper, equality $A(\pi) = A(\pi|_L)$ of commuting algebras shows, in the notation of Corollary 2.2, that $\pi|_L$ is multiplicity-free. So, for ν-almost-every α in the decomposition $\pi = \int_{\hat{G}} \pi_\alpha \, d\nu(\alpha)$ into irreducibles, the restriction $\pi_\alpha|_L$ determines π_α.

Non-positive polarizations and half-forms

by

J.H. Rawnsley

Introduction

We describe some of the geometry of non-positive Lagrangian subbundles of a symplectic vector bundle. For notation and terminology see [1,3,4]. The reader is warned however that sign conventions and normalizations do not always agree amongst the references. We use those of reference [1].

This is a report on joint work with R.J. Blattner and J.A. Wolf, and I am grateful to them for their hospitality at UCLA and Berkeley. Details and applications will appear elsewhere.

The linear theory

Let (V,ω) be a real symplectic vector space of dimension 2n. That is ω is a non-degenerate alternating bilinear form on V. If $V^{\mathbb{C}}$ denotes the complexification of V we can extend ω to be $\mathbb{C}$-bilinear and we denote its extension again by ω. Then $(V^{\mathbb{C}},\omega)$ is a complex symplectic vector space. Define a (pseudo-) Hermitian form H_ω on $V^{\mathbb{C}}$ by

$$H_\omega(v,w) = i\omega(v,\bar{w}), \qquad v,w \in V^{\mathbb{C}}$$

where $\bar{w}$ denotes the complex conjugate of w. This form has signature (n,n).

A complex Lagrangian subspace of (V,ω) is an n-dimensional complex subspace F of $V^{\mathbb{C}}$ such that

$$\omega(v,w) = 0, \quad \forall\, v,w \in F.$$

Consider the restriction of H_ω to F. It may be singular and it is easy to see that the kernel of H_ω on F is $F\cap\bar{F}$, the real part of F. We call the dimension of a maximal subspace of F on which H_ω is negative definite the index $i(F)$ of F. If $F\cap\bar{F}$ has dimension $r(F)$, then $n-i(F)-r(F)$ is the dimension of a maximal subspace of F on which H_ω is positive definite. If (p,q,r) is a triple of non-negative integers with $p+q+r = n$ we say F is of type (p,q,r) if $i(F) = p$, $r(F) = r$. Let $\mathcal{L}_{(p,q,r)}(V,\omega)$ denote the set of all complex Lagrangian subspaces of (V,ω) of type (p,q,r).

If (W,Ω) is a symplectic vector space over a field $\underline{k}$ we denote by $Sp(W,\Omega)$ the $\underline{k}$-linear endomorphisms of W which preserve   :

$$Sp(W,\Omega) \quad = \quad \{A \in \mathrm{End}W \mid \Omega(Ww_1,Aw_2) = \Omega(w_1,w_2)\ \forall\, w_1,w_2 \in W\}.$$

If $\mathcal{L}(V^{\mathbb{C}},\omega)$ denotes the set of all complex Lagrangian subspaces of (V,ω) then $Sp(V^{\mathbb{C}},\omega)$ acts transitively on $\mathcal{L}(V^{\mathbb{C}},\omega)$ and $\mathcal{L}(V^{\mathbb{C}},\omega)$ is a generalized flag manifold. $Sp(V,\omega)$ is a real form of $Sp(V^{\mathbb{C}},\omega)$ and

$$\mathcal{L}(V^{\mathbb{C}},\omega) \quad = \bigcup_{\substack{p,q,r\geq 0\\ p+q+r=n}} \mathcal{L}_{(p,q,r)}(V,\omega)$$

is a decomposition of $\mathcal{L}(V^{\mathbb{C}},\omega)$ into $Sp(V,\omega)$ orbits. See [5] for general results on the orbits of a real form on a generalized flag manifold.

The closed orbit corresponds with $p = q = 0$, $r = n$. A subspace F of this type is the complexification of $F\cap V$ which is a real Lagrangian subspace of (V,ω). The open orbits are those with $r = 0$ or equivalently $F\cap\bar{F} = 0$ so $V^{\mathbb{C}} = F\oplus\bar{F}$. There is then an endomorphism J_F of V with F and $\overline{F}$ as +i and -i eigenspaces respectively. Then $J_F^2 = -I$ and $J_F \in Sp(V,\omega)$. If $i(F) = 0$ F is said to be positive, which is the same as saying H_ω is non-negative on F. If H_ω is positive definite on F we say F is Kaehler. If only $r(F) = 0$ we say F is pseudo-Kaehler.

Symplectic and metaplectic vector bundles

Let (E,ω) be a real symplectic vector bundle over X. We call a subbundle F of $E^{\mathbb{C}}$ Lagrangian if $F_x \in \mathcal{L}(E_x^{\mathbb{C}}, \omega_x)$ for all x in X and $r(F_x)$ is constant for all x. Then $i(F_x)$ is also constant and we can then say F is of type (p,q,r) if F_x is for some x.

In [4] it was shown that any two positive Lagrangian subbundles are isomorphic. This is not true in general if we drop the positivity condition, but we have the following weaker result:

Theorem

Let F_1, F_2 be complex Lagrangian subbundles of (E,ω) then

$$c_k(F_1) \equiv c_k(F_2) \bmod 2, \quad k = 1, 2, \ldots, n$$

where $c_k(F)$ denotes the k-th Chern class of the complex vector bundle F.

To prove this we observe that from [4] we have $c_k(F_1) = c_k(F_2)$ for all k if F_1 and F_2 are both positive. Suppose F is of general type then we can split it $F = F_- + F'$ with H_ω negative definite on F_- and non-negative on F'. Then $\tilde{F} = \overline{F}_- + F'$ is positive Lagrangian. However

$$F + \overline{F}_- \cong \tilde{F} + F_- \qquad (*)$$

so

$$c(F)c(\bar{F}_-) \;=\; c(\tilde{F})c(F_-)$$

where

$$c(F) \;=\; 1 + c_1(F) + \ldots$$

is the total Chern class of F. Then for example

$$c_1(F) - c_1(F_-) \;=\; c_1(\tilde{F}) + c_1(F_-)$$

or

$$c_1(F) \;=\; c_1(\tilde{F}) + 2c_1(F_-).$$

This proves the result for c_1. The others follow by induction using similar arguments. We only need the result for c_1 in what follows.

<u>Corollary</u>

<u>(E,ω) is metaplectic if and only if some complex Lagrangian subbundle F is metalinear</u>.

This follows since (E,ω) is metaplectic if and only if $c_1(E,\omega)$ is even, but from [4]

$$c_1(E,\omega) = c_1(F) \equiv c_1(F) \bmod 2;$$

whilst F is metalinear if and only if $c_1(F)$ is even.

In fact arguing as in [4] it is easy to see that there is a 1-1 correspondence between metaplectic structures for (E,ω) and metalinear structures for F. If

$$i : \tilde{Q} \otimes \tilde{Q} \longrightarrow \Lambda^n \tilde{F}^o$$

gives half forms for $\tilde{F}$ then

$$Q \;=\; \tilde{Q} \otimes \Lambda^p F_- \qquad\qquad (**)$$

gives half-forms for F since (*) above implies

$$\Lambda^n F \otimes \Lambda^p \bar{F}_- \;\cong\; \Lambda^n F \otimes \Lambda^p F_-$$

and $F \cong F^o$.

So once we have constructed half-forms for positive Lagrangian subbundles we can obtain them for all Lagrangian subbundles by means of formula (**). However the choice of F_- is not canonical. Therefor a more canonical construction is required. There are two ways to proceed. We can use frame bundles as was done by Blattner [1] in the positive case or we can follow a suggestion of Kostant and generalize the symplectic spinors approach [2,3]. Here we describe the latter. Details of both will appear elsewhere.

Symplectic spinors

Let us recall how symplectic spinors are defined. If B denotes the bundle of symplectic frames of (E,ω) then a metaplectic structure on (E,ω) is a double covering of B by a principal $Mp(n,\mathbb{R})$ bundle $\tilde{B}$ so that

$$\begin{array}{ccccc} \tilde{B} \times Mp(n,\mathbb{R}) & \longrightarrow & \tilde{B} & \searrow & \\ \downarrow & & \downarrow & & X \\ B \times Sp(n,\mathbb{R}) & \longrightarrow & B & \nearrow & \end{array}$$

commutes, horizontal arrows being given by the right group actions. $\mathbb{R}^{2n}$ has a projective representation (the Schroedinger representation of the canonical commutation relations) on $L^2(\mathbb{R}^n)$ which is given infinitesimally on the Schwartz space $S(\mathbb{R}^n)$ by

$$(e_i.f)(q) = q_i f(q), \quad (e_{n+i}.f)(q) = -i\hbar \partial f/\partial q_i$$

and by duality also on the space $S'(\mathbb{R}^n)$ of tempered distributions. $Mp(n,\mathbb{R})$ acts also on L^2, S and S' as a group of self-equivalences of the representation. We can then form the bundle $S'(E,\omega) = \tilde{B} \times_{Mp(n,\mathbb{R})} S'(\mathbb{R}^n)$ over X whose fibre $S'(E,\omega)_x$ at x has a projective representation of the abelian Lie algebra E_x and of its complexification $E_x^{\mathbb{C}}$. One checks that

$$[v,w] = i\hbar\omega_x(v,w)\ I, \qquad v,w \in E_x^{\mathbb{C}}$$

as operators on $S'(E,\ )_x$.

If $F \subset E^{\mathbb{C}}$ is a Lagrangian subbundle let

$$(S'(E,\omega)^F)_x = \{ T \in S'(E,\omega)_x \mid v.T = 0\ \forall v \in F_x \}.$$

It is shown in [2,3] that if F is positive then $S'(E,\omega)^F$ is a line bundle dual to half-forms for F. However if F is not positive it is not hard to see that $S'(E,\omega)^F$ is zero. Kostant's suggestion is to regard $(S'(E,\omega)^F)_x$ as the zeroth cohomology group $H^0(F_x;\ S'(E,\omega)_x)$ of the abelian Lie algebra F_x with values in the F_x-module $S'(E,\omega)_x$. If this vanishes then one should examine the higher cohomology groups. This we have done and the result is the following:

Theorem

Let F be a complex Lagrangian subbundle of the metaplectic vector bundle (E,ω) then $H^k(F_x;\ S'(E,\omega)_x)$ is zero for all k except $k = i(F)$ and

$$H^p(F;\ S'(E,\omega))_x = H^p(F_x;\ S'(E,\omega)_x)$$

defines a line bundle over X whose dual gives half-forms for F.

We do not give a proof here but describe the main idea. Let F_-, F be defined as before and observe that since $\tilde{F}$ is positive then $\tilde{Q}^* = S'(E,\omega)^{\tilde{F}}$ is a line bundle and $\Lambda^p F_-^* \otimes \tilde{Q}^*$ embeds in $\Lambda^* F^* \otimes S'(E,\omega)$. The latter is the complex from which the Lie algebra cohomology groups are calculated. We construct a homotopy on this complex between the identity and a projection onto the subspace $\Lambda^p F_-^* \otimes \tilde{Q}^*$. This shows that the k-th cohomology group is zero unless $k = p = i(F)$. But

$\Lambda^p F^*_- \otimes \tilde{Q}^*$ is dual to $\Lambda^p F_- \otimes \tilde{Q}$ which we saw earlier was a square root of $\Lambda^n F^o$.

Remark

It is quite easy to self-pair the half-forms for F into the densities normal to $F \cap \bar{F}$ just as in the positive case, but we have no analogue at present of the pairing which exists for two positive Lagrangian subbundles F and G between $S'(E,\omega)^F$ and $S'(E,\omega)^G$ into the inverse densities on $F \cap \bar{G}$. The pointwise pairing of positive Lagrangian subbundles should probably be replaced by a spectral sequence for the cohomology groups but we don't yet know how to do this.

There are many situations where only non-positive polarizations of a symplectic manifold have some desired property such as invariance under a group (for example the minimal coadjoint orbits of su(p,q)) and applications of the above results and techniques to some of these problems are being studied in joint work with J.A. Wolf.

References

1. R.J. Blattner, The metalinear geometry of non-real polarizations. Lecture Notes in Math. vol. 570, Springer-Verlag, Berlin 1977.

2. B. Kostant, Symplectic spinors. Symposia Mathematica XIV, Academic Press, London 1974.

3. B. Kostant, On the definition of quantization. Colloque Symplectique (Aix-en-Provence), CNRS 1974.

4. J.H. Rawnsley, On the pairing of polarizations. Comm. Math. Phys. 58 (1978) 1-8.

5. J.A. Wolf, The action of a real semisimple Lie group on a complex flag manifold, I. Bull. Amer. Math. Soc. 75 (1969) 1121-1237.

CONNECTIONS ON SYMPLECTIC MANIFOLDS AND GEOMETRIC QUANTIZATION

Harald Hess

Freie Universität Berlin
FB 20, WE 4
Arnimallee 3
D-1000 Berlin 33

RESUME: Etant donné une variété symplectique munie des sous-fibrés Lagrangiens $F,G \subset T^{\mathbb{C}}M$ complémentaires, nous considérons des connexions symplectiques distinguées. Ces connexions ∇ sont utilisées pour quantifier les germes de fonctions dans les faisceaux $\underline{C}^k_F$ par des opérateurs différentiels dans un fibré quantique en lignes Q, qui est muni d'une connexion ∇^Q plate en direction de F. Aux propres choix de ∇, ∇^Q et d'une suite c de nombres réels, nous définissons des lois de quantification, qui généralizent les quantifications de Kostant-Souriau, de Weyl ou à l'ordre (anti-)normale.

ABSTRACT: Given a symplectic manifold equipped with complementary Lagrangian subtangent bundles $F,G \subset T^{\mathbb{C}}M$, we consider distinguished symplectic connections. These connections ∇ are used to quantize function germs in the sheaves $\underline{C}^k_F$ by differential operators on a quantum line bundle Q, which is equipped with a connection ∇^Q flat along F. For appropriate choices of ∇, ∇^Q and a sequence c of real numbers, we define quantization laws, generalizing the Kostant-Souriau, Weyl or (anti-)normal ordered quantizations.

For many geometric structures on manifolds, compatible connections have been used as a tool for investigating these structures. This applies to Riemannian, almost complex and almost hermitian manifolds, as well as to their indefinite variants. However, compatible connections on symplectic manifolds have hardly been considered in the literature, with the exception of a series of papers by Vey [18] and Lichnerowicz et al. [2,5], where such connections are used in the deformation approach to quantization (cf. also the contributions of Arnal and Lichnerowicz in these proceedings).

Apart from structural purposes, compatible connections on symplectic manifolds will be useful in geometric quantization, allowing to define a common generalization of one branch of the Kostant-Souriau theory and of older methods, as Weyl quantization and other factor ordering rules. Up to now, these older methods have been limited to symplectic vector spaces, if not even to $\mathbb{R}^{2n}$, with the exception of [15,16] where Weyl quantization is extended to cotangent bundles. In contrast, the quantization prescription we are going to describe applies to arbitrary symplectic manifolds.

Most of the material presented here will be treated in detail in the author's forthcoming doctor thesis; we also refer to [7] for some other aspects.

(M,ω) always denotes a 2n-dimensional symplectic manifold. As far as the definitions and elementary properties of symplectic connections are concerned, finite dimensionality of M is only imposed to simplify the presentation.

Underlined objects always denote sheaves over M. In particular, when V is a vector bundle over M, $\underline{V}$ denotes the sheaf of (germs of) sections in V.

We use a slightly widened notion of connection. A <u>connection</u> on a complex vector bundle $V \longrightarrow M$ will be a first order differential operator

$$\nabla : \underline{V} \longrightarrow \underline{\mathrm{Hom}}(T^{\mathbb{C}}M, V)$$

satisfying the derivation property

$$(1) \qquad \nabla_X(gs) = (\mathbb{L}_X g)s + g(\nabla_X s)$$

for all $X \in \underline{T^{\mathbb{C}}M}$, $g \in \underline{C}^{\infty}$ and $s \in \underline{V}$. If V is equipped with a real form, as e.g. when $V = T^{\mathbb{C}}M$, then a connection ∇ on V is called a <u>real connection</u> iff

$$(2) \qquad \overline{\nabla_X s} = \nabla_{\overline{X}} \overline{s} \qquad (X \in \underline{T^{\mathbb{C}}M},\ s \in \underline{V})$$

holds; in this case ∇ induces a connection in the usual sense on the real form of V. However, we do not use $\mathbb{R}$-vector bundles at all.

In addition, we have to refer to the notion of partial connections, cf. [10,13]. Therefore, consider a subtangent bundle $F \subset T^{\mathbb{C}}M$.

A partial connection on V along F then is a first order differential operator

$$\nabla_F : \underline{V} \longrightarrow \underline{\mathrm{Hom}}(F,V)$$

satisfying the derivation property (1) for all $X \in \underline{F}$.

As a motivation, we repeat some known facts of geometric quantization mostly oriented at Kostant [11], Rawnsley [13,14] with a few modifications. The subtangent bundle F is always supposed to be Lagrangian. Except for prequantization, the Kostant-Souriau theory only uses partial connections. However, our constructions are not based on prequantization in the usual way.

One of the most important concepts is that of a quantum bundle, which is a complex line bundle $Q \longrightarrow M$ equipped with a flat partial connection along F

$$\nabla_F^Q : \underline{Q} \longrightarrow \underline{\mathrm{Hom}}(F,Q)$$

such that Q is locally spanned by sections of the sheaf

(3) $$\underline{Q}_F := \mathrm{Ker}\ \nabla_F^Q .$$

In order that this is possible, F must at least be involutive, otherwise flatness of ∇_F^Q is not defined at all. We refer to [13,14] for a detailed discussion of this structure and its implications.

$\underline{Q}_F$ has to be interpreted as the sheaf of germs of pure states; it yields the global representation spaces as suitable cohomology spaces. To get Hilbert spaces, additional structures are needed, in particular Q must carry a suitable density-valued generalized hermitian structure. However, we do not consider the global representation spaces, and neglect the generalized hermitian structure for simplicity.

When F is involutive, there exists a canonical partial connection on F along F,

$$\nabla_F^F : \underline{F} \longrightarrow \underline{\mathrm{Hom}}(F,F)$$

given by

(4) $$X \lrcorner \nabla_F^F Y := \omega^{\#} \mathbb{L}_X \omega^{b} Y \qquad (X,Y \in \underline{F}),$$

which is always flat. In case $F = \overline{F}$, this partial connection has first been considered by Weinstein [19] and Kostant [12]. It is canonical for a number of reasons. Especially, ∇_F^F corresponds to the dual of the partial Bott connection [3,10] along F via the isomorphism $F \cong (T^{\mathbb{C}}M/F)^*$ induced by restricting ω^b. Another reason will be given below.

∇_F^F induces partial connections along F on all tensor bundles of F. Therefore, we may consider the (k+1)-fold symmetrized covariant derivative according to ∇_F^F

$$\check{\nabla}_F^{k+1} : \underline{C}^\infty \longrightarrow \underline{V^{k+1}F^*}$$

defined as follows:

$$(5) \qquad \check{\nabla}_F^{k+1} := S^{k+1}\nabla_F^{(k)} \cdots \nabla_F^{(1)} d_F ,$$

where d_F is the standard derivative followed by restriction to F,

$$\nabla_F^{(i)} : \underline{\otimes^i F^*} \longrightarrow \underline{\mathrm{Hom}}(F, \otimes^i F^*) = \underline{\otimes^{i+1}F^*}$$

is the partial connection along F induced by ∇_F^F , and S^{k+1} is the symmetrization operator (here normalized as (k+1)! times a projection).

Now we get subsheaves

$$(6) \qquad \underline{C}_F^k := \mathrm{Ker}\ \check{\nabla}_F^{k+1} \subset \underline{C}^\infty ,$$

and it follows from the definition of the $\check{\nabla}_F^{k+1}$ that these subsheaves form an increasing sequence

$$\underline{C}_F^0 \subset \underline{C}_F^1 \subset \ldots . \subset \underline{C}_F^k \subset \underline{C}_F^{k+1} \quad \ldots$$

We also consider the inductive limit

$$(7) \qquad \underline{C}_F^\bullet := \varinjlim_{k\to\infty} \underline{C}_F^k = \bigcup_{k=0}^\infty \underline{C}_F^k .$$

The definition (6) in particular means that $\underline{C}_F^0$ just consists of those function germs $f \in \underline{C}^\infty$ with $df|F = 0$.

F is called integrable (cf. [13]) iff F is locally spanned by locally Hamiltonian vector fields $\omega^\# dq_1, \ldots \omega^\# dq_n$, where the functions $q_1, \ldots q_n$ then are necessarily local sections of $\underline{C}_F^0$.

When F is integrable, the sheaves $\underline{C}_F^k$ coincide with those introduced by Kostant [11] under the same notation by defining $\underline{C}_F^0$ as above, and

$$(8) \qquad \underline{C}_F^k := \{ f \in \underline{C}^\infty | \bigwedge_{q_1, \ldots q_{k+1} \in \underline{C}_F^0} \mathrm{ad}q_1 \ldots \mathrm{ad}q_{k+1}(f) = 0 \},$$

where the adjoint representation is taken with respect to Poisson bracket.

In geometric quantization, it is always supposed that F and $F+\overline{F}$ are involutive subtangent bundles, because this condition seems necessary to construct Hilbertian representation spaces to $\underline{Q}_F$. In this case, F is automatically integrable [13].

However, integrability of F is not needed for our results, as long as the sheaves $\underline{C}_F^k$ are defined by (6).

Following Kostant [11], we want to quantize the germs in $\underline{C}_F^\bullet$, moreover the quantization should be given by a bidifferential operator

$$\delta_F : \underline{C}_F^{\bullet} \times \underline{Q}_F \longrightarrow \underline{Q}_F \; .$$

By a bidifferential operator, we mean a bilinear sheaf morphism, which is a differential operator in each variable. Alternatively, we may view δ_F as a differential operator valued differential operator

$$\delta_F : \underline{C}_F^{\bullet} \longrightarrow \underline{\mathrm{End}}(\underline{Q}_F) \; .$$

Such a δ_F will be called a <u>quantization map</u>.

Given a global function $f \in \Gamma(M, \underline{C}_F^{\bullet})$, $\delta_F(f)$ then is a global endomorphism of $\underline{Q}_F$, and therefore induces endomorphisms of the associated representation spaces arising in a functorial manner.

To construct quantization maps, it is very useful to have ordinary connections on Q and $T^{\mathbb{C}}M$ to our disposition, since then the construction essentially reduces to that of vector bundle morphisms. Note that even in the one variable case, for differential operators on Q, say, we need connections on both Q and $T^{\mathbb{C}}M$ to get this simplification.

Obviously, the ordinary connections ∇^Q on Q and ∇ on $T^{\mathbb{C}}M$ should extend the partial connections ∇_F^Q and ∇_F^F , respectively. Moreover, it is natural to require ∇ to be compatible with the symplectic structure.

It will turn out that ∇^Q and ∇ cannot be chosen independently. If ∇ is given in advance, the quantum bundle Q together with its connection ∇^Q in many cases may be constructed by some kind of generalized prequantization; one possibility of doing this is indicated in [7].

<u>Definition</u>: A connection ∇ on $T^{\mathbb{C}}M$ is called <u>symplectic</u> iff $\nabla\omega = 0$, or in more detail, iff

$$\mathbb{L}_X(\omega(Y,Z)) = \omega(\nabla_X Y, Z) + \omega(Y, \nabla_X Z) \qquad (X,Y,Z \in \underline{T^{\mathbb{C}}M}) \tag{9}$$

holds.

In virtue of $d\omega = 0$, torsion-free symplectic connections always exist, but they are far from being unique; in fact they build an infinite-dimensional affine space with difference vector space isomorphic to $\Gamma(M, \vee^3 (T^{\mathbb{C}}M)^*)$, cf. [2,18].

One method to single out a torsion-free symplectic connection ∇ is to fix an arbitrary torsion-free connection $\tilde{\nabla}$, then ∇ arises by defining

$$\nabla_X Y := \tilde{\nabla}_X Y + \tfrac{2}{3}\omega^{\#}(\tilde{\nabla}_X \omega)^{b} Y + \tfrac{1}{3}\omega^{\#}(\tilde{\nabla}\omega)(X,Y) \qquad (X,Y \in \underline{T^{\mathbb{C}}M}). \tag{10}$$

However, this method is not directly applicable to construct symplectic connections useful for geometric quantization. In fact, it seems that ultimately the appropriate connections need only satisfy a weaker torsion condition than being torsion-free. On the other hand, (10) does not help in constructing connections extending ∇^F_F.

Let Tor denote the torsion of ∇. It is easily seen that every symplectic connection ∇ satisfying

(11) $\nabla_U(\underline{F}) \subset \underline{F}$, $Tor(X,U) = 0$ $(U \in \underline{T^{\mathbb{C}}M},\ X \in \underline{F})$

automatically extends ∇^F_F (if such a ∇ exists at all, which is only possible for involutive F).

In general, the Kostant-Souriau quantization of classical observables does not only depend on one Lagrangian subtangent bundle F, but also on another Lagrangian subtangent bundle G, complementary to F:

$$F \oplus G = T^{\mathbb{C}}M .$$

Denote by pr_F , pr_G the projections according to this decomposition. Then we have

<u>Theorem 1</u>: There exists a unique symplectic connection ∇ satisfying

(12) $\nabla_U(\underline{F}) \subset \underline{F}$, $\nabla_U(\underline{G}) \subset \underline{G}$ $(U \in \underline{T^{\mathbb{C}}M})$,

(13) $Tor(X,Z) = 0$ $(X \in \underline{F},\ Z \in \underline{G})$.

Explicitly, ∇ is given by the following formulae:

(14)
$$\nabla_X Y = pr_F \omega^{\#} \mathbb{L}_X \omega^{b} Y \ , \quad \nabla_Z X = pr_F[Z,X]$$
$$\nabla_X Z = pr_G[X,Z] \qquad \nabla_Z W = pr_G \omega^{\#} \mathbb{L}_Z \omega^{b} W$$
$(X,Y \in \underline{F},\ Z,W \in \underline{G})$.

If either $F = \overline{F}$ and $G = \overline{G}$ or $F = \overline{G}$, then ∇ is real.

∇ is called <u>the bilagrangian connection associated to F and G.</u> □

No involutivity condition on F or G is needed in this theorem, it even holds if $d\omega$ does not vanish. ∇ is a generalization of the canonical connection on almost hermitian manifolds, cf. e.g. [17] for the latter. The proof of this theorem is very simple, cf. [7, 8]. In all of the following, we maintain the condition $d\omega = 0$.

The torsion of ∇ satisfies

(15) $Tor(X,Y) = -pr_G[X,Y]$ $(X,Y \in \underline{F})$.

In particular, the restriction of the torsion to F vanishes if and only if F is involutive. If this is true, then ∇ extends ∇^F_F and the projection on F in formula (14) for $\nabla_X Y$ is redundant.

Moreover, when F is involutive, ∇ is flat along F, i.e.

(16) $$\mathrm{curv}\nabla(X,Y) = 0 \qquad (X,Y \in \underline{F}).$$

The analogous assertions hold for G replacing F. Especially, if F and G are both involutive, the bilagrangian connection ∇ is torsion-free, and flat along both F and G, so that only the mixed components of the curvature may be different from zero.

Next, recall [11] that F and G are said to be Heisenberg related iff they are locally spanned by locally Hamiltonian vector fields $\omega^{\#}dq_i$ and $\omega^{\#}dp_i$, respectively, with functions q_i, p_i obeying the canonical Poisson bracket relations

(17) $$[q_i,q_j] = 0 = [p_i,p_j] \quad , \quad [q_i,p_j] = \delta_{ij} \qquad (i,j = 1,\dots n).$$

This notion admits the following characterization:

Theorem 2: F and G are Heisenberg related if and only if the associated bilagrangian connection ∇ is flat. □

A sketch of the proof has been given in [7], see [8] for a complete one; we only remark that one direction, namely to show that F and G are Heisenberg related if ∇ is flat, uses the complex Frobenius-Nirenberg theorem [9].

An example of this situation is given by the case where (M,ω) is a symplectic vector space and F,G are affine polarizations, i.e. are determined by parallel transport of complementary Lagrangian vector subspaces of $M^{\mathbb{C}}$. The associated bilagrangian connection then is always the standard connection given by ordinary derivation of vector space valued functions. In fact, locally this is the only example, up to symplectic diffeomorphisms, as long as the associated bilagrangian connection is real.

Now consider the case, where (M,ω) is a Kähler manifold, and $F = T^{1,0}M$, $G = T^{0,1}M$. It is an immediate consequence of theorem 1 and the usual characterization of the Levi-Cività connections that the bilagrangian connection with respect to F and G coincides with the Levi-Cività connection according to the Riemann metric derived from ω and the complex structure. Hence the bilagrangian connections in general are not flat.

Possibly, more types of symplectic connections are needed for geometric quantization. We shortly describe a type, which is especially suited for the case $M = T^{*}K$, K being a configuration space equipped with a Riemann metric, and F being the vertical polarization. More generally, let g be a Riemann metric on the normal bundle $T^{\mathbb{C}}M/F$, which is covariant constant along F according to the partial Bott connection. Then we have the

Theorem 3: Let F be involutive, then there exists a unique symplectic connection ∇ satisfying

(18) $\nabla_U(\underline{F}) \subset \underline{F}$, $\mathrm{Tor}(X,U) = 0$ $(U \in \underline{T^{\mathbb{C}}M},\ X \in \underline{F})$,

(19) $\nabla_Z(\underline{G}) \subset \underline{G}$, $\mathrm{Tor}(Z,W) \in \underline{F}$ $(Z, W \in \underline{G})$,

(20) $\nabla'' g = 0$,

where ∇'' is the induced connection on $T^{\mathbb{C}}M/F$. Moreover, when G is also involutive, then ∇ is torsion-free. □

Such a kind of connection may be needed to get the correct additional curvature terms to the Laplace-Beltrami operator when quantizing the free Hamiltonian.

Now we are going to describe the construction of quantization maps. Though it seems likely that the results on quantization maps could be extended to weaker assumptions, we suppose for simplicity that ∇ is a torsion-free symplectic connection satisfying

(21) $\nabla_U(\underline{F}) \subset \underline{F}$, $\nabla_Z(\underline{G}) \subset \underline{G}$ $(U \in \underline{T^{\mathbb{C}}M},\ Z \in \underline{G})$.

First, the connections ∇^Q and ∇ together induce connections

$$\nabla^{Q,(i)} : \underline{\mathrm{Hom}}(\otimes^i T^{\mathbb{C}}M, Q) \longrightarrow \underline{\mathrm{Hom}}(\otimes^{i+1} T^{\mathbb{C}}M, Q) ,$$

in particular $\nabla^{Q,(0)} = \nabla^Q$. Via these connections, ∇^Q and ∇ define an ℓ-fold symmetrized covariant derivative (for every $\ell \in \mathbb{N}_0$)

(22)
$$\check{\nabla}^{Q,\ell} : \underline{Q} \longrightarrow \underline{\mathrm{Hom}}(\vee^{\ell} T^{\mathbb{C}}M, Q)$$
$$\check{\nabla}^{Q,\ell} := S^{\ell}\, \nabla^{Q,(\ell-1)} \cdots \nabla^{Q,(0)} .$$

It is easily seen that every differential operator $D: \underline{Q} \longrightarrow \underline{Q}$ of order k may be decomposed uniquely as a sum

(23)
$$D = \sum_{\ell=0}^{k} \frac{1}{\ell!} \sigma^{\ell}(D)\, \check{\nabla}^{Q,\ell} ,$$

with suitable vector bundle morphisms

$$\sigma^{\ell}(D) : \mathrm{Hom}(\vee^{\ell} T^{\mathbb{C}}M, Q) \longrightarrow Q .$$

In the following, we essentially construct the $\sigma^{\ell}(\delta_F(f))$ which themselves become dependent on $f \in \underline{C}^{\bullet}_F$ by another differential operator, giving rise to a decomposition resembling (23). However, we arrange the occurring terms in another way.

Thus, let first $f \in \underline{C}^{\infty}$, and consider arbitrary $j, \ell \in \mathbb{N}_0$. Define a germ $f^{\ell}_j \in \underline{\vee^{\ell} G}$ by the composition of the following maps applied to f:

$$\underline{C}^\infty \xrightarrow{\overset{\vee}{\nabla}{}^{j+\ell}_F} \underline{\vee^{j+\ell}F^*} \xrightarrow{\vee^{j+\ell}\omega^\#} \underline{\vee^{j+\ell}G} \xrightarrow{\overset{\vee}{\nabla}{}^j_G} \underline{\mathrm{Hom}(\vee^j G,\vee^{j+\ell}G)} \xrightarrow{\overset{\vee}{tr}{}^j} \underline{\vee^\ell G}.$$

Here $\overset{\vee}{\nabla}{}^{j+\ell}_F$ is defined as in (5). Since F and G are complementary, F may be identified with $T^{\mathbb{C}}M/G$, and thus we have the isomorphism $\omega^\# : F^* \longrightarrow G$. Due to assumption (21), the restriction of ∇ to $\underline{G}$ in both arguments defines a partial connection denoted by ∇^G_G. It induces a partial connection on $\vee^{j+\ell}G$ along G. By analogy to (5) and (22), we then get the j-fold symmetrized covariant derivative $\overset{\vee}{\nabla}{}^j_G$. Finally, $\overset{\vee}{tr}{}^j$ will be the j-fold symmetrized contraction map defined by

(24) $$\overset{\vee}{tr}{}^j(\beta\otimes Y) := \beta \llcorner Y \qquad (\beta\in \vee^j G^*,\ Y\in \vee^{j+\ell}G).$$

Thus we have defined

(25) $$f^\ell_j := \overset{\vee}{tr}{}^j(\overset{\vee}{\nabla}{}^j_G(\vee^{j+\ell}\omega^\#)\overset{\vee}{\nabla}{}^{j+\ell}_F f).$$

It is a very important consequence of the definitions (6),(7) of $\underline{C}^k_F$, $\underline{C}^\bullet_F$, respectively, that f^ℓ_j vanishes for $f\in\underline{C}^\bullet_F$ if $j+\ell$ is great enough. Therefore, the following definition of a family of auxiliary bidifferential operators makes sense, because the summation only involves a finite number of terms:

$$D_j : \underline{C}^\bullet_F \times \underline{Q}_F \longrightarrow \underline{Q}$$

(26) $$D_j(f,s) := \sum_{\ell=0}^{\infty} (-i)^\ell \frac{1}{\ell!(j+\ell)!} f^\ell_j \lrcorner \overset{\vee}{\nabla}{}^{Q,\ell} s \qquad (f\in\underline{C}^\bullet_F,\ s\in\underline{Q}_F).$$

If j is great enough, $D_j(f) = D_j(f,\cdot)$ vanishes, too.

Hence for every sequence $c = (c_j)_{j\in\mathbb{N}_0}$ of real numbers, the following definition of a bidifferential operator makes sense:

$$\delta^c_F : \underline{C}^\bullet_F \times \underline{Q}_F \longrightarrow \underline{Q}$$

(27) $$\delta^c_F(f,s) := \sum_{j=0}^{\infty} (-i)^j \frac{1}{j!} c_j D_j(f,s) \qquad (f\in\underline{C}^\bullet_F,\ s\in\underline{Q}_F).$$

This is nearly a quantization map as desired. The only problem is whether the image of δ^c_F is contained again in the sheaf $\underline{Q}_F$ of germs of pure states. In general, this can only be expected if the connections ∇, ∇^Q and the sequence c together satisfy suitable conditions. In the following, we are going to give some sufficient conditions, and examples where these are satisfied.

For a germ $f\in\underline{C}^0_F$, $\delta^c_F(f) = \delta^c_F(f,\cdot)$ will only depend on c_0. In fact, $\delta^c_F(f)$ is the multiplication operator with $c_0 f$, which is an endomorphism of $\underline{Q}_F$ due to the definition of $\underline{C}^0_F$.

Next, when $f\in\underline{C}^1_F$, only c_0 and c_1 are relevant; moreover the expression for δ^c_F reduces to

(28) $$\delta_F^c(f) = c_0(f - i\nabla^Q_{\omega^\# d_F f}) - c_1 i\ \mathrm{tr}(\nabla^G_G \omega^\# d_F f)\ .$$

In this case, we have

Theorem 4: Suppose that

(29) $$c_0 \mathrm{curv}\nabla^Q(X,U) = c_0 i\omega(X,U) + c_1 \mathrm{tr}(\mathrm{curv}\nabla(X,U)|\underline{F})$$

for all vector fields $U \in \underline{T^{\mathbb{C}}M}$ and $X \in \underline{F}$. Then

(30) $$\delta_F^c(\underline{C}_F^1 \times \underline{Q}_F) \subset \underline{Q}_F$$

holds. □

The proof is fairly straightforward from the definition (3) of $\underline{Q}_F$.

If $c_0 = 1$, then in particular $\delta_F^c(1) = 1$ holds. When additionally relation (29) is satisfied even for arbitrary $X \in \underline{T^{\mathbb{C}}M}$, the quantization map δ_F^c maps Poisson brackets on $\underline{C}_F^1$ to commutators (up to a factor i), i.e. solves a restricted version of the Dirac problem.

Let us consider the two most important applications of this theorem to the Kostant-Souriau quantization theory. To this aim, denote by L the Kostant-Souriau prequantum bundle, and by ∇^L its connection, which by definition satisfies $\mathrm{curv}\nabla^L = i\omega$.

First, when $Q := L$ and $\nabla^Q := \nabla^L$, the assumption of theorem 4 holds for $c_0 = 1$ and $c_1 = 0$. In this case, the symplectic connection ∇ is irrelevant, and δ_F^c is just the prequantization map restricted to $\underline{C}_F^1$.

Secondly, let $\Lambda_{1/2}F$ be the bundle of half-forms over F, which now comes with an ordinary connection induced by the connection on F arising by restricting ∇ to F. Let $Q := L \otimes \Lambda_{1/2}F$, and ∇^Q be the tensor product (of connections) of ∇^L and the connection on $\Lambda_{1/2}F$. Then the assumption of theorem 4 is satisfied for $c_0 = 1$ and $c_1 = \frac{1}{2}$, and it turns out that

(31) $$\delta_F^c(f) = f - i\nabla^L_{\omega^\# df} \otimes 1 - i \otimes (\mathbb{L}_{\omega^\# df})^{1/2}$$

holds, which is just the Kostant prescription in this case, cf. [6].

For quantization of germs in $\underline{C}_F^\bullet$, not in $\underline{C}_F^1$, the best result we have at the time is the following one:

Theorem 5: Suppose that

(32) $$\mathrm{curv}\nabla^Q(X,U) = i\omega(X,U) \quad , \quad \mathrm{curv}\nabla(X,U) = 0$$

holds for all vector fields $U \in \underline{T^{\mathbb{C}}M}$ and $X \in \underline{F}$. Then

(33) $$\delta_F^c(\underline{C}_F^\bullet \times \underline{Q}_F) \subset \underline{Q}_F$$

is satisfied for an arbitrary sequence c. □

We only remark that under the above assumptions, even

(34) $$D_j(\underline{C}_F^{\bullet} \times \underline{Q}_F) \subset \underline{Q}_F$$

is true for every $j \in \mathbb{N}_0$, from which (33) follows at once. The proof is very technical.

In particular, this theorem applies, when ∇ is the bilagrangian connection associated to Heisenberg related subtangent bundles F and G, and $\mathrm{curv}\nabla^Q = i\omega$. Even more specially, if the former condition holds, the latter is satisfied e.g. when $Q = L$, $\nabla^Q = \nabla^L$, or when $Q = L \otimes \Lambda_{1/2}F$ and ∇^Q is the tensor product of ∇^L with the connection on the half-form bundle induced by ∇.
In case that F and G are Heisenberg related, ∇ is the associated bilagrangian connection, and $\mathrm{curv}\nabla^Q = i\omega$, the quantization maps δ_F^c admit simple coordinate representations, allowing to compare them with the quantization map of Kostant [11] as well as with older quantization prescriptions for various choices of c.

To write down these coordinate expressions, we use multi-indices, which as superscripts denote powers.

In terms of coordinates satisfying (17) such that $\omega^{\#}dq_i$, $\omega^{\#}dp_i$ span F,G respectively, every germ $f \in \underline{C}_F^k$ over the domain U of these coordinates admits a decomposition

(35) $$f = \sum_{0 \leq |r| \leq k} g_r p^r ,$$

where $g_r \in \underline{C}_F^0$, i.e. depend only on the q_i.
In addition, consider a local trivialization

$$\Phi : U \times \mathbb{C} \longrightarrow Q_{|U}$$

with

(36) $$\mathrm{Hom}(1,\Phi^{-1})\nabla^Q\Phi = d - i \sum_{\nu=1}^{n} p_\nu dq_\nu .$$

Then we get the following coordinate representation:

(37) $$\Phi^{-1}\delta_F^c(f)\Phi = \sum_{0 \leq |s+t| \leq k} (-i)^{s+t} c_{|s|} \binom{s+t}{s} \left(\frac{\partial^s g_{s+t}}{\partial q^s}\right) \frac{\partial^t}{\partial q^t} .$$

Up to normalization conventions, now Kostant's quantization map [11] (specialized to the Heisenberg related case) coincides with δ_F^c for the choice

(38) $$c_j = 1 - \tfrac{1}{2}j \qquad (j \in \mathbb{N}_0).$$

Other choices of the sequence c correspond to factor ordering rules. For comparison with the latter, we suppose that (M,ω) is the symplectic vector space $\mathbb{R}^{2n}$, and F, G are either given by parallel transport of the standard $\mathbb{R}^n$ summands, or are $T^{1,0}M$, $T^{0,1}M$ according to the standard complex structure. In the first case, the q_i,p_i are the standard position and momentum variables, respectively, while in the second case they are the standard complex variables $z_i,\overline{z}_i$

(up to a factor i).

With new coefficients derived from c via

(39) $$\tilde{c}_r^u := \sum_{0\leq h\leq u-r} (-1)^h \binom{u-r}{h} c_{|r+h|} ,$$

the inverse Leibniz rule applied to (37) yields another coordinate representation

(40) $$\Phi^{-1}\delta_F^c(f)\Phi = \sum_{0\leq|s+t|\leq k} (-i)^{s+t} \tilde{c}_s^{s+t} \binom{s+t}{s} \frac{\partial^s}{\partial q^s}\left(g_{s+t}\frac{\partial^t}{\partial q^t}\right) .$$

Now the quantization map δ_F^c for

(41) $$c_j = \delta_{j,0} \qquad (j\in\mathbb{N}_0)$$

extends the standard ordering [1] when $F = \overline{F}$, $G = \overline{G}$, and the normal ordering [1] when $F = \overline{G}$. Similarly, for

(42) $$c_j = 1 \qquad (j\in\mathbb{N}_0),$$

δ_F^c extends the antistandard or antinormal ordering [1], respectively. Finally, for the choice

(43) $$c_j = 2^{-j} \qquad (j\in\mathbb{N}_0),$$

the quantization map δ_F^c extends the Weyl quantization. In fact, then the right hand side of formula (40) is just the expression derived from Weyl's rule by McCoy, cf. e.g. [4].

We have erroneously stated in (at least the preprint to) [7] that the Born-Jordan quantization rule also could be obtained as a special case of δ_F^c. This can only be maintained if $\dim M = 2$. Otherwise, a geometric generalization of the Born-Jordan rule needs very specialized additional structures, which make it uninteresting, if it exists at all.

The quantization method described here still has many limitations. Some of them are caused by technical difficulties in proving that the image of δ_F^c again lies in the sheaf $\underline{Q}_F$. However, we have strong indications that theorems 4 and 5 do not exhaust the possibilities of getting quantization maps, and that also the general convention on ∇ is not fully necessary.

More serious are the limitations due to the purely local approach used here (once the connections are chosen), which only allows to answer questions on formal selfadjointness of the outcoming operators (if generalized hermitian structures are taken into account), but not those on essential selfadjointness. We hope to cure this defect by extending the definition of δ_F^c using parallel transport according to ∇^Q along the geodesics of ∇, if the latter is real.

The major advantage of our approach is to treat the extensively studied (anti-) normal ordered and Weyl quantizations as well as the Kostant-Souriau method all on the same geometric level.

ACKNOWLEDGEMENTS: I thank M.Forger, C.Günther, D.Kraußer and R.Schrader for helpful discussions.

REFERENCES:

1. G.S.Agarwal, E.Wolf Calculus for functions of noncommuting operators and general phase-space methods in quantum mechanics I Phys.Rev. D 2 (1970), 2161-2186
2. F.Bayen, M.Flato, C.Fronsdal, A.Lichnerowicz, D.Sternheimer Deformation theory and quantization I Ann.Phys. 111 (1978), 61-110
3. R.Bott in: Lectures on algebraic and differential topology Springer Lect.Notes in Math., vol.279
4. H.Daughaday, B.P.Nigam Function in quantum mechanics which corresponds to a given function in classical mechanics Phys.Rev. 139 B (1965), 1436-1442
5. M.Flato, A.Lichnerowicz, D.Sternheimer Crochet de Moyal-Vey et quantification C.R.Acad.Sc.Paris, sér.A 283 (1976) 19-24
6. K.Gawędzki Fourier-like kernels in geometric quantization Diss.Math. 128 (1976), 1-83
7. H.Hess On a geometric quantization scheme generalizing those of Kostant-Souriau and Czyż, to app. in: Proc.Inform.Meet. on Diff.Geom.Meth. in Physics, Clausthal 1978
8. H.Hess forthcoming thesis
9. L.Hörmander The Frobenius-Nirenberg theorem Arkiv för Matematik 5 (1965), 425-432
10. F.W.Kamber, P.Tondeur Foliated bundles and characteristic classes Springer Lect.Notes in Math., vol.493 (1975)
11. B.Kostant On the definition of quantization in: Coll.Int. du CNRS, Géométrie symplectique et physique mathématique, Aix-en-Provence 1974, ed. CNRS (1976)
12. B.Kostant Symplectic spinors in: Conv. di geom. simplett. e fisica matem., INDAM Rome 1973, Sympos.Math. XIV, Academic Press N.Y. (1974)

13. J.H.Rawnsley On the cohomology groups of a polarisation and diagonal quantisation
Trans.Am.Math.Soc. 230 (1977), 235-255

14. J.H.Rawnsley Flat partial connections and holomorphic structures in C^{∞} vector bundles
Proc.Am.Math.Soc. 73 (1979), 391-397

15. J.Underhill Quantization on a manifold with connection
J.Math.Phys. 19 (1978), 1932-1935

16. J.Underhill, S.Taraviras Weyl quantization on a sphere
in: Springer Lect.Notes in Phys., vol.50 (1976), 210-216

17. I.Vaisman Cohomology and differential forms
Dekker, N.Y. (1973)

18. J.Vey Déformation du crochet de Poisson sur une variété symplectique
Comment.Math.Helv. 50 (1975), 421-454

19. A.Weinstein Symplectic manifolds and their Lagrangian submanifolds
Adv.Math. 6 (1971), 329-346

GEOMETRIC ASPECTS OF THE FEYNMAN INTEGRAL

D. J. Simms

Mathematical Institute, Bonn[1]

In this talk I want to make some observations on the geometry involved in Feynman's path integral. We consider a dynamical system based on a smooth manifold X as configuration space, and a time-dependent Lagrangian function L. Thus L is a smooth real valued function on $TX \times R$ where TX is the tangent bundle of X. For each smooth path $\gamma : [s_1, s_2] \to X$,

$$S(\gamma) = \int_{s_1}^{s_2} L(\dot{\gamma}(t), t)\, dt$$

is called the action of the path γ, where $\gamma(t)$ is the velocity vector of γ at t.

Let ψ, a complex valued function on $X \times R$, be a time-dependent quantum mechanical wave function of the system. Then Feynman proposed that ψ should satisfy an equation of the form

$$\psi(y, s_2) = \int \left[\int \exp(2\pi i h^{-1} S(\gamma)) \mathcal{D}(z, s_1; y, s_2)(\gamma)\right] \psi(z, s_1)\, dz.$$

Here $\mathcal{D}(z, s_1; y, s_2)(\gamma)$ is to be some suitable measure on a space of paths $\gamma : [s_1, s_2] \to X$ such that $\gamma(s_1) = z$ and $\gamma(s_2) = y$, and h denotes Planck's constant. The factor $\exp(2\pi i h^{-1} S(\gamma))$ in the expression originated with Dirac.

This proposal of Feynman is closely related to the various geometric concepts introduced into quantisation theory by Kostant [1] and Souriau [4]. It is the purpose of this paper to describe these relationships.

We first note that, corresponding to a path γ in X we have a path $\tilde{\gamma}$ in the cotangent bundle M of $X \times R$, defined by the Legendre transformation associated with L. We then have $S(\gamma) = \int_{\tilde{\gamma}} \alpha$ where

[1] On leave of absence from Trinity College Dublin, supported by the Sonderforschungsbereich Theoretische Mathematik.

α is the canonical 1-form on M. For these facts see our other article in this volume. We see that the Dirac factor $\exp(2\pi i h^{-1} S(\gamma))$ is the multiplying factor given by a suitable parallel transport along $\tilde{\gamma}$. More precisely, we take the product complex line bundle $\mathcal{L}$ over M. We consider sections of $\mathcal{L}$ as smooth complex valued functions on M. For each vector field ξ on M we define a covariant derivative ∇_ξ, acting on the sections of $\mathcal{L}$, by

$$\nabla_\xi f = \xi f - 2\pi i h^{-1} \langle \alpha, \xi \rangle f$$

for each section f of $\mathcal{L}$. The associated connection on $\mathcal{L}$ has curvature form $-h^{-1}\omega$ where $\omega = d\alpha$ is the symplectic form on M. Thus $(\mathcal{L}, \nabla)$ is a Kostant-Souriau prequantisation line bundle. It is parallel transport with respect to this connection which gives the Dirac factor.

The direct approach to giving a meaning to Feynman's expression starts off by taking a partition of the time interval $[s_1, s_2]$ and replacing each path γ by a corresponding piecewise classical path. We now examine the geometrical aspects of the finite dimensional integral obtained in this kind of way.

The standard procedure, for the case $X = R^n$, may be described as follows. Fix a partition of $[s_1, s_2]$ into N equal segments, $s_1 = t_1 < t_2 < \ldots < t_{N+1} = s_2$. For x, y in X let $S_{jk}(x, y)$ denote the action of the classical path (supposed unique) from x at time t_j to y at time t_k. Let

$$\Delta_{jk}(x, y) = \left[\det \left(\frac{\partial^2 S_{jk}(x, y)}{\partial x^p \, \partial y^q} \right)_{p,\, q = 1,\ldots,n} \right]^{\frac{1}{2}}$$

If γ is a path in X such that $\gamma(s_1) = z$ and $\gamma(s_2) = y$ then we associate with it the sequence $(x_1, \ldots, x_{N+1})$ where $x_j = \gamma(t_j)$. In the Feynman expression we replace $S(\gamma)$ by $\sum_{i=1}^{N} S_{i,i+1}(x_i, x_{i+1})$ and we replace $\mathcal{D}(z, s_1; y, s_2)(\gamma)\,dz$ by $\prod_{i=1}^{N} \Delta_{i,i+1}(x_i, x_{i+1})\,dx_i$.

Instead of integrating over a space of paths γ, we integrate over $(x_1, \ldots, x_N)$. Thus we obtain the expression

$$\int \left[\prod_{i=1}^{N} (\exp 2\pi i h^{-1} S_{i,i+1}(x_i, x_{i+1}))\ \Delta_{i,i+1}(x_i, x_{i+1}) \right] \psi(x_1, t_1) dx_1 .. dx_N$$

which formally represents an integral over R^{nN}. This expression in turn may be represented as a composition of N integral transforms on R^n as

$$\left[T^{N+1,\,N} \circ \ldots \circ T^{2,1}\ \psi(\cdot\ ,\ t_1) \right] (y)$$

where T^{kj} is the transform given by

$$(T^{kj}\phi)(x') = \int (\exp 2\pi i h^{-1} S_{jk}(x, x'))\ \Delta_{jk}(x, x')\ \phi(x)\ dx.$$

We now consider the geometrical significance of the integral transform T^{kj}. We restrict ourselves to the case of a non-singular Lagrangian, where the range M_o of the Legendre transformation is a smooth submanifold of M of codimension 1. In this case the classical paths $\tilde{\gamma}$ in M_o give a 1-dimensional foliation of M_o.

Let ρ_j denote the map from M_o to X which assigns to each point m of M_o the position at time t_j of the classical path through m. Let S_j be the real valued function on M_o such that $S_j(m)$ is the integral of the canonical 1-form α along the classical path in M_o from time t_j up to the point m.

Let ϕ be a function on R^n and put $\phi_j = \phi \circ \rho_j$. For x, x' in R^n let m be the end-point in M_o of $\tilde{\gamma}$ where γ is the classical path (supposed unique) such that $\gamma(t_j) = x$ and $\gamma(t_k) = x'$. Then for part of the integrand of T^{kj} we have

$$(\exp 2\pi i h^{-1}\ S_{jk}(x, x'))\ \phi(x) = (\exp 2\pi i h^{-1}\ S_j)\ \phi_j \Big|_m .$$

The mapping ρ_j and the function $(\exp 2\pi i h^{-1}\ S_j)\phi_j$ on M_o have the following geometrical significance. i) The level sets of ρ_j are the leaves of a Lagrangian foliation (real polarisation) F of M_o. This means that the leaves of F are maximally isotropic with respect to the 2-form ω. Each leaf of F can be locally represented as the graph of an exact differential $d\Theta$ on $X \times R$. Such a function Θ is a solution of the Hamilton-Jacobi equation. Thus F itself represents a complete solution of the Hamilton-Jacobi equation.

ii) The function $(\exp 2\pi i h^{-1} S_j)\ \phi_j$ is a section of the pre-quantisation line bundle $\mathcal{L}$ over M_o which is covariant constant along the leaves of F.

To get a more complete understanding of the integrand appearing in T_{kj}, we take the inner product of $T^{kj}\phi$ with another function χ on R^n. We get

$$\int (T^{kj}\phi)(x')\ \overline{\chi(x')}\ dx'$$
$$= \int (\exp 2\pi i h^{-1} S_{jk}(x, x'))\ \phi(x)\ \overline{\chi(x')}\ \Delta_{jk}(x, x')\ dx\ dx'$$

If we transfer this integral to M_o using the map $\rho : M_o \to X \times X$ given by $\rho(m) = (\rho_j(m), \rho_k(m))$ then we get, formally.

$$\int (\exp 2\pi i h^{-1}(S_j - S_k))\ \phi_j\ \overline{\chi_k}\ \rho^*(\Delta_{jk}\ dx \wedge dx')$$

where here the integral is taken over M_o transverse to the foliation by the classical paths. As shown in [3] this is, up to a normalising factor, the expression for the Blattner-Kostant-Sternberg (BKS) pairing [2] of

$$(\exp 2\pi i h^{-1} S_j)\ \phi_j\ \nu_j \quad \text{with} \quad (\exp 2\pi i h^{-1} S_k)\ \chi_k\ \nu_k$$

where $\nu_j = \rho_j^*(dx)^{\frac{1}{2}}$.

Thus T_{kj} is given by the BKS pairing of $\mathcal{L}$-valued half-forms normal to, and covariant constant along, the level sets of ρ_j and ρ_k respectively.

References

1. B. Kostant, Quantisation and unitary representations. Lecture Notes in Mathematics 170, Springer, Berlin 1970
2. J. Rawnsley, On the pairing of polarisations. Comm. Math. Physics 58, (1978), 1 - 8.
3. D. J. Simms, Geometric quantisation the Feynman integral. Lecture Notes in Physics 106, Springer, Berlin 1979
4. J. M. Souriau, Structure des systèmes dynamiques, Dunod, Paris 1970.

RELATIVISTIC QUANTUM THEORY IN COMPLEX SPACETIME

Gerald Kaiser
Mathematics Department
University of Lowell
Lowell, Massachusetts 01854

1. INTRODUCTION

I wish to report here on some recent work which looks at relativistic quantum theory from a new geometric point of view. The main ingredient in this approach is the spectral condition, which states that the energy-momentum operators P_μ for an isolated (but possibly self-interacting) system have their joint spectrum in the closure of the forward light cone V_+. This eminently reasonable condition (which enters as an axiom into quantum field theory [1]) has an immediate and important consequence: it gives rise to a canonical complexification of spacetime, as follows: Let $y = (y_o, \vec{y})$ be an arbitrary future-pointing four-vector, i.e., $y_o > |\vec{y}|$. If $p = (p_o, \vec{p})$ is a point in the joint spectrum of (P_μ), then $p_o \geqslant |p|$, hence $yp \equiv y^\mu p_\mu \equiv y_o p_o - y \cdot p \geqslant o$. Thus the operator $yP \equiv y^\mu P_\mu$ is non-negative and the group of space-time translations $U(x) = \exp(-ixP)$ extends naturally to a holomorphic semigroup

$$U(z) = \exp(-izP) = \exp(-yP)\ U(x),$$

where $z = x-iy$ belongs to the forward tube

$$T = \{x-iy \in C^4 \mid y \in V_+\},$$

which is the complexification of spacetime referred to above.

All this is well-known and well-used. It leads to important results such as the holomorphy of the Wightman functions, which in turn provides the technical means for rigorously proving theorems in quantum field theory [1]. What is new in our approach is this: We show that quite generally, the imaginary coordinates y_μ can be given a direct physical interpretation as, roughly, energy-and momentum variables canonically associated with the theory, on equal footing with the space-time variables x_μ. This shows that T is much more than a mere technical device: it is an extended phase space somehow associated with the theory. It therefore becomes tempting to reformulate the theory from the beginning with T, rather than R^4, as the proper arena for physics.

The geometric structure of the new base space T must combine the Minkowskian structure of flat spacetime with the symplectic structure of classical phase space in a convariant way. Furthermore, to make the connection with quantum mechanics a solid one, the classical symplectic structure must play an important role in the quantum dynamics. For the simplest systems (single massive scalar particles) this has been accomplished in references [2-5], of which we now give a brief account. For a more detailed review, see [5].

The Hilbert space of quantum states is the set of positive-energy solutions of the Klein-Gordon equation, square integrable in momentum space (the positivity of the energy results from the spectral condition). Essentially by the argument above, these solutions extend as holomorphic functions to the forward tube T. The connection between classical states (points $z \in T$) and quantum states (holomorphic solutions $f(z)$) is furnished by a family $\{e_z | z \in T\}$ of "coherent states" of the Bargmann-Segal-Klauder type [6-8]. Solutions are given by inner products: $f(z) = \langle e_z | f \rangle$. The e_z are optimal wave-packets "focused" at x which have an expected energy-momentum proportional to y (where $z = x - iy$). (This provides the physical interpretation of y on which our approach is based.) The overlap between different e_z's is given by their inner product:

$$\langle e_z | e_w \rangle = -2i\Delta^+(z - \bar{w}),$$

where Δ^+ is the Wightman two-point function for the neutral scalar field [1], which turns out to be a reproducing kernel [9] for our Hilbert space. The synthesis of the Minkowskian and symplectic geometries is accomplished by giving T the (indefinite) Kähler metric

$$ds^2 = g_{\mu\nu} d\bar{z}^\mu dz^\nu ,$$

where $g_{\mu\nu}$ is the usual Minkowski metric. The associated (Kähler) 2-form

$$2i\alpha = g_{\mu\nu} d\bar{z}^\mu \wedge dz^\nu = 2i dy_\mu \wedge dx^\mu$$

then provides an invariant symplectic structure on T. However, T is too large for a phase space since it contains the time-and "energy" dimensions. A prospective phase space is a six-dimensional submanifold σ of T which has the product form $\sigma = S \times \Omega$, where S (configuration space) is a three-dimensional submanifold of spacetime R^4 and Ω (momentum space) is a hyperboloid in V_+. It turns out that the restriction of the 2-form α to σ will be symplectic if and only if S is a "space-or-light-like" submanifold of R_4 i.e., its normal n satisfies $n^2 \geq 0$. This is in conformity with intuition (the configuration space must be spacelike) and suggests that the two geometries have indeed been successfully combined. Convariance has been preserved in the sense that the Poincare group transforms different σ's (each with its own symplectic structure) into one another by canonical transformations. The role played by α in the quantum dynamics is as follows: The 6-form $\alpha \wedge \alpha \wedge \alpha$, restricted to any phase space σ, is a volume form (Liouville measure), and the inner product between states can be obtained by integration with respect to this form. (Incidentally, this gives a formalism quite analogous to classical statistical mechanics!) The squared norm $\| f \|_\sigma^2$ of a state f then automatically appears as the total flux of a (probability) current which is conserved (so that $\| f \|_\sigma^2$ is independent of σ) as a result of the dynamics (Klein-Gordon equation and positivity of the energy) satisfied by f. As a by-product, a consistent probabilistic interpretation for Klein-Gordon theory is obtained: one has a covariant probability current with a non-negative time component. This resolves a long-standing difficulty in the usual formalism (see the discussion in [4]).

These are the main results of [3] and [4]. In [5] an attempt is made to introduce external fields by replacing solutions $f(z)$ by holomorphic sections of a vector bundle equipped with a fiber metric, using ideas of gauge theory. The fiber metric then generates gauge fields of the Yang-Mills type, in much the same way as a Riemannian metric generates a connection.

2. TOWARD QUANTUM FIELD THEORY

Some of the basic elements of the above formalism can be generalized, in a simple and natural way, to the most advanced and general form of relativistic quantum mechanics: the theory of interacting systems of quantum fields. We now indicate how this can be done for the simplest such case: a neutral scalar field, possibly in self-interaction but otherwise isolated. Let $\phi(x)$ be such a field, operating on a Hilbert space F with a unique vacuum vector Ω (see [10] for background). Let P_μ be the energy-and momentum operators of the field, which satisfy the spectral condition and have the defining properties $P_\mu \Omega = o$ and

$$\frac{\partial \phi(x)}{\partial x^\mu} = i\,[P_\mu, \phi(x)].$$

Then

$$\phi(x) = \exp(ixP)\phi_o \exp(-ixP), \tag{1}$$

where $\phi_o \equiv \phi(o)$. (Equation (1) has only a formal significance since $\phi(x)$ is actually an operator-valued distribution [1], hence $\phi(o)$ cannot be defined.) We have seen that $\exp(-ixP)$ extends to $\exp(-izP)$, where $z=x-iy\varepsilon T$. Similarly, $\exp(ixP)$ extends to $\exp(i\bar{z}P)$ where $\bar{z}=x+iy$ is the complex conjugate of z. Thus define

$$\begin{aligned} \phi(z) &= \exp(i\bar{z}P)\phi_o \exp(-izP) \\ &= \exp(-yP)\phi(x) \exp(-yP)\,. \end{aligned} \tag{2}$$

Because $\exp(-yP)$ is a strongly smoothing operator, $\phi(z)$ is much better behaved than $\phi(x)$. In fact, it can be expected to be an operator-valued function rather than distribution. Note also that since $\phi(x)$ is hermitian, so is $\phi(z)$.

Now a field theory, if it satisfies certain minimal and physically reasonable axioms (among them the spectral condition), can be completely recovered from its Wightman functions [1], which in the usual formalism are distributions on R^{4n} given by

$$(3)\qquad \mathcal{W}_n(x_1,x_2,\ldots,x_n) \equiv \langle\Omega|\phi(x_1)\phi(x_2)\ldots\phi(x_n)\Omega\rangle.$$

Since $P_\mu\Omega = o$, eq. (1) implies that

$$\mathcal{W}_n(x_1,x_2,\ldots,x_n) =$$

$$\langle\Omega|\phi_o \exp(-i(x_1-x_2)P)\phi_o\ldots\phi_o \exp(-i(x_{n-1}-x_n)P)\phi_o\Omega\rangle$$

$$(4)\qquad \equiv W_n(x_1-x_2,x_2-x_3,\ldots,x_{n-1}-x_n).$$

The functions W_n are distributions on $R^{4(n-1)}$ which, by the spectral condition, are boundary values of holomorphic functions on T^{n-1}:

$$(5)\qquad W_n(\xi_1,\ldots,\xi_{n-1}) = \lim_{\eta_k\to o} W_n(\xi_1-i\eta_1,\ldots,\xi_{n-1}-i\eta_{n-1})$$

with $\eta_k \varepsilon V_+$, $k=1,\ldots,n-1$. Thus, in the usual formalism, the powerful condition of holomorphy is reached by continuing the auxiliary functions W_n. On the other, using (2) we can define $\mathcal{W}_n$ directly as a function on T^n by

$$\mathcal{W}_n(z_1,z_2,\ldots,z_n)=\langle\Omega|\phi(z_1)\phi(z_2)\ldots\phi(z_n)\Omega\rangle$$

$$=\langle\Omega|\phi_o\exp(-i(z_1-\bar z_2)P)\phi_o\ldots\phi_o\exp(-i(z_{n-1}-\bar z_n)P)\phi_o\Omega\rangle$$

$$(6)\qquad \equiv W_n(z_1-\bar z_2,z_2-\bar z_3,\ldots,z_{n-1}-\bar z_n),$$

which is holomorphic in T^{n-1} in the variables

$$(7)\qquad z_k-\bar z_{k+1} = (x_k-x_{k+1})-i(y_k+y_{k+1}).$$

To give the variables y a physical interpretation in the context of field theory, we need to generalize the "coherent states" e_z. Define

$$(8)\qquad e_z = \phi(z)\Omega \quad , \; z\varepsilon T.$$

If $\phi(x)$ is the free neutral scalar field, then e_z (which is then a one-particle state) turns out to be the previous e_z. In the general case, let us evaluate the expectation of P_μ in e_z. We have

$$\|e_z\|^2 = \langle\Omega|\phi_o \exp(-2yP)\phi_o\Omega\rangle, \tag{9}$$

which is invariant and hence a function only of $\lambda^2 \equiv y_\mu y^\mu$. Therefore

$$\begin{aligned}\langle P_\mu\rangle &= \|e_z\|^{-2} \langle e_z|P_\mu e_z\rangle \\ &= -\tfrac{1}{2} \|e_z\|^{-2} \frac{\partial}{\partial y^\mu} \|e_z\|^2 \\ &= -\tfrac{1}{2} \|e_z\|^{-2} \frac{\partial}{\partial\lambda^2}\|e_z\|^2 \cdot \frac{\partial\lambda^2}{\partial y^\mu} \\ &= -\frac{\partial}{\partial\lambda^2} (\ln \|e_z\|^2)\cdot y_\mu \\ &\equiv F(\lambda^2)\cdot y_\mu ,\end{aligned} \tag{10}$$

showing that $\langle P_\mu\rangle$ is proportional, by an invariant factor, to y_μ. This establishes the physical significance of y, hence of T, in the general framework of quantum field theory (for fields with spin, internal structure, etc., e_z simply acquires more indeces).

Of course, there is a price to be paid for working directly in T. The extended field is non-local in the sense that $\phi(z)$ fails to commute with $\phi(z')$ even if x and x' are spacelike with respect to one another. However, a somewhat weaker but still very reasonable condition does hold: Consider the vacuum expectation value

$$\begin{aligned}\langle\Omega|[\phi(z'),\phi(z)]\Omega\rangle &= \mathcal{W}_2(z',z)-\mathcal{W}_2(z,z') \\ &= 2i\,\mathrm{Im}\,\mathcal{W}_2(z',z) \\ &= 2i\,\mathrm{Im}\; W_2(z'-\bar{z}).\end{aligned} \tag{11}$$

By relativistic invariance, $W_2\,(z'-\bar{z})$ depends only on the variable

$$\begin{aligned}\zeta &= (z'-\bar{z})^2 \\ &= (x'-x)^2-(y'+y)^2-2i(x'-x)(y'+y)\end{aligned} \tag{12}$$

which runs over the complex plane cut along the positive real axis [10]. Suppose now that z and z' stand in the following relation, which we call

T-simultaneity for short: x and x' are simultaneous in the center-of-mass frame, i.e., in the frame in which $\vec{y} + \vec{y}' = 0$. Then the scalar $(x'-x)(y'+y)$ vanishes, hence ζ is real (and negative, since $x'-x$ is spacelike)*. Now by the Lehmann spectral representation (see [10], p. 471), W_2 is a real function of ζ in the cut plane, i.e., $\overline{W_2(\zeta)} = W_2(\bar{\zeta})$. Hence, for real ζ, the imaginary part of $W_2(\zeta)$ vanishes. It therefore follows from (12) that if z and z' are T-simultaneous, $\langle\Omega|[\phi(z'), \phi(z)]\Omega\rangle$ vanishes. This property of $\phi(z)$ is actually independent of the locality of $\phi(x)$, since the latter was not used in its derivation.

A possible advantage of the non-locality of $\phi(z)$ is that it would appear to deal directly and covariantly with extended particles: the asymptotic fields, being free, give rise to the "coherent states" e_z of references [2-5]. Thus some of the divergences plaguing the traditional theory, related to point-particles, may be avoided.

REFERENCES

1. R.F. Streater and A.S. Wightman, PCT, Spin and Statistics and All That (Benjamin, 1964).
2. G. Kaiser, Thesis, University of Toronto, 1977.
3. G. Kaiser, J. Math. Phys. 18, 952 (1977).
4. G. Kaiser, J. Math. Phys. 19, 502 (1978).
5. G. Kaiser, "Holomorphic Gauge Theory", in Geometric Methods in Mathematical Physics, G. Kaiser and J.E. Marsden, editors (Springer, 1980).
6. V. Bargmann, Commun. Pure Appl. Math. 14, 187 (1961).
7. I.E. Segal, Illinois J. Math. 14, 187 (1962).
8. J.R. Klauder, Ann. Phys. (N.Y.) 11, 123 (1960).
9. H. Meschkowski, Hilbertsche Räume mit Kernfunktion (Springer, 1962).
10. P. Roman, Introduction to Quantum Field Theory (Wiley, 1969).

*Conversely, if ζ is real, z and z' are T-simultaneous. Thus, unlike ordinary simultaneity, T-simultaneity does have an invariant meaning!

EXISTENCE ET EQUIVALENCE DE DEFORMATIONS ASSOCIATIVES
ASSOCIEES A UNE VARIETE SYMPLECTIQUE

André LICHNEROWICZ
Collège de France - Paris

Dans un programme commun avec M. Flato, D. Sternheimer, J. Vey et d'autres savants plus jeunes, nous avons étudié les propriétés et les applications des déformations de l'algèbre associative triviale et de l'algèbre de Lie de Poisson attachées à une variété symplectique. De telles déformations fournissent une nouvelle approche de la Mécanique quantique que nous avons développé ailleurs ([1],[2]) et sur laquelle je reviendrai dans ma conférence de Salamanca. La présente conférence est consacrée aux résultats récents concernant l'existence de l'équivalence des déformations associatives considérées (ou $*_\nu$-produits).

1 - Cohomologies d'Hochschild et de Chevalley.

a) Soit (W,F) une variété symplectique connexe, de dimension $2n$, classe C^∞ et 2-forme fondamentale F. Nous notons $b_k(W)$ les nombres de Betti de la variété W. Considérons l'isomorphisme de fibrés vectoriels $\mu : TW \to T^*W$ défini par $\mu(X) = - i(X)F$ (où $i(.)$ est le produit intérieur); cet isomorphisme s'étend naturellement aux tenseurs. Nous désignons par Λ (2-tenseur de structure) le 2-tenseur contravariant antisymétrique $\mu^{-1}(F)$.

L'espace $N = C^\infty(W;\mathbb{R})$ est naturellement muni de deux structures algèbriques

1) une structure d'algèbre associative donnée par le produit usuel des fonctions (qui est ici commutatif).

2) une structure d'algèbre de Lie donnée par le crochet de Poisson

$$(1\text{-}1) \qquad \{u,v\} = i(\Lambda)(du \wedge dv) = P(u,v) \qquad (u,v \in N)$$

où l'opérateur de Poisson P est un opérateur bidifférentiel, d'ordre 1 sur chaque argument, nul sur les constantes. Il est naturel d'étudier s'il est possible de déformer, en un sens convenable, ces deux lois de façon à obtenir un modèle isomorphe à la Mécanique quantique usuelle. La réponse apparaît comme positive.

b) Dérivations et déformations d'une algèbre associative procèdent d'une même cohomologie de l'algèbre à valeurs dans l'algèbre même et appelée cohomologie de Hochschild. Explicitons cette cohomologie sur le cas de l'algèbre $(N,.)$. Une p-cochaîne C est une application p-linéaire de N^p dans N, les 0-cochaînes étant identifiées aux éléments de N. Le cobord de Hochschild de la p-cochaîne C est la (p+1)-cochaîne $\tilde{\partial}C$ définie par :

$$(1\text{-}2) \qquad \tilde{\partial}C(u_o, \ldots, u_p) = u_o\, C(u_1,\ldots,u_p) - C(u_o u_1, u_2,\ldots,u_p) + C(u_o, u_1 u_2,\ldots,u_p) - \ldots + (-1)^p\, C(u_o,u_1,\ldots,u_{p-1}u_p) + (-1)^{p+1}\, C(u_o,\ldots,u_{p-1}) \cdot u_p$$

On a $\tilde{\partial}^2 = 0$ pour $p \geqslant 1$. Un 1-cocycle de (N,.) est une dérivation de cette algèbre, donc définie par un champ de vecteurs. Une p-cochaîne C est dite d-différentielle ($d \geqslant 0$) si elle est définie par un opérateur multidifférentiel d'ordre maximum d en chaque argument. Si T est une 1-cochaîne (d+1)-différentielle, $\tilde{\partial}T$ est d-différentielle. Inversement

Proposition - Si T est une 1-cochaîne de (N,.) telle que $C = \tilde{\partial}T$ soit d-différentielle ($d \geqslant 0$), T est elle-même (d+1)-différentielle. Si $\tilde{\partial}T$ est nulle sur les constantes, il en est de même pour T [5] .

Nous n'envisagerons ici que des cochaînes nulles sur les constantes. Soit $\tilde{H}^p(N;N)$ le p^{e} espace de la cohomologie de Hochschild correspondante; J. Vey [3] a établi :

Théorème (Vey). $\tilde{H}^p(N,N)$ est isomorphe à l'espace des p-tenseurs contravariants antisymétriques de W.

c) De manière symétrique, dérivations et déformations d'une algèbre de Lie procèdent d'une même cohomologie de l'algèbre de Lie à valeurs dans l'algèbre de Lie elle-même et correspondant à la représentation adjointe. Nous la nommons cohomologie de Chevalley; une p-cochaîne C de (N,P) est une application p-linéaire alternée de N^p dans N; le cobord de Chevalley de la p-cochaîne C est la (p+1)-cochaîne ∂C définie par :

$$(1\text{-}3) \qquad \partial C(u_o,\ldots, u_p) = \varepsilon^{\lambda_o\ldots\lambda_p}_{o\ldots p}\Big(\frac{1}{p!}\{u_{\lambda_o}, C(u_{\lambda_1},\ldots,u_{\lambda_p})\} - \frac{1}{2(p-1)!}\, C(\{u_{\lambda_o}, u_{\lambda_1}\}, u_{\lambda_2},\ldots,u_{\lambda_p})\Big)$$

où ε est l'indicateur antisymétrique de Kronecker et où $u_\lambda \in N$. L'espace des 1-cocycles de (N,P) est l'espace des dérivations, un 1-cocycle exact étant une dérivation intérieure. Même définition (pour $d \geqslant 1$) d'une p-cochaîne d-différentielle que dans le cas associatif. Si C est d-différentielle, ∂C l'est aussi. J'ai établi [3] .

Proposition - Si C est un 2-cocycle d-différentiel ($d \geqslant 1$) de Chevalley exact, il existe un opérateur d-différentiel T tel que $C = \partial T$.

Nous n'envisagerons encore que des cochaînes différentielles nulles sur les constantes et nous notons $H^p(N;N)$ le p^{e} espace de la cohomologie de Chevalley correspondante.

2 - Déformations formelles.

Je vais d'abord rappeler et étendre les éléments principaux de la théorie de Gerstenhaber [4] concernant les déformations des structures algèbriques, en particulier des algèbres associatives.

a) Soit $E(N;\nu)$ l'espace des fonctions formelles de $\nu \in \mathbb{C}$ à coefficients dans N; ν est dit le paramètre de déformation. Considérons une application bilinéaire $N \times N \to E(N;\nu)$ qui donne la série formelle

$$u *_\nu v = \sum_{r=0}^{\infty} \nu^r (u,v) = u.v + \sum_{r=1}^{\infty} \nu^r C_r(u,v) \qquad (u,v \in N) \tag{2-1}$$

où les C_r $(r \geq 1)$ sont des 2-cochaînes différentielles de $(N,.)$. Ces cochaînes s'étendent naturellement à $E(N;\nu)$. On a une déformation formelle de $(N,.)$ si (2-1) satisfait formellement la relation d'associativité. S'il en est ainsi, (2-1) définit sur $E(N;\nu)$ une structure d'algèbre associative formelle. Pour (2-1) arbitraire, on a pour $u,v,w \in N$

$$(u *_\nu v) *_\nu w - u *_\nu (v *_\nu w) = \sum_{t=1}^{\infty} \nu^t D_t(u,v,w) \tag{2-2}$$

où D_t est la 3-cochaîne

$$D_t(u,v,w) = \sum_{r+s=t}^{\infty} C_r(C_s(u,v),w) - \sum_{r+s=t}^{\infty} C_r(u,C_s(v,w) \qquad (r,s \geq 0) \tag{2-3}$$

Pour pouvoir raisonner par récurrence, nous posons :

$$E_t(u,v,w) = \sum_{r+s=t}^{\infty} C_r(C_s(u,v),w) - \sum_{r+s=t}^{\infty} C_r(u,C_s(v,w)) \qquad (r,s \geq 1) \tag{2-4}$$

On a l'identité

$$D_t = E_t - \tilde{\partial} C_t \tag{2-5}$$

Si (2-1) est limité à l'ordre q, on a une déformation d'ordre q si l'associativité est satisfaite à l'ordre (q+1) près. S'il en est ainsi, E_{q+1} est automatiquement un 3-cocycle de $(N,.)$. Pour qu'il existe une 2-cochaîne C_{q+1} vérifiant $D_{q+1} = E_{q+1} - \tilde{\partial} C_{q+1}$ il faut et il suffit que E_{q+1} soit exact; E_{q+1} définit une classe de cohomologie de Hochschild, élément de $\tilde{H}^3(N;N)$ qui est l'obstruction à l'ordre (q+1) à la construction d'une déformation. Une déformation d'ordre 1 est dite infinitésimale; on a $E_1 = 0$ et par suite seulement $\tilde{\partial} C_1 = 0$; C_1 est un 2-cocycle arbitraire de Hochschild.

b) Considérons une série formelle en ν

$$T_\nu = \sum_{s=0}^{\infty} \nu^s T_s = Id_N + \sum_{s=1}^{\infty} \nu^s T_s \tag{2-6}$$

où les T_s ($s \geqslant 1$) sont des opérateurs différentiels sur N; T_ν opère naturellement sur $E(N;\nu)$. Considérons une autre application bilinéaire $N \times N \to E(N;\nu)$ correspondant à la série formelle.

(2-7)
$$u \star'_\nu v = uv + \sum_{r=1}^{\infty} \nu^r C'_r(u,v)$$

où les C'_r sont encore des 2-cochaînes différentielles. Supposons que (2-6), (2-7) soient tels que l'identité suivante soit formellement satisfaite

(2-8)
$$T_\nu(u \star'_\nu v) = T_\nu u \star_\nu T_\nu v$$

En faisant usage de formules universelles, on peut prouver par récurrence la proposition suivante :

Proposition - La déformation formelle (2-1) de (N,.) étant donnée, toute série formelle (2-6) engendre une application bilinéaire unique (2-7) satisfaisant (2-8). Cette application définit une nouvelle déformation formelle qui est dite équivalente à (2-1). En particulier une déformation est dite triviale si elle est équivalente à la déformation identité ($C_r = 0$ pour tout $r \geqslant 1$)

Si deux déformations sont équivalentes à l'ordre q, il apparaît un 2-cocycle dont la classe, élément de $\tilde{H}^2(N;N)$ est l'obstruction à l'équivalence à l'ordre (q+1). En particulier deux déformations infinitésimales définies par les deux 2-cocycles C_1 et C'_1 sont équivalentes si $(C'_1 - C_1)$ est exact.

c) Soit $E(N;\lambda)$ l'espace des fonctions formelles de $\lambda \in \mathbb{C}$ à coefficients dans N. Une déformation de l'algèbre de Lie de Poisson (N,P) est définie par une application bilinéaire alternée $N \times N \to E(N;\lambda)$ donnée par :

(2-9)
$$[u,v]_\lambda = P(u,v) + \sum_{r=1}^{\infty} \lambda^r C_{2r+1}(u,v)$$

où les C_{2r+1} sont des 2-cochaînes différentielles de (N,P) telles que l'identité de Jacobi soit formellement satisfaite. La cohomologie de Chevalley joue exactement le même rôle pour les déformations d'algèbres de Lie que la cohomologie de Hochschild pour les déformations d'algèbres associatives; (2-9) définit sur $E(N;\lambda)$ une structure d'algèbre de Lie formelle.

3 - Les $\star_\nu$-produits.

a) Une 2-cochaîne C(u,v) de Hochschild est dite paire si elle est symétrique en u, v, impaire si elle est antisymétrique en u,v. Une 3-cochaîne B(u,v,w) de Hochschild est paire si elle est symétrique en u,v, impaire si elle est antisymétrique en u,w; si C est une 2-cochaîne, $\tilde{\partial} C$ est impaire pour C paire, paire pour C impaire. Soit B un 3-cocycle de Hochschild : si B est pair, il est exact et $B = \tilde{\partial} C^{(i)}$; si B est impair, $B = T + \tilde{\partial} C^{(p)}$, où $C^{(i)}$ (resp. $C^{(p)}$) est impaire (resp. paire) et où T est donné par un 3-tenseur antisymétrique. Des résultats analogues sont valables pour un 2-cocycle.

b) Sur la variété symplectique, P donné par le 2-tenseur antisymétrique Λ définit un 2-cocycle de Hochschild qui n'est jamais exact. Supposons qu'il existe sur (W,F) une déformation associative de la forme [5]

$$(3\text{-}1) \qquad u *_\nu v = u.v + \nu P(u,v) + \sum_{r=2}^{\infty} \nu^r C_r(u,v)$$

où les C_r nulles sur les constantes satisfont l'hypothèse suivante :

Hypothèse de parité. C_r est impaire pour r impair, paire pour r pair.

S'il en est ainsi, nous dirons que (3-1) définit un $*_\nu$-produit sur (W,F). Un $*_\nu$-produit engendre par antisymétrisation une algèbre de Lie formelle (2-9), où $\lambda = \nu^2$

$$[u,v]_\lambda = (2\nu)^{-1} (u *_\nu v - v *_\nu u)$$

J'ai établi [5] le théorème suivant qui est utile dans beaucoup de questions.

Théorème d'unicité - Si une algèbre de Lie formelle (2-9) est engendrée par un $*_\nu$-produit, ce $*_\nu$-produit est unique.

c) On vérifie aisément que, pour un $*_\nu$-produit, E_t est pair pour t impair, impair pour t pair. A partir de considérations de parité et en jouant simultanément sur les deux cohomologies Neroslavsky et Vlasov ont récemment établi dans un travail non encore publié [6] :

Théorème d'existence (N.V.). Sur toute variété symplectique (W,F) à $b_3(W) = 0$, il existe des $*_\nu$-produits.

4 - Les algèbres formelles de Vey.

a) Soit Γ une connexion symplectique définissant un opérateur de dérivation covariante ∇. Posons $P^0(u,v) = u.v$, $P^1 = P$ et introduisons les opérateurs bidifférentiels d'ordre maximum r en chaque argument, définis pour chaque domaine U d'une carte $\{x^i\}$ (i,j, ... = 1, ..., 2n) par l'expression suivante :

$$(4\text{-}1) \qquad P^r(u,v)\big|_U = P^r_\Gamma(u,v)\big|_U = \Lambda^{i_1 j_1} \ldots \Lambda^{i_r j_r} \nabla_{i_1 \ldots i_r} u \, \nabla_{j_1 \ldots j_r} v \qquad (u,v \in N)$$

Si Γ est plate, $\exp(\nu P)(u,v)$ définit un $*_\nu$-produit dit de Moyal.

Cette situation peut se généraliser de la manière suivante : soit Γ une connexion symplectique arbitraire; P et $P^2_\Gamma/2$ définissent toujours un $*_\nu$-produit à l'ordre 2. Pour $u \in N$, désignons par $\mathcal{L}(X_u)\Gamma$ le 3-tenseur covariant symétrique défini à partir de la dérivée de Lie de la connexion Γ par le champ hamiltonien $X_u = \mu^{-1}(du)$. La 2-cochaîne S^3_Γ donnée par

(4-2) $$S^3_\Gamma(u,v)\big|_U = \Lambda^{i_1j_1}\Lambda^{i_2j_2}\Lambda^{i_3j_3}(\mathcal{L}(X_u)\Gamma)_{i_1i_2i_3}(\mathcal{L}(X_v)\Gamma)_{j_1j_2j_3}$$

est un 2-cocycle de Chevalley ($\partial S^3_\Gamma = 0$) non exact admettant même symbole principal que P^3.

$$uv + \nu P(u,v) + (\nu^2/2!)\, P^2_\Gamma(u,v) + (\nu^3/3!)\, S^3_\Gamma(u,v)$$

définit un $*_\nu$-produit à l'ordre 3.

La 2-classe de cohomologie β de Chevalley définie par S^3_Γ est indépendante du choix de la connexion symplectique et est un invariant de la structure symplectique. On établit :

<u>Proposition - Le second espace $H^2(N;N)$ de cohomologie de Chevalley admet comme générateurs β et les classes définies par les images par μ^{-1} des 2-formes fermées de W.</u>

Ainsi $H^2(N;N)$ fait intervenir β et la cohomologie de G. de Rham de W en dimension 2.

b) Introduisons les notations suivantes : nous désignons par Q^r un opérateur bidifférentiel d'ordre maximum r en chaque argument, nul sur les constantes, satisfaisant l'hypothèse de parité et dont le symbole principal coïncide avec celui de P^r. En particulier, nous prenons $Q^o(u,v) = uv$, $Q^1 = P$. Nous sommes conduits à la définition suivante

<u>Définition - Un $*_\nu$-produit de Vey est un $*_\nu$-produit de la forme</u>

(4-3) $$u *_\nu v = \sum_{r=0}^{\infty} (\nu^r/r!)\, Q^r(u,v)$$

<u>Une algèbre de Lie de Vey est une algèbre de Lie formelle donnée par un crochet de la forme:</u>

(4-4) $$[u,v]_\lambda = \sum_{r=0}^{\infty} (\lambda^r/(2r+1)!)\, Q^{2r+1}(u,v)$$

<u>Un $*_\nu$-produit de Vey (resp. une algèbre de Lie de Vey) sont dits forts si $Q^3 \in \beta$</u> .

c) Considérons un $*_\nu$-produit de Vey à l'ordre 2; on montre qu'il existe une connexion symplectique <u>unique</u> Γ telle que :

(4-5) $$Q^2 = P^2_\Gamma + \tilde{\partial} H$$

où H est un opérateur différentiel <u>d'ordre maximum 2</u>.

Supposons que Q^3 soit un 2-cocycle de Chevalley ($\partial Q^3 = 0$); il existe de même une connexion symplectique <u>unique</u> Γ telle que :

(4-6) $$Q^3 = S^3_\Gamma + 3\,\partial H + T$$

où H est un opérateur différentiel d'ordre maximum 2 et T un 2-tenseur image d'une 2-forme fermée de W. Si Q^3 est donné , Q^2 défini, à partir de (4-6), par (4-5) est tel que :

(4-7) $uv + \nu P(u,v) + (\nu^2/2!)\, Q^2(u,v) + (\nu^3/3!)\, Q^3(u,v)$

donne un $*_\nu$-produit à l'ordre 3 et, par antisymétrisation, une déformation infinitésimale de l'algèbre de Lie de Poisson.

d) A partir d'une longue étude des types bidifférentiels, j'ai établi le lemme suivant :

Lemme - Supposons que (W,F) admette un $*_\nu$-produit de la forme

(4-8) $u *_\nu v = u.v + \nu P(u,v) + (\nu^2/2!)\, Q^2(u,v) + \ldots$

Ce $*_\nu$-produit est un $*_\nu$-produit de Vey.

On déduit de ce lemme [7] :

Théorème - Tout $*_\nu$-produit sur (W,F) est équivalent à un $*_\nu$-produit de Vey.

En effet, pour une connexion symplectique arbitraire, $C_2 - P_\Gamma^2/2$ est un 2-cocycle de Hochschild pair, donc exact : il existe un opérateur différentiel A tel que $C_2 - P_\Gamma^2/2 = \tilde{\partial}A$. En introduisant le $*_\nu$-produit déduit du $*_\nu$-produit donné à partir de $T_\nu = Id + \nu^2 A$, on obtient un $*_\nu$-produit de la forme (4-8), donc un $*_\nu$-produit de Vey.

On établit aussi [7] :

Théorème - Toute variété symplectique (W,F) telle que $b_3(W) = 0$ admet un $*_\nu$-produit de Vey fort.

En effet partons de (4-7) avec $Q^3 \in \beta$. En raffinant le raisonnement de Neroslavsky et Vlasov, on peut en déduire un $*_\nu$-produit complet de (W,F). D'après le lemme, ce $*_\nu$-produit est un $*_\nu$-produit de Vey fort.

5 - Le cas où $b_2(W) = 0$.

a) Pour une variété symplectique (W,F) telle que $b_2(W) = 0$, F est exacte et $H^2(N,N)$ admet β comme seul générateur. On établit :

Théorème - Sur une variété symplectique (W,F) à $b_2(W) = 0$, tous les $*_\nu$-produits existants sont équivalents entre eux et équivalents à un $*_\nu$-produit de Vey fort.

Considérons en effet sur (W,F) deux $*_\nu$-produits

(5-1) $$u *_\nu v = u.v + \nu P(u,v) + \sum_{r=2}^{\infty} \nu^r C_r(u,v)$$

(5-2) $$u *'_\nu v = u.v + \nu P(u,v) + \sum_{r=2}^{\infty} \nu^r C'_r(u,v)$$

Supposons maintenant que, par équivalence, on puisse faire en sorte que $C'_r = C_r$ jusqu'à un certain rang; tout 2-cocycle de Hochschild pair étant exact, on peut supposer ce rang pair. On a

$$C'_r = c_r \qquad (r = 1, \ldots, 2q)$$

où $q \geqslant 1$ et

$$C'_{2q+1} - C_{2q+1} = T$$

où T est un 2-tenseur contravariant antisymétrique vérifiant $\partial T = - [\Lambda, T] = 0$, donc exact si $b_2(W) = 0$.
Il existe ainsi un champ de vecteurs Z tel que $\mathcal{L}(Z)\Lambda = T$. Cela posé, considérons le nouveau produit, noté $*''_\nu$, déduit de (5-1) à partir de

$$\Gamma_\nu = Id_N + \nu^{2q}\mathcal{L}(Z)$$

On a, avec des notations évidentes $C''_r = C_r = C'_r$ pour $r = 1, \ldots, 2q$ et $C''_{2q+1} - C_{2q+1} = \mathcal{L}(Z)\Lambda = T$. Il en résulte $C''_{2q+1} = C'_{2q+1}$; on a ainsi établi par récurrence que tous les $*_\nu$-produits sont équivalents. D'après le théorème du §4,d, il y a équivalence avec un $*_\nu$-produit de Vey tel que, d'après (4-6), $Q^3 = S^3_\Gamma + 3\,\partial H$ pour une connexion convenable Γ. On a ainsi $Q^3 \in \beta$ et le $*_\nu$-produit de Vey est fort.

b) On a, en ce qui concerne les algèbres de Lie, les importants résultats suivants:
Lemme (S. Gutt [8]). Si $b_2(W) = 0$ les algèbres de Lie formelles

$$(5\text{-}3) \qquad [u,v]_\lambda = P(u,v) + \sum_{r=1}^{\infty} \lambda^r C_{2r+1}(u,v)$$

telles que $C_3 = Q^3/3!$ sont toutes équivalentes.
On en déduit [8] :
Proposition - Si $b_2(W) = 0$, toutes les algèbres de Lie de Vey sont équivalentes et fortes.

6 - Le théorème principal.

a) Considérons une variété symplectique arbitraire (W,F) admettant une algèbre de Lie formelle (5-3) telle que $C_3 = Q^3/3!$. Soit U un domaine contractile de W($b_2(U) = 0$) et introduisons la restriction à U de l'algèbre de Lie envisagée. D'après le lemme précédent, cette algèbre de Lie est équivalente à l'algèbre de Lie sur U (dite de Moyal) engendrée par le $*_\nu$-produit de Moyal-Vey sur U défini au §4,a. Il en résulte que notre algèbre de Lie $\{C_{2r+1}|_U\}$ sur U est engendrée par un $*_\nu$-produit $\{C_{2r(U)}, C_{2r+1}|_U\}$ sur U qui est un $*_\nu$-produit de Vey puisque $C_{2(U)} = Q^2_U/2$. On en déduit que l'algèbre de Lie formelle initiale (5-3) sur W est nécessairement une algèbre de Lie de Vey, c'est-à-dire que $C_{2r+1} = Q^{2r+1}/(2r+1)!$. On a établi
Proposition - Toute algèbre de Lie formelle (5-3) telle que $C_3 = Q^3/3!$ est nécessairement une algèbre de Lie de Vey.

b) D'après ce qui précède, $C_{2r(U)} = Q^{2r}_U/(2r)!$. Considérons deux domaines contractiles U,V de W d'intersection non vide; il résulte du théorème d'unicité que $Q^{2r}_U = Q^{2r}_V$ sur $U \cap V$. Il existe par suite des opérateurs bidifférentiels Q^{2r} sur W tels que

$Q_U^{2r} = Q^{2r}|_U$. Les Q^{2r}, Q^{2r+1} définissent sur W un $*_\nu$-produit de Vey qui engendre l'algèbre de Lie donnée.

Des considérations précédentes, on déduit

Théorème principal - Pour toute variété symplectique (W,F) les quatre propriétés suivantes sont équivalentes

1) Il existe sur (W,F) un $*_\nu$-produit

2) Il existe sur (W,F) une algèbre de Lie formelle telle que le 2-cocycle C_3 soit homologue au sens de Chevalley à $Q^3/3!$.

3) Il existe sur (W,F) un $*_\nu$-produit de Vey.

4) Il existe sur (W,F) une algèbre de Lie de Vey.

Pour qu'une algèbre de Lie formelle soit engendrée par un $*_\nu$-produit, il faut et il suffit que le 2-cocycle C_3 soit homologue, au sens de Chevalley, à $Q^3/3!$.

Références.

[1] M. Flato, A. Lichnerowicz, D. Sternheimer. C.R. Acad. Sci. Paris 283 A, (1976), p. 19-24.

[2] F. Bayen, M. Flato, C. Fronsdal, A. Lichnerowicz, D. Sternheimer. Lett. in Math. Phys. 1, (1977), p 521-530; Ann. of Phys. 111, (1978), p 61-152.

[3] J. Vey. Comm. Math. Helv. 50, (1975), p 421-454.

[4] M. Gerstenhaber. Ann. of Math. 79, (1964), p 59-103.

[5] A. Lichnerowicz. C. R. Acad. Sci. 286, (1978, p. 49-53; Ann. di Matem (à paraître).

[6] M. Neroslavsky et A. Vlasov. (à paraître).

[7] A. Lichnerowicz. Existence et équivalence de * -produits; C. R. Acad. Sci. Paris (à paraître).

[8] S. Gutt. Equivalence of deformations and associated * - products. Lett. in Math. Phys. 3, (1979), p 297-310.

A NEW SYMPLECTIC STRUCTURE OF FIELD THEORY

by

Jerzy Kijowski
Institute of Mathematical Methods in Physics,
University of Warsaw, ul. Hoża 74;
00 - 682 Warszawa, Poland

Usual canonical formulations of the field theory follow the general scheme of hamiltonian mechanics. The finite dimensional phase space of positions and momenta is replaced in field theory by the infinite dimensional symplectic space of Cauchy data on a given 1-parameter family of space-like surfaces in the space-time M /see [1] /. Dynamics is a /more or less regular/ ordinary differential equation in this infinite dimensional space. This approach has led to very important technical results /see e.g. Ebin - Marsden work on hydrodynamics or Choquet-Bruhat, Fisher, Marsden results on the Cauchy problem in General Relativity/. However, the physical insight into the field theory we get this way is very poor. The most important shortcoming of this "3+1" approach is that the relativistic invariance and the locality of the theory can not be described in terms of the symplectic structure which is used within this framework. Wheather the "hamiltonian vector field" defining the dynamics is given by the differential or by the integral operator acting on the space of Cauchy data is not relevant. The latter would even have been more natural from the point of view of regularity. This is why such a "canonical formalism" is not very useful for the purposes of the quantum field theory, where the relativistic invariance and the locality are fundamental.

There is, however, another canonical /i.e. symplectic/ formulation which is based on the notion of symplectic relation. The scope of validity of this approach is very large. It applies to both statical theories /as the theory of beams and membranes or the thermostatics/ and dynamical theories /mechanics of discrete systems and continuous media, field theory in given space-time and also General Relativity, i.e. the dynamics of geometry/. This approach is local and includes both "hamiltonian" and "lagrangian" formulations as special "control modes". The relativistic invariance of the theory obtains a natural symplectic formulation within this theory. It is also extremely adequate for the purposes of quantum theories.

The theory is based on ideas of W. Tulczyjew /see [10], [11], [12]/. Also the geometric theory of calculus of variations /[2], [3], [4], [5], [7]/ was useful to discover some fundamental symplectic structures of field theory. Our approach has been presented in the Springer Lecture Notes volume /[8]/ where the complete formulation can be found. The present paper is merely the "publicity" for this approach.

<u>Definition:</u> A symplectic relation is a lagrangian /maximal isotropic/ submanifold of a symplectic manifold.

It turns out that many different physical laws can be formulated in terms of symplectic relations.

<u>Examples:</u>

1° Let $\mathcal{P}=R^6$, with coordinates (q^i,f_i) and the symplectic form $\omega = df_i \wedge dq^i$, be the space of states of an elastically suspended small body in the three-dimensional physical space. Coordinates q^i describe the position of the body and coordinates f_i describe the force acting on the body because of the suspension /e.g. system of springs/. The physical elasticity laws describing the suspension define the relation between position and force:

$$f_j = \varphi_j(q^i) \tag{1}$$

which can be treated as a definition of the 3-dimensional surface

$$\lambda = \left\{ (q^i, f_j) \in \mathcal{P} \mid f_j = \varphi_j(q^i) \right\} \subset \mathcal{P} . \tag{2}$$

The relation λ is symplectic if the symplectic form ω vanishes when restricted to λ :

$$\omega|\lambda = d\,\varphi_j(q) \wedge dq^j = \frac{1}{2}\left(\frac{\partial \varphi_j}{\partial q^i} - \frac{\partial \varphi_i}{\partial q^j} \right) dq^i \wedge dq^j = 0. \tag{3}$$

We see that symplectic relations correspond to "reciprocal" /in the sense of Onsager/ laws of elasticity:

$$\frac{\partial \varphi_j}{\partial q^i} = \frac{\partial \varphi_i}{\partial q^j} . \tag{4}$$

2° Let $\mathcal{P} = R^4_+$ with coordinates (V,p,T,S) /all of them being positive/ and let the symplectic structure of $\mathcal{P}$ be given by the form

$$\omega = dV \wedge dp + dT \wedge dS. \tag{5}$$

Coordinates describe the volume, the pressure, the temperature and the entropy of the simple thermodynamical body. There are two "state equations" /i.e. equations of isotherms and adiabates/. Therefore, the subspace λ of states which are accessible for a given body is 4-2 = 2-dimensional. It is lagrangian because of the reciprocity usually expressed in terms of Maxwell equations :

$$\left(\frac{\partial S}{\partial V}\right)_T = \left(\frac{\partial p}{\partial T}\right)_V . \tag{6}$$

3° Let $\mathcal{P}^{t_2,t_1} = R^6 \times R^6$ with coordinates $(\underset{(1)}{q}{}^i, \underset{(1)}{p}{}_i, \underset{(2)}{q}{}^i, \underset{(2)}{p}{}_i)$ be a space of complete boundary positions and momenta /i.e. initial and final data/ of a classical particle moving in a given force field during the time interval $[t_1, t_2]$. Physical laws governing the system imply the existence of the resolvent mapping

$$R_{t_2, t_1} : R^6 \longrightarrow R^6 \tag{7}$$

from the space of initial data to the space of final data. Therefore the subspace λ of those points of $\mathcal{P}^{t_2, t_1}$ which are allowed by the dynamics of the particle is a relation:

$$\lambda = \left\{ (\underset{(1)}{q}^i, \underset{(1)}{p}_i, \underset{(2)}{q}^i, \underset{(2)}{p}_i) \in \mathcal{P}^{t_2, t_1} \middle| \, (\underset{(2)}{q}^i, \underset{(2)}{p}_i) = R_{t_2, t_1}(\underset{(1)}{q}^i, \underset{(1)}{p}_i) \right\} \subset \mathcal{P}^{t_2, t_1} . \tag{8}$$

The space $\mathcal{P}^{t_2, t_1}$ has a natural symplectic structure

$$\omega^{t_2, t_1} = \widetilde{\omega}_{t_2} - \widetilde{\omega}_{t_1} \tag{9}$$

where ω_{t_k} is the standard symplectic structure of the space of Cauchy data $\mathcal{P}_{t_k}$ at the time t_k:

$$\omega_{t_k} = d\underset{(k)}{p}_i \wedge d\underset{(k)}{q}^i \tag{10}$$

and $\widetilde{\omega}_{t_k}$ is its lift to $\mathcal{P}^{t_2, t_1}$:

$$\widetilde{\omega}_{t_k} = pr_k^* \, \omega_{t_k} \tag{11}$$

/pr_k is the projection of the Cartesian product $\mathcal{P}^{t_2, t_1}$ to its k'th component/. The minus sign in the formula (9) is of the homological character. The right-hand side is the integral of the "function"

$$t \longrightarrow \omega_t$$

over the zero-dimensional boundary $\partial[t_1, t_2]$ of the time interval $[t_1, t_2]$. Therefore, the point t_1 enters with the negative orientation. /In the field theory the time interval will be replaced by the volume $V \subset M$ and the zero-dimensional integration will be replaced by the actual integration over the boundary ∂V of V/. The relation λ is symplectic if the form

$$\omega|\lambda = pr_1^{*}\left(R^{*}_{t_2,t_1}\omega_{t_2} - \omega_{t_1}\right) \tag{12}$$

vanishes. This happens if and only if

$$R^{*}_{t_2,t_1}\omega_{t_2} = \omega_{t_1} \tag{13}$$

i.e. when the resolvent mapping is a symplectomorphism /a canonical transformation/.

4° In the previous example we put $t_1=t$, $t_2=t+\Delta t$. We change variables putting $\underset{(1)}{q}=q$, $\underset{(1)}{p}=p$, $\underset{(2)}{q}=q+\Delta t\cdot\dot{q}$, $\underset{(2)}{p}=p+\Delta t\cdot\dot{p}$. In coordinates $(q^i,p_i,\dot{q}^i,\dot{p}_i)$ the form $\omega^{t_2t_1}$ reads:

$$\omega^{t_2t_1} = \Delta t\left(dp_i\wedge d\dot{q}^i + d\dot{p}_i\wedge dq^i\right) + (\Delta t)^2\left(d\dot{p}_i\wedge d\dot{q}^i\right). \tag{14}$$

Now we "pass to the limit" with $\Delta t \longrightarrow 0$. This means that we replace the space of boundary data by the space $\mathcal{P}^I_t$ of first jets of histories of our particle. Jets are parametrized by positions, momenta and their first derivatives. The limiting procedure

$$\frac{1}{\Delta t}\omega^{t+\Delta t,t} = \frac{1}{\Delta t}\left(\tilde{\omega}_{t+\Delta t} - \tilde{\omega}_t\right) \xrightarrow[t \longrightarrow 0]{} \frac{d}{dt}\omega_t = \omega^I_t \tag{15}$$

can be given a precise geometrical meaning. In terms of coordinates

$$\omega^I_t = dp_i\wedge d\dot{q}^i + d\dot{p}_i\wedge dq^i . \tag{16}$$

The symplectic space $\left(\mathcal{P}^I_t,\omega^I_t\right)$ is called the infinitesimal phase space at the time t. The dynamics of the system is given by the vector field /possibly time-dependent/:

$$X = u^i\frac{\partial}{\partial q^i} + v_j\frac{\partial}{\partial p_j} \tag{17}$$

on each $\mathcal{P}_t$. Such a vector field is a relation in $\mathcal{P}^I_t$:

$$\lambda_t = \left\{ (q^i, p_i, \dot{q}^i, \dot{p}_i) \,\middle|\, \dot{q}^i = u^i(q,p),\ \dot{p}_i = v_i(q,p) \right\} \subset \mathcal{P}^I_t. \tag{18}$$

The relation λ_t is symplectic if the following form vanishes

$$\begin{aligned} \omega^I_t | \lambda_t &= dp_i \wedge \left(\frac{\partial u^i}{\partial q^j} dq^j + \frac{\partial u^i}{\partial p_j} dp_j \right) + \left(\frac{\partial v_j}{\partial q^i} dq^i + \frac{\partial v_j}{\partial p_i} dp_i \right) \wedge dq^j = \\ &= \left(\frac{\partial u^i}{\partial q^j} - \frac{\partial v_j}{\partial p_i} \right) dp_i \wedge dq^j + \frac{\partial u^i}{\partial p_j} dp_i \wedge dp_j + \frac{\partial v_j}{\partial q^i} dq^i \wedge dq^j = 0. \end{aligned} \tag{19}$$

This is equivalent to X being a hamiltonian vector field.

5° Let $\mathcal{P} = R^1 \times R^3 \times R^3$ be a fibre bundle over the last factor R^3 which is the physical space with coordinates (x^k), k=1,2,3. Let $\mathcal{P}_x = R^1 \times R^3$ be a fibre over $x \in R^3$. The coordinates (φ, p^k) in $\mathcal{P}_x$ describe the electrostatic potential and the electrostatic induction vector. The space $\mathcal{P}_x$ has no natural symplectic structure /there are 3 "momenta" p^k and only one "position" φ/. However, we can assign the following 2-form on $\mathcal{P}_x$:

$$\omega_x(\underline{n}) = (dp^k \wedge d\varphi)\, n_k \tag{20}$$

to each "hypersurface element", i.e. a two vector $\underline{n}$ attached at x:

$$\underline{n} = n_k dx^k \,\lrcorner \left(\frac{\partial}{\partial x^1} \wedge \frac{\partial}{\partial x^2} \wedge \frac{\partial}{\partial x^3} \right). \tag{21}$$

The form $\omega_x(\underline{n})$ is degenerate since it depends on the flux $\overset{n}{p} = p(\underline{n}) = p^k n_k$ of the induction p through the surface element $\underline{n}$ and not on the whole induction p. We may reduce $\mathcal{P}_x$ with respect to this degeneration. The resulting quotient space $\mathcal{P}_x(\underline{n})$ with coordinates $(\varphi, \overset{n}{p})$ is symplectic:

$$\omega_x(\underline{n}) = d\overset{n}{p} \wedge d\varphi\ . \tag{22}$$

This is the natural analogue of the space of initial data in particle mechanics. Similarily, the physical laws governing the electrostatical field are expressed in terms of symplectic relations. However, the relation is not a two-component relation /between initial and final data/ but a multi-component relation between states of the field in different points of the boundary ∂V of a dielectric body V which we describe. Therefore the 0-dimensional integral (9) has to be replaced by the direct integral

$$\mathcal{P}^{\partial V} = \int_{\partial V} \mathcal{P}_x \, (d\sigma) \tag{23}$$

where $d\sigma$ is a "surface element" of ∂V. More precisely: $\mathcal{P}^{\partial V}$ is the space of pairs $(\phi, P^{\partial V})$ of functions defined on ∂V. The functions describe the boundary value of the potential and the electrostatic induction flux through ∂V /i.e. the normal component of p/. The canonical 2-form

$$\omega^{\partial V} = \int_{\partial V} \omega_x(d\sigma) = \int_{\partial V} dp^{\underline{n}}(x) \wedge d\varphi(x) \tag{24}$$

is non-degenerate. Wheather it is strong or weak symplectic structure depends on the choice of the regularity class of functions ϕ and $P^{\partial V}$ /see [1], [8] /. The equations of electrostatics are elliptic:

$$\begin{aligned} p^k &= -\varepsilon^{kl}\partial_l\varphi \\ \partial_k p^k &= -4\pi\varrho \end{aligned} \tag{25}$$

where ε^{kl} is the dielectric tensor density and ϱ is a fixed distribution of the electric charge within the dielectric body V. This implies that the Dirichlet data ϕ and the Neumann data $P^{\partial V}$ are not independent. Given a Dirichlet data ϕ we can find the configuration φ in the whole V and consequently calculate P^V. The subspace $\lambda \subset \mathcal{P}^{\partial V}$

composed of those pairs which are related by the above procedure is the symplectic relation because of Green's formula /see [8]/.

6^0 For very small /infinitesimal/ body ΔV the Dirichlet data on ΔV can be approximated by the first jet of the potential, i.e. by the values (φ, φ_k) where

$$\varphi_k = \partial_k \varphi(x). \tag{26}$$

Similarily we could expect that the first jet $(p^k, p^k{}_l)$, where

$$p^k{}_l = \partial_l p^k \tag{27}$$

would be necessary to describe the Neumann data for the infinitesimal body. However, the formula (24) rewritten with use of Stokes theorem reads:

$$\begin{aligned} \omega^{\partial V} &= \int_{\partial V} (dp^k \wedge d\varphi)\, d\sigma_k = \int_V \partial_k (dp^k \wedge d\varphi)\, dV \\ &= \int_V (dp^k{}_k \wedge d\varphi + dp^k \wedge d\varphi_k) dV \approx (d\varkappa \wedge d\varphi + dp^k \wedge d\varphi_k)\cdot \Delta V \end{aligned} \tag{28}$$

where

$$\varkappa = p^k{}_k = \partial_k p^k. \tag{29}$$

This shows that only the divergence $\varkappa$ not all the derivatives $p^k{}_l$ are relevant. More precisely, the "scalar-density-valued" symplectic form

$$\omega^I_x = (d\varkappa \wedge d\varphi + dp^k \wedge d\varphi_k) \otimes (dx^1 \wedge dx^2 \wedge dx^3) \tag{30}$$

in the space $J^1_x \mathcal{P}$ of first jets of sections of $\mathcal{P}$ at the point x may be defined. This form is degenerate. Reducing $J^1_x \mathcal{P}$ with respect to this degeneracy we obtain the quotient space $\mathcal{P}^I_x$ which we call the infinitesimal phase space at x. The coordinates in $\mathcal{P}^I_x$ are $(\varphi, p^k, \varphi_k, \varkappa)$. The field equations (25) can be treated as a relation

$$\lambda_x = \left\{ (\varphi, p^k, \varphi_k, \varkappa) \;\middle|\; p^k = -\mathcal{E}^{kl}\varphi_l, \quad \varkappa = -4\pi\varrho \right\} \subset \mathcal{P}^I_x.$$

The relation is symplectic because of the symmetry of the dielectric tensor $\mathcal{E}^{kl}$.

7° Let Q be any fibre bundle over space-time M. We will use coordinate systems (φ^A, x^μ) in Q compatible with the fibration. This means that (x^μ) is a coordinate system in M and (φ^A) is a coordinate systems in fibres Q_x defined by equation x^μ = const. Given a fibre Q_x we define a space $\mathcal{P}_x$ of "vector-density-valued" covectors on Q_x. Every element $p \in \mathcal{P}_x$ is a combination of elements e^A_μ where

$$e^A_\mu = d\varphi^A \otimes \left[\frac{\partial}{\partial x^\mu} \rfloor (dx^0 \wedge .. \wedge dx^3)\right] . \tag{31}$$

The expansion coefficients

$$p = p^\mu_A e^A_\mu \tag{32}$$

of the field momentum p together with the values of field potentials φ^A form a coordinate system (φ^A, p^μ_A) in the space $\mathcal{P}_x$. We call $\mathcal{P}_x$ the phase space of the field at the point $x \in M$. This is not a symplectic space /there are more momenta p^μ_A than "positions" φ^A/. Similarily as in the example 5° we assign a symplectic space $\mathcal{P}_x(\underline{n})$ to each hypersurface-element $\underline{n}$ at $x \in M$. The symplectic form is equal to

$$\omega_x(\underline{n}) = (dp^\mu_A \wedge d\varphi^A)\, n_\mu = d\overset{n}{p}_A \wedge d\varphi^A \tag{33}$$

where n_μ are coordinates of $\underline{n}$ /cf. formula (21)/ and $\overset{n}{p}_A = p^\mu_A n_\mu$. Also to each volume $V \subset M$ we assign a symplectic space of complete boundary data

$$\mathcal{P}^{\partial V} = \int_{\partial V} \mathcal{P}_x(d\sigma) \tag{34}$$

with the symplectic form

$$\omega^{\partial V} = \int_{\partial V} \omega_x(d\sigma) = \int_{\partial V} dp_A^{\overline{n}}(x) \wedge d\varphi(x) . \tag{35}$$

A typical example of this situation is $V = [t_1, t_2] \times \Sigma$ where Σ is a 3-dimensional volume. Elements of $\mathcal{P}^{\partial V}$ describe thus initial Cauchy data at t_1, final Cauchy data at t_2 and the boundary data /incoming and outgoing radiation/. Field equations define a relation between those 3 components. For physically important field theories this relation is symplectic.

8^0 The infinitesimal version of this theory can be obtained when V is shrinked to a point. Using Stokes theorem we write

$$\begin{aligned} \omega^{\partial V} &= \int_V \partial_\mu (dp_A^\mu \wedge d\varphi^A) dV = \int_V (dp_{A\mu}^\mu \wedge d\varphi^A + dp_A^\mu \wedge d\varphi_\mu^A) \, dV \\ &\approx (dj_A \wedge d\varphi^A + dp_A^\mu \, d\varphi_\mu^A) \cdot \Delta V \end{aligned} \tag{36}$$

where

$$\begin{aligned} \varphi_\mu^A &= \partial_\mu \varphi^A \\ j_A &= p_{A\mu}^\mu = \partial_\mu p_A^\mu \end{aligned} \tag{37}$$

are components of the first jet of the section of the phase bundle $\mathcal{P} = \bigcup_x \mathcal{P}_x$. Similarily as in the example 6^0 we define "the infinitesimal phase space" $\mathcal{P}_x^I$ with the "scalar-density-valued" symplectic form

$$\omega_x^I = (dj_A \wedge d\varphi^A + dp_A^\mu \, d\varphi_\mu^A) \otimes (dx^0 \wedge \ldots \wedge dx^3) . \tag{38}$$

The field equations define a relation in $\mathcal{P}_x^I$. In the linear case the relation is symplectic if equations are self-adjoint. In the general case this may be taken as a definition of a self-adjointness.

Different approaches to the statics and dynamics of various physical systems /such as variational formulation, the hamiltonian formulation/ can be embedded into above general framework through the notion of a control mode.

Definition: By a /local/ control mode in the symplectic space $(\mathcal{P}, \omega)$ we call the /local/ symplectomorphism of $\mathcal{P}$ on a cotangent bundle:

$$\alpha : \mathcal{P} \longrightarrow T^*Q \tag{39}$$

where Q is any differential manifold.

A coordinate system (q^i) in Q induces a coordinate system (q^i, p_i) in $\mathcal{P}$, where p_i are expansion coefficients of covectors over Q with respect to the basis dq^i. We call coordinates q^i "the control parameters" and p_i "the response parameters". The choice of a control mode is equivalent to the choice of a 1-form Θ in $\mathcal{P}$ equal to the pull--back of the canonical 1-form

$$\vartheta = p_i dq^i \tag{40}$$

in T^*Q. Thus $\Theta = \alpha^* \vartheta$. The form Θ is a primary form for ω:

$$d\Theta = \omega \, . \tag{41}$$

Intuitively, the choice of a control mode consists in dividing the space $\mathcal{P}$ into control and response parameters.

Both variational and canonical formulations of dynamics are connected with special control modes. It is important to select those properties of the dynamics which are absolute and those which do depend on the control mode.

Given a control mode and a symplectic relation λ we have

$$d(\Theta|\lambda) = d\Theta|\lambda = \omega|\lambda = 0. \tag{42}$$

There exists /at least locally/ a function $\underline{S}$ defined on λ such that

$d\underline{S} = \Theta|\lambda$. The projection S of $\underline{S}$ to the space of control parameters is called a generating function of λ.

Examples:

Ad 2° The choice of V and S as control parameters and p together with T as response parameters in thermostatics gives rise to a 1-form

$$\Theta^U = -pdV + TdS. \qquad (43)$$

The generating function U of λ in this control mode is called the internal energy. It is defined by the equation $d\underline{S} = \Theta|\lambda$ which reads:

$$dU = -pdV + TdS. \qquad (44)$$

Other control modes define the entalphy H, the free energy F and the Gibbs function G as generating functions of the same state equations according to formulae:

$$\begin{aligned} dH &= Vdp + TdS \\ dF &= -pdV - SdT \qquad (45) \\ dG &= Vdp - SdT. \end{aligned}$$

Ad 3° Choosing boundary positions $(\underset{(1)}{q},\underset{(2)}{q})$ as control parameters and boundary momenta $(\underset{(1)}{p},\underset{(2)}{p})$ as response parameters we obtain a 1-form

$$\Theta = \underset{(2)}{p}_i d\underset{(2)}{q}^i - \underset{(1)}{p}_i d\underset{(1)}{q}^i . \qquad (46)$$

The generating function W of the dynamics in this control mode is the action defined by equations:

$$dW(\underset{(1)}{q},\underset{(2)}{q}) = \underset{(2)}{p}_i d\underset{(2)}{q}^i - \underset{(1)}{p}_i d\underset{(1)}{q}^i . \qquad (47)$$

Ad 4° Choosing position q and velocity $\dot{q}$ as control parameters and momentum p together with its derivative $\dot{p}$ as response parameters we obtain the "lagrangian control mode" with the 1-form

$$\Theta^L = p_i d\dot{q}^i + \dot{p}_i dq^i . \tag{48}$$

The generating function of the dynamics in this control mode is the Lagrangian L defined by equations

$$dL(q,\dot{q}) = p_i d\dot{q}^i + \dot{p}_i dq^i . \tag{49}$$

Choosing the Cauchy data (q^i, p_i) as control parameters and their derivatives as response parameters we obtain "the hamiltonian control mode" with the 1-form

$$\Theta^H = -\dot{q}^i dp_i + \dot{p}_i dq^i . \tag{50}$$

The corresponding generating function is equal to minus Hamiltonian:

$$d\big(-H(q,p)\big) = -\dot{q}^i dp_i + \dot{p}_i dq^i . \tag{51}$$

Ad 5° Choosing Dirichlet data $\bar{\phi}$ as control parameters and Neumann data as response parameters we obtain the "Dirichlet control mode" with the total energy of the electrostatic field as a generating function.

Ad 8° Choosing "field potentials" φ^A together with field strengths φ^A_μ as control parameters and field momenta p^μ_A together with "currents" j_A as response parameters we obtain the Lagrange control mode with 1-form

$$\Theta^L = \left(j_A d\varphi^A + p^\mu_A d\varphi^A_\mu\right) \otimes \left(dx^0 \wedge \ldots \wedge dx^3\right). \tag{52}$$

The corresponding generating function is the Lagrangian $\mathcal{L}$ which is a scalar density in space-time. Putting

$$\mathcal{L} = L dx^0 \wedge \ldots \wedge dx^3 \tag{53}$$

for given coordinate system (x^μ) in M we obtain

$$dL(\varphi^A, \varphi^A_\mu) = j_A d\varphi^A + p^\mu_A d\varphi^A_\mu . \qquad (54)$$

This is the Euler - Lagrange version of field equations:

$$p^\mu_A = \frac{\partial L}{\partial \varphi^A_\mu} , \qquad j_A = \partial_\mu p^\mu_A = \frac{\partial L}{\partial \varphi^A} . \qquad (55)$$

Choosing field potentials φ^A together with momenta p^μ_A as control parameters and "field strengths" φ^A_μ together with "currents" j_A as response parameters we obtain the control mode

$$\Theta^H = (j_A d\varphi^A - \varphi^A_\mu dp^\mu_A) \otimes (dx^0 \wedge \ldots \wedge dx^3)$$

which is extensively used in geometric theory of calculus of variations. This approach is closely related to the Poincaré - Cartan form /see [2], [3], [4], [5], [7], [8]/. The generating function /with negative sign/ is called there the generalized Hamiltonian but this is not the physical energy of the system. The latter is a generating function with respect to the control mode which we call "the Cauchy control mode" or "the evolution control mode". In order to define this control mode we have to choose a 3+1 decomposition of space-time. This is equivalent to the choice of a Cauchy hypersurface $\Sigma \subset M$ and a vector field X transversal to Σ. We choose Cauchy data on Σ as control parameters and their time derivatives as response parameters, where by time we mean a parameter of a 1-parameter family of diffeomorphisms generated by X. For the sake of notation convenience we can choose in M such coordinate system (x^μ) that $\Sigma = \{(x^\mu) \mid x^0 = \text{const.}\}$, $X = \frac{\partial}{\partial x^0}$. Now control parameters are $(\varphi^A, \varphi^A_k, p^0_A)$, where k=1,2,3 and response parameters are $(j_A, p^k_A, \varphi^A_0)$. The corresponding generating function is the energy /or momentum/ connected with the translation of the field along X. All the formulae are general covariant with respect to the choice of coordinates and therefore the energy has also the invariant, coordinate independent meanning /see [8] /.

The above short presentation of a new symplectic structure of field theories has only as a goal to show how natural and widely applicable is the notion of symplectic relation. We see that different theories can be obtained as special control modes within this framework. For new results obtained in this way see [8], [9], [6].

References

[1] Chernoff P.R., Marsden J.E. : Properties of infinite dimensional hamiltonian systems ; Springer Lecture Notes in Math., vol. 425

[2] Dedecker P. : Calcul des variations, formes differentielles et champs geodesiques ; In : Coll. internat. du C.N.R.S. Strassbourg 1953

[3] Dedecker P. : On the generalization of symplectic geometry to multiple integrals in the calculus of variations ; In : Diff. Geom. Methods in Math. Phys., Springer Lecture Notes in Math., vol. 570

[4] Kijowski J. : Comm. Math. Phys. 30 /1973/ p. 99

[5] Kijowski J. : Bull. Acad. Polon. Sci /math., astr., phys./ 22 /1974/ p. 1219

[6] Kijowski J. : G.R.G. Journal, 9 /1978/ p. 857

[7] Kijowski J., Szczyrba W. : Comm. Math. Phys. 46 /1976/ p. 183

[8] Kijowski J., Tulczyjew W. M. : A symplectic framework for field theories ; Springer Lecture Notes in Physics, vol. 107

[9] Kijowski J.,Pawlik B., Tulczyjew W.M. : A variational formulation of non-gravitating and gravitating hydrodynamics ; Bull. Acad. Polon. Sci. /in print/

[10] Tulczyjew W. M. : Seminar on Phase Spaces, Warsaw 1968 /unpublished/

[11] Tulczyjew W. M. : Symposia Math. 14 /1974/ p. 247

[12] Tulczyjew W. M. : Ann. Inst. H. Poincare 27 A /1977/ p. 101

CONFORMAL STRUCTURES AND CONNECTIONS

A.Z. Jadczyk
Institute of Theoretical Physics, University
of Wrocław, Cybulskiego 36, 50-205 Wroclaw, Poland

1. INTRODUCTION The primary aim of the present author was to understand "conformal invariant wave equation"

$$(\Box + \frac{R}{6})\varphi = \lambda\, \varphi^3 \qquad (0)$$

This equation appears in [1] without given any explicite reason as to "why" it is invariant. A conformal structure is best described by a tensor density $\gamma_{\mu\nu}$ [2]. If φ is a scalar density of dimension -1, then, with $g_{\mu\nu} = \varphi^2 \gamma_{\mu\nu}$, the equation (0) is nothing but $R(g) \sim \lambda$. Therefore its conformal invariance is obvious and "improved energy - momentum tensor" of [3] is automatic [4]. This observation can not be taken as satisfactory enough. One would like to have a machine /like a "covariant derivative" in a metric case/ which generates conformal invariant field equations. Long ago Cartan [5] considered connections more general than principal ones. A theory of conformal connections has been then developed to a highly sofisticated degree [6,7], and more recently it was shown, to be equivalent to a theory of twistor connections [8,9]. We prefer to work with O(4,2) vectors rather than spinors. The convenient mathematical apparatus is that of second order frames [7,10,11]. We give an interpretation of the bundle of O(4,2) -vectors in terms of jets of scalar densities and write down empty space field equations of gravitation in terms of the conformal connection. The theory is similar to one in [12]. More details can be found in [13].

2. CONFORMAL STRUCTURE OF SPACE - TIME Let M be a smooth, 4-dimensional manifold, thought of as being a set of space-time events. Let $B(M)$ be the bundle of linear frames over M. Then $B(M)$ is a principal bundle with the structure group $GL(4)$.

Let G be a Lie subgroup of $GL(4)$. A G-structure on M is a smooth subbundle of $B(M)$ with G as a structure group. In many interesting cases G can be identified as the stabilizer of some tensorial object on R^4. For example, to give M a pseudo - Riemannian structure is to give it an $O(1,3)$ -structure and $O(1,3)$ is the stabilizer of the standard metric tensor $\eta = (\eta_{ab}) = \mathrm{diag}(-1,1,1,1)$. Similarly, to give M a conformal structure is to give it CO 1,3 -structure, and $CO_+(1,3)$ is the stabilizer of /pseudo-/ tensor

$$\overset{\circ}{\chi}{}^{ab}_{cd} = \frac{1}{2}\,\epsilon_{cdef}\,\eta^{ea}\,\eta^{fb} \tag{1}$$

Let P be a conformal structure on M. The frames in P are called conformal frames /of the first order/. Take any coordinate system x^μ around $p \in M$, and let (e^μ_a) be a conformal frame over p. Then the formula

$$\chi^{\alpha\beta}_{\gamma\delta} = e^\alpha_a\, e^\beta_b\, e^c_\gamma\, e^d_\delta\, \chi^{ab}_{cd} \tag{2}$$

defines a /pseudo-/ tensor χ at p, which is independent of (e^μ_a). According to a general theorem /see [7]/, P is integrable /flat/ iff each point $p \in M$ admits a coordinate neighbourhood, with local coordinates x^μ, with respect to which the components of χ coincide with the standard ones (1). The tensor (2) is nothing but a Hodge - $*$ - operator restricted to 2-forms. Therefore, modulo topological subtleties, to give M a conformal structure is to give it a smooth $*$-operator acting linearly on the bundle of 2-forms and satisfying

$$(i) \qquad *^2 = -I,$$

(ii) $\star F \wedge G = F \wedge \star G$; $F, G \in \bigwedge^2(M)$.

/see [14] /

Although χ determines conformal structure completely, it is more convenient to deal with its "square root" i.e. with a tensor density $\gamma_{\mu\nu}$ uniquely defined by

$$\text{a)} \quad \chi^{\mu\nu}_{\sigma\rho} = \frac{1}{2} \epsilon_{\sigma\rho\alpha\beta} \gamma^{\alpha\mu} \gamma^{\beta\nu} ,$$

$$\text{b)} \quad \det(\gamma_{\mu\nu}) = -1 .$$

As it was above, a conformal structure is flat iff there are local coordinate systems in which $\gamma_{\mu\nu} \equiv \eta_{\mu\nu}$. Since this condition is known to be equivalent to vanishing of the Weyl conformal curvature tensor W , it should be possible to express W in terms of only.

Assume now that a conformal structure $\gamma_{\mu\nu}$ is given, and let Φ be a scalar density of dimension 1 , i.e.

$$\Phi'(x') = \left| \frac{\partial x'}{\partial x} \right|^{1/4} \phi(x).$$

If $1 = -1$, then $g_{\mu\nu} \doteq \Phi^2 \gamma_{\mu\nu}$ is a metric tensor on M . In this way one gets a correspondence between conformal structures and classes of conformally equivalent pseudo-Riemannian metrics on M . It follows in particular, that each scalar density of dimension -1 determines a symmetric affine connection which preserves the conformal structure. However, no such an affine connection is distinguished.

3. <u>CONFORMAL FIELD EQUATIONS</u> Usually field equations are considered on a flat Minkowskian background. What is a deeper meaning of conformal invariance in such a case? The group of all automorphisms of flat causal structure is the Weyl group - semidirect product of Poincaré transformations and dilatations. Special conformal transformations are singular, and conformal inversion $x^\mu \mapsto x^\mu / x^2$ does

not preserve causal relations. If so, then why should one require full conformal invariance and not only invariance with respect to the Weyl group? To answer this question it is necessary to consider what will happen after local deformation of the flat light-cone structure. In a flat space the very difference between local and global aspects can easily be lost. The total space can be naturally identified with its tangent space at a given point. A point can be identified with intersection of two infinite lines etc. On the contrary, in a generic Riemannian space no such identifications are possible, and no automorphisms of its causal structure exist. Here conformal structure exhibits its true meaning, Rimannian metric separates into a volume element /or length scale/ Φ^4 and a causal structure $\gamma_{\mu\nu}$. Conformal invariance means that the field equations are governed only by the causal structure, and not by the length scale. In consequence energy-momentum tensor is automatically traceless, and field equations are automatically invariant with respect to all Killing vector fields X of the causal structure. In particular they remain conformally invariant when specified to the flat case.

It follows that covariance of field equations under conformal inversions $x^{\mu} \mapsto x^{\mu} / x^2$ should be considered only as a hint that the equations are stable under local deformations of the conformal /flat/ structure. And only such stable systems are of physical interest. Given a transformation law of fields under conformal inversions it is then usually possible to deduce what kind of a geometrical object one is dealing with, and to generalize field equations to a curved background. However, the situation here is not as simple as in a Riemannian case, since no straightforward recipe like "replace derivatives by the covariant ones" is possible. /In fact, even in Riemannian case one meets umbiguity of curvature terms/.

4. SECOND ORDER CONFORMAL FRAME

a) The bundle $P^2(M)$

A second order frame at $p \in M$ is characterized /with respect to a coordinate, system x^μ / by a set of numbers $(e^\mu = x^\mu(p)\, , e^\mu_a ,$ $e^\mu_{ab} = e^\mu_{ba})$ with transformation laws

$$e^{\mu'} = x^{\mu'}(p)$$

$$e^{\mu'}_a = \frac{\partial x^{\mu'}}{\partial x^\mu}(p)\; e^\mu_a$$

$$e^{\mu'}_{ab} = \frac{\partial x^{\mu'}}{\partial x^\mu}\, e^\mu_{ab} + \frac{\partial^2 x^{\mu'}}{\partial x^\alpha \partial x^\beta}(p)\; e^\alpha_a\, e^\beta_b \, .$$

The coordinates e^μ_a can be interpreted as determining linear /1-st order/ frame at p , and e^μ_{ab} can be interpreted as determining connection coefficients

$$e^\mu_{\rho\sigma} \equiv -e^r_\rho\, e^s_\sigma\, e^\mu_{rs} \quad .$$

We denote by $P^2(M)$ the bundle of second-order frames.

The $G^2(4)$ set of all second-order frames at $0 \in R^4$ is a group with the multiplication law

$$(h^a{}_b , h^a{}_{bc})\; (k^a{}_b , k^a{}_{bc}) = (h^a{}_r\, k^r{}_b , h^a{}_{rs}\, k^r{}_b\, k^s{}_c + h^a{}_r k^r{}_{bc})$$

If $h^a{}_b , h^a{}_{bc}$ are replaced by e^μ_a , e^μ_{ab} , then the last formula gives natural action of $G^2(4)$ on $P^2(M)$ which makes $P^2(M)$ a principal bundle.

b) reduction of $P^2(M)$ induced by conformal structure.

Let a conformal structure C be given on M in terms of $\gamma_{\mu\nu}$. A general symmetric affine connection which preserves C is of the form

$$\Gamma^\alpha_{\mu\nu} = \hat{\Gamma}^\alpha_{\mu\nu} + (\delta^\alpha_\mu\, p_\nu + \delta^\alpha_\nu\, p_\mu - \gamma_{\mu\nu}\gamma^{\alpha\beta} p_\beta) \qquad (3)$$

where

$$\hat{\Gamma}^\alpha_{\mu\nu} = \frac{1}{2}\gamma^{\alpha\beta} (\partial_\mu \gamma_{\nu\beta} + \partial_\nu \gamma_{\mu\beta} - \partial_\beta \gamma_{\mu\nu}) \, .$$

Therefore $P^2(M)$ can be reduced to $P^2_C(M)$ defined as consisting of second order frames e such that e^{μ}_{a} are "orthonormal" conformal frames, and $e^{\mu}_{\sigma\rho}$ are coefficients of connections (3).

The structure group of $P^2_C(M)$ is $G = CO(1,3) \times R^{4*}$ with the multiplication law

$$(h^a{}_b, v_a)(k^a{}_b, w_a) = (h^a{}_r k^r{}_b, v_r k^r{}_a + w_a)$$

identified with a subgroup of $G^2(4)$ by

$$(h^a{}_b, v_a) \mapsto (h^a{}_b, h^a{}_{bc})$$

where

$$h^a{}_{bc} = h^a{}_r \left(\delta^r_b v_c + \delta^r_c v_b - \eta_{bc} \eta^{rs} v_s \right).$$

It is evident that, conversely, each reduction of $P^2(M)$ to G determines a conformal structure C. There is, however, a distinguished class of reductions, namely those of the form $P^2_c(M)$.

c) bundle of jets of scalar densities.

Let ϕ be a scalar density of dimension l.

The first jet of ϕ at p is parametrized by $1+4$ parameters:

$$\Phi = \Phi(p)$$
$$z_\mu = \partial_\mu \Phi(p)$$

If e is a conformal frame in P^2_c, then one defines coordinates of (ϕ, z) with respect to e by

$$\varphi(e) \doteq e^{1/4} \Phi \quad , \quad e = |e^{\mu}_{a}|$$

$$z^a(e) \doteq \eta^{ar} e^{\mu}_{r} e^{1/4} \left(z_\mu + \frac{1}{4} \Phi \Gamma_\mu \right)$$

where $\Gamma_\mu = e^{\sigma}_{\mu\sigma}$. From the transformation character of $(\varphi(e), z(e))$ one then finds that the bundle of jets of scalar densities of dimension l can be considered as an associated bundle $P^2_{c,l}$ of $P^2_c(M)$ corresponding to the following representation D_l of G on R^{1+4}:

$$D_1(\Lambda,\theta) = \theta^{(1-1)}\begin{pmatrix} \theta & , & 0 \\ 1\theta & , & \Lambda \end{pmatrix}$$

where $h^a{}_b = \theta\Lambda^a{}_b$, and $\Lambda\eta\Lambda^T = \eta$.

This representation admits no invariant bilinear form. If, however, a new parameter ψ is introduced, then the following quadratic form:

$$Z^2 = \begin{pmatrix}\varphi \\ z \\ \psi\end{pmatrix}^2 = z^2 - 2\varphi\psi$$

is scale invariant, provided D_1 is prolonged to $\tilde{D}_1$ given by

$$\tilde{D}_1 = \begin{pmatrix} & D_l & , 0 \\ & & , 0 \\ \frac{1}{2}^2\theta\, v^2 , & 1\, v\eta\Lambda , & \theta^{-1} \end{pmatrix}$$

The representation $\tilde{D}_1$ preserves the scalar product given by the matrix

$$S = \begin{pmatrix} 0 , & 0 , & -1 \\ 0 , & \eta , & 0 \\ -1 , & 0 , & 0 \end{pmatrix}$$

up to a factor:

$$D_1{}^T S D_1 = \theta^{2(1-1)} S .$$

In particular, for $1=1$, the quadratic form $Z^T S Z$ is invariant.

5. <u>CONFORMAL CONNECTIONS</u> The above prolongation $\tilde{P}^2_{c,1}$ of $P^2_{c,1}$ carries natural invariant bilinear form (Z,Z') with $O(2,4)$ as an invariance group (precisely speaking the stability group of S is <u>isomorphic</u> to $O(4,2)$). It carries also a natural one-dimensional isotropic line subbundle I defined by $\varphi = 0$, $z = 0$.

A <u>natural</u> linear connection in $\tilde{P}^2_{c,1}$ should preserve both. No such connection exists. There exists connection which preserve scalar product /but no I / . It is given by

$$\nabla_\mu Z = \partial_\mu Z + \Gamma_\mu Z$$

where

$$\Gamma_\mu = t_\mu D + \frac{1}{2}\omega_\mu{}^r{}_s M^s{}_r + v_{\mu r} K^r + w_\mu^r + P_r =$$

$$= \begin{pmatrix} t_\mu , & w_{\mu a} , & 0 \\ v_\mu{}^a , & \omega_\mu{}^a{}_b , & w_\mu{}^a \\ 0 , & v_{\mu a} & , -t_\mu \end{pmatrix}$$

and

$$\begin{aligned} w_\mu{}^a &= - e_\mu^a \\ \omega_\mu{}^a{}_b &= e^a{}_\sigma \nabla_\mu e_b^\sigma \\ v_{\mu a} &= \frac{1}{2}\left(R_{\mu\nu} - \frac{1}{6} R g_{\mu\nu}\right) e^\nu{}_a \\ t_\mu & \text{ - arbitrary one-form.} \end{aligned}$$

Here ∇_μ and $R_{\mu\nu}$ are covariant derivative and Ricci tensor of $g_{\mu\nu} = e_\mu{}^a \eta_{ab} e_\nu{}^b$. Since t_μ is a one-form, one can put $t_\mu = 0$ to get what is called "canonical normal Cartan connection".

It should be observed that this connection preserve the cone $Z^2 = 0$, and so defines a /nonlinear/ connection in a "compactified tangent bundle".

6. FIELD EQUATIONS FOR GRAVITATION. The equations $\nabla_\mu Z = 0$ are equivalent to the field equations of gravitation in empty space. In fact, $\nabla_\mu Z = 0$ implies Z^2 = const, and then $\nabla_\mu Z = 0$ is equivalent to $R(\tilde{g})$ = const, where $\tilde{g}_{\mu\nu} = \varphi^{-2} e_\mu{}^a \eta_{ab} e_\nu{}^b$ /observe that Z is always refered to section (e_a^μ) of P_c^1, which determines section of P_c^2 /. In particular Z^2 is proportional to a cosmological constant, and $R(\tilde{g})$ = const is nothing but the familar "wave equation" (0).

Equations of the above type can not be obtained from a Lagrangean. It is, however, interesting to notice that $D = K^\mu \nabla_\mu$ is an invariant operator /like the Dirac operator/, and Z satisfies $\nabla_\mu Z = 0$ iff it satisfies

i) $DZ = 0$

ii) Z^2 = const.

Therefore $Z^t D Z$ can be taken as proportional to Lagrangean density which, together with the constraint ii) gives us standard Einstein vacuum equations.

The connection Γ can be defined in every associated $O(2,4)$ bundle, in particular, it defines covariant derivative of twistors. However, in the twistorial /four-valued/ representation of $O(2,4)$ $K_\mu K_\nu \equiv 0$. Therefore the invariant operator D has no curvature terms in this case. Therefore $DZ = 0$ has a different meaning for a twistorial section Z.

REFERENCES

[1]. R. Penrose, Proceedings of the Royal Society, London, A284, 159 /1965/

[2]. F. Gürsey, Annals of Physics N.Y., 24, 211 /1963/

[3]. C.G. Callan, Jr., S. Coleman, and R. Jackiw, Ann.Phys.N.Y. 59, 1970, 42-73

[4]. A.Z. Jadczyk, Int.J. Theor.Phys., /1979/

[5]. E. Cartan, Bull.Soc.Math. France, 52, 1924, 205-241

[6]. N. Tanaka, Trans. Amer.Math.Soc. 92, 1959, 168-190

[7]. S. Kobayashi, Transformations Groups in Differential Geometry, Springer, New York 1972

[8]. H. Friedrich, Gen. Rel. Grav. 8, 1977, 303-312

[9]. K. Dighton, Int.J. Theor. Phys. 11, 1973, 31-43

[10]. J.P. Harnad, and R.B. Pettitt, Gauge Theories for Space-Time Symmetries II: Second Order Conformal Structures, Centre de Recherches Mathématique, Université de Montréal, Preprint, February 1978

[11]. J.D. Hennig, G-Structures and Space-Time Geometry - I: Geometric Objects of Higher Order, Preprint IC/78/46, ICTP, Trieste, 1978

[12]. S. Deser, Ann. Phys. N.Y. 59, 1970, 248-253

[13]. A. Jadczyk, Some Comments On Conformal Connections, Preprint, November 1978

[14]. A. Jadczyk, Bull. Pol. Acad. Sci. XXVII, no2, p. 91-94, 1979

Equilibrium Configurations of Fluids in General Relativity*

H.P. Künzle, Department of Mathematics

and

J.R. Savage, Department of Physics

University of Alberta, Edmonton
Alberta, Canada T6G 2G1

1. Introduction. This paper is a summary of investigation [11], [12] into the uniqueness problem of isolated relativistic, self-gravitating, perfect fluids in a static or stationary space-time. While the uniqueness of a spherically symmetric solution to a fixed equation of state and total mass of the static Einstein equations with suitable asymptotic conditions is considered physically evident, it has not yet been rigorously derived. In the stationary case, uniqueness for slow rotations corresponds to the classical result that the Maclaurin series of ellipsoids represents the only equilibrium configurations of such a fluid with constant density [14].

While the global analysis techniques we use, developed mainly by Fischer, Marsden and Cantor, do not let us show as general a result as may be hoped for, they do result in similar restricted uniqueness results for both the static and stationary (slow rotations close to spherical symmetry) cases. Our approach is an analysis of the linearized Einstein equations with a fixed equation of state $\rho = \rho(p)$ and a fixed total gravitational mass m on an arbitrary background solution over $\mathbb{R}^3$ and close to the unique spherically symmetric solution with the same $\rho(p)$ and m. We follow the spirit of Fischer and Marsden's work on lineraization stability [6],[7] in considering the solution set of Einstein's equations together with their boundary conditions as the inverse image of zero of a nonlinear differential operator $\mathcal{L}: X \to Y$ where X and Y are chosen to be isomorphic to suitable weighted Sobolev spaces as discussed in Cantor [1]-[3] and where an element of X uniquely specifies a space-time metric. (Moreover, all elements of X that are solutions of Einstein's equations have the same total mass.) The set of diffeomorphisms asymptotic

*Partially supported by the National Sciences and Engineering Research Council of Canada

to the identity at infinity operates in a natural way on X and the corresponding orbits are submanifolds of X. We would like to show that $\mathcal{L}^{-1}(0)$ is contained in the orbit through the spherically symmetric solution in the static case, and that $\mathcal{L}^{-1}(0) \cap U$ where U is some tubular neighborhood of the above orbit is contained in the orbits through a one-dimensional submanifold passing through the spherical solution and parametrized by the angular momentum in the stationary case.

We cannot prove this. However, in a neighborhood of a spherically symmetric solution $\underset{\sim}{g}$ in X, the set of equivalence classes of X with respect to the action of the diffeomorphism group is locally in one-to-one correspondence with a submanifold $\mathcal{S}$ transversal to the orbits which therefore represents the physically distinct space-time metrics. We show, that there can be no non-constant C^1-curve of solutions to Einstein's equations with suitable asymptotic conditions and fixed (small) angular momentum contained in the slice $\mathcal{S}$.

This result is analogous to the stationary black hole theorems [8], [4] and [17].

2. Space of solutions of Einstein's equations. Assume that the space-time (M,g) admits a globally time-like Killing vector field ξ and is a product manifold $R \times \Sigma$. There are local coordinates (t,x^i) such that $\xi = \partial_t$ and the Lorentz metric has the form

$$g_{00} = -e^{2U}, \quad g_{0i} = -e^{2U}\alpha_i, \quad g_{ij} = e^{-2U}\gamma_{ij} - e^{2U}\alpha_i\alpha_j \qquad (i,j = 1,2,3) \tag{2.1}$$

where γ is a positive definite Riemannian metric, U a scalar function and α a one-form on Σ. (See [10] for a rigorous development.) Einstein's equations for a perfect fluid

$$\overset{4}{R}_{\alpha\beta} - (\tfrac{1}{2})\overset{4}{R}\overset{4}{g}_{\alpha\beta} = T_{\alpha\beta} = (\rho+p)\overset{4}{u}_\alpha \overset{4}{u}_\beta + p\overset{4}{g}_{\alpha\beta} \tag{2.2}$$

then become in the 3-dimensional formalism, using γ to raise and lower indices,

$$\Delta U := \gamma^{ij}\nabla_i\partial_i U = \tfrac{1}{2}(\rho+3p)e^{-2U} + \tfrac{1}{2}e^{4U}h^2 + (\rho+p)e^{-4U}T^2\theta^2 \tag{2.3}$$

$$\nabla^r\nabla_r\alpha^i + \nabla^i\nabla_j\alpha^j - 4\partial_j U\, \nabla^{[j}\alpha^{i]} + R^{ik}\alpha_k = 2(\rho+p)e^{-6U}uT\theta^i \tag{2.4}$$

$$\overset{3}{R}_{ij} = 2\partial_i U\partial_j U + \tfrac{1}{2}e^{4U}h_i h_j + (\rho+p)e^{-4U}T^2\theta_i\theta_j - (2pe^{-2U} + (\rho+p)e^{-4U}T^2\theta^2)\gamma_{ij} \tag{2.5}$$

where $\theta^\alpha = T^{-1}u^\alpha$, $\theta^i = \theta^i$, T is the temperature, $h^i = 2\varepsilon^{ijk}\partial_{[j}\alpha_{k]}$, $h^2 = \gamma_{ij}h^ih^j$, $\upsilon = -u_0$ so that $\upsilon^2 = e^{2U} + T^2\theta^2$ and the equation of hydrodynamical equilibrium is $Tdp = (\rho+p)dT$. (M,g) is static iff $d\alpha = 0$ iff $h = 0$ iff $\theta = 0$ (by (2.4)). In the following we will therefore treat the static case as a special case of the stationary case by letting $d\alpha$ and θ vanish.

We assume space-time is asymptotically Euclidean in the sense of Lichnerowicz [13], i.e. that there exists a compact $K \subset \Sigma$ and a diffeomorphism $\varphi : \Sigma\setminus K \to \mathbb{R}^3 \setminus B$ where B is a closed ball centered at the origin so that with respect to the standard coordinate system in $\mathbb{R}^3$

$$\gamma_{ij} - \delta_{ij} \, , \; \alpha_i \, , \; U \in O(|x|^{-1}) \, , \; \partial_k\gamma_{ij} \, , \; \partial_k\alpha_i \, , \; \partial_u U \in O(|x|^{-2}) \tag{2.6}$$

where $|x|^2 = \Sigma(x^i)^2$. It is known that for a stationary vacuum space-time there is then an asymptotically cartesian coordinate system such that

$$\gamma_{ij} - \delta_{ij} \, , \; \alpha_i \in O(|x|^{-2}) \, , \; \partial_k\gamma_{ij} \, , \; \partial_k\alpha_i \in O(|x|^{-3}) \; . \tag{2.7}$$

Coordinate transformations which are asymptotically the identity will not destroy this behaviour.

Stationarity and thermodynamical equilibrium imply that θ^α is a Killing vector field inside matter, and Lindblom [15] has shown by analytic continuation that the space-time is globally axisymmetric. Although his argument cannot be readily adapted to our asymptotic conditions, we can with little loss of generality assume axisymmetry. Let $\eta^i\partial_i$ be the corresponding vectorfield on Σ where we fix the rotation axis by letting $\eta^i = \varepsilon^{3i}{}_k x^k + O(|x|^{-1})$ at infinity. Then there is a constant b such that $\theta = b\eta$.

In the following we study the set $S_{\rho(p)}$ $(\tilde{S}_{\rho(p)})$ of stationary (static) asymptotically Euclidean perfect fluid space-times with an a priori given equation of state $\rho(p)$ that is subject to the conditions (i) $p \leftrightarrow \rho(p)$ is a piece-wise C^1-function on a closed interval $I = [0,p_2]$, (ii) $\rho(p) \geq p$ on I , (iii) $0 \leq d\rho/dp < \infty$ on I and for which U has only one critical point, namely a nondegenerate minimum U_c at the centre (since we are close to the spherical solution). As shown in [11] these assumptions imply there is a compact region

$D \subset \Sigma$ with $\rho > p \geq 0$ and an exterior vacuum region with ρ dropping discontinuously from a finite value to 0 at the boundary ∂D of D defined by $p = 0$. (This is required for mathematical convenience.)

We will always consider $T_b = T(p = 0)$, fixed since there is no interest in changing the temperature by a global constant. An analysis of the spherically symmetric solutions [11] shows that the whole solution is fixed by m and $U_b = U(p=0)$ We conjecture that U_b is determined in terms of m by the condition that U_c is a critical point of U, as in fact it is for specific simple models such as the interior Schwarzschild solution. But since a proof is not simple for a general equation of state we regard U_b as an extra parameter and prove a linearized version of this statement. When regarding the pressure p, temperature T (and ρ) as functions on Σ we do so via their dependence on U and b, which is determined by $Tdp = (\rho+p)dT$ up to a constant a. We will define a through $\Theta^\alpha = a\xi^\alpha + b\eta^\alpha$ where we regard ξ^α and $\eta^\alpha = (0,\eta^i)$ as fixed by the previous argument so that $T^{-2} = e^{2U}(a^2+2u\alpha_i\theta^i+(\alpha_i\theta^i)^2) - e^{-2U}\theta^2$, i.e. $a^{-1} = T_c e^{U_c}$. In the static (or spherically symmetric) case this is equivalent to specifiying $U_b = -\ln(aT_b)$.

Although a priori we do not know how big $S_{\rho(p)}$ is or what kind of topology and differentiable structure it can be given we can regard it as the inverse image of a differentiable map on a bigger set that can be provided with a natural Banach manifold structure. There is a considerable arbitrariness in the choice of such a structure for a set of tensor fields on a noncompact manifold but physical and mathematical considerations combine to remarkably limit the choice if one restricts to the weighted Sobolev spaces $M^p_{s,\delta}$ introduced by Cantor.

<u>Definition</u>: $M^p_{s,\delta} = M^p_{s,\delta}(\mathbb{R}^n,\mathbb{R}^m)$ is the completion of $C_0^\infty(\mathbb{R}^n,\mathbb{R}^m)$ with respect to the norm

$$\|f\|_{p,s,\delta} := \sum_{|\alpha|\leq S} \|\sigma(x)^{(\delta+|\alpha|)} D^\alpha f\|_p \qquad (p>0,\ s\in\mathbb{N},\ \delta\in\mathbb{R}) \tag{2.8}$$

where $\| \ \|_p$ denotes the L^p-norm, $\sigma^2(x) := 1 + |x|^2$, $\alpha = (a_1,\ldots,a_n) \in \mathbb{N}^n$ and $|\alpha| = \Sigma\alpha_i$. For $f \in C^\infty(\mathbb{R}^n,\mathbb{R}^m)$ let $M^p_{s,\delta}(f) := \{g\colon \mathbb{R}^n \to \mathbb{R}^m \mid (g-f)\ M^p_{s,\delta}\}$.

For properties of these spaces see Cantor [1]-[3].

The asymptotic conditions (2.7) together with the Lichnerowicz [13] junction condition which state that U, γ_{ij} and α_i are C^1 and piecewise C^2 and C^3 indicate that $U + \frac{m}{|x|}$, $\gamma_{ij} - \delta_{ij}$ and α_i will be in $M^p_{s,1+\delta}$ with $0 \le \delta < 1 - 3/p$ and $s = 2$. $p > 3$ is required mathematically, e.g. for the application of Cantor's isomorphism theorem 1.4 of [3]. In order to obtain more than a C^0 action of the diffeomorphism group on the spherically symmetric $\underset{\sim}{\sigma} = (\underset{\sim}{\gamma}, \underset{\sim}{U}, 0, \underset{\sim}{\eta}, \underset{\sim}{a}, 0)$ we require $\underset{\sim}{\sigma} \in M^p_{3,1+\delta}$. By (2.5) we can choose $\underset{\sim}{\gamma}$ this way but (2.3) forces $\underset{\sim}{U}$ to be only C^1. To avoid this problem we introduce a modified $\hat{U} = U - f(\rho e^{2U})^2$, where $f = e^{2U}/(4(\underset{\sim}{\rho}+3\underset{\sim}{p})\underset{\sim}{\gamma}^{ij}\partial_i U \partial_j U)|_b = \text{const}$, which is C^2 in the spherical case to define the topology while for calculational ease we can use either U or $\hat{U}$. For the rest of the paper we fix $p > 3$, $s = 3$, $0 \le \delta < 1 - 3/p$ for the static case and $\delta = 0$ for the stationary case and define

$$\mathscr{R}^p_{s-1,\delta+1} := \{(h_{ij}, u, \beta, \zeta, A, B) \mid h_{ij} = h_{ji}, \; u, \beta_i \in M^p_{s-1,\delta+1}, \; \zeta^i \in M^p_{s-1,\delta}, \\ A, B \in \mathbb{R}, \; \gamma_{ij} + h_{ij} \text{ positive definite}\} \tag{2.9}$$

which is an open subset of a Banach space so that

$$\mathscr{P}^p_{s-1,\delta+1} := \{\sigma = (\gamma_{ij}, U, \alpha_i, \eta_i, a, b) \mid (\gamma_{ij} - \delta_{ij}, \hat{U} - \hat{\underset{\sim}{U}}, \alpha_i, \eta^i - \underset{\sim}{\eta}^i, a - \underset{\sim}{a}, b) \\ \in \mathscr{R}^p_{s-1,\delta+1}\} \tag{2.10}$$

is a Banach manifold with all elements which are also in $S_{\rho(p)}$ having the same mass. The static case $\mathscr{P}^p_{s-1,\delta+1}$ is obtained by letting $\alpha, \beta, \eta, \zeta, b, B$ vanish.

Definition: Letting $1\!\!1$ denote the indentity map on $\Sigma \simeq \mathbb{R}^3$ define the diffeomorphism group

$$\mathscr{D}^p_{s,\delta} := \{\varphi \in M^p_{s,\delta}(1\!\!1) \mid \varphi^{-1} \text{ exists and } \varphi^{-1} \in M^p_{s,\delta}(1\!\!1)\} . \tag{2.11}$$

By [6] this is an open submanifold of $M^p_{s,\delta}(1\!\!1)$ and a topological group. Let $\mathscr{G}^p_{s,\delta} = \mathscr{D}^p_{s,\delta} \times M^p_{s,\delta} = \{(\varphi, \chi) \mid \varphi \in \mathscr{D}^p_{s,\delta}, \chi \in M^p_{s,\delta}\}$ where $M^p_{s,}$ acts continuously on $X^p_{s-1,\delta+1}$ ($M^p_{s-1,\delta+1}$ vector fields or 1-forms) by $\alpha \to d\chi$.

A slight extension [12] of Cantor's theorem 5.3 and 5.5 of [3] gives the

Theorem 2.1: (i) $\mathcal{G}^p_{s,\delta}$ acts continuously on $\mathcal{P}^p_{s-1,\delta+1}$ by

$$A: \mathcal{G}^p_{s,\delta} \times \mathcal{P}^p_{s-1,\delta+1} \to \mathcal{P}^p_{s-1,\delta+1} : ((\varphi,\chi), (\gamma,U,\alpha,\eta,a.b)) \mapsto (\varphi^*\gamma, U\circ\varphi,\varphi^*\alpha+d\chi, \varphi_*^{-1}\eta, a, b)$$

Moreover $A_{(\varphi,\chi)}: \sigma \to A((\varphi,\chi),\sigma)$ is C^∞ and if $\sigma \in \mathcal{P}^p_{s-1+k,\delta+1}$, $A_\sigma: (\varphi,\chi) \to A((\varphi,\chi),\sigma)$ is C^k for $k = 0$ or 1.

(ii) If $\sigma \in \mathcal{P}^p_{s-1+k,\delta+1}$ (for $k = 0,1$) then the orbit $\mathcal{O}_\sigma := \{A((\varphi,\chi),\sigma) \mid (\varphi,\chi) \in \mathcal{G}^p_{s,\delta}\} \subset \mathcal{P}^p_{s-1,\delta+1}$ is a C^k-submanifold.

(iii) If $\sigma \in \mathcal{P}^p_{s;\delta+1} \exists$ a neighborhood V of $(1\!\!1,0)$ in $\mathcal{G}^p_{s,\delta}$ and a slice of the action, i.e. a submanifold $\mathcal{S}$ of $\mathcal{P}^p_{s-1,\delta+1}$ containing σ such that $((\varphi,\chi),\sigma') \mapsto A((\varphi,\chi),\sigma')$ is a homeomorphism of $V \times \mathcal{S}$ onto a neighborhood U of σ in $\mathcal{P}^p_{s-1,\delta+1}$ and $\mathcal{O}_\sigma \cap \mathcal{S} = \{\sigma\}$.

In the stationary case the proof of this theorem requires $\delta = 0$. (The obvious restriction of replacing $\mathcal{G}^p_{s,\delta}$ by $\mathcal{D}^p_{s,\delta}$ gives the appropriate theorem for the static case, where $0 \le \delta < 1 - 3/p$ is allowed.) We thus have that the tangent space to $\mathcal{P}^p_{s-1,\delta+1}$ at $\sigma \in \mathcal{P}^p_{s,\delta+1}$ splits, i.e.;

$$T_\sigma \mathcal{P}^p_{s-1,\delta+1} = T_\sigma \mathcal{O} \oplus T_\sigma \mathcal{S} \qquad (2.12)$$

We will let $X^p_{s,\delta}$ stand for $M^p_{s,\delta}$-vector fields and 1-forms and $S^p_{s,\delta}$ stand for $M^p_{s,\delta}$ twice convariant symmetric tensors.

3. Solution of the linearized Einstein equations. The set $Q := S_{\rho(p)} \cap \mathcal{P}^p_{s-1,\delta+1}$ can now be characterized as the inverse image of 0 under the map

$$\mathcal{L} : \mathcal{P}^p_{s-1,\delta+1} \to S^p_{s-3,\delta+3} \oplus M^p_{s-3,\delta+3} \oplus X^p_{s-3,\delta+3} \oplus X^p_{s-3,\delta+2} =: Z \qquad (3.1)$$

formed in the obvious manner from (2.5),(2.3),(2.4) and $\nabla_r\nabla^{(r}\eta^{i)} = 0$ since η is a Killing vector field. (The corresponding $\tilde{Q} := \tilde{S}_{\rho,m} \cap \tilde{\mathcal{P}}^p_{s-1,\delta+1}$ is characterized analogously by $\tilde{\mathcal{L}} : \tilde{\mathcal{P}}^p_{s-1,\delta+1} \to S^p_{s-3,\delta+3} \oplus M^p_{s-3,\ +3} =: \tilde{Z}$ by the restriction to

$d\alpha$, h, and η vanishing.)

We conjecture that $\tilde{\mathcal{L}}^{-1}(0) = \mathcal{O}_{\underset{\sim}{g}}$ and $\mathcal{L}^{-1}(0) \cap V = (\bigcup_{\sigma\in\mathcal{H}} \mathcal{O}_\sigma) \cap V$ for some neighborhood V of $\underset{\sim}{g}$ in $\mathcal{P}^P_{s-1,\delta+1}$, where $\mathcal{H}$ is a one-dimensional submanifold of $\mathcal{P}^P_{s-1,\delta+1}$ passing through $\underset{\sim}{g}$ and parameterized by the angular momentum. However we cannot show surjectivity of $\mathcal{L}'(\sigma) = T_\sigma\mathcal{L}$ at $\sigma = \underset{\sim}{g}$ nor transversality of $\mathcal{L}$ over a suitable submanifold of Z and so are unable to show that $\mathcal{L}'(0)$ is a submanifold. Also we do not try to prove existence of such a $\mathcal{H}$ in this paper.

We will first obtain an elliptic operator from $\mathcal{L}'(\sigma)$ by restricting it to the slice $\mathcal{S}$ so that a slightly modified [11] theorem of Nirenberg and Walker [16] can be applied to extend our investigation of $\mathcal{L}'(\underset{\sim}{g})$ to a neighborhood of $\underset{\sim}{g}$.

Let $(c_{ij} = \delta\gamma_{ij}, u = \delta U, \beta = \delta\alpha, \zeta = \delta\eta, A = \delta a\ , B = \delta b) \in T_\sigma\mathcal{P}^P_{2,\delta+1}$. For $\sigma = \underset{\sim}{g} \in \mathcal{P}^P_{3,\delta+1}$ (2.12) gives the unique decomposition $c_{ij} = \pounds_\xi\gamma_{ij} + \varphi_{ij}$ with $\nabla^i\varphi_{ij} = 0$, $u = \pounds_\xi U + \psi$ (for the relation between δU and $\delta\hat{U}$ see [11]), $\beta = \pounds_\xi\beta + df + \omega$ with $\nabla^i\omega_i = 0$ and $\zeta = \pounds_\xi\eta + \nu$, where $\xi \in X^P_{s,0}$. Since the equations $\mathcal{L}'(\sigma) : T_\sigma\mathcal{P}^P_{2,\delta+1} \to Z$ are very long and can be found in [12] we will only briefly sketch the construction of the elliptic operator.

We define $A(\sigma) : \hat{X} := S^P_{s-1,\delta+1} \oplus M^P_{s-1,\delta+1} \oplus X^P_{s-1,\delta+1} \oplus X^P_{s-1,\delta} \to S^P_{s-3,\delta+3} \oplus M^P_{s-3,\delta+3} + X^P_{s-3,\delta+3} \oplus X^P_{s-3,\delta+2} =: \hat{Y}$ by subtracting off divergence terms and the coefficients $b_1(\sigma)$ and $b_2(\sigma)$ of A and B, respectively, in $\mathcal{L}'(\sigma)$, so that

$$\mathcal{L}'(\sigma)(c,u,\beta,\zeta,A,B) = 0$$

is equivalent to

$$A(\sigma)(c,u,\beta,\zeta) = (b_1(\sigma),\ b_2(\sigma))(A,B)^T \tag{3.8}$$

provided (c,u,β,ζ,A,B) is in the tangent space to the slice $\mathcal{S}$. It is easily verified that $A(\sigma)$ is an elliptic operator. We therefore want to show that (3.8) implies that c,u,ζ and A vanish and that β is uniquely determined by B, vanishing for $B = 0$. We first investigate (3.8) on the spherical background $\underset{\sim}{g}$.

The splitting of $T_{\underset{\sim}{g}}\mathcal{P}^P_{s-1,\delta+1} = T_{\underset{\sim}{g}}\mathcal{O} \oplus T_{\underset{\sim}{g}}\mathcal{S}$ allows (3.8) to be investigated on the two subspaces separately.

Proposition 3.1: Equation (3.8) has no nonzero solution (c,u,β,ζ,A,B) on $T_\sigma \mathcal{O}_\sigma$. □

The proof is straightforward [11],[12], relying on the facts that there are no Killing vector fields on $\mathbb{R}^3$ that vanish at infinity and that $\mathrm{div}_\gamma \circ K_\gamma : X^p_{3,\delta} \to X^p_{1,\delta+2}$ and $\Delta : M^p_{3,\delta} \to M^p_{1,\delta+2}$ are isomorphisms [5],[3].

Although on the spherical background (3.8) decouples into $A_{1,2}(\underset{\sim}{\sigma})(c,u) = b_{1,2}(A,0)^T$ and $A_3(\underset{\sim}{\sigma})(\beta) = b_3(\sigma)\ (0,B)^T$ and $A_4(\underset{\sim}{\sigma})(c,\zeta) = 0$ the situation on $T_{\underset{\sim}{\sigma}}\mathcal{S}$ is considerably more complicated than on $T_{\underset{\sim}{\sigma}}\mathcal{O}_{\underset{\sim}{\sigma}}$ as the first equation amounts to several coupled ordinary second order equations, after an expansion into spherical harmonics. (Note that this is the only equation in the static case since there β,η,b,ζ and B are made to vanish.) This problem was solved more easily in [9] by using a $2+1$-dimensional formalism using U as a coordinate, but required that the central value of the potential U and the pressure remain constant. There the equation decoupled into a three-dimensional scalar second order equation which could be solved by maximum principle arguments and into equations on the compact equipotential surfaces which were easily solved afterwards. To reduce the present problem to that treated in [9] we need look only at the spherical variations, i.e. the $\ell = 0$ component of the spherical harmonics. However this system is still too complicated. By using the fact that with a suitable choice of coordinates the Einstein equations for a spherically symmetric space-time reduce to two first order ordinary differential equations for two quantities we could show that $\delta U_c = \delta p_c = A = 0$. By relating the linearizations in the 3-dimensional and 2 + 1-dimensional formalisms the result of [9] yielded that (3.8) implies c, u and A vanish (see [11]). Then also ζ vanishes since $\mathrm{div}_\gamma \circ K_\gamma : X^p_{s,\delta} \to X^p_{s-2,\delta+2}$ is an isomorphism. The equation $A_3(\underset{\sim}{\sigma}) = k_3 b_3(\underset{\sim}{\sigma})$ is identical to some of the linearized equations analyzed in [10] if one sets $K^i = \varepsilon^{ijk}\partial_i\beta_k$. The analysis there implies that there is a unique solution β which is asymptotically $O(|x|^{-2})$ or vanishes iff B vanishes. Since the variation of the angular momentum δJ can be shown to be

$$\delta J = \frac{1}{16\pi} \lim_{|x|\to\infty} \int_{|x|=\mathrm{const.}} \partial_{[i}\beta_{j]} n^j_\gamma{}^{i\ell} d\Sigma_\ell$$

it is then seen [12] that β vanishes iff $\delta J = 0$. We thus have the

Theorem 3.2: The operator equation $A(\underset{\sim}{\sigma})(c,u,\beta,\zeta) = b(\sigma)(A,B)^T$ with $(c,u,\beta,\zeta,A,B) \in T_{\underset{\sim}{g}}\mathcal{P}^p_{2,\delta+1}$ implies that c,u,ζ and A vanish and β is uniquely determined by δJ, vanishing if $\delta J = 0$. □

The presence of the inhomogenous term makes a direct application of Nirenberg and Walker's theorem impossible so we are forced to consider the larger space X so that Cantor's isomorphism theorem [3] can be used to show that $A(\underset{\sim}{g}) : X \to Y$ is a continuous linear operator with finite dimensional kernel and closed range. Nirenberg and Walker's theorem then yields the

Proposition 3.3: For σ in a neighborhood of $\underset{\sim}{g}$ in $\mathcal{P}^p_{2,\delta+1}$ the operator $A(\sigma) : X := S^p_{2,\delta} \oplus M^p_{2,\delta} \oplus X^p_{2,\delta} \to S^p_{0,\delta+2} \oplus M^p_{0,\delta+2} \oplus X^p_{0,\delta+2} =: Y$ is injective,

by noting that $A = 0$ implies $\delta m = 0$ [11]. Thus $A(\underset{\sim}{g})$ has a bounded inverse. This enables us to show that there are two (in the static case, one) unique solutions $x,y \in X$ of $A(\sigma)x = b(\sigma)(A,B)^T$ corresponding to the variables, A and B, one of which cannot lie in $\hat{X}$, the other must yield a non-zero variation of the angular momentum δJ. Thus we get

Theorem 3.4: (i) Static case: Let $c\colon [0,1] \to \tilde{\mathcal{S}} \cap \tilde{\mathcal{P}}^p_{2,\delta+1} \cap \mathcal{S}_{\rho(p)}$ be a C^1-curve such that $c(0) = \underset{\sim}{g}$ is a spherical solution. Then c is constant if the slice $\mathcal{S}$ is contained in a small enough neighborhood of $\underset{\sim}{g}$.

(ii) Stationary case: Let $c\colon [0,1] \to c(t) = (\gamma(t),U(t),\alpha(t),\eta(t),a(t),b(t)) \in \mathcal{S} \cap \mathcal{P}^p_{2,\delta+1} \cap S_{\rho(p)}$ be a C^1-curve of solutions with a constant angular momentum

Then c is constant if the slice $\mathcal{S}$ is contained in a small enough neighborhood of $\underset{\sim}{g}$.

Proof: Since $\mathcal{S}$ is a C^1-Banach manifold and $c(t) \in S_{\rho(p)}$, $\mathcal{L}(c(t)) = 0$ whence $\mathcal{L}'(c(t))(dc/dt) = 0$ for all t. But since c is also tangent to the slice $\mathcal{S}$ this implies $A(c(t))(dc/dt) = b(c(t))(da/dt,db/dt)^T$ which implies $dc/dt = 0$ for all t. □

We thank Werner Isreal, Pong Soo Jang and Jerry Marsden for some useful discussions.

References

[1] Cantor, M., Spaces of functions with asymptotic conditions on $\mathbb{R}^n$, Indiana Univ. Math. J., 24, 897-902, (1975).

[2] Canton, M., Perfect fluid flows over $\mathbb{R}^n$ with asymptotic conditions, J. Func. Anal. 18, 73-84, (1975).

[3] Cantor, M., Some problems of global analysis on asymptotically simple manifolds, Comp. Math., 38, 3-35 (1979).

[4] Carter, B., Axisymmetric black hole has only two degrees of freedom, Phys. Rev. Lett. 26, 331-333, (1971).

[5] Choquet, G. and Y. Choquet-Bruhat, Sur un problème lié à la stabilité des données initiales en relativité générale, C.R. Acad. Sc. Paris 287A, 1047-1049 (1978).

[6] Fischer, A.E. and J.E. Marsden, Linearization stability of nonlinear partial differential equations, Proc. Symp. Pure Math. Amer. Math. Soc., 27, 219-263 (1974).

[7] Fischer, A.E. and J.E. Marsden, Deformation of the scalar curvature, Duke Math. J., 42, 519-547, (1975).

[8] Isreal, W., Event horizons in static vacuum space-times, Phys. Rev. 164 1776-1779, (1967).

[9] Künzle, H.P., On the spherical symmetry of a static perfect fluid, Commun. Math. Phys. 20, 85-100 (1971).

[10] Künzle, H.P. and J.R. Savage, Equilibrium of slowly rotating relativistic fluids, J. Math. Phys. (to appear).

[11] Künzle, H.P. and J.R. Savage, A global analysis approach to the general relativistic fluid ball problem, G.R.G. (to appear).

[12] Künzle, H.P. and J.R. Savage, On the uniqueness of the equilibrium configurations of slowly rotating relativistic fluids (to be published).

[13] Lichnerowicz, A., Théories relativistes de la gravitation et de l'éléctromagnetism, Masson, Paris, 1955.

[14] Lichtenstein, L., Gleichgewichtsfiguren rotierender Flüssigkeiten, Springer, Berlin, 1933.

[15] Lindblom, L., Stationary stars are axisymmetric, Astrophys. J. 208, 873-880, (1976)

[16] Nirenberg, L. and H.F. Walker, The null spaces of elliptic partial differential operators in $\mathbb{R}^n$, J. Math. Anal. Appl. 42, 271-301 (1973).

[17] Robinson, D.C., Uniqueness of the Kerr black hole, Phys. Rev. Lett., 34, 905-906, (1975).

QUATERNIONIC AND SUPERSYMMETRIC σ - MODELS

Jerzy Lukierski
Institute of Theoretical Physics, University of Wrocław
Wrocław, ul. Cybulskiego 36, Poland

1. INTRODUCTION

Recent interest in σ-models is justified by their following properties:

a/ the interaction is introduced in geometric way as a characterization of the following mapping, defining M-valued bosonic σ-field

$$\varphi^A(x) : \quad x \in \mathcal{S} \longrightarrow \varphi^A \in \mathcal{M} \tag{1.1}$$

where

$\mathcal{S}$ describes the manifold of coordinates /Minkowski space-time $R^{d,1}$ with signature $(d-1,1)$, Euclidean space R^d, curved space-time manifold/

$\mathcal{M}$ denotes the "internal" manifold at point x, a curved Riemannian manifold with possible additional structure /complex or quaternionic/

The simplest choice of $\mathcal{M}$ is the homogeneous coset space $\mathcal{M} = \frac{G}{H}$ - a nonlinear Riemannian manifold with constant scalar curvature R. In such a case the nonlinearity / interaction / is paremetrized by a <u>numerical</u> coupling constant $\lambda = \frac{1}{R}$.

In the first part we shall restrict ourselves to the Riemannian manifolds which are the homogeneous symmetric coset spaces $\mathcal{M} = \frac{G}{H}$, where $G = U(n;F)$ / $F = R : U(n;R) = SO(n)$, $F = C : U(n;c) = U(n)$, $F = H : U(n;H) = Sp(n)$ /. If $H=1$, the corresponding σ-fields are called principal; the following two coset spaces are distinguished [1,2]

Stiefel manifolds
$$S_{n,m}(F) = \frac{U(n;F)}{U(n-m;F)} \tag{1.2a}$$

Grassmann manifolds
$$G_{n,m}(F) = \frac{U(n;F)}{U(n-m;F)\otimes U(m;F)} = \frac{F^{n\cdot m}}{GL(m;F)} \tag{1.2b}$$

The manifolds $G_{n+1,1}(F) = FP(n)$ are called projective planes; if $F=H$ we shall consider only the case of quaternionic projective plane $HP(n)$.

In the second half of this lecture we shall generalize the mapping (1.1) in order to include fermionic σ-fields.

b/ the $\frac{G}{H}$ σ-model describes G-symmetry broken spontaneously to its subgroup H, and can be used for the description of Higgs effect.

The $\frac{G}{H}$ σ-field transforms linearly under H and nonlinearly under the transformations from $\frac{G}{H}$. In particular if we couple the $\frac{G}{H}$ σ-model with the massless G-gauge fields, the $\frac{G}{H}$-valued components of the gauge potentials become massive [3,4]. Recently such a system with G=O(3) and H=1 /principal O(3) σ-model / has been discussed also in to quantized version by Arefeva and Slavnov[5] as an alternative to the conventional Higgs mechanism. In summary, one can call σ-fields the <u>geometric Higgs fields</u>, in contrast to the conventional ones, introduced in nongeometric way.

c/ The σ-fields have nontrivial topological solutions, depending for a given choice of $\mathcal{S}$ only on the topology of $\mathcal{M}$.

If we assume that $\mathcal{S} = R^d$ /Euclidean σ-fields/ and $\varphi^A(x) \xrightarrow[|x|\to\infty]{} \varphi^A_\infty$ = const., one can compactify the Euclidean d-dimensional space : $\mathcal{S} = S^d$ /d-dimensional sphere/ . The topological configurations are classified by the homotopy classes:

Instanton solutions
for $\varphi^A(x)$, $x \in S^d$: $\quad \pi_d(\mathcal{M}) \neq 0$ (1.3)

If we consider $\mathcal{M}$ for which $\pi_i(\mathcal{M}) = 0$ / $i = 0,1,2,\cdots d-1$ /, due to Hureviz theorem [6] one can express the topological charge of the field $\varphi^A(x)$ as the integral over S^d of a local density

$$Q = \int_{S^d} dx_1 \cdots dx_d \, \omega_d(x) = \pi_d(\mathcal{M}) \tag{1.4}$$

where $\omega_d \in H^d(\mathcal{M}, R)$ / d-th cohomology class of the manifold $\mathcal{M}$ / and $\omega_d(x)$ is the pull-back of ω_d from $\mathcal{M}$ on S^d, induced by the mapping (1.1) .

If d=4 /fourdimensional Euclidean theories / the $\mathcal{M}$-valued σ-fields have nontrivial topological configurations described by the local topological charge density (1.4) if

$$\pi_0(\mathcal{M}) = \pi_1(\mathcal{M}) = \pi_2(\mathcal{M}) = \pi_3(\mathcal{M}) = 0 \qquad \pi_4(\mathcal{M}) \neq 0 \tag{1.5}$$

The relations (1.5) are the ones which define the manifolds with the <u>quaternionic Kähler structure</u> [1/]; ω_4 in (1.4) is the fundamental nonzero harmonic 4-form on $\mathcal{M}$.

d/ In the formalism of σ-models one can introduce the local gauge invariance by the principle of minimal gauge coupling with composite gauge fields. Such a procedure recalls the elimination of connections / Cristoffel symbols/ in general relativity in terms of the metric or vierbein fields and can be also called the inverse Higgs effect [2].

It is known [10] that any connection /gauge field/ on differentiable principal $U(m;F)$ bundle over compact base $\mathcal{S}$ can be defined as a natural connection on the principal $S_{n,m}$ bundle

$$A_\mu^{ij} = \bar{\Phi}^{iJ} \partial_\mu \Phi^{Jj} \qquad i,j=1\ldots m \quad J,K=1\ldots n \; \left(n \text{ sufficiently large}\right) \tag{1.6}$$

where we use the following parametrization

$$S_{n,m}(F): \qquad \bar{\Phi}^{iJ} \Phi^{Jj} = \delta^{ij} \qquad \Phi^{Ji} \in F \tag{1.7}$$

and $\Phi \longrightarrow \bar{\Phi}$ denotes the principal involution in F. Such a formula has purely geometric origin, but it can be also obtained in $S_{N,m}(F)$ σ-models with $U(m;F)$ -covariant derivatives as a solution of the algebraic equation following from the action principle.

In $S_{N,m}(F)$ σ-models one introduces compensating $U(m;F)$ gauge fields without kinematic term in order to reduce locally the manifold of field values to $G_{N;m}(F)$. The "free" Yang-Mills theory of the composite compensating gauge fields can be introduced however as the generalized σ-model with four-linear Lagrangean. One of the aims of our talk is to consider such fourlinear actions for the σ-models [11,12].

e/ Using the supersymmetric extension of Riemannian geometry [13-16] /supersymmetric extensions of Lie groups [17-19] / one can generalize the σ-field formalism and include fermions.

Such a generalization implies that the mapping (1.1) should be supersymmetrized. There are the following three different ways of introducing supersymmetric σ-models:

- one supersymmetrizes the manifold $\mathcal{M}$ of field values, in particular by replacing the coset space of a Lie group G by a coset space of a Lie supergroup $\tilde{G}$:

$$\mathcal{M} = \frac{G}{H} \longrightarrow \tilde{\mathcal{M}} = \frac{\tilde{G}}{\tilde{H}} \tag{1.8}$$

where $\tilde{H}$ is in general an invariant sub-supergroup of $\tilde{G}$. In such a <u>supercoset approach</u> the supersymmetric σ-field is described by the set of bosonic and fermionic σ-fields $\{\varphi^A(x), \psi^\alpha(x)\}$, parametrizing $\tilde{\mathcal{M}}$:

$$\{\varphi^A(x), \psi^\alpha(x)\} \quad : \quad x \in S \rightarrow \{\varphi^A, \psi^\alpha\} \in \tilde{\mathcal{M}} \tag{1.9}$$

- one supersymmetrizes the coordinate manifold $\mathcal{S}$, by replacing it by a superspace $\tilde{\mathcal{S}}$. However $\tilde{\mathcal{S}}$ in principle can be curved, we shall consider here the case when $\tilde{\mathcal{S}}$ is a flat superspace, as it was introduced by Wess, Zumino [20] and Salam, Strathdee [21]. In such a superfield approach the field $\varphi^A(x)$ is replaced by the σ-superfield $\varphi^A(x,\theta)$:

$$\varphi^A(x,\theta) \quad : \quad (x,\theta) \in \tilde{\mathcal{S}} \longrightarrow \varphi^A \in \mathcal{M} \tag{1.10}$$

- one can consider also fully supersymmetrized σ-model: if both $\mathcal{S}$ and $\mathcal{M}$ become supermanifolds.

We see that the fourdimensional σ-models with interesting topological properties and fermions should be fourlinear, quaternionic and supersymmetrized. The aim of this lecture is to present the recent results on fourlinear, quaternionic and supersymmetric σ-models. We shall consider firstly these three features separately; only at the end of Sect. 5 /supersymmetric σ-models in supercoset approach/ and at the end of Sect. 6 /supersymmetric σ-models in superfield approach/ we shall indicate how to unify these three desired properties [3/]. The extensive analysis of quaternionic geometry and self-dual solutions in quaternionic HP(n) σ-models was presented recently by Tze and Gürsey[12], and we send an interested reader to this excellent detailed review. The second part /Sect 4-6/ concerning the supersymmetric σ-models contains several new results; in particular in this written text of the lecture we added /see Sect. 6 / a model obtained after the Conference [23] describing supersymmetric Yang-Mills theory [24,25] with composite gauge superfield.

We would like to mention that many aspects of quaternionic and super symmetric σ-models /e.g. local and nonlocal infinite number of conservation laws, application of inverse scattering method, relation to the topology of supermanifolds, the discussion of possible perturbative scheme/ are left out completely not only due to the lack of time, but also because many questions still remain unanswered.

2. DIFFERENT GEOMETRIES AND BOSONIC σ - MODELS.

It is well-known that any n-dimensional oriented Riemannian manifold can be described by the real symmetric metric tensor g_{ij} and zero torsion; its holonomy group is in SO(n) . Any classical group manifold G with the symmetric connection form is an example of Riemannian manifold with constant scalar curvature.

Here we shall be interested in Riemannian manifolds with additional complex or quaternionic structure. One can introduce

a/ 2n-dimensional complex manifolds with the holonomy group in U(n) .

Such a geometry is described by complex coordinate maps and nondegenerate Hermitean metric $h_{\bar{I}j} = \bar{h}_{\bar{J}i}$. The topological properties are determined via fundamental two-form

$$\omega_2 = \frac{i}{2} h_{\bar{I}j} d\bar{z}_i \wedge dz_j \tag{2.1}$$

If the two-form ω_2 is nondegenerate and closed

$$d\,\omega_2 = 0 \tag{2.2}$$

it describes complex Kähler manifold.

The examples of complex Kähler manifolds are provided by $G_{n,m}(c)$; another example is $G_{n+2,2}(R)$ which can be parametrized as complex quadric [2] .

c/ 4n- dimensional quaternionic manifolds, with the holonomy group in $Sp(n) \times Sp(1)$ / $Sp(n) = U(n;H)$ / .

Let us consider firstly the manifolds described by the quaternionic, coordinate maps and the quaternion - Hermitean metric $H_{\bar{I}j} = \bar{H}_{\bar{J}i}$, where the real line element is given by the formula

$$ds^2 = d\bar{q}_i H_{\bar{I}j} dq_j \qquad q_i = q^o + e_r q^r \tag{2.3}$$

and

$$H_{ij} = G_{ij} + e_r H_{ij}^r \qquad \begin{array}{l} G_{ij} = G_{ji}, \text{ real} \\ H_{ij}^r = -H_{ji}^r, \text{ real} \end{array} \tag{2.4}$$

$$e_r e_s = -\delta_{rs} + \epsilon_{rst} e_t \qquad Sp(e_r e_s) = 4\delta_{rs} \tag{2.5}$$

Introducing three two - forms

$$\omega_2^{(r)} = \frac{1}{4} Sp \left\{ e_r \operatorname{Im}(d\bar{q}_i H_{\bar{I}j} dq_j) \right\} \tag{2.6}$$

the topological properties of quaternionic manifolds are determined by the fundamental four-form

$$\omega_4 = \omega_2^{(1)} \wedge \omega_2^{(1)} + \omega_2^{(2)} \wedge \omega_2^{(2)} + \omega_2^{(3)} \wedge \omega_2^{(3)} \qquad (2.7)$$

If the four-form is nondegenerate and closed

$$d\,\omega_4 = 0 \qquad (2.8)$$

it describes quaternionic Kähler manifold.

The only known quaternionic manifolds which admit the quaternionic coordinatization (2.3) are $S_{n,m}(H)$ and $G_{n,m}(H)$; one can also show [26] that the only compact nonflat quaternionic Kähler manifold with the holonomy group equal to $Sp(n) \times Sp(1)$ is the projective quaternionic plane $HP(n)$.

In general the quaternionic coordinatization of the quaternionic manifold $\mathcal{M}$ does not exists. The quaternionic structure implies only that there is defined over every point of $\mathcal{M}$ a 3-dimensional $SO(3)$ -bundle describing the realization in $T_x(\mathcal{M})$ of the quaternionic algebra, and defining the two-forms $\omega_2^{(r)}$ in (2.7) with the $SO(3)$ index $r=1,2,3$ [27,28] . There are known several properties of the quaternionic manifolds:

a/ there is one-to-one correspondence between symmetric quaternionic manifolds and compact simply connected simple Lie groups [28], e.g. $SO(n+4) \rightarrow G_{n+4,4}(R)$ $SU(2+n) \rightarrow \frac{SU(2+n)}{S(U(2)\otimes U(n))}$ etc.

b/ Every homogeneous compact quaternionic manifold is symmetric [26]

c/ there are known nonsymmetric quaternionic manifolds [29] , described by quaternionic representations of Clifford algebras, classified by Atiyah, Bott and Shapiro [30] .

The formulae (2.1) and (2.6) express via the definition (1.3) the topological charges of Euclidean two-dimensional /d=2/ complex σ -fields and Euclidean four-dimensional /d=4/ quaternionic σ -fields; the relations (2.2) and (2.7) expresses their topological conservation law, without any use of the equations of motion.

If we wish to define $\frac{G}{H}$ σ-fields, one has to choose one of at least three possible parametrizations of the coset space $\mathcal{M} = \frac{G}{H}$:

i) one considers $\frac{G}{H}$ as the Riemannian manifold with independent coordinates. For example $G_{n,m}(F)$ manifold is described by its F - Hermitean metric defining the Riemannian line element as follows /if F=H , we put m = 1 / [31]

$$ds^2 = \mathrm{Tr}\left\{ (I + \varphi^+\varphi)^{-1} d\varphi^+ (I + \varphi\varphi^+)^{-1} d\varphi \right\} \qquad (2.9)$$

where $\varphi = \varphi^{ji}$ / i=1...m, =1...n-1 / describe local F-valued charts in $G_{n,m}(F)$.

Such a parametrization is advantageous if we wish to use the formulae of differential geometry / see e.g. [2]/.

ii) one considers $\frac{G}{H}$ as a linear manifold with constraints, and with possible equivalence class relations.

In order to define $G_{n,m}(F)$ manifold we parametrize firstly the Stiefel manifold $S_{n,m}(F)$ by the constrained rectangular matrices (1.7), and further one introduces the equivalence relations

$$\Phi^{ji} \sim \Phi^{jk} U_k^{\;i} \qquad U \in U(m;F) \tag{2.10}$$

Such an approach exhibits the role of gauge fields in the elimination of redundant degrees of freedom and introduces the global parametrization of the manifold.

iii) One can consider the coset space as a group manifold G restricted by the algebraic $\frac{G}{H}$ -orbit relations.

This method is the most general; it has been thoroughly studied for U(n) by Michel and Radicati [32] and applied to $G_{n,m}(c)$ σ-models by Eichenherr [33] ; the analogous discussion for Sp(n) is not known.

Such a parametrization is useful if we study the properties which do not depend on the algebraic constraints, e.g. the formulae for local currents.

In our formulation of dynamics we have chosen the second parametrization: in action integral the constraints (1.7) are imposed by introducing the Lagrange multipliers, and the equivalence relations are transformed into the local gauge invariance by the help of compensating gauge fields.

Finally we should add that the Euclidean "base" manifolds $\mathcal{S}$ if d=2 / $\mathcal{S} = S^2$/ and d=4 / $\mathcal{S} = S^4$ / describe respectively one-dimensional complex and quaternionic manifolds FP(1) /F=C,H/. If the mapping (1.1) becomes F-holomorphic, it describes the instanton solutions [5]

3. FOURLINEAR LAGRANGEANS AND QUATERNIONIC σ - MODELS.

The geometric framework, outlined in previous Section provides the explicite formulae for the topological charges for complex and quaternionic σ-fields. For example choosing $\mathcal{M}$ = CP(n) one obtains

$$\pi_0(CP(n)) = \pi_1(CP(n)) = 0 \qquad \pi_2(CP(n)) = Z \tag{3.1}$$

and the instanton solutions in two-dimensional CP(n) σ-models are classified by the <u>first Chern class</u> of the mapping $S_2=CP(1)\rightarrow CP(n)$. Using (1.4) and (2.1) one obtains [34]

$$Q = \frac{i}{2\pi}\int_{S_2} d^2x\, \epsilon_{\mu\nu} F_{\mu\nu} = \pi_2\,(CP(n)) \qquad \mu,\nu=1,2 \tag{3.2}$$

where $F_{\mu\nu} = \partial_\mu A_\nu - \partial_\nu A_\mu$ and A_μ is given by (1.6), with $F=C$ and $m=1$.

In fourdimensional HP(n) σ- model [11,12] the topological configurations are classified by the <u>second Chern class</u>, called also the Pontriagin index

$$Q = -\frac{1}{64\pi^2}\int_{S^4} d^4x\, Sp\,(F_{\mu\nu} F_{\varrho\tau})\,\epsilon_{\mu\nu\varrho\tau} \qquad \mu,\nu,..=1...4 \tag{3.3}$$

where $F_{\mu\nu}$ is the composite imaginary quaternion-valued field strenght, derived at the end of this Section, and Sp denotes quaternionic trace.

In order to determine the dynamics of $\frac{G}{H}$ σ-fields /field equations, self-duality equations/ we propose the actions which are

a) invariant under global G and local H transformations,

b) invariant under the conformal change of coordinates in $\mathcal{S}$,

c) lead to the field equations with the highest second order derivatives.

From the properties b) and c) it follows that in d dimensions the action should be d - linear in the first derivatives of the σ-field 7/; the dependence on the σ-field without derivatives is restricted by a) .

If d=2 we should choose the action bilinear; additionally if we require the presence of instanton configurations we should consider the σ-fields with values in complex Kähler manifold /see e.g. [34] /.

Let us consider firstly the principal U(n) σ-model in two dimensions . The action can be written as follows 8/

$$W_2 = \frac{1}{2}\int d^2x\, Tr\,(\mathcal{A}_\mu \mathcal{A}^\mu) = \frac{1}{2}\int d^2x\, Tr(\partial_\mu \bar{g}\ \partial^\mu g) \tag{3.4}$$

where $\mathcal{A}_\mu$ is the curvature-free U(n) - algebra - valued connection (1.6) with n=m /see [36,37] / and $g^+g = gg^+ = 1$ describes U(n) group manifold. The $S_{n,m}(C)$ σ-model can be obtained by the replacement in (3.4)

$$\partial_\mu \longrightarrow D_\mu = \partial_\mu - \mathcal{B}_\mu^{(1)} \tag{3.5}$$

where the decomposition of $n \times n$ connection matrix

$$\mathcal{A}_\mu = S_\mu + \mathcal{B}_\mu^{(1)} \tag{3.6}$$

is such that $\mathcal{B}_\mu \in U(n-m)$, and $S_\mu \in \frac{U(n)}{U(n-m)}$, or equivalently

$$W_2 = \frac{1}{2} \int d^2x \; \mathrm{Tr} \; (S_\mu \, S^\mu) \tag{3.7}$$

Using $(n-m)^2$ gauge conditions for the $U(n-m)$ gauge field $\mathcal{B}_\mu^{(1)}$, one can also write the $S_{n;m}(c)$ action as follows 9/

$$W_2 = \frac{1}{2} \int d^2x \; \mathrm{tr} \; (\partial_\mu \bar{\Phi} \, \partial^\mu \Phi) \tag{3.8}$$

where $\Phi = \{\Phi^{ji}\}$ is given by (1.7) . Finally $G_{n,m}(c)$ σ-model is obtained by the formula

$$W_2 = \frac{1}{2} \int d^2x \; \mathrm{tr} (\nabla_\mu \bar{\Phi} \, \nabla^\mu \Phi) = \frac{1}{2} \int d^2x \; \mathrm{Tr} \; (\mathcal{G}_\mu \mathcal{G}^\mu) \tag{3.9}$$

where $\nabla_\mu = \partial_\mu - A_\mu$, $A_\mu \in U(m)$, given by (1.6) , and

$$\mathcal{A}_\mu = \mathcal{G}_\mu + \mathcal{B}_\mu^{(1)} + \mathcal{B}_\mu^{(2)} \tag{3.10}$$

i.e. $\mathcal{G}_\mu$ is the $n \times n$ matrix realization of the $G_{n,m}(c)$ coset space algebra.

Let us consider now the case $d=4$, and consider the actions satisfying a) - c) . For principal $U(n)$ σ-fields one can introduce two fourlinear terms, satisfying our conditions

$$W_4 = \frac{\lambda_1}{4} \int d^4x \; \mathrm{Tr} \left([\mathcal{A}_\mu, \mathcal{A}_\nu][\mathcal{A}^\mu, \mathcal{A}^\nu] \right) = \tag{3.11}$$

$$= \frac{\lambda_1}{4} \int d^4x \; \mathrm{Tr} \left(\partial_{[\mu} \bar{g} \; \partial_{\nu]} g \;\; \partial^{[\mu} \bar{g} \; \partial^{\nu]} g \right)$$

and

$$W_4 = \frac{\lambda_2}{4} \int d^4x \left[\mathrm{Tr} (\mathcal{A}_\mu \mathcal{A}^\mu) \right]^2 = \frac{\lambda_2}{4} \int d^4x \left[\mathrm{Tr} (\partial_\mu \bar{g} \; \partial^\mu g) \right]^2 \tag{3.12}$$

The term (3.11) for the special choice $g \in G = SU(2)$ was introduced by Skyrme twenty years ago [38,39] ; the generalization to arbitrary compact group $U(n;F)$ was considered recently by Faddeev [40] and Romanov, Schwarz [41] . The generalization of action (3.11) to symmetric homogeneous compact space $\frac{G}{H}$ was discussed by Faddeev and Tian-Szanski [42] . Here we shall discuss only the first term; the second, which plays the role of fourlinear Higgs term, can be constructed as the square of the Lagrange density in bilinear σ-model /see (3.4-3.10) /.

We shall generalize now, the action (3.11) for the cosets (1.2) if F=C.

a) $S_{n,m}(c)$ σ-model

$$W_4 = \frac{\lambda_1}{4} \int d^4x \, Tr([S_\mu, S_\nu][S^\mu, S^\nu]) = \tag{3.13a}$$

$$= \frac{\lambda_1}{4} \int d^4x \, Tr \left(\overline{D_{[\mu} g} \, D_{\nu]} \, g \, \overline{D^{[\mu} g} \, D^{\nu]} \, g \right)$$

or

$$W_4 = \frac{\lambda_1}{4} \int d^4x \, tr \left(\partial_{[\mu} \bar{\Phi} \, \partial_{\nu]} \Phi \, \partial^{[\mu} \bar{\Phi} \, \partial^{\nu]} \Phi \right) \tag{3.13b}$$

b) $G_{n,m}(c)$ σ-model

$$W_4 = \frac{\lambda_1}{4} \int d^4x \, Tr \left([G_\mu, G_\nu] [G^\mu, G^\nu] \right) = \tag{3.14}$$

$$= \frac{\lambda_1}{4} \int d^4x \, tr \left(F_{\mu\nu} \, F^{\mu\nu} \right)$$

where

$$F^{ij}_{\mu\nu} = \nabla_{[\mu}{}^{ik} \, \bar{\Phi}^{kj} \, \nabla^{jl}_{\nu]} \, \Phi^{jl} \tag{3.15}$$

denotes composite U(m) -field strenght.

We see that the action (3.14) provides the Lagrangean for the composite U(n) gauge fields with familiar form of the Yang-Mills action. In particular due to the relation

$$\frac{\delta W_4}{\delta \Phi^{kl}} = \int d^4 y \, \frac{\delta S}{\delta A^{ij}_\nu} \, \frac{\delta A^{ij}_\nu}{\delta \Phi^{kl}} \tag{3.16}$$

the solutions of the Yang-Mills equation

$$\frac{\delta S}{\delta A^{ik}_\nu} = \nabla^{\mu ij} F^{jk}_{\mu\nu} = 0 \tag{3.17}$$

for composite gauge fields solves also the model (3.14) .

In order to use the formula (3.3) for the topological charge we write now the fourlinear HP(n) σ-model [11,12] . The HP(n) manifold is parametrized globally by (n+1) quaternionic homogeneous coordinates

$$\bar{q}_0 q_0 + \dots + \bar{q}_n q_n = 1 \qquad q_j \sim q_j a \tag{3.18}$$

where $a \in Sp(1) = SU(2)$ describes the unit quaternion. The imaginary quaternion-valued vector field $A_\mu = \bar{q}_j \partial_\mu q_j$ transforms under local $Sp(1)$ transformations as follows

$$A'_\mu = \bar{a} A_\mu a + \bar{a} \partial_\mu a \tag{3.19}$$

and $Sp(1)$ -algebra-valued $SU(2)$ field strenght $F_{\mu\nu}$ has a form

$$F_{\mu\nu} = \overline{\nabla_{[\mu} q} \nabla_{\nu]} q \tag{3.20}$$

The action with quaternionic trace

$$W_4 = \frac{\lambda_1}{4} \int d^4x \, Sp\,(F_{\mu\nu} F^{\mu\nu}) \tag{3.21}$$

leads to $SU(2)$ Yang-Mills action, and if $\mathcal{S} = R^4$ the topological charge of Euclidean $HP(n)$ σ-field is given by (3.3) with the substitution (3.20).

The selfduality equation for the Euclidean model (3.20) has the form

$$\overline{\nabla_{[\mu} q} \nabla_{\nu]} q = \epsilon_{\mu\nu\varrho\tau} \overline{\nabla_{[\varrho} q} \nabla_{\tau]} q \tag{3.22}$$

and its solutions describe the $HP(n)$ quaternionic column, parametrizing all the selfdual $SU(2)$ Yang-Mills solutions, occuring in ADHM construction [43]. One can also show that the relations (3.22) are the generalized Cauchy-Riemann relations for the quaternion-holomorphic mapping, with quaternionic analyticity defined in the sense proposed by Gürsey and Tze [12].

In order to derive the formulae (1.6) for $SU(n)$ instanton solutions in term of σ-fields one can use the $G_{n,m}(C)$ σ-model (3.14). Even $HP(n)$ σ-model and its $SU(2)$ instanton solutions can be expressed in terms of complex $S_{n+2,2}(C)$ σ-fields Φ^{Ji} / $i=1,2$; $J=1\dots n+2$ / if due to the known formula [45]

$$\frac{Sp(n+1)}{Sp(n)\otimes Sp(1)} = \frac{U(2n+2) \cap Sp(2n+2;C)}{(U(2n) \cap Sp(2n;C))\otimes SU(2)} \tag{3.23}$$

we impose besides the $S_{2n+2;2}(C)$ constraints (1.7) the following additional relations:

$$\Phi^T \sigma_2 \Phi = \sigma_2 \qquad \Phi \sim \Phi S \qquad S \in SU(2) \tag{3.24}$$

The $SU(m)$ instantons expressed in terms of suitably constrained

$G_{n,m}(C)$ matrices were recently given by Berezin [44] ; they are the solutions of selfduality equation in suitably constrained four--linear $G_{n,m}(C)$ σ-model [46] .

4. DIFFERENT SUPERGEOMETRIES AND COSET SPACES FOR SUPERGROUPS.

In order to geometrize the fermionic degrees of freedom one adds to the bosonic coordinates $(x_1 \ldots x_n)$ the anticommuting fermionic Grassmann variables $(\theta_1 \ldots \theta_m)$, where

$$\{\theta_\alpha, \theta_\beta\} = 0 \qquad [\theta_\alpha, x_\mu] = [x_\mu, x_\nu] = 0 \tag{4.1}$$

The vector in R^n is defined by n real numbers, transforming under the change of linear frame by $GL(n,R)$ transformations . Similarly the supervector $Y = (x, \theta) \in R^{n,m}$ is the set of n bosonic and m fermionic variables, transforming by the Z_2 - graded matrix supergroup $GL(n,m;R)$, where

$$GL(n;m;R): \quad \mathfrak{F} = \begin{pmatrix} B_1 & F_1 \\ F_2 & B_2 \end{pmatrix} \begin{matrix} n \\ m \end{matrix} \qquad \begin{matrix} \deg B_i = 0 & i=1,2 \\ \deg F_i = 1 & i=1,2 \end{matrix} \tag{4.2}$$

(columns: n, m)

One can define therefore the real superspace as the Z_2 - graded $GL(n,m;R)$ -module, where for the nonsingular graded matrices $\mathfrak{F} \in GL(n,m;R)$ we have $\operatorname{Ber} \mathfrak{F} \neq 0$ 10/
The notion of real Euclidean space is generalized in the presence of even number of fermionic variables by introducing the orthosymplectic transformations which preserve the "flat" superRiemannian norm

$$(Y, Y') = Y_A^T G^{(o)}_{AB} Y_B \tag{4.3}$$

where

$$G^{(o)} = \begin{pmatrix} 1 & 0 \\ 0 & c \end{pmatrix} \begin{matrix} n \\ m \end{matrix} \qquad \begin{matrix} c^2 = -1 \\ c^T = -c \end{matrix} \tag{4.4}$$

(columns: n, m)

We obtain

$$OSp(n;2k;R): \qquad \mathfrak{F}^{\tilde{T}} G^{(o)} \mathfrak{F} = G^{(o)} \tag{4.5}$$

where graded transposition is defined as follows

$$\mathfrak{F}^{\tilde{T}} = \begin{pmatrix} B_1^T & -F_2^T \\ F_1^T & B_2^T \end{pmatrix} \tag{4.6}$$

We introduce also

i) Linear superspaces $F^{n,m}$ with complex and quaternionic structure.

Let us introduce respectively in $R^{2n,2m}$ and $R^{4n,4m}$ the complex and quaternionic superspace coordinates:

$$C^{n,m} : \quad z_i = x_i + ix_{i+1} \qquad t_\alpha = \theta_\alpha + i\,\theta_{\alpha+k}$$

$$H^{n,m} : \begin{cases} q_i = x_i + x_{i+r \cdot k} \cdot e_r \\ \xi_\alpha = \theta_\alpha + \theta_{\alpha+rk} \cdot e_r \end{cases}$$

One can generalize the norm (4.3) as follows

$$(Y, Y')_F = \overline{Y}_A^T \, G_{AB}^o (F) \, Y_A \tag{4.7}$$

where Y_A are the F-valued superspace coordinates, and the flat metric $G_{AB}^o(F)$ is F-Hermitean in bosonic and F-antiHermitean in the fermionic sector. The general choice in fermionic sector is

$$G_{AB}^o (C) = \begin{pmatrix} \overset{n}{1} & & \overset{m}{0} & \\ & i_{\ddots i} & & 0 \\ 0 & & -i_{\ddots} & \\ & 0 & & -i \end{pmatrix} \tag{4.8}$$

$$G_{AB}^o (H) = \begin{pmatrix} 1 & & 0 \\ & e_{\ddots} & \\ 0 & & \ddots_e \end{pmatrix} \qquad \begin{array}{l} e = n_i e_i \\ n_i^2 = 1 \end{array} \tag{4.9}$$

and one can define the following supergroups:

$$U(n;m-s,s) : \qquad \overline{Y}_A^T \, G_{AB}^o(C) \, Y_B = \text{inv.} \tag{4.10a}$$

$$UU\alpha(n;m;H) : \qquad \overline{Y}_A^T \, G_{AB}^o(H) \, Y_B = \text{inv.} \tag{4.10b}$$

which provide the supersymmetric extensions of $U(n,F)$ $(F=R,C,H)$:

i) $\quad F = R \qquad O(n) \longrightarrow OSp(n;2k)$

ii) $\quad F = C \qquad U(n) \longrightarrow UU\alpha(n;m)$ $\qquad\qquad (4.11)$

iii) $\quad F = H \qquad Sp(n) = U(n;H) \longrightarrow UU\alpha(n;m;H)$

One can denote these three supergroups by $UU\alpha(n;m;F)$, with the algebra in bosonic sector $U(n;F) \times U\alpha(m;F)$, where $U(m;F)$ denotes the group of antiunitary matrices, leaving the F-antiHermitean metric invariant. In particular we obtain [45,48]

$$
\begin{array}{llll}
\text{i)} & F = R & U\alpha(m;R) = Sp(m;R) & m \text{ even} \\
\text{ii)} & F = C & U\alpha(m;C) = U(m-s,s) & m \text{ arbitrary},\ 0 \leq s \leq m \\
\text{iii)} & F = H & U\alpha(m;H) = O(m;H) & m \text{ arbitrary}
\end{array}
\qquad (4.12)
$$

We see that only the complex supergroups can have the compact bosonic subgroup $U(n) \times U(m)$.

ii) Curved supergeometries.

One can define the general nonlinear supermanifold by an atlas of supermaps [49,50] ; the superRiemannian structure is obtained by the following graded extension of the Riemannian metric tensor g_{ij}

$$
G_{AB} = \begin{pmatrix} g_{ij} & g_{i\beta} \\ g_{\alpha j} & g_{\alpha\beta} \end{pmatrix}
\qquad
\begin{array}{l} g_{ij} = g_{ji} \\ g_{\alpha j} = g_{j\alpha} \\ g_{\alpha\beta} = - g_{\beta\alpha} \end{array}
\qquad (4.13)
$$

The real superRiemannian manifold may carry additional complex or quaternionic structure; after the introduction of complex or quaternionic supercoordinate maps one can define the graded extensions of the F- Hermitean metrics (2.1) and (2.3) . Because the graded extension of the calculus of exterior F - valued forms in known [51,52] one can introduce the graded extensions of Chern and Pontriagin classes, and define the notion of complex and quaternionic Kähler supermanifolds 53 .

The supergroups $UU\alpha(n;m;F)$ are the examples of special superRiemannian manifolds with constant scalar supercurvature and additional complex F= C or quaternionic F=H structure. We shall distinguish here the following three classes of supercoset spaces with additional F - structure (F=C or H) :

a) F - valued supersymmetric projective planes

$$
F\,\tilde{P}(n;m) = \frac{UU\alpha(n+1;m;F)}{UU\alpha(n;m;F) \otimes U(1;F)} \qquad (4.14)
$$

The supermanifold (4.14) has n bosonic and m fermionic F-valued dimensions. If n=0 , we get the purely fermionic projective plane $F\tilde{P}(m)$ with F - valued Grassmann coordinates, which may be used in the formalism of σ -models for the geometrization of purely fermionic interaction.

b) Stiefel supermanifolds

$$\tilde{S}^{b}_{n,m}(F) = \frac{UU_\alpha(n;m;F)}{U(n;F)} \tag{4.15a}$$

$$\tilde{S}^{f}_{n,m}(F) = \frac{UU\alpha(n;m;F)}{U\alpha(m;F)} \tag{4.15b}$$

c) F - valued fermionic Grassmanians

$$\tilde{G}_{n,m}(F) = \frac{UU\alpha(n;m;F)}{U(n;F)\otimes U\alpha(n;F)} \tag{4.16}$$

with $n\cdot m$ F -valued fermionic coordinates.

These supercoset spaces, similarly like in bosonic case /see Sect.2/ , can be parametrized at least in three different ways:

i) by using independent local supercoordinate maps. For example for the purely fermionic complex projective plane $\tilde{CP}(m)$ the super-Kähler metric has the Fubini - Study form /see [54] /

ii one can use global parametrization by introducing flat superspace with constraints and equivalence class relations. For example the manifold $\tilde{S}^{f}_{n,m}(C)$ is described by the supercoordinates satisfying the constraints

$$\bar{\Psi}^{i\alpha}\Psi^{\alpha k} + \bar{\Phi}^{ij}\ \Phi^{jk} = \delta^{ik} \qquad \begin{array}{l} i,j,k=1\ldots n \\ \alpha=1\ldots m \end{array} \tag{4.17}$$

where the first term in (4.17) is the $U\alpha(n;C)$ inner scalar product. The manifold $\tilde{G}_{n,m}(C)$ is obtained if we introduce additionally the following equivalence

$$\{\Psi^{\alpha k}, \Phi^{jk}\} \sim \{\Psi^{\alpha i}\ U^{ik},\ \ \Phi^{ji}\ U^{ik}\} \tag{4.18}$$

where $U^{ik} \in U(n)$. It is easy to check that using the freedom (4.18) one can gauge away from $\{\Psi^{\alpha k}, \Phi^{jk}\}$ all the bosonic degrees of freedom.

iii) it is also possible to define supercosets by imposing on supergroup manifold the orbit relations.

5. SUPERSYMMETRIC σ - MODELS : SUPERCOSET APPROACH.

We introduce the following classes of super-coset σ - models:

a) real orthosymplectic $R\tilde{P}(n-1;2m)$ σ-models.

$$\mathcal{M} = \frac{O(n)}{O(n-1)} \longrightarrow \tilde{\mathcal{M}} = \frac{OSp(n;2m)}{OSp(n-1;2m)} \qquad (5.1)$$

If $n=1$, and $Osp(0;2m) \equiv Sp(2m;R)$, one obtains the σ-model considered in [55] ; putting $n=3$ one obtains the super-coset generalization of the $O(3)$ σ-model. The global parametrization by the constraints in $R^{3,2m}$

$$x_1^2 + x_2^2 + x_3^2 = 1 \longrightarrow Y_A\, G^{(o)}_{AB}\, Y_B = Y_A\, Y^A = 1 \qquad (5.2)$$

leads to the model

$$S = \frac{1}{2}\int dx_1 \dots dx_d \left\{ \partial_\mu Y_A\ \partial^\mu Y^A + \lambda\left(Y_A\, Y^A - 1\right)\right\} \qquad (5.3)$$

which is conformal - invariant and has nontrivial topological configurations of Euclidean σ-field for $d=2$. Similarly one can show that the replacement (5.1) for $n=5$ leads to the supercoset extension of $O(5)$ σ-model which is conformal -invariant and has instanton solutions if $d=4$.

b) complex fermionic-Grassmannian $\tilde{G}_{n,m}(C)$ σ-models.

Let us consider the simplest case: the fermionic complex projective plane $C\tilde{P}(m) = \tilde{G}_{1;m}(C)$. In order to relate the model with possible physical applications we choose $m=4$ and $U\alpha(4) = U(2,2)$. In such a case the global parametrization (4.17-18) is ($\alpha=1,2,3,4$)

$$|\Phi|^2 + \bar{\Psi}_\alpha \Psi^\alpha = 1 \qquad (\Phi, \Psi^\alpha) \sim (e^{i\beta}\Phi\,,\ e^{i\beta}\Psi^\alpha) \qquad (5.4)$$

where $\bar{\Psi}_\alpha =$ C and C denotes antihermitean $U(2,2)$ metric /real antisymmetric or symmetric purely imaginary with signature $(+\,+\,,\,-\,,\,-)$/. The five variables in (5.4) transform linearly under the graded superconformal group $SU(2,2;1)$ and were called fermionic supertwistors /see [54] /

Using global parametrization (5.4) one can introduce the following two $C\tilde{P}(4)$ σ-models:

i) bilinear, for $d=2$ /see [56] /

$$W_2 = \frac{1}{2}\int d^2x \left(\overline{\nabla_\mu \Phi}\, \nabla^\mu \Phi + \overline{\nabla_\mu \Psi_\alpha}\, \nabla^\mu \Psi^\alpha \right) \qquad (5.5)$$

where $\nabla_\mu = \partial_\mu - A_\mu$, and the $U(1)$ gauge field A_μ calculated from the action (5.5) with constraints (5.4) has the form

$$A_\mu = \bar{\Phi}\,\partial_\mu \Phi + \bar{\Psi}_\alpha\, \partial_\mu \Psi^\alpha \qquad (5.6)$$

which is an example of the generalization of the Narasimhan-Ramanan formula (1.6) to the case of particular Stiefel supermanifold $\tilde{S}^{f}_{1,m}$ /see (4.15b) /.

Let us assume that the model (5.5) is Euclidean. Because $\widetilde{CP}(m)$ is a Kähler supermanifold, the graded fundamental two-form ω_2 pulled back on $R^2 \wedge \infty = S^2$ provides the formula for the topological charge, which in our case has the form

$$Q = \frac{i}{2\pi}\int d^2x \ \epsilon_{\mu\nu}\left(\overline{\nabla_{[\mu}\Phi}\,\nabla_{\nu]}\Phi + \overline{\nabla_{[\mu}\Psi_\alpha}\,\nabla_{\nu]}\Psi^\alpha\right) = \tag{5.7}$$

$$= \frac{i}{2\pi}\int d^2x \ \epsilon_{\mu\nu} F_{\mu\nu} \qquad \mu,\nu = 1,2$$

where

$$F_{\mu\nu} = \partial_{[\mu}\overline{\Phi}\,\partial_{\nu]}\Phi + \partial_{[\mu}\overline{\Psi}_\alpha\,\partial_{\nu]}\Psi^\alpha \tag{5.8}$$

denotes the U(1) curvature on the fibre $\tilde{S}^{f}_{1,4} \longrightarrow \widetilde{CP}(4)$, which is the fermionic generalization of the Hopf fibering $S_{5,1} = S^9 \longrightarrow CP(4)$ [56].

It is easy to check /see [57]/that the selfduality equations, following from (5.5) and (5.7)

$$\nabla_\mu \Phi = \epsilon_{\mu\nu}\nabla_\nu\Phi \qquad \nabla_\mu\Psi^\alpha = \epsilon_{\mu\nu}\nabla_\nu\Psi^\alpha \tag{5.9}$$

imply that the gauge-independent inhomogeneous σ-field coordinates

$$\chi^\alpha = \Psi^\alpha \cdot (\Phi)^{-1} \tag{5.10}$$

satisfy the free massless two-dimensional Dirac equation.

ii) fourlinear, for d=4

$$W_4 = -\frac{1}{4}\int d^4x \quad F_{\mu\nu}F^{\mu\nu} \tag{5.11}$$

where $F_{\mu\nu} = \partial_\mu A_\nu - \partial_\nu A_\mu$ is given by the formula (5.8).

Such a model describes free electrodynamics with composite e.m. field A_μ, with fundamental fermionic σ-fields transforming as the conformal spinor/twistor/.

The models (5.5) and (5.11) can be extended with the help of simply generalized formulae from Sect. 3 to the case of $\tilde{G}_{4;m}(C)$ σ-model, with the fundamental σ-fields describing the coset space

$$\tilde{G}_{4;m}(C): \qquad \frac{SU(2,2;n)}{SU(2,2)\otimes U(n)} \qquad \begin{pmatrix}\text{quark-twistor}\\ \text{variables [58]}\end{pmatrix} \tag{5.12}$$

Generalization of the formula (5.6) provides the U(n) composite gauge field, and obvious generalization of the action (5.11) provides free U(n) gauge theory, invariant under the extended graded conformal supergroup $SU(2,2;n)$.

c) quaternionic supercoset σ-models $H\tilde{P}(n;m)$

The projective quaternionic space $HP(n)$ can be supersymmetrized as follows

$$\frac{Sp(n+1)}{Sp(n)\otimes Sp(1)} \longrightarrow \frac{UU\alpha(n+1;m;H)}{UU\alpha(n;m;H)\otimes Sp(1)} \qquad (5.13)$$

The quaternionic independent supercoset coordinates transform linearly under the product of three bosonic groups $U(n;H)\otimes U\alpha(m;H)\otimes Sp(1)$. If we choose $n = 2$ because $U(2;H)$ is the spinor form of $O(5)$ the $H\tilde{P}(2;m)$ σ-field is an Euclidean de-Sitter spinor, transforming additionally under the following internal symmetries:

$$\begin{array}{lll} m = 1: & O(1,1) \times SU(2) & \\ m = 2: & SU(1,1) \times SU(2) \times SU(2) & \\ m = 3: & SU(3,1) \times SU(2) & \\ m = 4: & O(6,2) \times SU(2) & \text{etc.} \end{array} \qquad (5.14)$$

The supersymmetric generalization of the fourlinear Euclidean $HP(n)$ σ-model 3.21 describing $SU(2)$ composite Yang-Mills theory is straightforward , with $F_{\mu\nu}$ containing besides the bosonic part (3.15) an additional term obtained from the fermionic quaternionic fields $\xi_\alpha(x) = \xi_\alpha^0(x) + e_r$ x . The formula (3.3) for the topological charge remains valid.

6. SUPERSYMMETRIC σ - MODELS: SUPERFIELD APPROACH.

In the superfield approach one assumes that the "coordinate manifold" is a superspace. One can also interprete the mapping as the particular extension of the internal manifold $\mathcal{M}$ in (1.1) by the exterior product of spinor representations T_α in $\mathcal{S}$ /spinor bundles over $\mathcal{S}$ /

$$\mathcal{M} \longrightarrow \mathcal{M} \otimes \bigoplus_{i=0}^{d} \wedge T_\alpha \qquad (6.1)$$

i.e. the σ-superfield is described by the set of antisymmetric spinor fields $\bigoplus_{i=0}^{d} \varphi^A_{\alpha_1 \ldots \alpha_i}$ with values in $\mathcal{M}$.

We shall consider here only the case $d=2$ and $d=4$. For $d=2$ one uses

— real superspace $(x_1, x_2, \theta_\alpha)$, θ_α real, $\alpha = 1,2$;

the σ -superfield

$$\varphi^A(x,\theta) = \varphi^A(x) + \theta^\alpha \psi_\alpha{}^A + \bar\theta\theta f^A \qquad (6.2)$$

can be real or complex.

— O(2) complex superspace $(x_1, x_2, \theta_\alpha, \bar\theta_\alpha)$, θ_α complex, $\alpha = 1,2$; we shall consider complex chiral σ -superfields

$$\bar D_\alpha \varphi^A(x,\theta,\bar\theta) = 0 \qquad \bar D_\alpha = \frac{\partial}{\partial\bar\theta_\alpha} - \frac{i}{2}(\bar\theta\sigma)^\mu\partial_\mu \qquad (6.3a)$$

$$D_\alpha \bar\varphi^A(x,\theta,\bar\theta) = 0 \qquad D_\alpha = \frac{\partial}{\partial\theta_\alpha} - \frac{i}{2}(\sigma^\mu\theta)\partial_\mu$$

For d=4 we shall consider only real superspace (x_μ, θ_α), θ_α real /Majorana/ $\alpha = 1,2,3,4$, and complex chiral σ -superfields $\Phi_\pm(x,\theta)$:

$$D^\pm_\alpha \Phi_\pm(x,\theta) = 0 \qquad D^\pm_\alpha = \frac{1}{2}(1 + i\gamma_5)D_\alpha \qquad (6.3b)$$

$$D_\alpha = \frac{\partial}{\partial\theta_\alpha} - \frac{i}{2}(\not\partial\theta)_\alpha \qquad \alpha = 1\ldots 4$$

The first σ -superfields introduced in the literature [58,59] were real ; they defined O(3) supersymmetric σ -models for d=2 in real superspace. If the internal manifold $\mathcal{M}$ is complex, one can use simpler /less components ! / chiral σ -superfields.

Introducing σ -superfields one should remember that the cosets $\frac{G}{H}$ can be parametrized

— locally by independent coordinate charts leading to independent -superfields,

— globally e.g. by the relations (1.7), (1.10). The relation (1.7) leads to the constrained superfields, and the equivalence relation (1.10) implies the supergauge invariance,

— one can also use the orbit relations [32]. For d=2 $\mathcal{M}$ -valued chiral independent σ -superfields $z_i(x,\theta,\bar\theta)$, if $\mathcal{M}$ is a Kähler manifold with the metric expressed in terms of Kähler potential

$$h_{\bar i j}(\bar z, z) = \frac{\partial\Sigma(\bar z, z)}{\partial\bar z_i\,\partial z_j}, \qquad (6.4)$$

it has been shown by Zumino [60] that the action is given by the formula

$$W_2 = \frac{1}{2}\int d^2x\, d^2\theta\; d^2\bar\theta\; \Sigma(\bar z, z) \qquad (6.5)$$

This elegant compact form permits to write down e.g. the supersymmetric $G_{n,m}(C)$ σ-model if we only observe that the metric (2.9) can be obtained from the formula (6.5) by putting [2]

$$\Sigma(\varphi^+,\varphi) = \ln\det(I + \varphi\varphi^+) \tag{6.6}$$

If we use the superfields (6.2) with complex components, one can introduce the description of $G_{n,m}(C)$ σ-model in terms of the constrained superfields. The relations

$$\bar{\varphi}^{ij}(x,\theta)\,\varphi^{jk}(x,\theta) = \delta^{ij} \tag{6.7}$$

imply

$$\begin{aligned} &\bar{\varphi}^{ij}\varphi^{jk} = \delta^{ij} \\ &\bar{\psi}_\alpha^{ij}\varphi^{jk} + \bar{\varphi}^{ij}\psi^{ji} = 0 \\ &\bar{\varphi}^{ij}F^{jk} + \bar{F}^{ij}\varphi^{jk} = \bar{\psi}^{ij}\psi^{jk} \end{aligned} \tag{6.8}$$

The action, written down for $m=1$ /CP(n) supersymmetric σ-model/ by d'Adda, di Vecchia , Lüscher [61] and Michailov, Perelomov 62 has the form

$$W_2 = \frac{1}{2}\int d^2x\, d\theta_1\, d\theta_2\, (\overline{\nabla_\alpha\varphi})^{ji}(\nabla^\alpha\varphi)^{ji} \tag{6.9}$$

where / $\sigma^\mu = (\sigma^1, \sigma^2)$, real /

$$\nabla_\alpha = \mathcal{D}_\alpha + A_\alpha \qquad \mathcal{D}_\alpha = \frac{\partial}{\partial\theta_\alpha} + i(\sigma^\mu\theta)_\alpha\partial_\mu \tag{6.10}$$

and

$$A_\alpha^{ij} = \bar{\varphi}^{ij}\mathcal{D}_\alpha\varphi^{jk} \tag{6.11}$$

In order to write the supersymmetric $G_{n;m}(C)$ σ-model for $d=4$ we shall extend the fourlinear $G_{n,m}(C)$ bosonic model (3.14) by considering composite SU(n) gauge superfields in supersymmetric SU(n) Yang-Mills theory [24,25] . The supersymmetric Lagrangean

$$\mathcal{L} = \frac{1}{8}\,\mathrm{tr}\,\bar{D}_-D_+[\bar{\Psi}_{--}\Psi_{++}] + \text{h.c.} \tag{6.12}$$

where the chiral spinor superfields $\Psi_{\pm\pm}$ are defined as follows

$$\Psi_{\alpha\,\pm\pm} = -\frac{1}{2\sqrt{2}}\,\bar{D}_\pm D_\mp\left[e^{\mp 2V} D_{\alpha\,\pm}\, e^{\pm 2V}\right] \tag{6.13}$$

can be written in terms of $G_{n;m}(C)$ chiral σ-fields $\Phi_\pm$ if we postulate the relation

$$\Phi_+\Phi_- = e^{-2V} \tag{6.14}$$

The formula (6.14) expresses the $n\times n$ vector superfield V as composite in terms of the independent components φ^{Ji}, $\psi^{Ji}_{\alpha\pm}$ and F^{Ji}. In particular if we assume that

$$\bar{\varphi}^{iJ}\varphi^{Jk} = \delta^{ik} \qquad \bar{\varphi}^{iJ}\psi^{Jk} = 0 \qquad \bar{\varphi}^{iJ}F^{Jk} = 0 \tag{6.15}$$

the composite superfield V can be written in Wess-Zumino gauge

$$V^{ij} = \frac{1}{4}\bar{\theta}\gamma^{\mu}\gamma_5\theta\, V^{ij}_{\mu} + \frac{1}{4}(\bar{\theta}\theta\bar{\theta})^{\alpha}\chi^{ij}_{\alpha} + \frac{1}{32}(\bar{\theta}\theta)^2 D^{ij} \tag{6.16}$$

where

$$\begin{aligned} V_{\mu} &= i\varphi^{+}\partial_{\mu}\varphi + \frac{1}{2}\psi_{+}^{T}C\gamma_{\mu}\psi_{-} = A_{\mu} + j_{\mu} \\ \chi_{+} &= -i\partial^{\mu}\varphi^{+}\gamma_{\mu}\psi_{-} - \psi_{+}F \end{aligned} \tag{6.17}$$

$$D = -2(\nabla_{\mu}\varphi)^{+}\nabla^{\mu}\varphi + i\psi_{+}^{T}C(\overleftarrow{\nabla} - \overrightarrow{\nabla})\psi_{-} + 2j_{\mu}j^{\mu} - 2F^{+}F$$

and

$$\begin{aligned} (\nabla_{\mu}\varphi)^{+} &= (\partial_{\mu} - iA_{\mu})\varphi^{+} \\ \nabla_{\mu}\varphi &= \partial_{\mu}\varphi + i\varphi A_{\mu} \end{aligned}$$

The action (3.14) is generalized as follows

$$\begin{aligned} \mathcal{L} = &-\frac{1}{4}\left(F_{\mu\nu} + \nabla_{[\mu}j_{\nu]} - i[j_{\mu}, j_{\nu}]\right)^2 \\ &-\frac{i}{2}\chi_{+}^{T}C(\not{\nabla}\chi_{-} - i\gamma^{\mu}[j_{\mu}, \chi_{-}]) + \frac{1}{2}D^2 \end{aligned} \tag{6.18}$$

It can be shown [23] that the Lagrangean (6.18) is invariant under local U(n) gauge transformations and the supersymmetry transformations which are obtained as the superposition of superconformal and supergauge transformations leaving the σ-superfield constraints (6.15) invariant.

The model is formulated in d=4 Minkowski space. Because the O(4) group does not have 4-dimensional real spinor representation, the Euclidean modification of the model is not trivial. It can be easily checked [30] that the fundamental spinor representations of all d=4 Euclidean groups O(4), O(4,1), O(5) and O(5,1) are quaternionic. It seems therefore appropriate to consider in d=4 Euclidean supersymmetric formalism the quaternionic superfields with real Majorana spinors in Minkowski case replaced by quaternionic spinors /a pair of one-dimensional fermionic quaternions in O(4) case/ .

7. FINAL REMARKS

The map of σ-models is large; our aim was not to present many detailed results but provide sufficiently wide perspective. We would like to mention finally these aspects which according to our opinion especially merit further investigations:

a) the physical properties of the fourlinear Lagrangeans and their possible role in explaining dynamically the confirement of quarks.

It should be stressed that the fermionic σ-fields with quark indices in the supersymmetric Lagrangeans (5.11) and (6.18) do not have bilinear kinematic terms and are subcanonical. Due to the conformal invariance the subcanonical dimensions imply long range interaction /using naive arguments: $\frac{1}{K^4}$ behaviour in momentum space/.

b) the role of σ-models in extended supergravity theories.

It has been shown long time ago [63] that Einstein gravity is the $\frac{GLA(4,R)}{O(3,1)}$ σ-model /GLA(4,R) denotes 20-parameter general affine group / with constraints following from the conformal invariance. On the other hand in N=8 extended supergravity it has been shown [64] that there are internal symmetry sectors governed by the dynamics on coset spaces $\frac{E_7}{SU(8)}$ and $\frac{SU(8)}{O(8)}$.

c) The role of <u>fully supersymmetrized</u> σ-models in elementary particle physics.

It seems plausible that spin degrees of freedom should be put into the coordinate monifold $\mathcal{S}$; if colour, electroweak and flavour degrees of freedom describe $\mathcal{M}$ it is very enticing to treat them as the bosonic sectors of the internal symmetry supergroup. Some efforts in this direction recently have been made e.g. by considering the SU(5;1) unification scheme [65,66].

FOOTNOTES :

1. For the quaternionic manifolds the nonvanishing Betti numbers B^k = dimension of $H^k(\mathcal{M};R) \neq 0$ have index k modulo four /k=4n, n, natural / [7,8]. From this property follow the relations (1.5).

2. The notion of inverse Higgs effect is due to Ogiewetski /see [10]/. In the conventional Higgs mechanism the σ-fields are eaten up by

gauge fields and gauge fields become massive; inverse Higgs effect means that the gauge fields are eaten up by σ-fields, and σ-fields become selfinteracting in a way imitating the presence of gauge fields.

3. More detailed presentation of fourlinear supersymmetric σ-models will be given in [22] .

4. We denote $\mathrm{Im}\, q_i = e_r q^r$, $\mathrm{Re}\, q = q^o$.

5. The notion of quaternionic analyticity is thoroughly investigated in [12] .

6. This conclusion was also reached independently by dr D. Maison /private communication/.

7. It should be mentioned that if we remove the requirement c) one can retain the bilinearity in terms of σ-field derivatives by allowing the higher order field equations. If the dimension d is even one obtains the field equations with the kinetic part described by $\frac{d}{2}$ -iterated Laplace /in Euclidean case/ or d'Alambert /in Minkowski case/ operator. Such a possibility for d=4 was considered by some authors /see e.g. [35] / but will not be considered here.

8. For generality we use covariant and contravariant indices, embracing Euclidean as well as Minkowski case. For simplicity we shall omit the obvious Lagrange multiplier terms.

9. We denote by Tr the summ 1... n , and by tr the summ over m / m = 1... n-1 / .

10. For the definition of Berezinian, called also superdeterminant, see [47] .

REFERENCES:

1. S. Helgason, "Differential geometry and symmetric spaces", Academic Press, New York, 1962
2. S. Kobayashi and K.Nom zu, "Foundations of Differential Geometry" , Vol. II, Interscience Publ. New York, 1968
3. A.A. Slavnov, TMF 10, 305 /1972/
4. F. Gürsey and L. Marchildon, Phys.Rev. D 17, 2038 /1978/
5. I.J. Arefeva and A.A. Slavnov, "Geometric origin of Higgs mechanism " /in Russian/ , to be published
6. E.H. Spanier , "Algebraic Topology", Mc Graw-Hill Inc., New York 1966

7. V.Y. Kraines, Trans. Am. Math. Soc. 122, 357 /1966/
8. S. Ishihara, Journ. Diff. Geom. 9, 483 /1974/
9. E.A. Ivanov and V.I. Ogievetski, TMF 25, 164 /1975/
10. M.S. Narasimhan and S. Ramanan , Am. J. Math. 83, 356 /1961/; 85, 223 /1963/
11. J. Lukierski , CERN preprint TH-2678, May 1979; improved version to be publ. in Proc. of Summer Institute, Kaiserslautern, August 1979, Springer Verlag
12. F. Gürsey and H.C Tze , Yale preprint, August 1979 to be published in Annals of Physics.
13. R. Arnowitt and P. Nath, Phys. Lett. 56 B, 177 /1975/
14. C. Fronsdal , Letters Math. Phys. 1, 165 /1976/
15. J. Wess and B. Zumino , Phys. Lett. 56B, 361 /1977/
16. V.I. Ogievetski and E. Sokhatchev, Dubna preprints , 1979
17. P.G.O.Freund and I. Kaplansky , J. Math. Phys. 17, 228 /1976/
18. V. Rittenberg, in "Group - Theoretic Methods in Physics", Proc. of V Int. Symposium, Tübingen, July 1977, publ.Springer Verlag
19. J. Lukierski, "Quaternionic superspaces and supersymmetric extensions of quaternionic groups", to be publ. in Proc. of Intern. Seminar of Group - Theoretical Methods, Zvenigorod /USSR/ , November 1979
20. S. Ferrara, J. Wess and B. Zumino, Phys. Lett. B51, 239 /1974/
21. A. Salam and J. Strathdee, Phys. Rev. D11, 1521 /1975/
22. J. Lukierski, to be published as a lecture at Karpacz Winter School, February 1980
23. J. Lukierski and B. Milewski , to be published
24. S. Ferrara and B. Zumino, Nucl.Phys. B79, 413 /1974/
25. A. Salam and J. Strathdee, Phys. Lett. B51, 353 /1974/
26. D.W. Alekseevski, Funkc. Anal. i ego Prim.,2 , 11/1968//in Russian/
27. A. Gray, Michigan Math. J. 16, 125 /1969/
28. I.A. Wolf, J. Math. Mech. 14, 1033 /1965/
29. D.W. Alekseevski, Izv. AN USSR, Ser. Mat.39, 315 /1975//in Russian/
30. M.F. Atiyah, R. Bott and A. Shapiro, Topology, 3, /suppl.I /, 3 /1964/
31. Y.C. Wong, Proc. Acad. Sci. USA 57, 589 /1967/
32. L. Michel and L.A. Radicati in Coral Gable Conf., 1968
33. H. Eichenherr , Nucl. Phys. B146, 215 /1968/
34. A.M. Perelomov, Comm.Math.Phys. 63, 237 /1978/

35. V.de Alfaro, S.Fubini and G. Furlan, Nuovo Cim. 50A,523 /1979/
36. W.E. Zacharov and A.W. Michailov, JETP 74, 1953 /1978/
37. A.T. Ogielski,Phys. Rev. D , in press
38. T.H.R. Skyrme, Proc. Roy. Soc. A260, 127 /1961/
39. N.K. Pak and H.C.Tze, Ann. of Phys. 117,164 /1979/
40. L.D. Faddeev, Lett. in Math. Phys. 1, 289 /1976/
41. W.N. Romanov and A.C. Schwarz, TMF,37, 305 /1978/ /in Russian/
42. M.A. Semenov-Tian-Szanski and L.D. Faddeev, Westnik LGU,13,81 /1977/
43. M.F. Atiyah, N.J. Hitchin,V.G. Drinfeld and Yu.I. Manin, Phys. Lett. 65A, 185 /1978/
44. F.A. Berezin, Funkc.Analiz i ego pril. 11,n.2/1979//in Russian/
45. R. Gilmore, "Lie groups ,Lie algebras and some of its applications", Wiley-Interscience Publ.,New York 1974
46. F.A. Berezin and G.I. Kac,Mat.Sbornik, 82, 314 /1970/
48. J. Tits, "Tabellen zu den einfachen Lie Gruppen und ihre Darstellungen",Lect.Notes in Math.,Vol.40,Springer Verlag,Berlin,1967
49. F.A. Berezin, Jad.Fiz. 29, 1970 /1979/
50. A. Rogers, Imperial College prepreint, March 1979
51. B. Kostant, in Proc.of the Conf.on Differential-Geom.Methods. in Math.Phys.,Bonn, July 1975, publ. in Springer 1976
52. F.A. Berezin, Jad.Fiz. 30, 1168 /1979/
53. J. Lukierski, in "Supergravity" ed. P.van Nieuwenhuizen and D.Z. Freedman, North-Holland, 1979, p. 301
54. J. Lukierski, ICTP preprint 78/82, J. Math. Phys., in press
55. J. Lukierski, Lett. Math. Phys. 3, 135 /1979/
56. J. Lukierski, "Quarks and Fermionic Geometry",Lecture at IV Hadronic Workshop, Erice, October 1978;in Proceedings, in press; see also Czech. Journ. Phys. 29, 44 /1979/
57. A. Trautman, Int. J. Theor. Phys. 16, 561 /1977/
58. P.di Vecchia and S. Ferrara, Nucl. Phys. B130, 93 /1977/
59. E. Witten, Phys. Rev. 16, 2991 /1977/
60. B. Zumino, CERN preprint TH - 2733, August 1979
61. A. d'Adda, P. di Vecchia and M. Lüscher, Nucl. Phys.B152,125/1979/
62. A.W. Michailov and A.M. Perelomov, JETP Letters 29, 445/1979/
63. A.B. Borisov and V.I. Ogievetski,TMF 21, 329 /1974/
64. E.Cremmer and B. Julia , Nucl.Phys. B159, 141 /1979/
65. J.G. Taylor, Phys. Rev. Lett. 43, 824 /1979/
66. A. Salam, unpublished

SUPERGRAVITY AS THE GAUGE THEORY OF SUPERSYMMETRY

S. Ferrara
Laboratoire de Physique Théorique de l'Ecole Normale Supérieure, Paris, France
and
Laboratori Nazionali di Frascati, INFN, Frascati, Italy

We review different approaches to supergravity as a geometrical theory of local supersymmetry.

INTRODUCTION

The aim of the present discussion is to review equivalent approaches to the theory of supergravity [1] which are based on different gauge groups as well as on different base manifolds.

The first approach [1] [2] considers supergravity as the gauge theory of the graded Poincaré group in ordinary Minkowski space-time. This approach can be viewed as the natural generalization of the Weyl-Cartan formulation of Einstein theory of gravitation. It has the advantages of being very simple and straighrforward, but the disadvantage of being incomplete and limited in application. In particular, it is difficult to explain the existence of the auxiliary fields [3] i.e. fields which vanish when the classical equations of motion of (pure) supergravity are fulfilled. These fields play a crucial role in the construction of supergravity models and in the quantization of supergravity because they couple to matter and to the Faddeev-Popov ghosts.

The second approach uses the concept of superspace first introduced by Salam and Strathdee [4] in the framework of global (rigid) supersymmetry.

The early approach [5] to gauged supersymmetry in superspace used the orthosymplectic groups as gauge groups in superspace. It is very similar to Einstein theory because it is based on a "Riemannian" superspace but it has the drawback of reproducing supergravity only in a limiting situation [6] . The late approach [7] [8] to gauged supersymmetry is based on a more complicated affine (non metrical) geometry in superspace. The gauged group is simply the Lorentz group and there exists no metric tensor in superspace. This approach is even more non conventional for the very reason that the gauged group acts in a reducible way on the supertetrad fields which transform as the (I/2, 1/2) + (1/2, 0) + (0, 1/2) representation of the spinor group

SL(2,C). This reducibility is the origin of the superspace constraints. These constraints are restrictions on the torsion components which enable us to solve the supertetrads (supervielbien) in terms of some prepotential superfields which are the true gauge fields of the theory.

This constrained geometry is required by the fact that the dynamical gauge fields must describe massless excitations with appropriate helicity content. This is not the case in the Riemannian superspace of Nath and Arnowitt [5] [6] in which massless exotic states with helicity higher than two propagate unless a suitable limit is taken.

The superspace approach appears also to be preferable in the case of extended supergravity, i.e. in theories in which the N gauged Fermi generators are labelled by an internal symmetry index. Here the gauged theory in Minkowski space seems to work [9] only for N = 2,and for N$\geq$3 the superspace formulation seems to be unavoidable.

In section II we will review the first approach previously discussed while in section III we will make a survey over superspace. In section IV, we will finally compare the two approaches. Due to limitation in space and time, our presentation will be limited. We will not be able to discuss new schemes which have been developed and which could be convenient alternatives to the approaches discussed in this review. We have in mind the approaches by Ogievetsky, Sokatchev[10] and Siegel and Gates [11]. They have the advantage of using an unconstrained geometry in superspace. Another interesting approach [12] uses dimensional reduction from a higher dimensional space-time. This approach could be particularly useful for a geometrical formulation of extended supergravity models.

II. SUPERGRAVITY AS THE GAUGE THEORY OF THE GRADED POINCARE GROUP

Let us consider the 14-dimensional graded Poincaré algebra whose even part is the usual Poincaré algebra with displacement generators P_r and Lorentz generators M_{rs}

$$[M_{rs}, M_{tu}] = \delta_{st} M_{ru} - \delta_{rt} M_{su} - \delta_{su} M_{rt} + \delta_{ru} M_{st}$$

$$[M_{rs}, P_t] = \delta_{st} P_r - \delta_{rt} P_s \qquad [P_r, P_s] = 0 \tag{1}$$

and the odd generators Q_a obey the following commutation and anticommutation relations

$$[P_r, Q_a] = 0 \qquad [M_{rs}, Q_a] = (\gamma_{rs})_a{}^b Q_b \tag{2}$$

$$\{Q_a, Q_b\} = -\tfrac{1}{2} (\gamma^r C)_{ab} P_r \tag{3}$$

where $C\gamma_r C^{-1} = -\gamma_r^T$ and thus C is the charge conjugation matrix. (We use the conventions and notations of the first of ref.[3]).

We would like to consider now supergravity as the gauge theory of the graded

Poincaré algebra (1) (2) (3) alluded above.
Denoting the (anti) commutation of the 14 generators $X_A = (P_r, M_{rs}, Q_a)$ by

$$[X_A, X_B\} = f_{BA}^{\ \ C} X_C \tag{4}$$

where the symbol $[\ \ \}$ denotes the graded Poisson bracket

$$X_A X_B - (-)^{AB} X_B X_A \qquad A = 0 \text{ for } A = r \text{ (vector)}, \quad A = 1 \text{ for } A = a \text{ (spinor)}$$

one defines a Lie algebra valued vector field

$$h_\mu^{\ A} X_A = e_\mu^{\ r} P_r - \omega_\mu^{\ rs} M_{rs} + \bar{\psi}_\mu^{\ a} Q_a \tag{5}$$

and a parameter field

$$\eta = \epsilon^A X_A = \xi^r P_r + \lambda^{rs} M_{rs} + \bar{\epsilon}^a Q_a \tag{6}$$

and curvatures

$$R_{\mu\nu}^{\ A} = \partial_\nu h_\mu^{\ A} - \partial_\mu h_\nu^{\ A} + f_{BC}^{\ \ A} h_\nu^{\ B} h_\mu^{\ C} \tag{7}$$

The explicit expressions of curvatures defined by (7) are

$$R_{\mu\nu}^{\ rs}(M) = \partial_\mu \omega_\nu^{\ rs} + \omega_\mu^{\ rt} \omega_\nu^{\ ts} - \mu \leftrightarrow \nu \tag{8}$$

$$R_{\mu\nu}(P) = D_\nu e_\mu^{\ r} - D_\mu e_\nu^{\ r} + \tfrac{1}{2} \bar{\psi}_\mu \gamma^r \psi_\nu = \partial_\nu e_\mu^{\ r} + \omega_\nu^{\ rs} e_{s\nu} + \tfrac{1}{4} \bar{\psi}_\mu \gamma^r \psi_\nu - (\mu \leftrightarrow \nu) \tag{9}$$

$$R_{\mu\nu}(Q) = D_\nu \psi_\mu - D_\mu \psi_\nu = (\partial_\nu + \tfrac{1}{2} \omega_\nu^{\ rs} \sigma_{rs}) \psi_\mu - (\mu \leftrightarrow \nu) \tag{10}$$

(more precisely $R_{\mu\nu}$ would be the $\bar{R}_{\mu\nu}$ but we use here a simplified notation)
From the gauge transformations

$$\delta(h_\mu^{\ A} X_A) = \partial_\mu \eta + [\eta, h_\mu^{\ A} X_A] \tag{11}$$

one finds

$$\delta h_\mu^{\ A} = (D_\mu \epsilon)^A = \partial_\mu \epsilon^A + f_{BC}^{\ \ A} h_\mu^{\ B} \epsilon^C \tag{12}$$

hence

$$\delta e^r_{\ \mu} = \tfrac{1}{2} \bar{\epsilon} \gamma^r \psi_\mu \quad , \quad \delta \psi_\mu = D_\mu \epsilon \tag{13}$$

We have not written the gauge variation of the spin connection $\omega_\mu^{\ rs}$ because in gauging the graded Poincaré group one must impose the additional constraint on the P curvature

$$R_{\mu\nu}^{r}(P) = 0 \qquad \text{(torsion free space)} \tag{14}$$

which allows us to solve ω_μ^{rs} in terms of the other gauge fields $e_{r\mu}, \psi_{\mu a}$

$$\omega_\mu^{rs} = \omega_\mu^{rs}(e) + \frac{1}{4}(\bar{\psi}_\mu \gamma^r \psi^s - \bar{\psi}_\mu \gamma^s \psi^r + \bar{\psi}^r \gamma^\mu \psi^s) \tag{15}$$

The constraint (14) is needed in order not to count the translations twice.
We also note that general coordinate transformations are not the same as P-gauge transformations, since

$$\begin{aligned}\delta_{GC} h_\mu^A = \partial_\mu \xi^\nu h_\nu^A + \xi^\nu \partial_\nu h_\mu^A &= \partial_\mu(\xi^\nu h_\nu^A) + f_{BC}^A h_\mu^B (\xi^\nu h_\nu^C) \\ &+ \xi^\nu(\partial_\nu h_\mu^A - \partial_\mu h_\nu^A + f_{BC}^A h_\nu^B h_\mu^C) = \\ &= D_\mu(\xi^\nu h_\nu^A) + \xi^\nu R_{\mu\nu}^A \end{aligned} \tag{16}$$

the term $D_\mu(\xi^\nu h_\nu^A)$ is a sum of gauge transformations including P -gauges which are not symmetries of the theory.
The pure supergravity Lagrangian is [1] (we put the gravitational constant κ = 1)

$$\mathcal{L}_{SG} = -\frac{1}{2} e\, e_r^\mu e_s^\nu R_{\mu\nu}^{rs}(M) - \frac{1}{4}\varepsilon^{\mu\nu\rho\sigma}\bar{\psi}_\mu \gamma_5 \gamma_\nu R_{\sigma\rho}(Q) \tag{17}$$

We note incidentally that if we happened to vary (17) with respect to ω_μ^{rs} we would precisely get the constraint (14) i.e.

$$\delta\mathcal{L}_{SG}/\delta\omega_\mu^{rs} = 0 \quad \Longrightarrow \quad R_{\mu\nu}^{r}(P) = 0 \tag{18}$$

Then because (18) is equivalent to (14) we see that it is correct to use $\delta_\epsilon \omega_\mu^{rs} = 0$ because this variation is multiplied by (18) which vanishes in virtue of the constraint(14). This property is called "1.5 order formalism".

It is amazing to observe that the spin connection given in (15) by solving the group-theoretical constraint (14) coincides with the Weyl-Cartan connection in a theory of gravitation with torsion minimally coupled to spinor matter.

We conclude this section by pointing out that the above property is true only for pure supergravity and for supergravity coupled to (conformal) spinor-vector matter. It is not true for general matter couplings and indeed an explanation of this peculiar property can be found in the existing literature [13].

III. SUPERSPACE SUPERGRAVITY AS THE GAUGE THEORY OF THE LORENTZ GROUP

Some years ago, Salam and Strathdee [4] enlarged ordinary space-time to superspace, a manifold with spinning degrees of freedom. The base manifold of superspace has points parametrized by coordinates

$$Z^\Lambda = (x^\mu, \theta^\alpha) \qquad \mu, \alpha = 1 \ldots 4 \tag{19}$$

(Greek indices denote world tensors, early letters refer to spinors and late letters to vectors). x^μ are (c -number commuting) space-time coordinates while θ^α are (anticommuting) Grassmann variables

$$[x^\mu, x^\nu] = [x^\mu, \theta^\alpha] = \{\theta^\alpha, \theta^\beta\} = 0 \tag{20}$$

In extended superspace - suitable to describe extended supersymmetry - the spinorial coordinates are supplemented by an additional (internal symmetry) Yang-Mills index $\theta^{\alpha i}$. In this lecture we will confine ourselves to the simplest case with no internal symmetry.

Group-theoretically, superspace is the quotient space G/H in which G is the (14-dimensional) graded Poincaré group defined by the graded commutation relations (1) (2) and (3) and H is the Lorentz group. The generators P^r , Q^α act as translations and supertranslations in superspace, respectively.

In the infinitesimal we have

$$\delta z^\Lambda = (\delta x^\mu, \delta\theta^\alpha) = (a^\mu + \tfrac{1}{4}\bar{\epsilon}\gamma^\mu\theta \,,\, -\epsilon^\alpha) \tag{21}$$

The composition rule of supertranslations is obtained by performing the commutation of two infinitesimal transformations

$$[\delta_2, \delta_1] z^\Lambda = ([\delta_2,\delta_1]x^\mu, [\delta_2,\delta_1]\theta^\alpha) = (\tfrac{1}{2}\bar{\epsilon}_2\gamma^\mu\epsilon_1, 0) \tag{22}$$

Relations (21) and (22) show that superspace gives a realization of P^r and Q^α (as well as M^{rs}) in terms of differential operators

$$P_r = i\frac{\partial}{\partial x^r} \qquad\qquad Q^\alpha = \frac{\partial}{\partial\bar{\theta}^\alpha} - \frac{1}{4}(\gamma^\mu\theta)_\alpha \frac{\partial}{\partial x^\mu} \tag{23}$$

A superfield is a function $f(z)$ in superspace. Due to the anticommuting nature of the spinor coordinates, $f(z)$ is equivalent to a finite collection of ordinary fields in Minkowski space-time

$$f(z^\Lambda) = \sum_{n=0}^{4} \frac{1}{n!}\theta^{\alpha_1}\cdots\theta^{\alpha_n} f_{\alpha_1\ldots\alpha_n}(x^\mu) \tag{24}$$

Therefore, $f(z)$ unifies eight Bose and eight Fermi fields in a single object.

A scalar superfield transforms as follows under (global) supersymmetry transformations

$$f'(x',\theta') = f'(x + \tfrac{1}{4}\bar{\epsilon}\gamma^\mu\theta, \theta - \epsilon) = f(x,\theta) \tag{25}$$

From (25) we can deduce supercovariant derivatives

$$\mathcal{D}_A = \left(\frac{\partial}{\partial x^r}, \; D_a = \frac{\partial}{\partial\bar{\theta}^a} + \frac{1}{4}(\gamma^\mu\theta)_a\frac{\partial}{\partial x^\mu}\right) \tag{26}$$

which commute with supertranslations and fulfill the algebra

$$\left[\frac{\partial}{\partial x^r},\frac{\partial}{\partial x^s}\right] = \left[\frac{\partial}{\partial x^k},\mathcal{D}_a\right] = \left\{\mathcal{D}_a,\mathcal{D}_b\right\} - \frac{1}{2}(\gamma^r C)_{ab}\frac{\partial}{\partial x^r} = 0 \tag{27}$$

Relations (27) show that flat superspace has non vanishing supertorsion even though its supercurvature vanishes.

We now turn to curved superspace, following the approach pursued by Wess, Zumino [7] and Brink, Gell-Mann, Ramond and Schwarz [8] . The superspace is the base-manifold, with points labeled by coordinate $z^\Lambda = (x^\mu, \theta^\alpha)$. These points undergo general coordinate transformations

$$z^\Lambda \rightarrow z^\Lambda + \Xi^\Lambda(z) \tag{28}$$

At each point of superspace one erects a local tangent frame and one defines supertetrads (supervielbein field)

$$E^A_\Lambda(z) \qquad A = (a, r) \;,\; \Lambda = (\mu, \alpha) \tag{29}$$

(Latin letters refer to flat indices) with inverse

$$E^\Lambda_A E^B_\Lambda = \delta^B_A \;,\qquad E^A_\Lambda E^\Sigma_A = \delta^\Sigma_\Lambda \tag{30}$$

In the tangent space the gauge group is assumed to be the Lorentz group or more precisely its spinor (universal covering) group SL(2,C) . This is the main difference from an earlier approach made by Nath and Arnowitt [5] [6] in which the gauge group was assumed to be the full orthosymplectic group $Osp(4/1)$. We will see in a moment that this difference in the identification of the gauge groups in superspace is the very reason for the possibility of correctly describing massless particles with a limited number of helicity states which correspond to the known representations of massless one-particle states of global supersymmetry.

In the tangent space of superspace there exists no invertible metric in contrast with the case envisaged by Nath and Arnowitt. This is due to the fact that the Lorentz group acts in a reducible way on the supertetrads. The two tensors

$$\eta_{AB} = \begin{pmatrix} \eta_{rs} & 0 \\ 0 & 0 \end{pmatrix} \;,\qquad \tilde{\eta}_{AB} = \begin{pmatrix} 0 & 0 \\ 0 & C_{ab} \end{pmatrix} \tag{31}$$

are both invariant tensors but they do not have inverses. From this peculiar property it follows that we cannot construct a metric tensor $g_{\Lambda\Sigma}$ out of the vielbein and the resulting space is not only non-Riemannian but also non-metric. As usual, as in any general affine space, we can define covariant derivatives

$$\mathcal{D}_\Lambda = \frac{\partial}{\partial x^\Lambda} + \Omega_\Lambda \tag{32}$$

where Ω_Λ is the Lie algebra valued superconnection

$$\Omega_\Lambda = \Omega_\Lambda^{rs} X_{rs} \tag{33}$$

and X_{rs} are the Lorentz generators.
Covariant derivatives with tangent space indices

$$\mathcal{D}_A = E_A^\Lambda \mathcal{D}_\Lambda = E_A^\Lambda \partial_\Lambda + \Omega_A \tag{34}$$

satisfy the graded commutation relations

$$[\mathcal{D}_A, \mathcal{D}_B\} = R_{AB} - 2 T_{AB}^{\ C} \mathcal{D}_C \tag{35}$$

where R_{AB} is the Lie algebra valued supercurvature

$$R_{AB} = (-)^{\Pi(B+\Lambda)} E_A^\Pi E_B^\Lambda \left(\partial_\Pi \Omega_\Lambda - (-)^{\Lambda\Pi} \partial_\Lambda \Omega_\Pi + [\Omega_\Pi, \Omega_\Lambda\}\right) \tag{36}$$

and $T_{AB}^{\ C}$ is the supertorsion tensor

$$T_{AB}^{\ C} = (-)^{\Lambda(B+\Pi)} E_A^\Lambda E_B^\Pi \left(\mathcal{D}_\Lambda E_\Pi^C - (-)^{\Lambda\Pi} \mathcal{D}_\Pi E_\Lambda^C\right) \tag{37}$$

Covariant derivatives are defined as follows

$$\mathcal{D}_\Lambda E_\Pi^r = \partial_\Lambda E_\Pi^r + \Omega_\Lambda{}^r{}_s E_\Pi^s$$

$$\mathcal{D}_\Lambda E_\Pi^a = \partial_\Lambda E_\Pi^a + \frac{1}{4} \Omega_{\Lambda rs} (\gamma^{rs})^a{}_b E_\Pi^b \tag{38}$$

From the structure of the tangent space-group (Lorentz group) it follows that we can extend Ω_A^{rs} and R_{AB}^{rs} to tensors with components

$$\Omega_A{}^C{}_D = \left(\Omega_A{}^r{}_s, \Omega_A{}^c{}_d = \tfrac{1}{4}\Omega_A^{rs}(\gamma_{rs})^c{}_d, \Omega_A{}^c{}_s = \Omega_A{}^r{}_b = 0\right) \tag{39}$$

(idem for R_{AB}). From (37) we can also solve the connection Ω_A in terms of $T_{AB}^{\ C}$ and E_A^Λ, E_Λ^A as follows

$$2\Omega_{A,CB} = \tilde{T}_{AB,C} + (-)^{CA+BC} \tilde{T}_{CA,B} - (-)^{BA+CA} \tilde{T}_{BC,A}$$

$$\tilde{T}_{AB,C} = \tilde{\eta}_{CC'} \left[T_{AB}^{\ C'} - (-)^{\Lambda(B+\Pi)} E_A^\Lambda E_B^\Pi \left(\partial_\Lambda E_\Pi^{C'} - (-)^{\Lambda\Pi} \partial_\Pi E_\Lambda^{C'}\right)\right]$$

$$(\tilde{\eta}_{CC'} : \eta_{rs}, C_{ab}) \tag{40}$$

From (35) and the Jacobi identities $[[\mathcal{D}_A, \mathcal{D}_B\}, \mathcal{D}_C\} = 0$ it follows that R_{AB} and $T_{AB}^{\ C}$ satisfy two sets of Bianchi identities

$$2\mathcal{D}_{[A} T_{BC]}^{\ \ D} + 4 T_{[AB}^{\ \ C'} T_{C'C]}^{\ \ D} + R_{[AB,C]}^{\ \ \ \ D} = 0 \tag{41}$$

$$\mathcal{D}_{[A} R_{BC]} + 2 T_{[AB}^{\ \ D} R_{DC]} = 0 \tag{42}$$

(where $[ABC]$ means graded cyclic permutation on the three indices A, B, C).

We now come to the main point of the superspace geometry, namely the further constraints on the torsion coefficients which are needed to reproduce the dynamics of supergravity theory correctly. These constraints are [14]

$$T_{ts}^{\;t} = T_{ab}^{\;c} = 0 \tag{43}$$

$$T_{ab}^{\;r} = \frac{1}{4}(C\gamma^{r})_{ab} \tag{44}$$

$$T_{a\;r}^{\;s} = 0 \tag{45}$$

Constraints given by (43) are essentially equivalent to the torsion free condition of Einstein theory. They are sufficient, as easily seen by (40), to solve completely the superconnection Ω_A in terms of the supervielbein E_Λ^A (or its inverse). Conditions (44) and (45) give further relation on the vielbein itself. They state that the inverse vielbein E_A^Λ is only a function of its spinorial vertical component E_a^Λ and moreover they give further constraints on the field E_a^Λ itself. These properties have been widely discussed in refs. (14-15-16). Condition(44) can also be regarded as a sort of equivalence principle for supergravity. It is the condition needed in order for the flat vielbein with components $E^{0A}_\Lambda = (\delta_\mu^r, \delta_\alpha^a, 0, -\frac{1}{4}(\bar{\theta}\gamma^r)_\alpha)$ to be a particular solution of the general constraints (43) (44) (45) with $R_{AB}(E^0) = 0$ and $T_{rs}^{\;a} = T_{rb}^{\;c} = 0$

We would like to stress that the possibility of having constraints like(43)(44) and (45) is closely related to the fact that the tangent space group acts in a reducible way on boson and spinor components. This would be impossible with the orthosymplectic group as a gauge group. In this latter case, the only consistent constraint would be $T_{AB}^{\;C} = 0$, i.e. a torsion free (Riemannian) space. However, this solution would be inconsistent with global (rigid) supersymmetry described at the beginning of this section.

IV RELATION BETWEEN THE SUPERSPACE AND THE SPACE-TIME FORMULATION OF SUPERGRAVITY

The relation between supergravity, as formulated in ordinary space-time, and the superspace version comes by identification of the lowest ($\theta = 0$) components of superspace geometrical quantities with fields defined over space-time [8] [17] [18] . This must be done both for superfields and superparameters. More precisely one requires consistency of the composition rules of the gauge algebra over Minkowski space with parameters $\xi^\mu(x), \epsilon^\alpha(x), \lambda^{rs}(x)$ with the composition rules of the gauge algebra over superspace with parameters $\Xi^\Lambda(z), \Lambda^{rs}(z)$. Moreover, one requires consistency of the transformation rules of the supervielbein under general coordinate and Lorentz transformation in superspace with the transformation rules of the vierbein e_μ^r and Rarita-Schwinger field ψ_μ^a in Minkowski space. This consistency procedure can be carried out provided the following identifications are made as starting points :

$$E_\mu^r(x,\theta=0) = e_\mu^r(x) \quad , \quad E_\mu^a(x,\theta=0) = \psi_\mu^a(x) \tag{46}$$

$$\Omega_\mu^{rs}(x,\theta=0) = \omega_\mu^{rs}(e,\psi) \tag{47}$$

$$\Xi^\alpha(x,\theta=0) = \epsilon^\alpha(x) \, , \; \Xi^r(x,\theta=0) = \xi^r(x) \; , \; \Lambda^{rs}(x,\theta=0) = \lambda^{rs}(x) \tag{48}$$

Then, this consistency procedure enables us to reconstruct all superspace quantities to all orders in θ , once the $\theta = 0$ components are known. However, in order for the procedure to be consistent, the gauge algebra in Minkowski space must close, possibly with field dependent parameters. This is possible only if auxiliary fields [3] are added in the formulation given in section II. These auxiliary fields turn out to be a scalar S , a pseudoscalar P and an axial vector A_μ.
In presence of S , P , A_μ the e_μ^r transformation rule is unchanged but the $\delta\psi_\mu^a$ given by (13) acquires additional terms given [3] by

$$\delta\psi_\mu^a = D_\mu\epsilon^a + \frac{i}{2}A_\mu\gamma_5\epsilon + \frac{1}{6}\gamma_\mu(S - i\gamma_5 P - i \not{A}\gamma_5)\epsilon \tag{49}$$

In superspace, the auxiliary fields turn out to be the irreducible components of the following torsion component [18]

$$T_a{}^c{}_r(x,\theta=0) = \frac{1}{12}\left[\gamma_r(S - i\gamma_5 P - i\not{A}\gamma_5)\right]^c{}_a + \frac{i}{4} A_r(\gamma_5)^c{}_a \tag{50}$$

It is easy to see that if we had put $T_a{}^c{}_r = 0$, then the Bianchi identities (41) and (42) would have implied the Einstein and Rarita-Schwinger equations. This is clearly related to the fact that when $S = P = A_\mu = 0$ the gauge algebra of supergravity closes on the supergravity mass-shell only [19] .
We would like to conclude this discussion by finally giving the geometrical explanation of the extra terms present in $\delta\psi_\mu^a$ as dictated by (49) in contrast with the group-theoretical variation of a gauge field as given by (12) and (13).
The very reason is that Q gauges, like P gauges discussed in the previous sections, are not the same as local supersymmetry transformations but they differ from them by curvature terms according to the law [18]

$$\delta_G(\Xi^\Lambda) h_\Lambda^A = (D_\Lambda^G H)^A - \Xi^\Pi R_{\Pi\Lambda}^A \tag{51}$$

where $H = \Xi^\Lambda h_\Lambda^A$
Now for $\psi_\mu^a = E_\mu^a(x,\theta=0)$ the law (51) gives the gauge law as in (13) plus an extra term proportional to $T_a{}^b{}_\mu$. With the result given in (50) one may verify that the curvature term in (51) indeed reproduces the field variation as given by (49).

REFERENCES

[1] D.Z. Freedman, P. Van Nieuwenhuizen and S. Ferrara, Phys. Rev.D13 (1976) 3214
S. Deser and B. Zumino, Phys. Lett. 62B (1976) 335
[2] S.W. Mac Dowell and F. Mansouri, Phys. Lett. 38 (1977) 739
A. Chamseddine and P.C. West, Nucl. Phys. B129 (1977) 34
[3] S. Ferrara and P. Van Nieuwenhuizen, Phys. Lett. 74B (1978) 333
K. Stelle and P.C. West, Phys. Lett. 74B (1978) 330
[4] A. Salam and J. Strathdee, Nucl. Phys. B76 (1974) 477
[5] R. Arnowitt and P. Nath, Phys. Lett. 56B (1975) 117
[6] P. Nath and R. Arnowitt, Phys. Lett. 65B (1976) 63
[7] J. Wess and B. Zumino, Phys. Lett. 66B (1977) 361
[8] L. Brink, M. Gell-Mann, P. Ramond and J.H. Schwarz, Phys. Lett. 74B (1978) 336
[9] P.K. Townsend and P. Van Nieuwenhuizen, Phys. Lett. 67B (1977) 439
[10] V.I. Ogievetsky and F. Sokatchev, Phys. Lett. 79B (1978) 22
[11] W. Siegel, Nucl. Phys.B142 (1978) 301
W. Siegel and S.J. Gates, Nucl. Phys. (1979) 77
[12] A. D'Adda, R. D'Auria, P.Fré and T. Regge, Turin preprint (1979)
[13] S. Ferrara and P. Van Nieuwenhuizen, Phys. Rev.D20 (1979) 2079
[14] R. Grimm , J. Wess and B. Zumino, Nucl. Phys. B152 (1979) 255
[15] S.J. Gates and W. Siegel, HUTP 79/A034, Harvard Preprint
[16] K. Stelle and P.C. West ICTP/79-80/5
S.J. Gates, K. Stelle and P.C. West ICTP 79-80/8, Imperial College preprint
[17] J. Wess and B. Zumino, Phys. Lett. 79B (1978) 394
[18] S. Ferrara and P. Van Nieuwenhuizen, ITP-SB-79-78, to appear in Annals of Physics
[19] D.Z. Freedman and P. Van Nieuwenhuizen, Phys. Rev. D14 (1976) 912

HYPERGRAVITIES

S. Deser*

Department of Physics
Brandeis University
Waltham, Massachusetts 02254

The gravitational couplings of higher spin gauge fields are considered. We exhibit, for fermions of spin $\geq 5/2$ (hypergravity) and for a spin 2 field, the strong restrictions imposed by the consistency requirements on the matter equations. For spin 2, they are constraints on the field itself, while for fermions they require vanishing of the Weyl tensor.

Introduction

This is a report on current work carried out in collaboration with C. Aragone on the properties of higher spin ($>3/2$) fields and their interactions with gravitation, particularly spins 2 and 5/2.

One motivation for this study is to see whether it is possible to extend the framework of supergravity, in which spin 3/2 coupling is made consistent, to spin 5/2 interactions. This is of importance since there is a direct link in supersymmetric theories between the highest spin of a supermultiplet and the internal symmetry it can carry, while the supergravity multiplet with its maximal O(8) invariance seems too narrow to accomodate the known groups of particle physics. Since supergravity [1,2] itself is the unification of the free global spin (2+3/2) multiplet into a locally symmetric interacting theory [3], one might hope that the same extension can be carried out for the a priori similar (5/2,2) model. In any case, being able to "bound" the possible interaction of matter and gravity will give a better understanding of what makes supergravity exceptional.

The study of the bosonic system of spin 2 matter coupled to gravity is also closely related to this program. On the one hand, if higher spin supersymmetry is possible, it will require other spin 2 components than the graviton, in order to build up larger multiplets than just the basic O(1) spin (3/2,2) combination. On the other hand, earlier negative conclusions [4] regarding the spin 2-gravity system were carried out before the "miracles" of supergravity were understood, and it might be hoped that similar miracles could occur here (e.g. by including torsion and suitable nonminimal couplings). In fact, we shall see that although some miracles do occur (and in particular that one occurs only for spin 2), they are not sufficient to ensure consistency. The conclusions reached here are then essentially negative and tend to reinforce the preferred role of supergravity as the only consistent theory which involves gravity and spins greater than 1, at least in a finite particle multiplet.

Spin 2-Gravity [5]

We show here that despite some miracles, spin 2-gravity coupling cannot be consistent without excessive constraints on the spin 2 field's dynamics, and that these results hold for minimal or nonminimal coupling and with or without torsion.

Let us first recall the problems of higher spin couplings to gravity and the miracles whereby they are avoided by spin 3/2. This will give us a comparison for treating the spin 2 and S>3/2 fermion cases. The basic consistency requirement is that the divergence of the matter field equation vanish, if not identically, then at least by virtue of the Einstein equations. This means in particular that the Weyl tensor must not appear, since it is not determined by the field equations. [For spins 0 or ½ the problem does not arise at all since their field equations have no free indices.] For spin 1, the identity $\partial_\mu(\partial_\nu F^{\mu\nu}) \equiv 0$ does not change form in curved space, and there is no problem with the Ricci identity $[D_\mu, D_\nu] \sim$ curvature. For spin 3/2, the Rarita-Schwinger equation has the general form $R^\mu \sim \varepsilon^{\mu\nu\alpha\beta}\gamma_s\gamma_\nu D_\alpha\psi_\beta$ and its divergence is of the schematic form

$$2D_\mu R^\mu \sim iG_{\mu a}\gamma^a\psi^\mu \equiv (G_{\mu a}-T_{\mu a}(\psi))i\gamma^a\psi^\mu + T_{\mu a}i\gamma^a\psi^\mu . \tag{1}$$

The Weyl tensor is indeed absent ("miracle" 1) and the leftover $T\gamma\psi$ term actually vanishes identically by Fierz rearrangements (miracle 2). The fact that $D_\mu R^\mu$ is proportional to the Einstein equation (G-T) implies the existence of a local fermionic gauge invariance [2] of the interacting system, since $R^\mu = \delta I/\delta\psi_\mu$, $(G-T)=\delta I/\delta e_{\mu a}$ and (1) is equivalent to the well-known identity of supergravity $\int[\delta I/\delta\psi_\mu 2D_\mu\alpha(x)+\delta I/\delta e_{\mu a} i\bar{\alpha}\gamma_a\psi_\mu] \equiv 0$. The latter means that $\delta\psi_\mu = 2D_\mu\alpha$ $\delta e_{\mu a} = i\bar{\alpha}\gamma_a\psi_\mu$ is a (local) invariance. Conversely, we shall see that consistency problems arise because of the absence - or breakdown - of the corresponding identity in the coupled higher spin case. The first "miracle", absence of the Weyl tensor $C_{\mu\nu\alpha\beta}$ in the analog of (1) does occur for spin 2, and indeed only for spin 2 (for spin 3/2 it is in fact not a miracle, since C simply cannot be present by index counting). The second miracle could occur for bosonic fields. For example, for a scalar field, there is an identity of the type we would need,

$$[T^{\mu\nu}(\phi) - g^{\mu\nu}T^\alpha_\alpha(\phi)]\partial_\nu\phi \equiv 0 \tag{2}$$

bu unfortunately it doesn't quite work for spin 2.

The action of spin 2 minimally coupled to gravity has the form

$$I_2(\psi_{\mu\nu};g) = \int[h^{\mu\nu}(\gamma_{\mu\nu;\alpha}{}^{\alpha}-\gamma_{\mu\alpha}{}^{\alpha}{}_{;\nu}) + \sqrt{-g}\, g^{\mu\nu}(\gamma_{\mu\alpha}{}^{\beta}\gamma_{\nu\beta}{}^{\alpha}-\gamma_{\mu\nu}{}^{\alpha}\gamma_{\alpha\beta}{}^{\beta})] \tag{3}$$

where $h^{\mu\nu} \equiv \psi^{\mu\nu} - \frac{1}{2}g^{\mu\nu}\psi^\alpha_\alpha$ is a useful choice of field variable, and $\gamma_{\mu\nu}{}^{\alpha}$

is the conjugate variable in first order formulation. All index operations and covariant differentiations are with respect to the metric $g_{\mu\nu}$. It may be considered as (one possible) linearization about a given background of the Einstein action in terms of the variations $h^{\mu\nu} \equiv \delta(\sqrt{-g}\, g^{\mu\nu})$, $\gamma_{\mu\nu}{}^{\alpha} = \delta\Gamma_{\mu\nu}{}^{\alpha}$. The combined field equations are

$$G_L^{\mu\nu}(\psi,g) = 0 \qquad (4a) \qquad ; \qquad G_{\mu\nu}(g) = \kappa^2 T_{\mu\nu}(\psi) \qquad (4b)$$

where $G_L^{\mu\nu}(\psi)$ is the linearized Einstein tensor formed from the second covariant derivatives of the ψ field. However, $D_\mu G_L^{\mu\nu}$ fails to vanish; it takes the form

$$D_\mu G_L^{\mu\nu} = R^{\nu}{}_{\alpha} D_\beta h^{\beta\alpha} + [R_{\nu\alpha;\beta} + R_{\nu\beta;\alpha} - R_{\alpha\beta;\mu}]h^{\alpha\beta} \quad . \qquad (5)$$

In the flat space (free field) case, $\partial_\mu G_L^{\mu\nu}$ is indeed identically conserved, being linked to the abelian invariance of $I_2(\psi)$ under $\delta\psi_{\mu\nu} = \partial_\mu \xi_\nu + \partial_\nu \xi_\mu$. The breakdown embodied in (5) is due to the fact that the generalization $\delta\psi_{\mu\nu} = D_\mu \xi_\nu + D_\nu \xi_\mu$ is not an invariance, even if linked to any ξ-dependent variation one might simultaneously make on the metric. Note that there is no dependence in (5) on the Weyl tensor, which may be understood because small excitations about a given background are consistently described by a spin 2-like field, and this could not occur were $C_{\mu\nu\alpha\beta}$ unavoidably present in any spin 2 formulation. Unfortunately, insertion of the Einstein equation (4b) into (5) does not lead to complete cancellation (except in the approximation $R_{\mu\nu} = 0$ which represents the linearization about an Einstein background).

This situation is quite analogous to that of small excitations about a Yang-Mills background as well, whose action is consistent only if the latter satisfies the source-free Yang-Mills equations. What goes wrong specifically when (4b) is inserted into (5) is that it leaves a number of terms involving the pure helicity two parts of $h^{\mu\nu}$ and its derivatives, and the requirement takes on the generic form $hh\,\partial^3 h = 0$. These terms cannot be removed either by gauge conditions or by use of the matter field equations (4a).

Now even for supergravity, torsion was required at this stage to cancel the leftover terms in the analog of (5). It is therefore natural to introduce it here as well. Basically torsion is proportional to the variational derivative of the spin 2 action with respect to the covariant derivative (or $\Gamma_{\mu\nu}{}^{\alpha}$), i.e. it is proportional to $h\gamma$ or $h\,\partial\, h$, and leads to effective contact terms in the action $\sim (h\,\partial\, h)^2$, which are of the generic type needed to cancel the constraints we obtained. Nevertheless, although it too generates $hh\,\partial^3 h$ terms, it is not possible to achieve cancellation.

The remaining possibility is to include non-minimal couplings. We have also tried all possible such couplings, $\sim R_{\mu\nu}\psi^{\mu\alpha}\psi^{\nu}{}_{\alpha} + R_{\mu\nu\alpha\beta}\psi^{\mu\beta}\psi^{\nu\alpha} + R\psi_{\mu\nu}\psi^{\mu\nu} + \ldots$ and again complete cancellation of the extra terms does not occur, even in the optimal form $G_{\alpha\beta}\gamma^{\alpha\beta}{}_{\mu} = 0$ one can obtain for the consistency condition by appropriate choice of the non-minimal terms. In addition, non-minimal coupling runs into additional causality problems on the spin 2 field propagation. Likewise, any other choice of basic spin 2 variables (e.g. vierbein-like) does not help (one choice corresponds to relative non-minimality with respect to any other [6]).

The underlying difficulty is that the original flat space abelian invariance of the spin 2 field, which ensured that it had only helicity ± 2 excitations, cannot be maintained in curved space, nor can additional variations of the metric compensate for this loss. There is no new algebra linking matter and gravity, in other words, as there is in supergravity. This failure is also related to the consistency requirement on the Einstein equations, that the stress tensor be (covariantly) conserved in order to comply with the Bianchi identity $D_\mu G^{\mu\nu} \equiv 0$. Under a coordinate transformation,

$$\delta_c g_{\mu\nu} = D_\mu \xi_\nu + D_\nu \xi_\mu \quad , \quad \delta_c \psi_{\mu\nu} = D_\mu(\xi^\alpha \psi_{\alpha\nu}) + D_\nu(\xi^\alpha \psi_{\alpha\mu}) + 2\xi^\alpha \gamma_{\mu\nu\alpha} \tag{6}$$

the total action is a coordinate invariant, which transforms explicitly as

$$\begin{aligned} 0 = \delta_c\left[I_E + I_2(\psi, g)\right] &= \int \{\delta_c g_{\mu\nu}[\xi^{\mu\nu} - T^{\mu\nu}] + \delta_c \psi_{\mu\nu} \delta I/\delta\psi_{\mu\nu}\} \\ &= \int \xi^\alpha [D_\mu T^\mu{}_\alpha + \psi_{\alpha\nu} D_\mu G_L^{\mu\nu} + \gamma_{\mu\nu\alpha} G_L^{\mu\nu}] \quad . \end{aligned} \tag{7}$$

It is clear that for conservation of the stress tensor, the matter field equations $G^L_{\mu\nu} = 0$ do not suffice; the consistency condition $D_\mu G_L^{\mu\nu} = 0$ is also needed.

We conclude then, that a linear spin 2 system coupled to the Einstein field minimally or not, with or without torsion is subjected to very strong constraints on its dynamics despite the "miraculous" absence of any Weyl tensor dependence in the constraints.

We also note that, although it is not within our present aims, one can also investigate the massive spin 2 case. It too, despite not having gauge problems, also seems to be too strongly constrained by gravitational coupling.

S > 3/2 Fermions and Gravity

As stated in the introduction, coupling of higher spin fermions, particularly the other nearest neighbor, S = 5/2, of the graviton is of importance in understanding the limitations of supersymmetry, in attempting to extend supergravity beyond its present bounds, and to reconsider the problems of higher spin gravity couplings in the light of the successes of supergravity. For details and further references, see [7] for the properties of free higher spin fermions, [8] for their various formulations in tensor and "vierbein" form and [8,9] for their coupling to gravity.

We will concentrate here on the gravity coupling problem. At the free field level, there is not apparent distinction between the (2,3/2) and (5/2,2) multiplets. Both have the appropriate abelian invariances enduring that only helicities $\pm S$ are present, the global version of coordinate invariance and global boson-fermion supersymmetry. The basic difference is that the Noether current associated with the latter is a vector-spinor (like ψ_μ) of spin 3/2, whereas spin 5/2 is represented by a symmetric tensor-spinor $\psi_{\mu\nu}$ or by a "vierbein" nonsymmetric $\psi_{\mu\tilde{a}}$ with $\gamma^a\psi_{\mu\tilde{a}} \equiv 0$. Its source would therefore have to be of the same type; but it would be the Noether current not of the grading $G_{1/2}(P_\psi)$, but rather of $G_{3/2}(P_\psi)$. Unfortunately, although this graded extension of the Poincare algebra P_ψ can be given [10], it is known [11] that it does not admit particle representations. This means that there is no appropriate conserved current to which $\psi_{\mu\nu}$ can couple. We shall see this in a more explicit way by considering the coupled equations directly.

That the consistency conditions for higher spin fermions involve the full Riemann tensor (and hence cannot be removed by use of the Einstein equation) was already noted in the first two papers of [4], and independently of [9] in [12]. Here we will see in detail just how strong the constraints really are. We shall omit torsion for simplicity, because it cannot help with the basic Weyl tensor problem. In terms of the symmetric tensor representation, the spin 5/2 action has the form [13]

$$I_{5/2}[\psi;e_{\mu\alpha}] = -\frac{i}{2}\int [\bar{\psi}^{\mu\nu}\not{D}\psi_{\mu\nu} + \frac{1}{2}\bar{\psi}\not{D}\psi - 2\bar{\psi}D_\nu\psi^{\nu'} - 2\bar{\psi}_{\mu\nu}\gamma^{\nu\kappa\lambda}D_\kappa\psi^\mu{}_\lambda] \tag{8}$$

where $\psi \equiv \psi_\alpha{}^\alpha$, $\psi'_\mu \equiv \gamma^\alpha\psi_{\alpha\mu}$, $\gamma^{\nu\kappa\lambda} \equiv \epsilon^{\nu\kappa\lambda\sigma}\gamma_\sigma\gamma_5$, and all contractions, differentiations etc are understood to be covariantly with respect to the vierbeins. Whereas the free action is invariant under

$$\delta\psi_{\mu\alpha} = \partial_\mu\xi_{\tilde{\alpha}} + \partial_\alpha\xi_{\tilde{\mu}} \quad , \quad \gamma^\mu\xi_{\tilde{\mu}} \equiv 0 \tag{9}$$

this is not longer the case when $\partial \to D$.

The consistency constraints which result may, after appropriate use of γ-algebra be recast essentially in the form

$$C^{+}_{\mu\nu\alpha\beta}\gamma^{\alpha}\psi_{+}^{\nu\beta} + C^{-}_{\mu\nu\alpha\beta}\gamma^{\alpha}\psi_{-}^{\nu\beta} + R_{\nu\lambda}\gamma\psi = 0 \tag{10}$$

where $C_{\mu\nu\alpha\beta}$ is the Weyl tensor, $R\gamma\psi$ represents the Ricci dependent terms and $C^{\pm}$ and $\psi_{\pm}$ are the pure self dual and helicity components of the fields:

$$C^{\pm}_{\mu\nu\alpha\beta} \equiv \frac{1}{2}(C_{\mu\nu\alpha\beta} \pm iC^{*}_{\mu\nu\alpha\beta}) \quad , \quad \psi^{\mu\nu}_{\pm} \equiv \frac{1}{2}(1 \pm i\sqrt{5})\psi^{\mu\nu} \quad . \tag{11}$$

Then (10) requires either that the full Weyl tensor vanish, or that (C^{+}, ψ_{-}) or (C^{-}, ψ_{+}) do. But in real Minkowski signature, $\psi_{\pm} = 0$ implies $\psi = 0$, and likewise $C^{+} = 0$, means $C = 0$. These requirements agree with those obtained earlier from consideration [14] of propagation of general irreducible representations of the Lorentz group in a background gravitational field in harmonic matter gauge.

Exactly the same constraints arise also from the vierbein formulation [8] in which the action has the Rarita-Schwinger form

$$I_{5/2} = \frac{i}{2}\int \bar{\psi}_{\rho\tilde{a}}\gamma^{\rho\mu\nu}f_{\mu\nu\tilde{a}} \quad , \quad \gamma^{\alpha}\psi_{\rho\tilde{a}} \equiv 0 \tag{12}$$

with $f_{\mu\nu\tilde{a}} \equiv D_{\mu}\psi_{\nu\tilde{a}} - D_{\nu}\psi_{\mu\tilde{a}}$. Here the flat-space invariance under $\delta\psi_{\mu\tilde{a}} = \partial_{\mu}\xi_{\tilde{a}}$ breaks down when $\partial_{\mu} \to D_{\mu}$ and the same calculation as above, i.e. taking the divergence of the field equations once again leads to (10).

Our negative results cannot be improved by non-minimal coupling or torsion, as is clear on dimensional grounds. However, it may be that they are weaker when viewed in Euclidean signature where self-duality is perfectly acceptable. This question, as well as the independent one of whether the Ricci-dependent terms can be made to vanish (possibly with torsion) is open at present.

It would be surprising if going to still higher spin fermions would improve matters, and indeed it does not [8]. We will sketch the general form here. To a spin S + 3/2 particle, we associate the field $\psi_{\mu a_1 \dots a_s}$, which is symmetric and γ-transverse in all the tangent space a-indices. The action may again be put in the Rarita-Schwinger form in which the a_i act essentially as internal indices:

$$I^{S+3/2} = \frac{i}{2}\int \bar{\psi}_{\rho a_1 \dots a_s}\gamma^{\rho\mu\nu}f_{\mu\nu a_1 \dots a_s} \quad ,$$
$$f_{\mu\nu a_1 \dots a_s} \equiv D_{\mu}\psi_{\nu a_1 \dots} - D_{\nu}\psi_{\mu a_1 \dots} \tag{13}$$

The free action is invariant under $\delta\psi_{\mu a_1 \dots} = \partial_{\mu}\xi_{a_1 \dots}$, where ξ is also symmetric and γ-traceless. But because $[D_{\mu}, D_{\nu}] \neq 0$, we again find constraints of the form (10). More precisely, the leading helicity parts of $\psi_{\mu a_1 \dots}$ appear in them through

$$C^{+}_{a_1\lambda\mu\nu}\gamma^{\nu}\psi^{\lambda\mu}_{-\,a_2 \dots} + C^{-}_{a_1\lambda\mu\nu}\gamma^{\nu}\psi^{\lambda\mu}_{+\,a_2 \dots} + \dots = 0 \tag{14}$$

where we have omitted the additional lower spin representations contained in ψ as well as Ricci-tensor terms. The $\psi_{\pm}$ again represents the pure helicity parts of ψ, and $C^{\pm}$ the pure dual parts of the Weyl tensor.

As mentioned earlier, the various representations, vierbein or symmetric tensor, for the spinors are equivalent as far as consistency is concerned, and the form (14) would also appear in the tensor represnetation. It is also clear that (as for spin 5/2) one must either have vanishing Weyl tensor or vanishing ψ-field in order to fulfill (14).

The history of supergravity has shown that it is dangerous to state that consistency can never be achieved. It is conceivable that cancellations might occur with an infinite tower of fields, for example, but it is difficult to avoid the conclusion that a simple extension of supergravity to include spin 5/2 as an elementary excitation (let alone higher spin fermions) cannot be constructed within the normal field theoretical framework. There is clearly a challenge in extending the O(8) limit of supergravity given these results.

REFERENCES

* Supported in part by the NSF under grant no. PHY78 09644 A01.

1. D.Z. Freedman, P. van Nieuwenhuizen, and S. Ferrara, Phys. Rev. D13, 3214 (1976).
2. S. Deser and B. Zumino, Phys. Lett. 62B, 335 (1976).
3. D.G. Boulware, S. Deser, and J.H. Kay, Physica 96A, 141 (1979).
4. H. Buchdahl, Nuov. Cim. 10, 96 (1958); 25, 486 (1962). C. Aragone and S. Deser, Nuovo. Cim. 3A, 709 (1971).
5. C. Aragone and S. Deser, Nuov. Cim. (in press).
6. C. Aragone, A. Restuccia, and R. Siri, Gen. Rel. and Grav. 5, 643 (1974).
7. C. Aragone and S. Deser, Phys. Rev. D (1980).
8. C. Aragone and S. Deser, Nucl. Phys. (1980).
9. C. Aragone and S. Deser, Phys. Lett. 86B, 161 (1979).
10. B.G. Konopel'chenko, Sov. J. Nucl. Phys. 23, 478 (1976).
11. R. Haag, J.T. Lopuszanski, and M. Sohnius, Nucl. Phys. B88, 275 (1975).
12. F.A. Berends, J.W. van Holten, B. de Wit and P. van Nieuwenhuizen, Stony Brook preprint ITP-SB-79-13.
13. J. Schwinger, Particles, Sources and Fields (Addison Wesley 1970).
14. S.M. Christensen and M. Duff, Nucl. Phys. (1979).

P A R T II

Proceedings of the Conference

Held at Salamanca

September 10 - 14, 1979

Edited by P.L. García and A. Pérez-Rendón

P R E F A C E

The second part of the Conference on "Differential Geometrical Methods in Mathematical Physics" of 1979 was held in the *Facultad de Ciencias de la Universidad de Salamanca* from September 10th to 14th, 1979.

The members of the International Advisory Committee were: K. Bleuler (Bonn), H. D. Doebner (Clausthal), D. Galletto (Torino), B. Kostant (M.I.T.), A. Lichnerowicz (College de France), K. Maurin (Warsaw) and S. Sternberg (Harvard & Tel-Aviv).

The second part of this volume includes 21 talks grouped into four Chapters: Gauge Theories; Quantization and Symplectic Structures; General Relativity; Classical Field Theory and Analytical Mechanics.

The organizers would like to express their gratitude to all participants for their contributions and cooperation and to Springer-Verlag for its readiness to publish these proceedings.

We would like to acknowledge the financial support of the *Ministerio de Universidades e Investigación*, the *Consejo Superior de Investigaciones Científicas*, the *Rectorado de la Universidad de Salamanca* and the *Dirección de los Cursos Internacionales de la Universidad de Salamanca*, which have made the celebration of this Conference possible for the first time in Spain.

Salamanca, June 1980 P.L. García A. Pérez-Rendón

List of Participants

Abad, J. - Physics Dept., Univ. Zaragoza, Spain.

Aguirre, E. - Institut fur Theoretische Physik, Clausthal, Germany.

Alonso, J.M. - Math. Dept., Univ. Salamanca, Spain.

Alves, A.S. - Math. Dept., Univ. Coimbra, Portugal.

Ancochea, G. - Real Academia de Ciencias, Madrid, Spain.

Andersson, S. - Institut fur Theoretische Physik, Clausthal, Germany.

Angermann, B. - Institut fur Theoretische Physik, Clausthal, Germany.

Asorey, M. - Physics Dept., Univ. Zaragoza, Spain.

Baddou, J. - Institut National de Statistique et D'Economie Appl., Rabat, Marruecos.

Banyaga, A. - Math. Dept., Harvard Univ., Cambridge, U.S.A.

Bellaiche, A. - Universite Paris VII - U.E.R. Mathematiques, Paris, France.

Benenti, S. - Istituto di Fisica Matematica, Univ. di Torino, Italy.

Berg, H.P. - Institut fur Theoretische Physik, Clausthal, Germany.

Binz, E. - Lehrstuhl fur Mathematik I, Univ. Mannheim, Germany.

Bleuler, K. - Institut fur Theoretische Kernphysik der Universitat Bonn.

Bobo, M. - Math. Dept., Univ. Salamanca, Spain.

Bott, R. - Math. Dept., Harvard Univ., Cambridge, U.S.A.

Boyero, A. - Math. Dept., Univ. Salamanca, Spain.

Bueno, A. - Math. Dept., Univ. Salamanca, Spain.

Calderón, J. - Math. Dept., Univ. Simón Bolivar, Caracas, Venezuela.

Cariñena, J.F. - Physics Dept., Univ. Zaragoza, Spain.

Carmeli, M. - Department of Physics, Univ. Ben Gurion, Beer Sheva, Israel.

Carvalho, M^{a} - Escola Secundaria D. Luisa de Gusmao, Lisboa, Portugal.

Castro, R. - Math. Dept., Univ. Santiago de Compostela, Spain.

Cembranos, M^{a}. L. - Math. Dept., Univ. Salamanca, Spain.

Cid, L. - Math. Dept., Univ. Carabobo, Valencia, Venezuela.

Chinea, F.J. - Physics Dept., Univ. Madrid, Spain.

Cobos, J.A. - Math. Dept., Univ. Salamanca, Spain.

Curras, C. - Math. Dept., Univ. Barcelona, Spain.

Cushman, R. - Rijksuniversiteit te Utrecht, The Netherlands.

Dedecker, P. Inst. of Math., Univ. Lovaina, Belgium.

Dedene, G. - Katholieke Universiteit Leuven, Dept. Wiskunde, Leuven, Belgium.

Deser, S. - Physics Dept., Univ. Brandeis, Waltham, U.S.A.

Diaz, A. - Math. Dept., Univ. Barcelona, Bellatera, Spain.

Doebner, H.D. - Institut fur Theoretische Physik, Clausthal, Germany.

Ferraris, M. - Istituto di Fisica Matematica, Torino, Italy.

Forger, M. - Institut fur Theoretische Physik, Berlin.

Galián , R. - Math. Dept., Univ. Salamanca, Spain.

Gambi, J. - Physics Dept., Univ. Valladolid, Spain.

Gårding, L. - Lunds Universitets Matematika Institution, Lund, Swedeen.

García, P. - Math. Dept., Univ. Salamanca, Spain.

Gaspar, M^{a}. - Math. Dept., Univ. Complutense, Madrid, Spain.

González-Tablas, R. - Math. Dept., Univ. Salamanca, Spain.

Guillemin, V. - Massachusetts Institute of Technology, Cambridge, MA, U.S.A.

Günther, Ch. - Technische Universitat, Berlin.

Heß, H. - Fachbereich Physik, Freier Univ. Berlin.

Hernández, L. - Math. Dept., Univ. Salamanca, Spain.

Hernández Ruipérez, D. - Math. Dept., Univ. Salamanca, Spain.

Hoyos, J.J. - Univ. Alcalá de Henares, Madrid, Spain.

Kaiser, G. - Math. Dept., Univ. Lowell, U.S.A.

Kerner, R. - Mecanique Dept., Univ. Pierre et Marie Curie, Paris, France.

Kijowski, J. - Inst. of Math. Methods in Physics, Univ. Warsaw, Poland.

Kopczynski, W. - Inst. Theoretical Physics, Univ. Warsaw, Poland.

Kostant, B. - Math. Dept., M.I.T., Cambridge,MA, U.S.A.

Lancho, M. - Math. Dept., Univ. Salamanca, Spain.

León, M. - Math. Dept., Univ. Santiago de Compostela, Spain.

Libermann, P. - Universite Paris VII, Paris, France.

Lichnerowicz, A. - College de France, Paris, France.

Lorenzo, J. - Math. Dept., Univ. Salamanca, Spain.

Losada, R. - Escuela Superior de Ingenieros Industriales, Bilbao, Spain.

Lozano, M^{a}. T. - Math. Dept., Univ. Zaragoza, Spain.

Lucas, D. - Faculdade de Ciencias e Tecnologia, Univ. Coimbra, Portugal.

Marle, C.M. - Academie de Paris, Universite Paris VI, Paris, France.

Martin, M^{a}. I. - Physics Dept., Univ. Valladolid, Spain.

Maurin, K. - University of Warsaw, Poland.

Mendes, A. - Math. Dept., Univ. Coimbra, Portugal.

Moncrief, V. - Physics Dept., Yale Univ., New Haven, Connecticut, U.S.A.

Montesinos, A. - Math. Dept., Univ. Complutense, Madrid, Spain.

Montesinos, R. - Math. Dept., Univ. Salamanca, Spain.

Moreira, R.A. - C.F.M.C. Gama Pinto, Lisboa, Portugal.

Moroz, B.Z. - Physics Dept., Univ. Ben Gurion, Israel.

Muriel, F.J. - Math. Dept., Univ. Salamanca, Spain.

Muñoz, J. M^{a}. - Math. Dept., Univ. Salamanca, Spain.

Muñoz Masqué, J. - Math. Dept., Univ. Salamanca, Spain.

Navarro, J.A. - Math. Dept., Univ. Salamanca, Spain.

Ne'eman, Y. - Physics Dept., Univ. Tel Aviv, Israel, Univ. Texas, U.S.A.

Padró, J.R. - Math. Dept., Univ. Puerto Rico, Rio Piedras, Puerto Rico.

Pascual, J.F. - Physics Dept., Univ. Valladolid, Spain.

Peñalba, I. - Math. Dept.,Univ. Salamanca, Spain.

Pereira da Silva, J.A. - Math. Dept., Univ. Coimbra, Portugal.

Pérez-Rendón, A. - Math. Dept., Univ. Salamanca, Spain.

Pérez, S. - Math. Dept., Univ. Salamanca, Spain.

Petry, H. Institut fur Theoretische Kernphysik der Universitat Bonn.

Pham Mau Quan - Université Paris-Nord, Paris, France.

Recht, L. - Math. Dept., Univ. Simón Bolivar, Caracas, Venezuela.

Rica, A.J. - Math. Dept., Univ. Lisboa, Portugal.

Ribeiro, A. - Math. Dept., Univ. Coimbra, Portugal.

Rio, J. - Math. Dept., Univ. Salamanca, Spain.

Rodríguez, G. - Math. Dept., Univ. Salamanca, Spain.

Romo, M^{a}. R. - Math. Dept., Univ. Salamanca, Spain.

Sancho de Salas, J. - Math. Dept., Univ. Salamanca, Spain.

Santos, J.R. - C.F.M.E., Math. Dept., Univ. Lisboa, Portugal.

Schiappacasse, O. - Math. Dept., Univ. Puerto Rico.

Segal, I.E. - M.I.T., Math. Dept., Cambridge, MA, U.S.A.

Sen, R.N. - Math. Dept., Ben Gurion Univ., Israel.

Seredynska, M. - Institute of Fundamental Tech. Research, Warsaw, Poland.

Serrano, M^{a}. G. - Math. Dept., Univ. Salamanca, Spain.

Simms, D.J. - University of Bonn and Trinity College Dublin.

Soilán, An. - Math. Dept., Univ. Complutense, Madrid, Spain.

Sternberg, S. - Harvard University, Cambridge, MA, U.S.A., Tel Aviv Univ., Israel.

Suárez, L. - Math. Dept., Univ. Salamanca, Spain.

Szapiro, T. - Physics Dept., Univ. Warsaw, Poland.

Taborda, J. - Centro Fisica da Materia Condensada, Portugal.

Trias, - Math. Dept., Univ. Simón Bolivar, Caracas, Venezuela.

Turiel, F.J. - Math. Dept., Univ. Autónoma, Madrid, Spain.

Uhlenbrock, D.A. - Math. Dept., Univ. Wisconsin, Madison, WI, U.S.A.

Ungar, T. - Mathematisches Institute der Universitat Bonn.

Vázquez, M^{a}. J. - Math. Dept., Univ. Salamanca, Spain.

Visus, I. - Math. Dept., Univ. Salamanca, Spain.

Wolf, J.A. - Math. Dept., Univ. California, Berkeley, CA, U.S.A.

MORSE THEORY AND THE YANG-MILLS EQUATIONS

Raoul Bott*

University of Harvard

1. Introduction.

In the Variational problems of mathematical physics one is usually interested only in the extremals which afford an absolute minimum to the functional being varied. The other, the unstable solutions, therefore tend to be forgotten. On the other hand, in the Morse theory the totality of the extrema satisfy relations which are forced by the topology of the space underlying the variational problem. There is thus a certain amount of feedback between the stable and unstable critical sets which enables one to derive information about the topology of the absolute minimum by studying the sometimes more accessible unstable critical sets. The Yang-Mills functional for a principal bundle P over a compact Riemann surface M gives a striking illustration of this principle, and the purpose of this talk is to sketch in the main steps of this application of the Morse theory for you.

I am reporting on joint work with M. Atiyah still very much in progress.

2. The nondegenerate Morse theory.

Let f be a smooth function on a manifold M, and let Cr(f) be the set of its critical points, that is, the set of points when $df = 0$. On the restriction of the tangent bundle of M to Cr(f) the Hessian of f, Hf, now becomes a well defined quadratic form and we call a component $N \subset Cr(f)$ a nondegenerate critical manifold if and only if the following conditions are satisfied:

2.1) N is an imbedded submanifold.

2.2) Hf is nondegenerate on the orthogonal complement to N.

Under these conditions one may choose a subbundle $\nu_N^-(f)$ of $TM|N$ on which Hf is negative definite, and this negative bundle is unique up to isomorphism. We call its fiber dimension the index of N, and say that f is orientable along N if and only if $\nu_N^-(f)$ is orientable.

In the nondegenerate Morse theory one deals with functions whose critical sets consist entirely of nondegenerate critical manifolds: $Cr(f) = \{N\}$; and one associates to such an f the following "Morse-Series":

$$m_t(f) = \sum_{\{N\}} t^{\lambda(N)} P_t(N) \tag{2.3}$$

* This work was supported in part through funds provided by the National Science Foundation under the grant 33-966-7566-2.

where $P_t(N)$ is the Poincaré Polynomial of N relative to some fixed coefficient field, twisted by the orientation of $\nu_-(f)$ along N when f fails to be orientable along N.

Remarks. When the critical set of f consists of points alone, $m_t(f)$ simply counts these points weighted with $t^{\lambda p}$. Thus at = 1 $m_t(f)$ provides one with an honest count of the critical set. In our more general context the constant function is of course also nondegenerate and clearly has for its Morse-Series $P_t(M)$, the Poincaré Polynomial of M itself. The fundamental theorem of the Morse theory now asserts that in some sense such a constant function has the smallest Morse Series. Precisely, one has the following slight extension of the basic Morse inequalities, see [B1].

Let f be nondegenerate on a compact manifold M. Then there exists a polynomial $Q(t) = a_0 + a_1 t + \dots$ with coefficients ≥ 0, such that

$$m_t(f) - P_t(M) = (1+t)Q(t) \ . \tag{2.4}$$

In view of this inequality we will call a nondegenerate function perfect if $m_t(f) = P_t(M)$. Such a function then exhibits the minimal critical behavior forced on it by the topology of the underlying space. Note by the way, that because of the $(1+t)$ factor in (2.3) a function is automatically perfect if the product of any two consecutive coefficients, $m_i \cdot m_{i+1}$, of $m_t(f)$ vanish - in particular therefore if all indices are even. As an illustration of this principle let me indicate how one can use it to compute the Poincaré Series of the complex projective space $\mathbb{C}P_n$.

Indeed the function

$$\tilde{f} = a_0|z_0|^2 + a_1|z_1|^2 + \dots + a_n|z_n|^2$$

with $a_0 < a_1 < \dots a_n$ real, gives rise to the function

$$f(z) = \tilde{f}(z)/|z|^2$$

on $\mathbb{C}P_n$ whose critical points are seen to be the coordinate axes and whose indices go up by two each time. Thus

$$m_t(f) = 1 + t^2 + \dots t^{2n} , \tag{2.5}$$

and therefore $P_t(\mathbb{C}P_n)$ is also given by this formula. Let me remind you in passing that (2.5) is also the correct formula for $n = \infty$:

$$P_t(\mathbb{C}P_\infty) = 1 + t^2 + t^4 + \dots$$
$$= 1/1 - t^2 \ .$$

3. The treatment of symmetries.

We now ask the following question: How do symmetries of a function affect this Morse calculus? Consider then the case where the nondegenerate function f is assumed to be invariant under on group G of diffeomorphisms of M. Clearly, if G acts freely on M, all one has to do is to pass from f on M to the function f_G induced by f on the quotient M/G, and then apply the usual Morse theory. However when the action is not free M/G is not a manifold and this procedure gives quite the wrong results in general. A historical case where many people including myself, have foundered by trying to apply this simple quotient construction is the case of closed geodesics. Here the functional is the energy $E(\mu)$ of a map μ of S^1 into a Riemann manifold W. Thus $E(\mu)$ is defined on the space of maps: $Map(S^1;W)$ and E is invariant under the action of S^1 obtained by rotating the source space. This action is not free near closed paths which are circumnavigated several times.

The "correct" way of treating this situation, which Atiyah and I propose is to take a leaf out of the book of the homotopy theorists and to relate the critical behavior of f on M, to the cohomology - not of M/G - but of the "homotopy quotient" M_G of the action. Recall that the space M_G is obtained as follows. We first choose a classifying action of G on a manifold W. Thus W must be (1) contractible and (2) the action of G must be free.

Remark. These W's are usually infinite dimensional. Eg. for $G = S^1$, W is the unit sphere in an infinite dimensional complex Hilbert space with $S^1 = \{z \mid |z| = 1\}$ acting by multiplication. Note also that therefore the quotient $W/G = \mathbb{C}P_\infty$! The quotient of such a universal action is called the classifying space of G and is a very important space in modern topology. It is usually denoted by BG.

In any case in terms of such a W, the homotopy quotient of an action of G on M is simply defined as the ordinary quotient of the diagonal action on $W \times M$:

(3.1) $$M_G \doteq W \times M/G .$$

Thus M_G fibers over BG with fiber W, and its cohomology can therefore be treated by standard methods. One often refers to this cohomology as the equivariant cohomology of M, and uses the notation

(3.2) $$H^*_G(M) \equiv H^*(M_G) .$$

Similarly we call the Poincaré series of M_G the equivariant Poincaré series of M and denote it by

(3.3) $$P^G_t(M) \equiv \sum t^k \dim H^k_G(M) .$$

It goes without saying, of course, that when the action is free then M_G is of the same homotopy type as M/G, so that $P_t^G(M)$ then just reproduces $P_t(M/G)$. On the other hand when the action fails to be free then $P_t^G(M)$ will in general be quite a different kettle of fish from $P_t(M/G)$.

With all this understood we can now proceed to the equivariant Morse-Inequalities. Consider then a situation where f is invariant under the compact group G acting on the compact manifold M and that the critical sets of f fall into nondegenerate manifolds {N}, each a single orbit with stability group H_N:

$$N = G/H_N \ . \tag{3.4}$$

Then define the equivariant Morse Series of f by:

$$m_t^G(f) = \sum t^{\lambda_N} P_t(BH_N) \ . \tag{3.5}$$

Thus in the equivariant theory we count each orbit with t^{λ_N} augmented by the cohomology of the classifying space of the stability group of the orbit N.

Theorem. Under our hypotheses we have the "Morse inequalities":

$$m_t^G(f) - P_t^G(M) = (1 + t)Q^G(t) \tag{3.6}$$

where $Q^G(t)$ is some formal series with non-negative coefficients.

4. The Yang-Mills Theory Over Riemann Surfaces.

Whenever the Morse equalities hold in (3.6), that is,

$$m_t^G(f) = P_t^G(M) \tag{4.1}$$

we will speak of a perfect Morse function in the equivariant sense. These then exhibit the minimal critical behavior compatible with both the symmetry of f and the topology of M.

The interesting fact which M. Atiyah noted two years ago and whose proof we have been perfecting ever since, is now the following:

Theorem. The Yang-Mills Functional

$$S(A) = \int_M \|F_A\|^2 \tag{4.2}$$

defined on the space of connections $\mathcal{a}(P)$ of a principal U(n) bundle P over a compact Riemann Surface defines a Perfect Morse Function in the equivariant sense, relative to its group of symmetries: G(P): the group of "local" gauge transformations of P.

Remark. For the mathematicians in the audience let me recall that G = G(P) is simply the group of bundle automorphisms of P, which cover the identity. In the case of the trivial bundle $P = M \times G$, the automorphisms amount to maps of M to G. Note also that the space of connections $\mathcal{a} = \mathcal{a}(P)$, is an affine space and hence contractible. Thus in our instance

$$\mathcal{a}_G \simeq BG ,$$

so that

$$P_t^G(\mathcal{a}) = P_t(BG) , \tag{4.3}$$

and this is then the pertinent object which forces the critical behavior of the Yang-Mills functional.

It is of course a highly nontrivial matter to show that the Morse theory "works" in this infinite dimensional setting. In fact it does not, unless the dimension of the base manifold is less than or equal to three. This is a recent result of Karen Uhlenbeck. We are of course in this range and here this result can also be established by other means. The fact that Yang-Mills is perfect in this case is more technical - for instance it is not true when the structure group of P is say O(n)! - and I really cannot attempt to explain this phenomenon to you here. Rather, let me use it to illustrate the point I made at the start of this lecture, that the unstable critical points furnish us with information about the stable ones.

We apply our theorem to the case of Yang-Mills for a principal U(2)-bundle P over a Riemann Surface (compact) of genus g. Assume also that $c_1(P) = 1$. On the one hand $P_t(BG)$ can then be computed by standard methods of topology. The result is

$$P_t(BG) = \frac{\{(1 + t)(1 + t^3)\}^{2g}}{(1 - t^2)^2(1 - t)^4} \tag{4.4}$$

On the other hand the extrema of Yang-Mills can be analyzed, and this much can be found out about them rather easily:

a) The extrema of Yang-Mills fall into a collection of smooth varieties of $\mathcal{G}$-orbits.

b) The variety corresponding to the minimum of Yang-Mills, is isomorphic to the variety of stable bundles of dimension 2 over M, in the sense of algebraic geometry. We call this variety Min. This variety is of high dimension and its topology is obscure.

c) The other (unstable critical sets are isomorphic to the square of the Jacobian of the curve M, and thus are Tori of dimension 4g.

d) The stability groups of these extrema are: the center S^1 - for Min, and $S^1 \times S^1$ for all the others.

Using these facts, and after computing the relevant indices, one can compute the Equivariant Morse Series for Yang-Mills to find:

$$m_t^G(Y-M) = \frac{P_t(\mathrm{Min})}{1-t^2} + \sum_{k>0} \frac{t^{2g+4k}}{(1-t^2)^2} \cdot (1+t)^{4g} . \tag{4.5}$$

The summation is here over the unstable extrema, and can clearly be carried out to yield

$$\frac{t^{2g}}{(1-t^2)^2} \; \frac{(1+t)^{4g}}{(1-t^4)} .$$

Finally now, according to our theorem - that is the perfection of Yang-Mills in the equivariant sense - we may equate (4.4) and (4.5) to find the following formula for the Poincaré Polynomial P_t of the variety of stable extrema Min:

$$P_t(\mathrm{Min}) = \frac{[(1+t)(1+t^3)]^{2g}}{(1-t^2)^2(1-t^4)} - \frac{t^{2g}(1+t)^{4g}}{(1-t^2)(1-t^4)} . \tag{4.6}$$

When Michael Atiyah and I first wrote down this formula we had quite a hard time convincing ourselves that (4.6) actually reduces to a polynomial. At that time we were checking (4.6) against known results of Newstead who had explicity computed $P_t(\mathrm{Min})$ for low g. Later we actually found (4.6) in the literature - but derived in a completely different way.

Indeed in [2] Harder had tackled the problem of computing $P_t(\mathrm{Min})$ purely in the context of algebraic geometry. Namely he defined these varieties over finite fields and found the number of rational points on them with the aid of deep results in number theory going back to C.L. Siegel. He then applied the Weil conjectures - which at that time had not yet been proved to be valid - and this led him to precisely the formula (4.6).

My time is up, so I will stop here. Although our main theorem has no direct application to Physics so far, we have hopes that its techniques and conceptual background will someday. We suspect for instance, that although the Morse theory does not work over S^4, say, still it works better and better as the structure group gets larger i.e. as $n \longrightarrow \infty$ in $U(n)$, and this could well be of relevance in the $1/n$ expansion.

I append the following bibliography in case someone is interested in following some aspect of the subject. A paper, scheduled for the London Math. Journal, is under preparation by Atiyah and myself.

Bibliography.

1 R. Bott and H. Samelson, Applications of the theory of Morse to symmetric spaces, Amer. J. of Math. vol. 80 (1968), pp. 964-1029.

2 G. Harder, Eine Bemerkung zu einer Arbeit von P.E. Newstead, Jour.für Math. 242 (1970), 16-25.

3 G. Harder and M.S. Narasimhan, On the cohomology groups of moduli spaces of vector bundles over curves, Math. Ann. 212 (1975), 215-248.

4 D. Mumford and P.E. Newstead, Periods of a moduli space of bundles on curves, Amer. J. Math. 90 (1968), 1201-1208.

5 M.S. Narasimhan and S. Ramanan, Moduli of vector bundles on a compact Riemann surface, Ann. of Math. 89 (1969), 19-51.

6 M.S. Narasimhan and S. Ramanan, Vector bundles on curves, Proceedings of the Bombay Colloquium of Algebraic Geometry, 335-346, Oxford University Press, 1969.

7 M.S. Narasimhan and C.S. Seshadri, Stable and Unitary vector bundles on a compact Riemann surface, Ann. of Math. 82 (1965), 540-576.

8 P.E. Newstead, Topological properties of some spaces of stable bundles, Topology 6 (1967), 241-262.

9 ______, Stable bundles of rank 2 and odd degree over a curve of genus 2, Topology 7 (1968), 205-215.

10 ______, Characteristic classes of stable bundles of rank 2 over an algebraic curve, Trans. Amer. Math. Soc. 169(1972), 337-345.

11 ______, Rationality of moduli spaces of stable bundles, to appear.

12 C.S. Seshadri, Space of unitary vector bundles on a compact Riemann surface, Ann. of Math. 85 (1967), 303-336.

REDUCTION OF THE YANG MILLS EQUATIONS

Vincent Moncrief
Department of Physics
Yale University
New Haven, Connecticut 06520

I. INTRODUCTION

In this paper we shall discuss the reduction of the classical Yang-Mills equations, regarding these equations as a constrained Hamiltonian system with an infinite dimensional (gauge) symmetry group. Roughly speaking this reduction consists of:

(i) solving the Yang-Mills constraint equations,

(ii) taking the quotient of the resulting space by the action of the gauge group, and

(iii) showing that the dynamics projected to the quotient space is in fact Hamiltonian.

We shall carry out this reduction explicitly (locally in a suitable function space) for the case of Yang-Mills fields propagating in a flat spacetime with compact spacelike (Cauchy) hypersurfaces.

We first introduce a suitable (weak symplectic) function space of (unconstrained) canonical variables for the Yang-Mills system and describe within it the conventional formulation of Yang-Mills dynamics. We next discuss the geometry of the constraint subset of this phase space and show that this subset is in fact a submanifold except near points which represent symmetrical solutions of the Yang-Mills equations (i.e., solutions with non-trivial, continuous (gauge) isotropy groups). We (temporarily) cut out the symmetric, singular points and work with the manifold of (generic) non-symmetric solutions. We show that the intersection of this constraint submanifold with a suitable slice for the gauge group action is in fact a symplectic submanifold of the original phase space. We give an explicit coordinate system for this submanifold (locally, near any non-singular point) and show how Hamilton's equations may be expressed in this local chart.

In physics terminology the restriction to a slice for the gauge group action corresponds to a particular choice of gauge. Such a choice gives an explicit (though highly arbitrary) way to quotient out the gauge group action on the constraint submanifold. The particular slice that we shall use might be called a "generalized Coulomb gauge" since it reduces at the trivial, zero field solution of the constraint equation to the conventional Coulomb gauge condition.

In the final section we shall reinstate the singular points of the constraint subset and discuss the geometry of the constraint set near such points. We sketch a proof (following a pattern developed by Fischer, Marsden and Moncrief[1] in their treatment of the Einstein equations) that, near any point with a one dimensional isotropy group, the constraint subset is homeomorphic to a (manifold x cone). The manifold points represent nearby solutions with the "same (one-dimensional) symmetry" as the given one whereas the cones represent the branching of the constraint set to solutions of lower (i.e., no continuous) symmetry. Arms[2] has essentially completed the characterization of the singular points of Yang-Mills theory by showing how one can handle the case of higher dimensional isotropy groups. Arms, Marsen and Moncrief[3] have extended this method to study the zero sets of momentum mappings of general Hamiltonian systems and Arms, Fischer, Marsden and Moncrief[4] have applied it to treat the case of many (spacelike) Killing fields in gravity.

II. GEOMETRY OF THE CONSTRAINT SET

We first introduce some notation and then recall some results from the literature on the geometry of the solution set of the constraint equations.

A. Notation and basic formulas

For the background spacetime we shall take a flat, Lorentzian metric on $T^3 \times R$. In suitable coordinates the metric is (putting $x^0 = t$)

$$ds^2 = -dt^2 + dx^i dx^i \tag{2.1}$$

where the x^i are periodic coordinates on the torus. The choice of a spacially compact spacetime is made primarily to simplify the analysis. The elliptic theory used in our reduction is simpler in the compact case since there are no asymptotic conditions to worry about.

We let G be a compact, semi-simple (matrix) Lie group and write $\mathcal{G}$ for the Lie algebra of G. We let $\{\theta_a\}$ designate a Hermitian basis for $\mathcal{G}$ so that

$$[\theta_a, \theta_b] = i f^{abc} \theta_c \tag{2.2}$$

for suitable constants f^{abc}. In the geometrical approach to Yang-Mills theory one constructs a principal fiber bundle (B, π, G) over spacetime with fiber G (so that $\pi^{-1}(x) = G$) and considers connection fields ${}^{(4)}a$ in this bundle. (For simplicity we consider only the case of trivial (product) bundles). The connection fields may be regarded as $\mathcal{G}$-valued one forms over spacetime and expanded as

$${}^{(4)}a = {}^{(4)}a_\mu^{(a)} \theta_a dx^\mu \tag{2.3}$$

The group $\mathcal{G}$ of (sufficiently smooth) antomorphisms of the bundle acts on the connections in a well known way. This group action generates the "gauge transformations" of ${}^{(4)}a$.

As a configuration manifold $\mathcal{A}$ for the Yang-Mills equations we take the space of H_s $(s \geq 2)$, $\mathcal{G}$-valued one forms over T^3. An element $a \in \mathcal{A}$ may be regarded as the potential induced on a $t=$ constant hypersurface of spacetime by the spacetime potential ${}^{(4)}a$.

(Here H_s designates the Sobolev space of functions in L^2 with derivations up to order s also in L^2). We define $\mathcal{E}$ = "the space of electric fields" to be the space of H_{s-1}, $\mathfrak{G}$ -valued vector densities over T^3. We may regard $\mathcal{A} \times \mathcal{E} \approx T^*\mathcal{A}$ (the L^2 cotangent bundle of $\mathcal{A}$) as an appropriate phase space for the Yang-Mills equations. The particular choice of function spaces was made with a view towards the local existence problem for the classical evolution equations. It also ensures the smoothness of the constraint map introduced below.

If ${}^{(4)}F$ is the curvature of a spacetime connection ${}^{(4)}a$, then the projections of ${}^{(4)}F$ onto a given $t = t_o =$ constant hypersurface define the "electric" and "magnetic" fields $e(t_o)$ and $b(t_o)$ of that surface. ${}^{(4)}F_{\perp\parallel}(t_o)$ corresponds to $e(t_o)$ and ${}^{(4)}F_{\parallel\parallel}(t_o)$ corresponds to (the dual of) $b(t_o)$ where $\perp$ and $\parallel$ designate the projections normal and tangent to the surface. $b(t_o)$ may also be regarded as the dual of the curvature ${}^{(3)}F(a(t_o))$ since ${}^{(4)}F({}^{(4)}a) = {}^{(3)}F(a(t_o))$.

The constraint map $\Phi : T^*\mathcal{A} \to \Lambda^0_d$ (where Λ^0_d is the space of $\mathfrak{G}$ -valued scalar densities over T^3) is defined by the formula

$$\begin{aligned} \Phi(a,e) &= \delta \cdot e + i[e\cdot, a] \\ &= \partial_j e^j + i[e^j, a_j] \end{aligned} \tag{2.4}$$

We wish to study the constraint subset $\Phi^{-1}(0) \subset T^*\mathcal{A}$.
The equation $\Phi(a(t), e(t)) = 0$ is equivalent to the normal projection of the Yang-Mills equations, i.e., to

$$\left(\nabla_{{}^{(4)}a} \cdot {}^{(4)}F({}^{(4)}a) \right)_\perp = 0 \tag{2.5}$$

where $\nabla_{{}^{(4)}a}$ represents the covariant derivative operator. This constraint is preserved in time by the evolution equations

$$\left(\nabla_{{}^{(4)}a} \cdot {}^{(4)}F({}^{(4)}a) \right)_\parallel = 0 \tag{2.6}$$

which in turn may be written in Hamiltonian form on $T^*\mathcal{A}$.
The normal projection ${}^{(4)}a_\perp$ is undetermined by this system and occurs as a "Lagrange multiplier" in the equations for $(a(t), e(t))$. This arbitrariness corresponds to the gauge invariance of the Yang-Mills equations.

Let (a_0, e_0) be any particular solution of $\Phi(a,e) = 0$ and consider $D\Phi(a_0,e_0)$, the derivative of Φ at (a_0, e_0). This operator is given by

$$D\Phi(a_0,e_0)\cdot(a',e') = \delta\cdot e' + i[e'\cdot, a_0] + i[e_0\cdot, a'] \tag{2.7}$$

where (a', e') is an arbitrary tangent vector at (a_0, e_0). The L^2 adjoint operator $D\Phi(a_0,e_0)^*$ is defined through

$$\mathrm{Tr}\int_{T^3} \eta\cdot D\Phi(a_0,e_0)\cdot(a',e') = \langle D\Phi(a_0,e_0)^*\cdot\eta\, , (a',e')\rangle \tag{2.8}$$

where η is an arbitrary $\mathcal{G}$-valued function on T^3 and $\langle\ ,\ \rangle$ is the (weak) Riemannian metric on $T^*\mathcal{A}$ given by

$$\langle(a',e'),(a',e')\rangle = \mathrm{Tr}\int_{T^3}\left(\mu\, a'\cdot a' + \mu^{-1} e'\cdot e'\right) \tag{2.9}$$

Here μ is the volume element of the hypersurface (recall that e' is a density), Tr represents a trace over the group indices (we may assume that $\mathrm{Tr}\,\theta_a\theta_b = \delta_{ab}$) and $\cdot$ represents contraction using the metric of the hypersurface.

The appropriate (weak) symplectic form ω on $T^*\mathcal{A}$ is given by

$$\omega((a',e'),(a'',e'')) = \mathrm{Tr}\int_{T^3}(a'\cdot e'' - e'\cdot a'') \tag{2.10}$$

and it is useful to define a complex structure J on each tangent space $T_{(a,e)}T^*\mathcal{A}$ by

$$\langle(a',e'),(a'',e'')\rangle = \omega((a',e'), J\cdot(a'',e'')) \tag{2.11}$$

J is symplectic and $J^2 = -\mathrm{Id}$.

B. Symmetries and singularities of the constraint set

The geometry of the constraint set $\Phi^{-1}(0)$ has been studied by Moncrief[5,6] and Arms[7] using methods developed by Fischer and Marsden[8,9] Moncrief[10,11] (see also Fischer, Marsden and Moncrief[1] and Arms, Marsden and Moncrief [3]) in their study of the Einstein equations. (The paper by Arms actually treats the more general case of gauge fields coupled to gravity.)

A key result is given by:

Theorem 1: If $\Phi(a_0, e_0) = 0$ and $\ker D\Phi(a_0,e_0)^* = \{0\}$ then $\Phi^{-1}(0)$ is a manifold near (a_0, e_0).

Sketch of proof: From the implicit function theorem one knows that if $D\Phi(a_0, e_0)$ is surjective then $\Phi^{-1}(0)$ is a manifold near (a_0, e_0). Using elliptic theory one shows that $D\Phi(a_0, e_0)$ is surjective if and only if $D\Phi(a_0,e_0)^*$ is injective. ∎

One can show by a straightforward computation that Φ is a moment map for the action of the gauge group $\mathcal{G}$ on $T^*\mathcal{A} \approx \mathcal{A} \times \mathcal{E}$. In other words the subspace

$$\text{range}\ J \cdot D\Phi(a_0,e_0)^* \subset T_{(a_0,e_0)} T^*\mathcal{A} \tag{2.12}$$

may be identified with the infinitesimal gauge transformations of (a_0, e_0) (i.e., with the tangent space to the $\mathcal{G}$-orbit of (a_0, e_0)). Thus $\ker D\Phi(a_0,e_0)^*$ coincides with the infinitesimal symmetries of (a_0, e_0). If the evolution equations (2.6) are taken into account one can show that the infinitesimal symmetries of a spacetime solution ${}^{(4)}a_0$ of (2.5) and (2.6) are isomorphic to the infinitesimal symmetries of Cauchy data (a_0, e_0) for this solution. More precisely if ${}^{(4)}\eta$ is a $\mathcal{G}$-valued function on spacetime for which

$$d\,{}^{(4)}\eta + i[{}^{(4)}\eta, {}^{(4)}a_0] = 0 \tag{2.13}$$

then ${}^{(4)}\eta$ induces on the $t = \text{constant}$ surface with data (a_0, e_0) an element ${}^{(4)}\eta(t) = \eta \in \ker D\Phi(a_0,e_0)^*$. Conversely any element of $\ker D\Phi(a_0,e_0)^*$ may be propagated to give an infinitesimal symmetry of ${}^{(4)}a_0$. In turn the infinitesimal symmetries may be exponentiated to yield finite dimensional subgroups of the gauge group $\mathcal{G}$.

From the above remarks and theorem 1 we see that points of $\Phi^{-1}(0)$ having no infinitesimal symmetries are in fact all manifold points. One can show conversely that any point with (a continuous) symmetry is singular in the sense that $\Phi^{-1}(0)$ fails to be a manifold near such a point (see Arms[7,2]). However this conclusion that symmetric points are singular depends crucially upon the compactness of the spacelike hypersurfaces. In the non-compact case (i.e., on Minkowski space) the constraint subset may be shown to be a manifold provided function spaces with suitable asymptotic conditions are used (see Moncrief[6]).

To simplify the analysis in the compact case we shall simply cut out all the symmetrical points of $\Phi^{-1}(0)$ and consider only the manifold of (generic) non-symmetrical points in our reduction. At such points $D\Phi(a,e)^*$ has trivial kernel and the operator $D\Phi(a,e)\cdot D\Phi(a,e)^*$ is an isomorphism of Λ^0 to Λ^0_d.

C. Decompositions of $T^*\mathcal{A}$

We show now how the linear space $T^*\mathcal{A}$ may be coordinatized in a particularly convenient way. To do this we introduce a decomposition of $T^*\mathcal{A}$ which generalizes the familiar transverse-longitudinal splitting of the vector potential and electric field of Maxwell theory.

For any point $(a_0,e_0)\in T^*\mathcal{A}$ we have the decomposition

$$T^*\mathcal{A} = \{(a_0,e_0)\} + \left(\text{range } D\Phi(a_0,e_0)^* \oplus \ker D\Phi(a_0,e_0)\right) \tag{2.14}$$

and if $(a_0,e_0)\in\Phi^{-1}(0)$ we may refine this to

$$T^*\mathcal{A} = \{(a_0,e_0)\} + \Big(\text{range } D\Phi(a_0,e_0)^* \oplus \text{range } \mathbb{J}\circ D\Phi(a_0,e_0)^* \oplus \left(\ker D\Phi(a_0,e_0)\cap \ker D\Phi(a_0,e_0)\circ\mathbb{J}\right)\Big) \tag{2.15}$$

The three summands in parentheses in (2.15), regarded as subspaces of $T_{(a_0,e_0)}T^*\mathcal{A}$, are mutually orthogonal relative to the metric $\langle\ ,\ \rangle$ and have the following interpretations:

(i) range $\mathbb{J}\circ D\Phi(a_0,e_0)^*$ = tangent space to the $\mathcal{G}$-orbit of (a_0,e_0),

(ii) range $D\Phi(a_0,e_0)^*$ = orthogonal complement to the tangent space ($\ker D\Phi(a_0,e_0)$) of the constraint submanifold at (a_0,e_0),

(iii) ker $D\Phi(a_0,e_0)\cap\ker D\Phi(a_0,e_0)\circ\mathbb{J}$ = a parameterization for the space of "true degrees of freedom" of the Yang-Mills field.

Thus any $(a,e)\in T^*\mathcal{A}$ has, for any fixed reference point $(a_0,e_0)\in\Phi^{-1}(0)$ the decomposition

$$(a,e) = (a_0,e_0) + D\Phi(a_0,e_0)^*\cdot\eta + \mathbb{J}\cdot D\Phi(a_0,e_0)^*\cdot\nu + (a',e')^{TT} \tag{2.16}$$

where η and ν are uniquely determined elements of Λ^0 (the space of $\mathcal{G}$-valued function on T^3) and where $(a',e')^{TT}$ is a uniquely determined element of the "transverse-transverse" space (iii) of true degrees of freedom. The proof of existence of this decomposition follows the same pattern used in the gravitational case (see Moncrief[12,6]). One makes use of the elliptic character of $D\Phi(a,e)^*$ and the fact that range $\mathbb{J}\circ D\Phi(a,e)^*\subset\ker D\Phi(a,e)$ provided $\Phi(a,e)=0$.

To suggest the usefulness of this decomposition we now show that the constraint submanifold $\Phi^{-1}(0)$ may be regarded as a graph over the affine submanifold

$$\mathcal{C}^T \equiv \{(a_0,e_0)\} + \ker D\Phi(a_0,e_0) \tag{2.17}$$

on a sufficiently small neighborhood of (a_0,e_0). Using (2.14) to expand an arbitrary (a,e) as

$$(a,e) = (a_0,e_0) + D\Phi(a_0,e_0)^*\cdot\eta + (a',e')^T \tag{2.18}$$

where $(a',e')^T\in\ker D\Phi(a_0,e_0)$ we claim that $\Phi(a,e)=0$ is uniquely soluble for η on a neighborhood of (a_0,e_0). This follows from the implicit function theorem and the fact that **the operator** $D\Phi(a_0,e_0)\circ D\Phi(a_0,e_0)^*$ is an isomorphism.

III. REDUCTION

Consider the affine space $S_{(a_0,e_0)}$ given by

$$S_{(a_0,e_0)} = \{(a_0,e_0)\} + \Big(\text{range } D\Phi(a_0,e_0)^* \oplus \big(\ker D\Phi(a_0,e_0) \cap \ker D\Phi(a_0,e_0)\circ J \big)\Big) \qquad (3.1)$$

By construction the tangent space to $S_{(a_0,e_0)}$ at (a_0,e_0) is the $\langle\ ,\ \rangle$ – orthogonal complement to the tangent space ($\text{range } J\circ D\Phi(a_0,e_0)^*$) of the $\mathcal{G}$ -orbit of (a_0,e_0). Using methods developed by Ebin[13] and Palais[14] one can show that a sufficiently small ball in $S_{(a_0,e_0)}$ about (a_0,e_0) is in fact a slice $\mathcal{S}_{(a_0,e_0)}$ for the gauge group action on $T^*\mathcal{A}$. (The corresponding construction of a slice for gravity was given by Fischer, Marsden and Moncrief.[1])

Now consider the set

$$\mathcal{P}_{(a_0,e_0)} \equiv \Phi^{-1}(0) \cap \mathcal{S}_{(a_0,e_0)} \qquad (3.2)$$

We have

Theorem 2: $\mathcal{P}_{(a_0,e_0)}$ is a submanifold of $T^*\mathcal{A}$. This submanifold is a graph over the affine space

$$\mathcal{C}^{TT} \equiv \{(a_0,e_0)\} + \big(\ker D\Phi(a_0,e_0) \cap \ker D\Phi(a_0,e_0)\circ J \big) \qquad (3.3)$$

on a sufficiently small neighborhood of (a_0,e_0).

Sketch of Proof: That $\mathcal{P}_{(a_0,e_0)}$ is a manifold follows from the transversality of the intersection of the manifolds $\Phi^{-1}(0)$ and $\mathcal{S}_{(a_0,e_0)}$. At any point $(a,e) \in \mathcal{P}_{(a_0,e_0)}$ we have, from the properties of a slice, that

$$T_{(a,e)} T^*\mathcal{A} = T_{(a,e)} \mathcal{S}_{(a_0,e_0)} + \text{range } J\circ D\Phi(a,e)^* \qquad (3.4)$$

(where $\text{range } J\circ D\Phi(a,e)^*$ is the tangent space to the $\mathcal{G}$ -orbit through (a,e)). However, since $\Phi(a,e) = 0$ we have

$$\text{range}\ \mathbb{J}\circ D\Phi(a,e)^{*} \subset \ker D\Phi(a,e) = T_{(a,e)}(\Phi^{-1}(0)) \qquad (3.5)$$

Thus

$$T_{(a,e)}T^{*}A = T_{(a,e)}\mathcal{S}_{(a_0,e_0)} + T_{(a,e)}(\Phi^{-1}(0)) \qquad (3.6)$$

and the intersection is transversal at every point of $\mathcal{P}_{(a_0,e_0)}$.

That $\mathcal{P}_{(a_0,e_0)}$ is a graph over the affine space $\{(a_0,e_0)\} + (\ker D\Phi(a_0,e_0) \cap \ker D\Phi(a_0,e_0)\circ\mathbb{J})$ near (a_0,e_0) may be seen by restricting the graph of $\Phi^{-1}(0)$ over $(\{(a_0,e_0)\} + \ker D\Phi(a_0,e_0))$ to the space of true degrees of freedom and showing directly that this graph preserves the slice.

The submanifold $\mathcal{P}_{(a_0,e_0)}$ consists of all data sufficiently near (a_0,e_0) satisfying the constraints and the gauge condition implicit in the choice of the slice. It is thus a natural candidate for a (local) reduced phase space for the dynamics. To refine this idea we show that $\mathcal{P}_{(a_0,e_0)}$ is in fact a symplectic manifold.

Proposition 3: $\mathcal{P}_{(a_0,e_0)}$ is a symplectic submanifold of the (weak) symplectic manifold $(T^{*}A, \omega)$ on a neighborhood of (a_0,e_0).

Sketch of proof: Clearly the two-form induced by ω on $\mathcal{P}_{(a_0,e_0)}$ will be closed (it is the pull-back $i_{*}\omega$ of the closed form ω by the inclusion map). To prove non-degeneracy we first note that if $(a',e') \in (\ker D\Phi(a_0,e_0) \cap \ker D\Phi(a_0,e_0)\circ\mathbb{J})$ then $\mathbb{J}\cdot(a',e')$ lies in this same space (the tangent space to $\mathcal{P}_{(a_0,e_0)}$ at (a_0,e_0)). For any such (a',e') we have, using (2.11),

$$\omega((a',e'),\mathbb{J}\cdot(a',e')) = \langle (a',e'),(a',e')\rangle \qquad (3.7)$$

The vanishing of the left hand side thus implies that $(a',e') = 0$ which gives the desired result. ■

In the following we shall restrict the definition of $\mathcal{P}_{(a_0,e_0)}$ to a sufficiently small neighborhood of (a_0,e_0) that the restriction is a graph over $(\ker D\Phi(a_0,e_0) \cap \ker D\Phi(a_0,e_0)\circ\mathbb{J}) + \{(a_0,e_0)\}$ and that $i_{*}\omega$ is non-degenerate on the restriction. We use the same symbol $\mathcal{P}_{(a_0,e_0)}$ for this restriction.

To complete the description of the dynamics in $(\mathcal{P}_{(a_0,e_0)}, i_*\omega)$ we need to specify the Hamiltonian function H. We claim that the appropriate Hamiltonian is simply

$$H = \frac{1}{2} \mathrm{Tr} \int_{T^3} (e \cdot e + b \cdot b) \Big|_{\mathcal{P}_{(a_0,e_0)}} \tag{3.8}$$

i.e., the restriction of the usual energy function on $T^*\mathcal{A}$ to $\mathcal{P}_{(a_0,e_0)}$. This may be verified by computing the Hamilton equations on $(\mathcal{P}_{(a_0,e_0)}, i_*\omega, H)$ and comparing them with the conventional equations of motion on $(T^*\mathcal{A},\omega)$. One must restrict the latter, imposing a suitable gauge condition on ${}^{(4)}a_\perp$ to ensure that the dynamical flow on $(T^*\mathcal{A},\omega)$ is tangent to $\mathcal{P}_{(a_0,e_0)}$ at every point on this submanifold. This restriction may be imposed by solving a suitable elliptic equation for ${}^{(4)}a_\perp$. One can write out the Hamiltonian vector field explicitly in the chart provided by the space of true degrees of freedom

$$C^{TT} = \{(a_0,e_0)\} + \left(\ker D\Phi(a_0,e_0) \cap \ker D\Phi(a_0,e_0)\cdot \mathbb{J}\right)$$

We remark that if $(a_0,e_0) = (0,0)$ then our gauge condition reduces to the familiar Coulomb condition that $a(t)$ have zero divergence. For that reason we call the gauge condition implicit in the choice of $\mathcal{S}_{(a_0,e_0)}$ a "generalized Coulomb gauge." It was pointed out by Gribov[15], for the non-compact case, that the usual Coulomb gauge condition does not define a global cross section of $T^*\mathcal{A}/\mathcal{G}$. (Strictly speaking Gribov considered $\mathcal{A}/\mathcal{G}$ but one can lift his result to $T^*\mathcal{A}/\mathcal{G}$ as discussed by Chodos and Moncrief[16].) For $SU(2)$ gauge theory on a Euclidean manifold Singer[17] has extended Gribov's idea to a general theorem that global, continuous cross sections of $\mathcal{A}/\mathcal{G}$ do not exist.

It seems clear that one should not expect the slices $\mathcal{S}_{(a_0,e_0)}$ considered here to define more than local cross sections of $T^*\mathcal{A}/\mathcal{G}$. If one considers the evolution of some initial data $(a,e) \in \mathcal{P}_{(a_0,e_0)}$ then the solution curve $(a(t),e(t))$ may eventually wander "off the edge" of $\mathcal{P}_{(a_0,e_0)}$. To continue such a solution one would generally need to construct a sequence $\{\mathcal{P}_{(a_i,e_i)}\}$ of local phase spaces and patch together segments of the solution curve as it passes from one local phase space to another.

Using this patchwork scheme for continuing solutions, the author has carried out some preliminary work on the global existence problem for the Yang-Mills equations in (2 + 1)-dimensional spacetime (1979, unpublished). One can show, using a simpler variant of the gauge conditions discussed above, that the blow-up time τ of the $H_2 \times H_1$ norm of a solution $(a(t), e(t))$ is <u>independent</u> of the reference point $(a_0, e_0) = (a(t_0), e(t_0))$ used in the construction of the local phase space $\mathcal{P}_{(a_0, e_0)}$. This result strongly suggests that any solution may be continued to arbitrarily long time intervals by simply passing to new local phase spaces $\mathcal{P}_{(a_i, e_i)}$ at a sequence of instants $\{t_i\}$ for which $(t_{i+1} - t_i) < \tau$ (taking $(a_i, e_i) = (a(t_i), e(t_i))$ at each transition). The finite blow-up time associated with any single local phase space would merely represent the the failure of the given local slice to define a global choice of gauge. Some additional work is needed to justify this interpretation of the aforementioned result.

IV. SINGULARITIES OF THE CONSTRAINT SET

In this section we shall reinstate the singular points and discuss the geometry of $\Phi^{-1}(0)$ near those points which have one dimensional isotropy groups. The argument sketched here follows that of Fischer, Marsden and Moncrief[1] for the case of one-dimensional (Killing) symmetries in gravity. Arms[2] has shown how to treat isotropy groups of higher dimension in Yang-Mills theory and Arms, Marsden and Moncrief[3] have extended this method to study the zero sets of momentum mappings of general Hamiltonian systems. In particular Arms, Fischer, Marsden and Moncrief[4] have applied this extension to treat the case of many (spacelike) Killing fields in gravity.

Let (a_0,e_0) be a point satisfying

$$\Phi(a_0,e_0) = 0\ , \quad \dim\left(\ker D\Phi(a_0,e_0)^*\right) = 1 \tag{4.1}$$

so that (a_0,e_0) is a solution with one dimensional isotropy group. We want to show that near (a_0,e_0) the set $\Phi^{-1}(0)$ is homeomorphic to a (manifold $\times$ cone). The manifold points will correspond to nearby solutions with the "same symmetry" (i.e., conjugate isotropy groups) as (a_0,e_0). From each such symmetrical solution there branches a cone of solutions of lower (i.e., no continuous) symmetry. The argument uses techniques of bifurcation theory together with the decomposition and slice results discussed above. We shall just sketch the argument here since the general theory (including the extension to higher dimensional symmetries has been given elsewhere (see Ref. (3)).

We first remark that the decomposition given in (2.15) and the construction of the affine slice $\mathcal{S}_{(a_0,e_0)}$ discussed in Section III work equally well when (a_0,e_0) has symmetries. One shows that the isotropy group $I_{(a_0,e_0)}$ of (a_0,e_0) preserves $\mathcal{S}_{(a_0,e_0)}$ by noting that the operator $D\Phi(a_0,e_0)\circ \mathbb{J}$ commutes with the action of $I_{(a_0,e_0)}$ on $T_{(a_0,e_0)}T^*\mathcal{A}$.

Let η span $\ker D\Phi(a_0,e_0)^*$ and consider the points (a,e) which satisfy

$$D\Phi(a,e)^*\cdot\eta = 0\ . \tag{4.2}$$

From the linearity of (4.2) in the variables (a,e) one can show that the solution set of this equation is an affine submanifold of $T^*\mathcal{A}$. Furthermore this submanifold intersects the (affine) slice $\mathcal{S}_{(a_0,e_0)}$ in a manifold we shall call $\mathcal{B}_{(a_0,e_0)}$. By definition $\mathcal{B}_{(a_0,e_0)}$ consists of all points of $T^*\mathcal{A}$ within the slice which have η as an infinitesimal symmetry.

We introduce an L^2 inner product $\langle\!\langle\ ,\ \rangle\!\rangle$ in the space Λ^0_d and define $P\Phi$ to be the projection of Φ onto $\mu\cdot(\ker D\Phi(a_0,e_0)^*) \subset \Lambda^0_d$, i.e., onto $\mu\cdot\eta$ (recall that μ is the volume element of a t = constant spacelike hypersurface). One may identify $P\Phi$ with

$$
\begin{aligned}
f &\equiv \int_{T^3} \mathrm{Tr}(\eta\cdot\Phi) \\
&= \langle\!\langle \mu\eta, \Phi \rangle\!\rangle
\end{aligned}
\tag{4.3}
$$

We let $\Lambda^0_{d,\eta}$ denote the $\langle\!\langle\ ,\ \rangle\!\rangle$ - projection of Λ^0_d orthogonal to $\mu\cdot\eta$.

Within the slice we define the map

$$
\begin{aligned}
&\Gamma : \mathcal{S}_{(a_0,e_0)} \longrightarrow \Lambda^0_{d,\eta} \ ; \\
&\Gamma(a,e) = (I - P)\,\Phi(a,e) \ .
\end{aligned}
\tag{4.4}
$$

It is straightforward to show that Γ has surjective derivative at (a_0,e_0) and thus that $\Gamma^{-1}(0)$ is a submanifold of $\mathcal{S}_{(a_0,e_0)}$ near (a_0,e_0). In a similar way one shows that $\Gamma^{-1}(0) \cap \mathcal{B}_{(a_0,e_0)}$ is a manifold. Furthermore f vanishes identically on $\Gamma^{-1}(0) \cap \mathcal{B}_{(a_0,e_0)}$ so that this space is in fact a manifold of solutions of the constraints.

To complete the determination of $(\Phi^{-1}(0) \cap \mathcal{S}_{(a_0,e_0)})$ we must impose the condition $f = 0$ within the manifold $\Gamma^{-1}(0)$. One first shows that $(\Gamma^{-1}(0) \cap \mathcal{B}_{(a_0,e_0)})$ is a manifold of zeros and critical points of the function f. To see this we compute

$$
Df(a,e)\cdot(a',e') = \mathrm{Tr}\int_{T^3} \eta\cdot D\Phi(a,e)\cdot(a',e')
$$

$$= \operatorname{Tr} \int_{T^3} \left(D\Phi(a,e)^{*} \cdot \eta , (a',e') \right) \tag{4.5}$$
$$= 0$$

the last equality following from the fact that $D\Phi(a,e)^{*}\cdot\eta = 0$ on $\mathcal{B}_{(a_0,e_0)}$. Thus f is constant on $(\Gamma^{-1}(0) \cap \mathcal{B}_{(a_0,e_0)})$ and thus vanishes since $f(a_0,e_0) = 0$. In addition one can show that the degeneracy space of the quadratic form $d^2 f(a_0,e_0)$ coincides with the tangent space to $\Gamma^{-1}(0) \cap \mathcal{B}_{(a_0,e_0)}$ at (a_0,e_0).

One can now appeal to a generalization of the Morse lemma due to Bott[18] (see Tromba[19] for the infinite dimensional case) to show that the zero set of f near (a_0,e_0) is homeomorphic to a (manifold x cone). The manifold is simply the space $\Gamma^{-1}(0) \cap \mathcal{B}_{(a_0,e_0)}$ of nearby solutions with the same symmetry as (a_0,e_0). From each such symmetrical point there branches a cone which is homeomorphic to the zeros of $d^2 f(a_0,e_0)$ restricted to a complement of its degeneracy space.

One can remove the gauge condition implicit in the choice of the slice by sweeping the (manifold x cone) structure out of the slice by the action of the coset space $\mathcal{G}/I_{(a_0,e_0)}$.

Since the full constraint set $\Phi^{-1}(0)$ is not a manifold one cannot extend the reduction scheme of Section III to it in any simple way. However sets of solutions of fixed symmetry type (e.g., the set $\Gamma^{-1}(0) \cap \mathcal{B}_{(a_0,e_0)}$ discussed above) do form manifolds. Since the evolution cannot change the symmetry type of any given set of Cauchy data (see the remarks in Section II) these submanifolds of fixed symmetry type are ruled by solutions on the Hamilton equations (supplemented by the gauge condition needed to preserve $\mathcal{S}_{(a_0,e_0)}$). This suggests that one can extend the procedure of reduction to these manifolds of fixed symmetry type. Indeed it is straightforward to show that the manifold $\Gamma^{-1}(0) \cap \mathcal{B}_{(a_0,e_0)}$ is in fact a symplectic submanifold of $(T^*\mathcal{A}, \omega)$. It is a natural phase space in which to formulate the dynamics of these solutions with the same symmetry as (a_0,e_0). A more complete treatment of this extension of the reduction procedure is given in Ref.(3).

REFERENCES

1. A. Fischer, J. Marsden and V. Moncrief, "The structure of the space of solutions of Einstein's equations. I. One Killing Field," (1979) to appear.
2. J. Arms, "The structure of the solution set for the Yang-Mills equations", (1980), in preparation.
3. J. Arms, J. Marsden and V. Moncrief, "Bifurcations of momentum mapping", (1980), in preparation.
4. J. Arms, A. Fischer, J. Marsden and V. Moncrief, "The structure of the space of solutions of Einstein's equations. II. Many Killing fields", in preparation.
5. V. Moncrief, Ann. Phys. 108, 387 (1977).
6. V. Moncrief, J. Math. Phys. 20, 579 (1979).
7. J. Arms, J. Math. Phys. 20, 443 (1979).
8. A. Fischer and J. Marsden, Bull. Am. Math. Soc. 79, 997 (1973).
9. A. Fischer and J. Marsden, Proc. Symp. Pure Math. 27, 219 (1975).
10. V. Moncrief, J. Math. Phys. 16, 493 (1975).
11. V. Moncrief, J. Math. Phys. 17, 1893 (1976).
12. V. Moncrief, J. Math. Phys. 16, 1556 (1975).
13. D. Ebin, Symm. Pure Math., Amer. Math. Soc. 15, 11 (1970).
14. R. Palais (unpublished) has constructed an affine slice for the action of the diffeomorphism group on the space of Riemannian metrics of a compact manifold.
15. V. N. Gribov, "Quantization of non-Abelian gauge theories," Leningrad Nuclear Physics Institute preprint (1977).
16. A. Chodos and V. Moncrief, "Geometrical gauge conditions in Yang-Mills theory: some non-existence results", (1978) to appear.
17. I. M. Singer, Commun. Math. Phys. 60 , 7 (1978).
18. R. Bott, Ann. of Math. 60, 248 (1954).
19. A. Tromba, Canad. J. Math. 28, 640 (1976).

Note: Research for this paper was supported in part by NSF Grant PHY76-82353.

TANGENT STRUCTURE OF YANG-MILLS EQUATIONS
AND
HODGE THEORY

Pedro L. García
Universidad de Salamanca

Abstract

The geometry of Yang-Mills fields is placed in the general frame of the Hamilton-Cartán formalism of the Calculus of Variations. The pre-symplectic structure of the space of solutions of the linearization of field equations at one of their solutions is studied. Using results of the Hodge theory for harmonic forms, the radical of the pre-symplectic metric of a Yang-Mills field is determined and conclusions are drawn about the corresponding manifolds of moduli with respect to the gauge group. The procedure to be followed in order to generalize the method to "minimal interactions" is illustrated with an elementary example, and finally some possible ways for further generalizations are pointed out.

Introduction

It is a well known fact that to every lagrangian system defined by a variational problem one can associate a "formal hamiltonian structure" (V,ω_2,A), where V is the manifold of solutions of the field equations, ω_2 is a 2-form on V defined by the classical Poincaré-Cartán invariant of the system, and A is a Lie algebra of functions on V to which all interesting dynamical quantities (energy, linear and angular moments etc.) belong.

In certain examples V has been endowed with a differentiable structure, in such a way that ω_2 becomes a closed 2-form and the elements of A become differentiable functions (finite-dimensional manifolds for systems with a finite number of degrees of freedom and infinite-dimensional manifolds for classical fields). From the dynamic viewpoint, the interesting case is when ω_2 is non-singular; in particular, the functions on V can be endowed with a Lie algebra structure (the Poisson algebra of the system), such that a becomes A sublagebra thereof. In this case V is the so-called "phase space" of the system, while ω_2 is an example of a symplectic metric. Nevertheles, there are important examples where ω_2 is singular. In such cases, if the dimension of the radical of ω_2 at each point is constant along V, it will define a distribution on V, which is involutive because $d\omega_2 = 0$. If the set $\tilde{V}$ of maximal integral submanifolds of this distribution can be endowed with a differentiable structure relative to which the natural projection $\pi: V \longrightarrow \tilde{V}$ is differentiable, then π would project ω_2 on a symplectic metric $\bar{\omega}_2$ on $\bar{V}$. The new structure $(\bar{V},\bar{\omega}_2)$ is the authentic "phase space" of the corresponding system. The case where a Lie group G acts on V its orbits being the above said integral submanifolds, is of special interest (this is, for example, the system defined by the General Relativity, where V is the manifold of Einstein metrics and G is the group of diffeomorphisms of space time). In a general way, a gauge group is a Lie group of transformations of V whose orbits are tangent at each point to the radical of the metric ω_2.

The Yang-Mills field theory is a typical example of the above situation. In this case the points of V are connections on a principal bundle satisfying the Yang-Mills equations, and the gauge group is the group of vertical automorphisms of the bundle. With the aim of establishing for such fields a programme similar to the one above refering to General Relativity, we could start with the study of the situation at the level of tangent space; i.e. trying to establish whether the radical of the 2-form ω_2 at each point coincides with the tangent space to the orbit of the gauge group through the said point.

In this paper it is proved, by using results of the Hodge theory for harmonic forms, that this is the case for Yang-Mills fields over a Lorentz manifold, while the result is false for the elliptic version of the electromagnetic field. An important point in order to be able to develop our method is to have on hand on the base manifold of an adequate intrinsic expression of the linearization of the field equations in one of its solutions, which we give on several occasions in this paper. The method is, in principle, generalizable to the "minimal interactions" as is illustrated with an elementary example (Klein-Gordon field over a complex line bundle in minimal interaction with the electromagnetic field generated by itself).

Some of the results of this paper were presented at the NSF - CBMS Conference on Geometric Methods in Mathematical Physics, University of Lowell, March 19th-23th.

The author wishes to thank to Professor Juan Sancho Guimerá for his numerous orientations during the carrying out of this paper, and to Professor Irving Ezra Segal for his useful correspondence.

1. The Hamilton-Cartán formalism of the Calculus of Variations.

1.1. The starting point of this formalism (three recent references are P. García [9], H. Goldschmidt-S. Sternberg [13], J. Kijowski-W. Szczyrba [14]) is a variational problem defined by a lagrangian density $\mathcal{L}\eta$ on the fiber bundle J^1E of the 1-jets of local sections of a fibered manifold $\pi: E \longrightarrow X$, where X is an orientable manifold endowed with a volume element η.

If $j^1s: X \longrightarrow J^1E$ is the 1-jet extension of a section $s: X \longrightarrow E$, one can define a functional $L:\Gamma(X,E) \longrightarrow \mathbb{R}$ by the formula:

$$L(s) = \int_{j^1s} \mathcal{L}\eta$$

Now, a fundamental problem in the Calculus of Variations consists in giving an adequate definition of extremal with respect to L with the further idea of characterising them in a handy way.

There are two classical methods of attacking the problem. One of them consists in endowing $\Gamma(X,E)$ (or an adequate extension thereof) with an infinite-dimensional differentiable structure in such a way that L becomes a differentiable function; the extremals are the zeros of dL. In the other method one exploits, as much as possible, the geometric structure of the problem data giving a direct definition of extremal, with the further idea of characterising them as solutions of some sort of partial differential equations. The Hamilton-Cartán formalism is a typical example of the last method.

If D is a π-projectable vector field on E and j^1D its 1-jet extension, one can give the following:

Definition 1. A section $s:X \longrightarrow E$ is extremal when for any π-projectable vector field D on E one has:

$$\int_{j^1 s} L_{j^1 D} \mathcal{L}\eta = 0$$

In the different versions of the so-called "Hamilton-Cartán formalism of the Calculus of Variations" a global n-form Θ on the manifold J^1E is introduced $(n = \dim X)$, in such a way that one has the following characterization of extremals:

Theorema 1 (Cartán). A section $s:X \longrightarrow E$ is extremal if and only if for any vector field Y on J^1E one has:

$$(1) \qquad i\, Y\, d\, \Theta \Big|_{j^1 s} = 0$$

This condition is locally equivalent to the classical Euler-Lagrange equations. In the case $X = \mathbb{R}$, the 1-form Θ was introduced and Theorem 1 proved by Elie Cártan in the well-known book Leçons sur les invariants intégraux. Thus the present formalism is a generalization of Cartán Theory to an arbitrary fibered manifold.

The differences between the various versions of the theory lie essentially on the different ways of introducing the Cartán form. In particular, the treatment followed in [9] is based on the systematic use of a particular 1-form which is canonically associated to every jet fiber bundle. It is defined starting from the notion of vertical differential of a section of E, and it is a 1-form θ on the manifold J^1E valued on the module M of sections of the induced vector bundle $p^*V(E)$, where V(E) is the vector bundle of vertical vector fields on E and p is the canonical projection $J^1E \longrightarrow E$. The idea of the geometrization of the Calculus of Variations proposed in [9] consists in the formulation of all concepts and manipulations of the theory in terms of differential forms and operations of differential calculus on the manifold J^1E with values in the modules M, M*, Hom(M,M) etc. For instance, the 1-jet extension j^1s of a section $s:X \longrightarrow E$ is thus characterized as the only section $\bar{s}:X \longrightarrow J^1E$ such that $p \circ \bar{s} = s$ and $\theta|_{\bar{s}} = 0$; the 1-jet extension j^1D of a π-proyectable vector field on E is characterized as the only vector field $\bar{D}$ on J^1E which is p-projectable on D and such that $L_{\bar{D}}\theta = \phi \circ \theta$, where $\phi \in Hom(M,M)$ and the product "$\circ$" is taken with respect to the natural bilinear product $Hom(M,M) \times M \longrightarrow M$. Proceding in this way, the introduction of the form Θ and the proof of Theorem 1 reduce to easy-going operations of vector bundle-valued exterior differential calculus.

1.2. The set V of extremals of a variational problem can be thought of as a sort of "infinite-dimensional manifold" on which the following definition of tangent vector is given:

Definition 2. A Jacobi field along an extremal s is a π-vertical vector field D_s on E, defined along s, and such that for any vector field Y on J^1E one has:

$$i\,Y\,L_{j^1D_s}\,d\,\Theta\Big|_{j^1s} = 0 \tag{2}$$

where j^1D_s is the 1-jet extension of D_s (i.e., the only vector field on J^1E, defined along j^1s, which is p-projectable on D_s and satisfies $L_{j^1D_s}\theta|_{j^1s} = 0$).

The real vector space defined by all these fields will be called tangent space to V at s. It will be denoted by T_sV. V will be called, as is usual, the manifold of solutions of the variational problem.

Condition (2) defines a system of linear partial differential equations in the fields D_s which is the geometric version of the "linearization" of the equation (1) in one of its solutions. When V is a differentiable manifold, every Jacobi field D_s is the tangent vector at s to a differentiable curve $\{s_t\}$ of solutions of (1) through s (except, possibly, for some points of V called "linearization unstables" A. Fischer - J. Marsden [7]). This gives the geometrical interpretation of this concept.

There are two basic notions in the Calculus of Variations admitting an inmediate interpretation in terms of the manifold of solutions; they are the so-called infinitesimal symmetries and their associated Noether invariants.

Definition 3. An infinitesimal symmetry of a variational problem is a π-projectable vector field D on E such that $L_{j^1D}\mathcal{L}\eta = 0$.

Theorem 2 (Noether). If D is an infinitesimal symmetry then, for every extremal s, the (n-1)-form defined on X by $i(j^1D)\Theta|_{j^1s}$ is closed.

The (n-1)-forms $i(j^1D)\Theta$ on J^1E associated to infinitesimal symmetries are a generalization of the first integrals for the one variable Euler-Lagrange equations. They were introduced by Emmy Noether in 1912 with the then usual terminology for the Calculus of Variations. Therefore they are called Noether invariants.

If one takes the cohomology class in $H^{n-1}(X,\mathbb{R})$ defined by the closed (n-1)-form $i(j^1D)\Theta|_{j^1s}$, then the Noether invariants can be interpreted as $H^{n-1}(X,\mathbb{R})$-valued functions on the manifold of solutions. This suggests taking this cohomology space as "scalars" in place of real numbers. In particular, when one deals with one-variable problems, $H^{n-1}(X,\mathbb{R}) = \mathbb{R}$, thus obtaining ordinary real functions.

On the other hand, if D is an infinitesimal symmetry, the vertical component D_s^v of D with respect to an extremal s is a Jacobi field along itself. In this way, infinitesimal symmetries can be considered as vector fields on the manifold of solutions.

All this suggests the definition of tensors at a point $s \in V$ as $\mathbb{R}$-multilinear mappings from the tangent space $T_s(V)$ into $H^{n-1}(X,\mathbb{R})$. In particular, if D and D' are two Jacobi fields along the extremal s, $i_{D'}\, i_D\, d\Theta|_{j^1 s}$ is closed, thus allowing us to define an $\mathbb{R}$-bilinear hemisymmetric mapping:

$$T_s(V) \times T_s(V) \xrightarrow{(\omega_2)_s} H^{n-1}(X,\mathbb{R}) \tag{3}$$

In this way, the manifold of solutions of a variational problem is canonically endowed with a 2-form ω_2, which we shall call its pre-symplectic metric.

For a broad class of one-variable variational problems (V,ω_2) is a symplectic manifold of finite dimension on which every infinitesimal symmetry and the differential of its corresponding Noether invariant are in the "pole-polar" relation with respect to ω_2, and where infinitesimal symmetries and Noether invariants are respectively subalgebras of the Lie algebra of vector fields on X and the Poisson algebra of differentiable functions on X.

For $n > 1$, if S is an oriented (n-1)-dimensional compact submanifold of X, the map $\int_S : H^{n-1}(X,\mathbb{R}) \longrightarrow \mathbb{R}$ allows us to consider the above introduced tensors on V as ordinary tensors. By adopting this viewpoint there are examples where (V,ω_2) is an symplectic infinite-dimensional differentiable manifold, I. Ségal [16], while there exist other examples where V is an infinite-dimensional differentiable manifold on which ω_2 is closed though singular A. Fischer - J. Marsden [8].

In any case (V,ω_2) can be thought of as an adequate geometric language unifying many notions and results in Calculus of Variations. Moreover, such a language has revealed itself as a very useful one to define the differentiable structure of the manifold of solucions in some examples where this has been achieved.

2. Yang-Mills fields.

2.1. The origin of this notion is found in the theory of electromagnetic field. In the most simple case, an electromagnetic field is defined by a variational problem whose fibered manifold $\pi : E \longrightarrow X$ is the cotangent bundle $T^*(X)$ of a pseudoriemannian orientable manifold X, volume element η the one canonically associated to the metric, and lagrangian $\mathcal{L}$ the real function on J^1E given by:

$$\mathcal{L}(j_x^1\omega) = 1/4 \, \|d\omega\|_x^2$$

where $\| \ \|$ is the norm defined by the metric on 2-forms.

Equation (1) becomes for this case the well known Maxwell equations:

$$\delta \, d \omega = 0$$

where $\delta = *^{-1} d*$ is the codifferential with respect to the metric. One has a system of linear partial differential equations, hence in this case, the manifold of solutions is a real vector space.

On the other hand, as the vertical tangent bundle V(E) can be identified with the induced vector bundle $\pi^* T^*(X)$, π-vertical vector fields on E defined along a solution can be identified with 1-forms on X. By this identification, equation (2) reproduces Maxwell equations as one could expect from their linearity.

Infinitesimal vertical symmetries of this variational problem are of special interest. They coincide locally with the infinitesimal generators D^f of the one-parameter groups of automorphisms τ_t^f on $T^*(X)$ defined by:

$$\tau_t^f(\omega_x) = \omega_x + t(df)_x$$

where f runs over the set of differentiable functions on X. The interest lies on the fact that they belong to the radical of the pre-symplectic metric ω_2 on the space of solutions. More precisely, one has an involutive subdistribution of $\mathrm{rad}\,\omega_2$ whose maximal integral submanifolds are:

$$\omega + df$$

ω being a solution and f an arbitrary function. Therefore the set of such submanifolds is the quotient $\bar{V}$ of V with respect to the subspace of exact 1-forms.

The idea of the recent geometric version of the Yang-Mills fields is based on the following reformulation of the preceding theory.

If $p: P \longrightarrow X$ is the U(1)-principal bundle direct product $X \times U(1)$ and σ_0 is the canonical flat connection of this product, then 1-forms ω on X can be identified with connections σ on P by the formula:

$$\sigma = \sigma_0 + \omega$$

By means of this identification the lagrangian of the electromagnetic field can be re-writen as follows:

$$\mathcal{L}(j_x^1 \sigma) = 1/4 \, \|F\|_x^2 , \quad F = \mathrm{Curv}\,\sigma$$

So Maxwell equations appear under the new form:

(1') $$F = \mathrm{Curv}\,\sigma , \quad \delta F = 0$$

The manifold of solutions thus loses its vector space structure, though its affine structure is preserved. In this way the tangent space $T_\sigma(V)$ at a point σ to the manifold of solutions V is identified with the real vector space of 1-forms ω on X, verifying the ordinary version of Maxwell equations:

(2')
$$\delta\, d\, \omega = 0$$

With respect to this parametrization of $T_\sigma(V)$ the pre-symplectic metric $(\omega_2)_\sigma$ is given by:

(3')
$$(\omega_2)_\sigma(\omega,\omega') = \langle\omega, iF'^2\eta\rangle - \langle\omega', iF^2\eta\rangle$$

where F^2 is the contraction of $d\omega$ with the metric, $iF^2\eta$ is the contraction of the first index in F^2 with the first one in η and $\langle\omega', iF^2\eta\rangle$ denotes the contraction of the covariant index in ω' with the contravariant one in $iF^2\eta$.

The crucial point of this new formulation is that the maximal integral submanifolds of the above considered subdistribution of rad ω_2 are precisely the orbits produced in V by the natural action of the group of vertical automorphisms of P (gauge group) on the connections. In fact, as far as one deals with a direct product principal bundle $P = X \times U(1)$, these automorphisms can be identified with the differentiable maps $c: X \longrightarrow U(1)$, which act on the connections of P by:

$$\sigma \longmapsto \sigma + \frac{1}{2\pi i}\,\frac{dc}{c}$$

so, if $c = e^{2\pi i f}$, the 1-form corresponding to the transformed connections will be $\omega + df$. On the other hand, this proves that the subspace of $\mathrm{rad}(\omega_2)_\sigma$ defined by the tangents to the orbits of the gauge group passing through σ is that of exact 1-forms $\{df\}$.

In this way, by using the usual terminology for this type of geometrical problems, the quotient $\bar{V}$ is the "moduli space" of the affine space V with respect to the gauge group. In particular, the pre-symplectic metric ω_2 on V is projected upon a metric $\bar{\omega}_2$ on the moduli space.

2.2. Now the natural generalization becomes immediate, A. Pérez-Rendón [15], T.T. Wu - C.N. Yang [19], P. García [10], W. Drechsler - M. Mayer [6] etc. If $p: P \longrightarrow X$ is a principal bundle with structural group G with Lie algebra L, the connections on P can be identified with the global sections of the affine bundle $\pi: E \longrightarrow X$ corresponding to the vector bundle $\mathrm{Hom}(T(X), \mathrm{Ad}\,P)$, where Ad P is the bundle associated to P with respect to the adjoint representation of G on L. The Lie group of vertical automorphisms on P (gauge group), when acting in the natural way on the connections on P, induces a representation $s \in \Gamma(X, \mathrm{Ad}\,P) \longrightarrow D^s$ of its Lie algebra by vector fields on E. By analogy with the electromagnetic field case, if X is an orientable manifold endowed with a volume element η, one can define a Yang-Mills field as a variational problem on E whose lagrangian density admits the vector fields D^s as infinitesimal symmetries.

In particular, if X is pseudo-riemannian and G is semisimple, the metric on X adn the Cartân-Killing metric on L allow one to define a natural norm on the Ad P-valued differential forms on X. This allows one to introduce a gauge-invariant lagrangian analogous to the one of electromagnetic field by the formula:

$$\mathcal{L}(j^1_x\sigma) = 1/4\ \|F\|^2_x\ ,\quad F = \mathrm{Curv}\,\sigma$$

Yang-Mills fields defined by the above lagrangian have been up to now the only ones dealt with in the literature, and they will be called ordinary.

Equation (1) corresponding to such fields becomes the Yang-Mills equations under the global version as they have recently been presented in the literature:

$$(1'')\qquad F = \mathrm{Curv}\,\sigma\ ,\quad \delta_\sigma F = 0$$

where δ_σ is the operator $\overset{-1}{*}d_\sigma *$, d_σ being the exterior differential for Ad P-valued forms on X with respect to the connection σ.

In general, (1") defines a system of nonlinear partial differential equations, the cause of this "non linearity" being in the "non commutativity" of the group G.

On the other hand, as in this case the vertical tangent bundle V(E) is identified with the induced vector bundle $\pi^*\mathrm{Hom}(T(X),\mathrm{Ad}\,P)$, the π-vertical vector fields on E defined along a section $\sigma: X \longrightarrow E$ can be identified with the Ad P-valued 1-forms on X. By way of this identification, the tangent space $T_\sigma(V)$ at a point σ of the manifold of solutions V of the Yang-Mills equations becomes the real vector space of Ad P-valued 1-forms ω on X satisfying the equation:

$$(2'')\qquad \delta_\sigma d_\sigma \omega - [\langle F_\sigma,\omega\rangle] = 0$$

where F_σ is the curvature of the connection σ contravaried with the metric on X and $[\langle F_\sigma,\omega\rangle]$ denotes the contraction of the first contravariant index in F_σ with the covariant one in ω with respect to the bilinear product defined by the Lie module structure of $\Gamma(X,\mathrm{Ad}\,P)$.

Equation (2") constitutes an intrinsic expresion for the "linearization" of Yang-Mills equations in one of its solutions, which show very clearly the σ-dependence of the tangent space $T_\sigma(V)$.

With respect to this parametrization of $T_\sigma(V)$, the pre-symplectic metric $(\omega_2)_\sigma$ is given by the formula:

$$(3'')\qquad (\omega_2)_\sigma(\omega,\omega') = \langle\omega, iF'^2\eta\rangle - \langle\omega', iF^2\eta\rangle$$

where F^2 is the contraction of $d_\sigma\omega$ with the metric of X, $iF^2\eta$ is the contraction of the first index in F^2 with the first one in η and $\langle\omega', iF^2\eta\rangle$ is the contraction of the covariant index in ω' with the covariant one in $iF^2\eta$ with respect to the bilinear product $\Gamma(X,\mathrm{Ad}\,P) \times \Gamma(X,\mathrm{Ad}\,P) \longrightarrow C^{\infty}(X)$, defined by the Cartan-Killing metric.

As in the theory of electromagnetic field, it can be proved with all generality, P. García [11], P. García - A. Pérez-Rendón [12], that the orbits of the gauge group in the manifold of solutions are tangent to the radical of the pre-symplectic metric ω_2. In [11] this result is given for an arbitrary (free) Yang-Mills field, and in [12] this is proved for a Yang-Mills field in "minimal interaction" with another field, both being arbitrary. In the free case, the subspace of $\mathrm{rad}(\omega_2)_\sigma$ defined by the tangents at σ to the orbits of gauge group is that of d_σ-exact 1-forms, i.e., $\{d_\sigma s\}$, $s\varepsilon\Gamma(X,\mathrm{Ad}\,P)$.

Now a fundamental problem is the study of the structure of the manifold of moduli $\bar{V}$ with respect to the gauge group. In particular, an important question is to see in which cases and under what circumstances the metric $\bar{\omega}_2$ on $\bar{V}$, upon which the pre-symplectic metric ω_2 is projected, is already irreducible. In what follows we shall deal with this last question.

3. Determination of the radical of the pre-symplectic metric of a Yang-Mills field.

3.1. The elliptic variant of the electromagnetic field theory provides an elementary example where the radical of ω_2 at each point $\sigma\varepsilon V$ is bigger than the subspace defined by the vectors tangent at σ to the corresponding orbit of the gauge group. In fact, with the notations of §2.1 and under the hypothesis of X being an orientable, compact, connected riemannian manifold, one has the following:

Lemma. The subspace of $T_\sigma(V)$ defined by the 1-forms $\omega\varepsilon T_\sigma(V)$ satisfying the (Lorentz) condition $\delta\omega = 0$ is complementary to the subspace $\{df\}$ of exact 1-forms.

Proof. If df satisfies $\delta df = 0$, i.e. $\Delta f = 0$, then f is constant, therefore $df = 0$. Let $\omega\varepsilon T_\sigma(V)$, then by the Hodge decomposition theorem, $\omega = H\omega + \delta\omega' + df$, $H\omega$ being the harmonic component of ω, ω' a 2-form and f a function. Then $H\omega + \delta\omega'\varepsilon T_\sigma(V)$ and $\delta(H\omega + \delta\omega') = \delta H\omega + \delta^2\omega' = 0$ //

Theorem 3. The pre-symplectic metric ω_2 on the space of solutions V of an electromagnetic field on an orientable, compact, connected riemannian manifold is identically zero.

Proof. By the above Lemma it suffices to prove that for every $\sigma\varepsilon V$, $(\omega_2)_\sigma = 0$ on the subspace of 1-forms $\omega\varepsilon T_\sigma(V)$ such that $\delta\omega = 0$. But for these 1-forms one has $\Delta\omega = \delta d\omega + d\delta\omega = 0$, so $d\omega = 0$ and then $F^2 = 0$ //

Then, the space of moduli $\bar{V}$ of V with respect to the gauge group can be identified in this case, with the affine space associated to the real vector space of harmonic 1-forms on X ($= H^1(X, \mathbb{R})$), and the general theory does not provide a pre-symplectic metric on the space of solutions.

3.2. Let us now consider the ordinary electromagnetic field theory (i.e. X is a 4-dimensional manifold endowed with a Lorentz metric ${}^{(4)}g$). Let $S \subset X$ be a spacelike compact hypersurface. If one takes a normal gaussian coordinate system along S, then in a tubular neighbourhood of S one has $X = (-\varepsilon,\varepsilon) \times S$, and ${}^{(4)}g$ becomes:

$$(4) \qquad {}^{(4)}g = -dt^2 + {}^{(3)}g_t$$

where "t" is the natural coordinate in $(-\varepsilon,\varepsilon)$ and ${}^{(3)}g_t$ is the riemannian metric defined on S by the restriction of ${}^{(4)}g$ to the hypersurface $\{t\} \times S$.

With respect to the local decomposition $X = (-\varepsilon,\varepsilon) \times S$, a 1-form ω and its exterior differential $d\omega$ can be expressed as follow: $\omega = \phi_t dt + A_t$, $d\omega = dt \wedge E_t + H_t$, where ϕ_t A_t E_t and H_t are, respectively, a function, two 1-forms and a 2-form on S defined by the restriction to the hypersurface $\{t\} \times S$ of $\omega(\frac{\partial}{\partial t})$, ω, $i_{\frac{\partial}{\partial t}} d\omega$ and $d\omega$. In physical terminology, they are the scalar and vector potencials and the electric and magnetic fields. In terms of these new objects, Maxwell equations become:

$$(5) \qquad \begin{cases} \dfrac{dA_t}{dt} = E_t + d\phi_t & H_t = dA_t \\[2ex] \dfrac{dE_t}{dt} = \delta_t H_t & \delta_t E_t = 0 \end{cases}$$

where the operators d and δ_t appearing in the second members are, respectively, the exterior differential and the codifferential with respect to the metric ${}^{(3)}g_t$ on the hypersurface S.

The solution of the Cauchy problem for the Maxwell equations establishes that, up to gauge transformations, there exists a unique solution (ϕ_t A_t E_t H_t) of (5) with arbitrarily given initial conditions $\phi_0 = \phi$, $A_0 = A$, $E_0 = E$, $H_0 = H$ in the space defined by the constraint equations $H = dA$, $\delta E = 0$. Here δ denotes the codifferential with respect to the metric ${}^{(3)}g = {}^{(3)}g_0$. More precisely, we can enunciate the following:

<u>Proposition</u>. Let $T_\sigma(V)$ be the tangent space to the space of solutions of the Maxwell equations on $X = (-\varepsilon,\varepsilon) \times S$ at one point σ, let $\{df\}_S^{(1}$ be the subspace of $\{df\} \subset T_\sigma(V)$ defined by the exact 1-forms which are zero along S, and let E be the space of vectors (ϕ,A,E) such that $\delta E = 0$. Then one has:

$$E = T_\sigma(V) \,/\, \{df\}_S^{(1}$$

According to this, if π is the canonical projection from $T_\sigma(V)$ to E, then the pre-symplectic metric $(\omega_2)_\sigma$ is projected by π on a metric $\hat{\omega}_2$ on E whose radical contains $\pi\{df\}$. A simple calculus proves that $\pi\{df\}$ is the subspace of E defined by the vectors $(\phi, d\psi, 0)$, where ϕ and ψ are arbitrary functions on S. On the other hand, $\hat{\omega}_2$, when interpreted as a real 2-form via the map $\int_S : H^{n-1}(X, \mathbb{R}) \longrightarrow \mathbb{R}$, is given by the formula:

$$\hat{\omega}_2((\phi\, A\, E), (\phi'\, A'\, E')) = \int_S (\langle A, E'\rangle - \langle A', E\rangle)\, {}^{(3)}\eta \tag{6}$$

where $\langle A, E'\rangle$ denotes the scalar product with respect to ${}^{(3)}g$ of the 1-forms A and E', and ${}^{(3)}\eta$ is the volume element on S canonically associated to ${}^{(3)}g$.

This reduction, based on a convenient statement of the solution of the Cauchy problem for Maxwell equations, allows one to prove the following:

Theorem 4. The radical of the pre-symplectic metric ω_2 on the space of solutions V of an ordinary electromagnetic field is generated at each point $\sigma \varepsilon V$ by the tangents at such a point to the orbit of the gauge group. Consequently ω_2 is projected upon an irreducible metric on the corresponding space of moduli $\bar{V}$.

Proof. It will be enough to prove that every vector in the radical of $\hat{\omega}_2$ is of type $(\phi, d\psi, 0)$. Let $(\phi, A, E) \varepsilon \text{ rad } \hat{\omega}_2$, i.e., for every $(\phi', A', E') \varepsilon E$ one has:

$$0 = \hat{\omega}_2((\phi\, A\, E), (\phi'\, A'\, E')) = \int_S (\langle A, E'\rangle - \langle A', E\rangle)\, {}^{(3)}\eta$$

By taking $A' = 0$ then, for every 1-form E' with $\delta E' = 0$, one must have:

$$\int_S \langle A, E'\rangle\, {}^{(3)}\eta = 0$$

By the Hodge decomposition theorem, $A = HA + \delta\omega + d\psi$, where HA is the harmonic componet of A, ω is a 2-form and ψ is a function. Then, for every E' with $\delta E' = 0$, one must have:

$$\int_S \langle HA, E'\rangle\, {}^{(3)}\eta + \int_S \langle \delta\omega, E'\rangle\, {}^{(3)}\eta = 0$$

By taking $E' = HA$, which is closed and coclosed because it is harmonic, one has:

$$\int_S \langle HA, HA\rangle\, {}^{(3)}\eta = 0$$

from where $HA = 0$.

Now, if one takes $E' = \delta\omega$, which is coclosed for $\delta^2 = 0$, one has:

$$\int_S \langle \delta\omega, \delta\omega\rangle\, {}^{(3)}\eta = 0$$

so $\delta\omega = 0$.

Thus $A = d\psi$, in turn implying that for every 1-form A', one has:

$$\int_S \langle E, A'\rangle\, {}^{(3)}\eta = 0$$

so finally one has $E=0$ //

3.3. Last, we shall deal with the case of an ordinary Yang-Mills field on a Lorentz manifold $(X, {}^{(4)}g)$ in a way identical to the one used for the case of the electromagnetic field.

First of all one must establish the way for equation (2") to be expressed as a system of equations analogous to Maxwell with respect to the local decomposition $X=(-\varepsilon,\varepsilon)\times S$. This can be achieved as follows.

The parallel translation with respect to the connection σ along the curves $\{y=\text{const.}, y\in S\}$ of $X=(-\varepsilon,\varepsilon)\times S$, allows one to identify the bundles $P|_{\{t\}\times S}$ and $P|_S$ and, consequently, $\mathrm{Ad}\,P|_{\{t\}\times S}$ and $\mathrm{Ad}\,P|_S$. By way of this identification, a $\mathrm{Ad}\,P$-valued form ω on X and its exterior differential $d_\sigma\omega$ with respect to σ can be expressed as follows: $\omega=\phi'_t dt+A'_t$, $d_\sigma\omega=dt\wedge E'_t+H'_t$, where ϕ'_t A'_t E'_t and H'_t are, respectively, a function, two 1-forms and a 2-form on S with values in $\mathrm{Ad}\,P$, defined by the restriction to $\{t\}\times S$ of $\omega(\frac{\partial}{\partial t})$, ω, $i\frac{\partial}{\partial t}d_\sigma\omega$ and $d_\sigma\omega$. In terms of these new objects equation (2") can be expressed as follows:

$$(7)\qquad \begin{cases} \dfrac{dA'_t}{dt}=E'_t+\tilde{d}_t\phi'_t & H'_t=\tilde{d}_tA'_t \\ \dfrac{dE'_t}{dt}=\tilde{\delta}_tH'_t+[E_t,\phi'_t]+[<H_t,A'_t>] & \tilde{\delta}_tE'_t=-[<E_t,A'_t>] \end{cases}$$

where $\tilde{d}_t$ is the exterior differential on $\mathrm{Ad}\,P|_S$-valued forms on S with respect to the connection defined on $P|_S\simeq P|_{\{t\}\times S}$ by the restriction of σ to $P|_{\{t\}\times S}$, $\tilde{\delta}_t$ is the codifferential with respect to $\tilde{d}_t$ and to the metric ${}^{(3)}g_t$, E_t and H_t are the $\mathrm{Ad}\,P|_S$-valued 1-form and 2-form on S defined, respectively, by the restriction to $\{t\}\times S$ of $i\frac{\partial}{\partial t}\mathrm{Curv}\,\sigma$ and $\mathrm{Curv}\,\sigma$, and finally the products $[\ ,\ ]$ and $[<\ ,\ >]$ are defined in the usual way by the contraction with ${}^{(3)}g_t$ and the bilinear product defined by the Lie module structure in $\Gamma(S,\mathrm{Ad}\,P|_S)$.

The solution of the Cauchy problem for the system of first order linear partial equations (7) establishes that, up to gauge transformations, there exists a unique solution $(\phi'_t\ A'_t\ E'_t\ H'_t)$ of (7) with arbitrarily given initial conditions $\phi'_0=\phi'$, $A'_0=A'$, $E'_0=E'$, $H'_0=H'$ in the space defined by the constraint equations:

$$H'=\tilde{d}A'\quad,\quad \tilde{\delta}E'=-[<E,A'>]$$

where, now, $\tilde{d}$ is the exterior differential on $\mathrm{Ad}\,P|_S$-valued forms on S with respect to the restriction to $P|_S$ of the connection σ, $\tilde{\delta}$ is the codifferential with respect to $\tilde{d}$ and to the metric ${}^{(3)}g$, and E is the restriction to S of the 1-form $i\frac{\partial}{\partial t}\mathrm{Curv}\,\sigma$.

In a more precise way one has:

$$E=T_\sigma(V)\,/\,\{d_\sigma s\}_S^{(2}$$

E being the space of vectors $(\phi' A' E')$ such that $\tilde{\delta}E' = -[<E,A'>]$ and $\{d_\sigma s\}_S^{(2}$ is the subspace of $\{d_\sigma s\} \subset T_\sigma(V)$ defined by the elements such that $d_\sigma s$ and $d_\sigma^2 s$ are zero along S.

Remark. The Cauchy problem for Yang-Mills equations on a Lorenz manifold has been recently dealt with by I. Segal [17]. The above statement is merely a geometric version of the solution of the said problem for the "linearization" of such equations at one of their solutions. In particular, the condition $d_\sigma^2 s = 0$ along S is an algebraic one because $d_\sigma^2 s = [\text{Curv}\,\sigma, s]$.

By following the analogy with the case of electromagnetic field, if π is the canonical projection from $T_\sigma(V)$ to E, then the pre-symplectic metric $(\omega_2)_\sigma$ is projected by π upon a metric $\hat{\omega}_2$ on E whose radical contains $\pi\{d_\sigma s\}$. A simple calculus shows that $\pi\{d_\sigma s\}$ is the subspace of E defined by the vectors $(\phi', \tilde{d}\psi', [E,\psi'])$, where ϕ' and ψ' are arbitrary sections of $\text{Ad}\,P|_S$. On the other hand, $\hat{\omega}_2$, when interpreted as a real 2-form via the map $\int_S : H^{n-1}(X, \mathbb{R}) \longrightarrow \mathbb{R}$, is given by the formula:

$$(8) \qquad \hat{\omega}_2((\phi' A' E'),(\bar{\phi}' \bar{A}' \bar{E}')) = \int_S (<A',\bar{E}'> - <\bar{A}',E'>)\,{}^{(3)}\eta$$

where $<A',\bar{E}'>$ denotes the scalar product of the $\text{Ad}\,P|_S$-valued 1-forms A' and $\bar{E}'$ on S with respect to the metric ${}^{(3)}g$ on S and to the riemannian structure on $\text{Ad}\,P|_S$ defined by the Cartán-Killing metric.

Having in mind the idea of generalizing Theorem 4 to a Yang-Mills field, it would be advisable to apply to the bundle $\text{Ad}\,P|_S$ the "Hodge decomposition theorem" for differential forms with values in a riemannian vector bundle endowed with a connection (e.g. see S. Bochner [4]). This is precisely the case for the bundle $\text{Ad}\,P|_S$, which is endowed with the riemannian structure defined by the Cartân-Killing metric and with the connection defined by the restriction of σ to $P|_S$. Let us see how Hodge theory is established in this case.

In order to symplify the notation, let $\text{Ad}\,P|_S = E$. One can define a scalar product on the real vector space $\Lambda^p(S,E)$ of E-valued p-forms on S by the formula:

$$\omega \cdot \omega' = \int_S <\omega,\omega'>\,{}^{(3)}\eta$$

where < , > is the above defined product by means of the metric ${}^{(3)}g$ and the riemannian structure of E. By ortogonality, one can extend this scalar product to $\Lambda(S,E) = \bigoplus_p \Lambda^p(S,E)$, and one can easily establish that the operators $\tilde{d}$ and $\tilde{\delta}$ are adjoint to each other, i.e.:

$$\tilde{d}\omega \cdot \omega' = \omega \cdot \tilde{\delta}\omega'$$

As in the ordinary theory, one defines the Laplace operator by the formula $\tilde{\Delta} = \tilde{d}\tilde{\delta} + \tilde{\delta}\tilde{d}$, and one says that a form ω is harmonic when $\tilde{\Delta}\omega = 0$. This is equivalent to ω being closed and coclosed. The only point one must take care of is that now $\tilde{d}^2$

and $\tilde{\delta}^2$ are not zero, in general, because of the curvature of the given connection.

The Hodge decomposition theorem now takes the following form:

$$\Lambda^p(S,E) = H^p(S,E) \oplus \left[\tilde{d}\,\Lambda^{p-1}(S,E) + \tilde{\delta}\,\Lambda^{p+1}(S,E)\right] \tag{9}$$

where $H^p(S,E)$, the space of harmonic p-forms, is orthogonal to the other two summands.

If one tries to prove Theorem 4 for a Yang-Mills field using the decomposition (9), one does not get the desired result. This is due, precisely, to the fact that $\tilde{\delta}^2 \neq 0$. Nevertheless, one can obtain the desired generalization as follows.

Lemma.

$$\Lambda^1(S,E) = \tilde{d}\Lambda^0(S,E) \oplus \ker(\tilde{\delta}:\Lambda^1(S,E) \longrightarrow \Lambda^0(S,E)) \tag{10}$$

Proof. Because $(\mathrm{Sim}_\xi \tilde{d})\omega = \xi \wedge \omega$, $\xi \varepsilon T^*(S)$, $\omega \varepsilon \Lambda^p(S,E)$, the symbol of the operator $\tilde{d}:\Lambda^0(S,E) \longrightarrow \Lambda^1(S,E)$ is injective. Thus it is enough to apply the Fredholm alternative //

Theorem 5. The radical of the pre-symplectic metric ω_2 on the manifold of solutions V of an ordinary Yang-Mills field on a Lorentz manifold is generated at each point $\sigma \varepsilon V$ by the tangents at such a point to the orbit of the gauge group. Consequently ω_2 is projected upon an irreducible metric on the corresponding manifold of moduli $\bar{V}$.

Proof. It will be enough to prove that every vector in the radical of $\hat{\omega}_2$ is of type $(\phi',\tilde{d}\psi',[E,\psi'])$. Let $(\phi',A',E') \varepsilon \operatorname{rad} \hat{\omega}_2$, i.e. for every $(\bar{\phi}',\bar{A}',\bar{E}')\varepsilon E$ one has:

$$0 = \hat{\omega}_2((\phi' A' E'),(\bar{\phi}' \bar{A}' \bar{E}')) = \int_S (\langle A,\bar{E}'\rangle - \langle \bar{A}',\bar{E}'\rangle)^{(3)}\eta$$

By taking $\bar{A}' = 0$ then, for every 1-form $\bar{E}'$ with $\tilde{\delta}\bar{E}' = 0$, one must have:

$$\int_S \langle A',\bar{E}'\rangle^{(3)} = 0$$

By the decomposition formula (10), $A' = \tilde{d}\psi' + \omega$, where ψ' is a 0-form and ω is an 1-form with $\tilde{\delta}\omega = 0$. Then, for every $\bar{E}'$ with $\tilde{\delta}\bar{E}' = 0$, one must have:

$$\int_S \langle \omega,\bar{E}'\rangle^{(3)}\eta = 0$$

in particular for $\bar{E}' = \omega$. Therefore $\omega = 0$.

Thus $A' = d\psi'$, and this in turn implies that for every $(\bar{\phi}' \bar{A}' \bar{E}')\varepsilon E$ one has:

$$0 = \int_S (\langle \tilde{d}\psi',\bar{E}'\rangle - \langle E',\bar{A}'\rangle)^{(3)}\eta = \int_S (\langle \psi',\tilde{\delta}\bar{E}'\rangle - \langle E',\bar{A}'\rangle)^{(3)}\eta =$$

$$= \int_S (\langle \psi', -[\langle E,\bar{A}'\rangle]\rangle - \langle E',\bar{A}'\rangle)^{(3)}\eta$$

From the invariance of Cartān-Killing metric with respect to the adjoint representation, one obtains:

$$<[E,\psi'],\bar{A}'> + <\psi',[<E,\bar{A}'>]> = 0$$

Then, the last integral can be written in the form:

$$\int_S (<[E,\psi'],\bar{A}'> - <E',\bar{A}'>^{(3)}\eta = \int_S <[E,\psi'] - E',\bar{A}'>^{(3)}\eta$$

But the 1-form $\bar{A}'$ is arbitrary, so finally one has $E' = [E,\psi']$ //

The theorem which has just been proved has been recently enunciated by I.E. Segal [18]. It may also be found in the Ph.D. Thesis of T.P. Branson [5].

Remark. Formula (10) is satisfied by ordinary forms of any order as a consequence of the Hodge decomposition theorem and of the identity $\delta^2 = 0$. If one had applied such a formula in the proofs of Theorems 3 and 4, such proofs would have been more direct and essential. This suggests the consideration of the decomposition (10) rather than the Hodge theorem as "basic formula" to prove, in Yang-Mills theory, results of the type we are dealing with here.

4. Generalization to minimal interactions.

The method we have just employed for determining the radical of the pre-symplectic metric of a Yang-Mills field can be generalized to "minimal interactions" as we shall illustrate with an example: Klein-Gordon field over a complex line bundle in minimal interaction with the electromagnetic field generated by itself.

Let L be a complex line bundle over an orientable pseudoriemannian manifold with pseudoriemannian volume element η, endowed with a hermitian metric h and a hermitian connection σ with respect to h. The metric on X and the hermitian metric define a natural norm $\|\ \|$ over the L-valued forms on X. This allows us to globalize the ordinary notion of a complex scalar field by considering the variational problem defined on $J^1(L)$ by the lagrangian density $\mathcal{L}\eta$, where $\mathcal{L}$ is the function given by:

$$\mathcal{L}(j^1_x s) = 1/2(\|d_\sigma s\|^2_x - m^2\|s\|^2_x)$$

The corresponding field equations then become:

$$(\Box_\sigma - m^2)s = 0$$

where $\Box_\sigma$ is the Laplace operator for L-valued forms on X relative to the pseudoriemannian metric of X and to the connection σ on L. Physically, one has a "Klein-Gordon field over a complex line bundle in the external electromagnetic field defined by the connection σ".

As is known, these connections are precisely the connections over the principal U(1)-bundle P defined by the elements $e \in L$ such that $h(e,e) = 1$. Let E be the affine bundle whose global sections are these connections. The Yang-Mills field defined on E in the sense of §2 can be interpreted as "the electromagnetic field generated by the electric particle described by the above Klein-Gordon field".

Now the minimal interaction between these two fields is defined by the variational problem on $J^1(L \times_X E)$ whose lagrangian is the sum of the corresponding lagrangians, i.e.:

$$\mathcal{L}(j^1_x(s,\sigma)) = 1/2\,(\|d_\sigma s\|^2_x - m^2\,\|s\|^2_x) + 1/4\,\|F\|^2_x\,, \quad F = \mathrm{Curv}\,\sigma \tag{11}$$

Equation (1) for this variational problem becomes the coupled Maxwell-Klein-Gordon equations in the following global version:

$$\left\{\begin{array}{l}(\Box_\sigma - m^2)s = 0 \\ \delta F = \mathrm{Re}\langle is, d_\sigma s\rangle \quad , \quad F = \mathrm{Curv}\,\sigma\end{array}\right. \tag{1'''}$$

where $\mathrm{Re}\langle is, d_\sigma s\rangle$ is the real part of the product of the section is and the 1-form $d_\sigma s$ with respect to the hermitian product defined on $\Gamma(X,L)$ by the metric h.

Remark. The term $\mathrm{Re}\langle is, d_\sigma s\rangle$ is the current 1-form associated to this minimal interaction. It depends on both interacting fields s and σ and on the hermitian metric. It furnishes us with a very simple answer to the question of the so called Wu - Yang dictionary [19]. This notion is defined in general in P. García - A. Pérez-Rendón [12] (§2.2. "Current tensor associated to a Minimal Interaction").

In this case, the vertical tangent bundle of $L \times_X E$ can be identified with the induced bundle of the Whitney sum $L \oplus T^*(X)$. Thus vertical vector fields on $L \times_X E$ along a section (s,σ) can be identified with the couples (s',ω), where s' is a section of L and ω is a 1-form on X. By way of this identification, the tangent space $T_{(s,\sigma)}V$ at a point (s,σ) to the solutions manifold V of equations (1''') is the real vector space of all couples (s',ω) such that:

$$\left\{\begin{array}{l}(\Box_\sigma - m^2)s' = i((\delta\omega)s - 2\langle\omega, d_\sigma s\rangle) \\ \delta d\omega = \mathrm{Re}(\langle is', d_\sigma s\rangle + \langle is, d_\sigma s'\rangle) - \|s\|^2\,\omega\end{array}\right. \tag{2'''}$$

where $\langle\omega, d_\sigma s\rangle$ denotes the scalar product of the ordinary 1-form, ω , and the L-valued 1-form, $d_\sigma s$, defined by the metric of X.

Equations (2''') are an intrinsic expression for the "linearization" of the coupled Maxwell-Klein-Gordon equations in one of its solutions, which show very clearly the (s,σ)-dependence of the tangent space $T_{(s,\sigma)}V$.

With respect to this parametrization of $T_{(s,\sigma)}V$, the pre-symplectic metric $(\omega_2)_{(s,\sigma)}$ is given by the formula:

$$(3''')\quad (\omega_2)_{(s,\sigma)}((s'\omega),(\bar{s}'\bar{\omega})) = -\mathrm{Re}(<\bar{s}',d_\sigma s'> - <s',d_\sigma \bar{s}'> + <is,s'>\bar{\omega} - <is,\bar{s}'>\omega)\cdot\eta +$$

$$+ <\omega,\bar{F}^2\cdot\eta> - <\bar{\omega},\bar{F}^2\cdot\eta>$$

where the second part is the pre-symplectic metric (3') of the electromagnetic field, where i is changed for a dot "•" in order to avoid confusion with the imaginary unit, and the 1-form Re( ) in the first part is considered contravariated with the metric of X, the dot means contraction with the first covariant index of η.

On acting in the natural way on the connections of P and on the sections of its associated vector bundle L, the Lie group of vertical automorphisms of P (gauge group) induces a representation of its Lie algebra (differentiable functions on X with Lie product zero) by vertical vector fields of the bundle $L\times_X E$. The vector fields thus obtained are infinitesimal symmetries of the variational problem under consideration. This can be seen directly by observing that the lagrangian (11) is invariant with respect to vertical automorphisms of P.

By the general theory (§1), such infinitesimal symmetries define vector fields on the solutions manifold. This can be seen directly, too, by observing that the vertical vector field of $L\times_X E$ defined along a section (s,σ) by one of these infinitesimal symmetries corresponding to a function f on X is (ifs,df). As remarked in §2, by the general theory of Minimal Interactions P. García-A. Pérez-Rendón [12], all these vector fields belong to the radical of the pre-symplectic metric (3'''). We shall prove this directly for our case as an application of the explicit formulas we have just obtained.

If (ifs,df) is a vector tangent to the orbit of the gauge group through the point (s,σ) and $(s',\omega)\in T_{(s,\sigma)}V$ is arbitrary, one has:

$$(\omega_2)_{(s,\sigma)}((ifs,df),(s'\omega)) = -\mathrm{Re}(<s',id_\sigma(fs)> - <ifs,d_\sigma s'> + <is,ifs>\omega -$$

$$- <is,s'>df)\cdot\eta + <df,F^2\cdot\eta> = -\mathrm{Re}(<s',is>df - f<is',d_\sigma s> - f<is,d_\sigma s'> +$$

$$+ f\|s\|^2\omega - <is,s'>df)\cdot\eta + \;\cdot d^\nabla(fF^2\cdot\eta) - f\left[\cdot d^\nabla(F^2\cdot\eta)\right]$$

where $\cdot d^\nabla$ indicates contraction of the covariant index of the differentiation with the unique contravariant index of the tensor on which it acts.

Bearing in mind that $\cdot d^\nabla(F^2\cdot\eta) = (\cdot d^\nabla F^2)\cdot\eta = (\mathrm{div}\,F^2)\cdot\eta = (\delta d\omega)\cdot\eta$, where the 1-form $\delta d\omega$ is considered contravaried with the metric, the above expression becomes:

$$- f\left[\delta d\omega - \mathrm{Re}(<is',d_\sigma s> + <is,d_\sigma s'>) + \|s\|^2\omega\right]\cdot\eta + \cdot d^\nabla(fF^2\cdot\eta)$$

The first term is zero, by the second of the equations (2'''), which must be fulfilled by the tangent vector (s',ω).

The result can be now obtained from the following:

Lemma. The (n-1)-form $\cdot d^{\nabla}(fF^2 \cdot \eta)$ is exact.

Proof.

$$\cdot d^{\nabla}(fF^2 \cdot \eta) = \left[\cdot d^{\nabla}(fF^2)\right] \cdot \eta = \mathrm{div}(fF^2) \cdot \eta = *\delta(fd\omega) =$$

$$= **^{-1} d*(fd\omega) = d*(fd\omega) \ //$$

We now see how one can achieve the determination of the radical of $(\omega_2)_{(s,\sigma)}$ for the case of a Lorentz manifold $(X, {}^{(4)}g)$.

With the same notations and under the same hypotheses in §3, the parallel translation with respect to the connection σ along the curves $\{y = \text{const.}, y \in S\}$ of $X = (-\varepsilon,\varepsilon) \times S$, allows one to identify the bundles $L|_{\{t\} \times S}$ and L_S. Via this identification, let s_t and $\dot{s}_t$ be the sections of L_S defined by restricting a section s of L and $\frac{\partial^{\nabla}}{\partial t} s$ to $\{t\} \times S$. By following the procedure in §3, equations (2''') can be rewritten as follows:

The first group becomes two evolution equations

$$(12) \quad \begin{cases} \dfrac{ds'_t}{dt} = \dot{s}'_t \\[2ex] \dfrac{d\dot{s}'_t}{dt} - is_t \dfrac{d\phi_t}{dt} = (\tilde{\Delta}_t - m^2)s'_t - i(\delta_t A_t)s_t + 2i\phi_t \dot{s}_t - 2i\langle A_t, \tilde{d}_t s_t\rangle \end{cases}$$

while the second group gives rise to two evolution equations:

$$(13) \quad \begin{cases} \dfrac{dA_t}{dt} = E_t + d\phi_t \\[2ex] \dfrac{dE_t}{dt} = \delta_t H_t + \mathrm{Re}(\langle is'_t, \tilde{d}_t s_t\rangle + \langle is_t, \tilde{d}_t s'_t\rangle) - \|s_t\|^2 A_t \end{cases}$$

and two constraint equations:

$$(14) \qquad H_t = dA_t \ , \quad \delta_t E_t = -\mathrm{Re}(\langle is'_t, \dot{s}_t\rangle + \langle is_t, \dot{s}'_t\rangle) + \|s_t\|^2 \phi_t$$

where δ_t is the codifferential for ordinary forms on S with respect to the metric ${}^{(3)}g_t$, $\tilde{d}_t$ is the exterior differential for L-valued forms on S with respect to the connection defined on $L_S \simeq L_{\{t\} \times S}$ by restricting σ to $L_{\{t\} \times S}$, and $\tilde{\Delta}_t$ is the Laplace operator with respect to $\tilde{d}_t$ and ${}^{(3)}g_t$.

The solution of the Cauchy problem for the first order partial differential equations system (12) (13) (14) allows us to establish a canonical projection $\pi : T_{(s,\sigma)} V \longrightarrow E$ from the tangent space $T_{(s,\sigma)} V$ upon the space E defined by the elements on S, $(s' \ \dot{s}' \ \phi \ A \ E)$ such that:

(15) $$\delta E = -\mathrm{Re}(\langle is', \dot{s}\rangle + \langle is, \dot{s}'\rangle) + \|s\|^2 \phi$$

The subspace {ifs,df} of $T_{(s,\sigma)}V$ defined by the vectors tangent to the orbit of the gauge group through (s,σ) is projected by π on the subspace of E defined by the vectors of type:

(16) $$(i\psi s, i(\phi s + \psi\dot{s}), \phi, d\psi, 0)$$

ϕ and ψ being arbitrary functions on S.

On the other hand, the pre-symplectic metric $(\omega_2)_{(s,\sigma)}$, when interpreted as a real 2-form via the mapping $\int_S : H^{n-1}(X, \mathbb{R}) \longrightarrow \mathbb{R}$, is projected by π on the following 2-form $\hat{\omega}_2$:

(17) $$\hat{\omega}_2((s'\dot{s}'\phi AE), (\bar{s}'\dot{\bar{s}}'\bar{\phi}\bar{A}\bar{E})) = -\int_S \mathrm{Re}(\langle\bar{s}', \dot{s}'\rangle - \langle s', \dot{\bar{s}}'\rangle + \bar{\phi}\langle is, s'\rangle - \phi\langle is, \bar{s}'\rangle)^{(3)}\eta + $$
$$+ \int_S (\langle A, \bar{E}\rangle - \langle\bar{A}, E\rangle)^{(3)}\eta$$

We can now prove the following:

<u>Theorem 6.</u> The radical of the pre-symplectic metric ω_2 on the manifold of solutions V of the coupled Maxwell-Klein-Gordon equations on a Lorentz manifold is generated at each point $(s,\sigma)\in V$ by the tangents at such a point to the orbit of the gauge group. Consequently ω_2 is projected upon an irreducible metric on the corresponding manifold of moduli $\bar{V}$.

<u>Proof.</u> It will be enough to prove that every vector in the radical of $\hat{\omega}_2$ is of the type (16). Let $(s'\,\dot{s}'\,\phi\,A\,E) \in \mathrm{rad}\,\hat{\omega}_2$, i.e. for every $(\bar{s}'\,\dot{\bar{s}}'\,\bar{\phi}\,\bar{A}\,\bar{E})\in E$ one has:

(*) $$-\int_S \mathrm{Re}(\langle\bar{s}', \dot{s}'\rangle - \langle s', \dot{\bar{s}}'\rangle + \bar{\phi}\langle is, s'\rangle - \phi\langle is, \bar{s}'\rangle)^{(3)}\eta + \int_S (\langle A, \bar{E}\rangle - \langle\bar{A}, E\rangle)^{(3)}\eta = 0$$

By taking $\bar{s}' = \dot{\bar{s}}' = 0$, $\bar{\phi} = 0$ and $\bar{A} = 0$ then, for every 1-form $\bar{E}$ with $\delta\bar{E} = 0$, one must have:

$$\int_S \langle A, \bar{E}\rangle^{(3)}\eta = 0$$

thus implying (see proof of Th. 4) $A = d\psi$. Carrying this to (*) and taking $\bar{s}' = \dot{\bar{s}}' = 0$ and $\bar{\phi} = 0$, it follows that $E = 0$.

Then, for every $(\bar{s}'\,\dot{\bar{s}}'\,\bar{\phi}\,\bar{A}\,\bar{E})\in E$, one must have:

(**) $$-\int_S \mathrm{Re}(\langle\bar{s}', \dot{s}'\rangle - \langle s', \dot{\bar{s}}'\rangle + \bar{\phi}\langle is, s'\rangle - \phi\langle is, \bar{s}'\rangle)^{(3)}\eta -$$
$$-\int_S \psi\,\mathrm{Re}(\langle i\bar{s}', \dot{s}\rangle + \langle is, \dot{\bar{s}}'\rangle - \|s\|^2\bar{\phi})^{(3)}\eta = 0$$

Taking $\bar{s}' = 0$ and $\bar{\phi} = 0$, for every section $\dot{\bar{s}}'$ we get:

$$\int_S \mathrm{Re}\langle -i\psi s + s', \dot{\bar{s}}'\rangle^{(3)}\eta = 0$$

from which $s' = i\psi s$ follows because the hermitian metric h is irreducible.

Finally, by substituting in (**), we have for every section $\bar{s}'$:

$$-\int_S \mathrm{Re}(\langle \bar{s}', \dot{s}'\rangle + \bar{\phi}\langle is, i\psi s\rangle - \phi\langle is, \bar{s}'\rangle)^{(3)}\eta - \int_S \psi\, \mathrm{Re}(\langle i\bar{s}', \dot{s}\rangle - \|s\|^2\bar{\phi})^{(3)}\eta = 0$$

$$= \int_S \mathrm{Re}\langle -\dot{s}' + i\phi s + i\psi\dot{s}, \bar{s}'\rangle^{(3)}\eta = 0$$

and, again by the irreduciblity of h, we have $\dot{s}' = i(\phi s + \psi s)$ //

5. Final comments.

It would be convenient to generalize theorems 4,5 and 6 to an arbitrary Yang-Mills field and a minimal interaction on a Lorenz manifold respectively. Such research would imply, on one hand, a study of the Cauchy problem for the linearization of the corresponding field equations at one of its solutions, on the other, an adequate application of the theory of elliptic operators on the space-like hypersurface which carries the inicial data. In the cases wherely the said theorems are generalized with no change, we would have the agreeable result that the "phase space" of the corresponding variational problem would be the "moduli" of the solutions of the field equations with respect to the natural action of the gauge group.

The fact that the pre-symplectic metric of the elliptic version of electromagnetic field is identically zero (Theorem 3) does not imply that this is the case for a general Yang-Mills field. The study of this question could also be of some interest. Of course, for the pre-symplectic metric not to be zero, the cohomology space $H^{n-1}(X, \mathbb{R})$ of the base manifold must be non zero too. In particular, this already discards the case of Yang-Mills theory on the sphere S^4, which has been so actively studied in recent years, M. Atiyah, N. Hitchin, I. Singer, R. Ward, etc. [1], [2], though the problem is open when the base manifold is a closed surface of arbitrary genus, which has been recently dealt with from the view point of Morse theory, M. Atiyah - R. Bott [3].

The first two of the proposed generalizations, which correspond to the hyperbolic case, are those which present a very clear physical interest. One must realize that the symplectic formalism of variational problems was introduced, fundamentally, in order to describe the classical fields on the Minkowski space as hamiltonian systems with the aim of their further quantization. Nevertheless, the pre-symplectic structure of the manifold of solutions can be defined for any variational problem, and the fact that it is zero for certain elliptic problems should not mean that it is always zero.

Bibliography.

[1] M. Atiyah - N. Hitchin - I. Singer, Deformations of Instantons, Proc. Nat. Acad. Sci. U.S.A., 1977.

[2] M. Atiyah - R. Ward, Instantons and Algebraic Geometry, Comm. Math. Phys. 55, 1977.

[3] M. Atiyah - R. Bott, to appear.

[4] S. Bochner, Curvature and Betti numbers in real and complex vector bundles, Rendiconti del Seminario Matematico di Torino, Vol. 15, 1955.

[5] T.P. Branson, The Yang-Mills equations: quasi-invariance, special solutions, and Banach manifold geometry (Thesis), M.I.T., 1979.

[6] W. Drechsler - M. Mayer, Fiber bundle techniques in gauge theories, Lect. Not. in Math. Phys. Springer-Verlag, 67, 1977.

[7] A. Fischer - J. Marsden, Linearization stability of nonlinear partial differential equations, Proc. Symp. Pure Math. A.M.S., 27 (2), 1975.

[8] A. Fischer - J. Marsden, General Relativity as a Hamiltonian System, Symposia Mathematica, Vol. 14, 1974.

[9] P. García, The Poincare-Cartán Invariant in the Calculus of Variations, Symposia Mathematica, Vol. 14, 1974 .

[10] P. García, Gauge Algebras, Curvature and Symplectic Structure, J. Diff. Geometry, 12, 1977.

[11] P. García, Reducibility of the Symplectic Structure of classical fields with gauge-simmetry, Lect. Not. in Math. Springer-Verlag, 570, 1977.

[12] P. García - A. Pérez-Rendón, Reducibility of the Symplectic Structure of Minimal Interactions, Lect. Not. in Math., Springer-Verlag, 676, 1978.

[13] H. Goldschmidt - S. Sternberg, The Hamilton-Cartán formalism in the Calculus of Variations, Ann. Inst. Fourier, 23, 1973.

[14] J. Kijowski - W. Szczyrba, Multisimplectic manifolds and the geometrical construction of the Poisson Brackets in the classical field theory, Colloq. Inter. CNRS, 237, 1975.

[15] A. Pérez-Rendón, A Minimal Interaction Principle for Classical fields, Symposia Mathematica, Vol. 14, 1974.

[16] I. Segal, Differential Operators in the manifold of solutions of a non linear differential equation, J. Math. Pures et Apl., XLIV, 1965.

[17] I. Segal, The Cauchy Problem for the Yang-Mills equations, to appear in J. Funct. Anal.

[18] I. Segal, General properties of the Yang-Mills equations in physical space (nonlinear wave equation /gauge-invariance/ Cauchy problem / symplectic structure), Proc. Natl. Acad. Sci. U.S.A., Vol. 75, n° 10, 1978.

[19] T.T. Wu - C.N. Yang, Concept of non integrable phase factors and global formulation of gauge fields, Phys. Rev., 12 D, 1975.

CLASSIFICATION OF GAUGE FIELDS AND GROUP REPRESENTATIONS*

M. Carmeli

Department of Physics
Ben Gurion University of the Negev
Beer Sheva 84120, Israel

and

B.Z. Moroz

Department of Mathematics
Hebrew University, Jerusalem, Israel

ABSTRACT

The problem of classification of SU(2) gauge fields is reviewed. Previous work on classification has been done using spinor and matrix methods. In this paper the classification problem is studied in terms of the theory of orbits of the representations of Lie groups. The $SL(2,C) \times SU(2)$ case is treated in some detail.

---

* Paper presented at the Conference on Differential Geometrical Methods in Mathematical Physics held at the University of Salamanca, Spain, 10-14 September 1979.

A gauge field F can be considered (locally) as a function $F: M \to V$ from the Minkowski space M into a complex finite-dimensional vector space V. It is assumed that V is a representation space for a Lie group G, so that a homomorphism $D: G \to GL(V)$ is given. The equations of motion of the gauge field are supposed to be invariant under the gauge transformations $F(x) \to (D(g_x)\cdot F)(x)$, where $x \in M$ and $g_x \in G$.

In the last few years several authors have studied the problem of the classification of gauge fields; namely, they tried to find for some classical group G a canonical form for $F(x)$ with respect to the action of G.[1-13] The exact statement of the problem is as follows:

1) To classify all the orbits of the representation D; and
2) To find a convenient parametrization for the set of the orbits in V.

This problem has been treated in general[14] and independently on its applications to physics. In this paper an example which can be discussed in an elementary fashion is given.

Let G be the product group, $G = SL(2,C) \times SU(2)$, and $D: G \to GL(V)$ is an irreducible finite-dimensional representation of G. One can then decompose the representation D into a tensor product $D = D_1 \times D_2$, where D_1 is a representation of $SL(2,C)$ and D_2 is a representation of $SU(2)$. Thus it is enough to classify the orbits of D_1 and D_2 separately.

It is well-known[15-17] that all the complex finite-dimensional irreducible representations of $SL(2,C)$ can be given by the formula

$$(D(g)f)(u,v) = (\gamma u + \delta)^m (\gamma^* v + \delta^*)^n f\left(\frac{\alpha u+\beta}{\gamma u+\delta}, \frac{\alpha^* v+\beta^*}{\gamma^* v+\delta^*}\right) \tag{1}$$

where f is a polynomial of degree not larger than m in u and not larger than n in v,

$$g = \begin{pmatrix} \alpha & \beta \\ \gamma & \delta \end{pmatrix}$$

and a^* denotes the complex conjugate of $a \in C$. The dimension of the representation (1) is equal to $(m + 1)(n + 1)$. Representations with $n = 0$ are called analytical and their restrictions to $SU(2)$ gives *all* the irreducible representations of this compact group. We first classify the orbits of the analytical representations of $SL(2,C)$ and the orbits of $SU(2)$, and then generalize our considerations to treat the representations of $SL(2,C)$ with $n \neq 0$.

Let us denote the right-hand side in (1) by $f_g(u,v)$; then the orbit $\mathcal{O}_f^G$ containing the element f has the form $\mathcal{O}_f^G = \{f_g | g \in G\}$, where G is one of the groups $SL(2,C)$ or $SU(2)$. In the $SU(2)$ case as well as for analytical representations of SL(2,C) the polynomial f_g depends only on one variable which we denote by z, so that (1) has the form

$$(D(g)f)(z) = (\gamma z + \delta)^m f\left(\frac{\alpha z+\beta}{\gamma z+\delta}\right) \overset{\text{def}}{=\!=} f_g(z) . \tag{2}$$

We note that for any $a \in C$ there exists an element $g \in SU(2)$ such that

$$g\cdot a = \frac{\alpha a+\beta}{\gamma a+\delta} = 0 , \tag{3}$$

and if

$$g = \begin{pmatrix} \alpha & \beta \\ \gamma & \delta \end{pmatrix} \in SU(2)$$

has zero as a fixed point (namely $g\cdot 0 = 0$), then

$$g = \begin{pmatrix} e^{i\varphi} & 0 \\ 0 & e^{-i\varphi} \end{pmatrix} \tag{4}$$

On the other hand, $g \in SL(2,C)$ is uniquely determined (up to a scalar factor ± 1) by its action on *three* distinct complex numbers.

We can now argue as follows: f is determined by its roots and the coefficient of the highest term, and $f_g(a) = 0$ if and only if $f(ga) = 0$. Thus to choose a "canonical" representative $f_g \in \mathcal{O}_f^G$ one fixes three roots in case of the analytical SL(2,C) representation, and a root and the highest coefficient in the SU(2) case. These arguments lead to the following proposition.

Lemma 1. 1) Every SU(2)-orbit contains a polynomial of the form $f(z) = az\varphi(z)$, where $a \in R$ and $a \geqslant 0$, and the leading coefficient of $\varphi(z)$ is equal to 1; 2) in any analytical representation of SL(2,C) every orbit contains a polynomial f having one of the following forms:

a) $f(z) = z(z-1)(z+1)\varphi(z)$;

b) $f(z) = a(z+1)^k(z-1)^\ell$ with $a = 1$ when $k \neq \ell$;

c) $f(z) = z^n$; or

d) $f = 0$.

Moreover, it can be easily shown that any orbit contains no more than a finite number of canonical representatives described in the lemma and these representatives can be actually calculated as soon as the roots of *some* point on the orbit are known.

In a similar fashion one can prove the following lemma about general SL(2,C) orbits (i.e. for $m,n \neq 0$).

Lemma 2. If the function $h_f(z) = f(z,z^*)$ has at least three but no more than a finite number of roots, then the orbit $\mathcal{O}_f$ contains a polynomial f_g such that $f_g(0,0) = f_g(\pm 1,0) = 0$ and no more than a finite number of polynomials satisfying this condition.

However, since there are polynomials [for example, $f(u,v) = u\cdot v + 1$] whose diagonalization h_f has no roots as well as polynomials f [for example, $f(u,v) = u^2 + v^2$, or $f(u,v) = u + v$] for which h_f has an infinite number of roots, lemma 2 does not solve the problem.

To treat the general case one notices that an orbit $\mathcal{O}_f$ is completely determined by a three parameters set (more precisely, not more than three parameters) of algebraic curves $\{f_g(u,v) = 0 \mid g \in SL(2,C)\}$. To every curve $f_g(u,v) = 0$ there corresponds an algebraic function $u = \varphi_g(v)$ on the Riemannian surface and we have to pick a function φ_{g_0} from this family of functions. Itis sufficient for this aim to fix its value in three (or less) points on this surface.

For let us consider a polynomial f and the corresponding algebraic curve $f(u,v) = 0$. Let now S_f be the Riemann surface uniformizing this curve so that $u = \varphi(\tau)$ provide the general solution of $f(u,v) = 0$, where the parameter τ runs over S_f. It follows from the representation formula Eq. (1) that the solutions of

$$f_g(u,v) = 0 \tag{5}$$

are given by

$$\begin{aligned} u &= g\cdot\varphi(\tau) , \\ v &= g\cdot\psi(\tau) , \end{aligned} \tag{6}$$

where $\tau \in S_f$ and

$$g\cdot z = \frac{\alpha z+\beta}{\gamma z+\delta} \tag{7}$$

with $z \in C$ and

$$g = \begin{pmatrix} \alpha & \beta \\ \gamma & \delta \end{pmatrix} . \tag{8}$$

In particular the genus p_f of S_f is a two-polynomial invariant of an orbit.[18-19] Alternatively, given a Riemann surface S one can pick a pair of meromorphic functions φ and ψ on S and consider the orbit

$$\mathcal{O}_f = \{(g\cdot\varphi,\ g\cdot\psi \mid g \in SL(2,C)\} \tag{9}$$

where f is the irreducible polynomial such that

$$f(\varphi(\tau),\ \psi(\tau)) = 0 \tag{10}$$

for any $\tau \in S$.

In summary, we see that in the case of the group SU(2) and analytical (or purely antianalytical) representations of the group SL(2,C) one parametrizes the orbits $\mathcal{O}_f$ by the sets $(\alpha_1,\ldots,\alpha_n)$ of the roots of f, so that the invariants of the Yang-Mills fields can be expressed as functions $P(\alpha_1,\ldots,\alpha_n)$ such that

$$P(g\cdot\alpha_1,\ldots,g\cdot\alpha_n) = P(\alpha_1,\ldots,\alpha_n)\ . \tag{11}$$

In the general case of the SL(2,C) representations the orbits are parametrized by pairs (φ,ψ) of meromorphic functions on a compact Riemann surface and the invariants are complex valued functions defined on the sets of these pairs which are invariant under the action

$$(\varphi,\psi) \to (g\cdot\varphi,\ g\cdot\psi)$$

where $g \in SL(2,C)$.

REFERENCES

1. T. Eguchi, Phys. Rev. D 13, 1561 (1976).
2. R. Roskies, Phys. Rev. D 15, 1722 (1977).
3. M. Carmeli, Phys. Rev. Lett. 39, 523 (1977).
4. L.-L. Wang and C.N. Yang, Phys. Rev. D 17, 2687 (1978).
5. M. Carmeli, in: *Differential Geometrical Methods in Mathematical Physics*, Eds. K. Bleuler et al., Springer-Verlag, Heidelberg 1978.
6. M. Carmeli, Phys. Lett. 77B, 188 (1978).
7. J. Anandan and K.P. Tod, Phys. Rev. D 18, 1144 (1978).
8. J. Anandan and R. Roskies, Phys. Rev. D 18, 1152 (1978).
9. J. Anandan and R. Roskies, J. Math. Phys. 19, 2614 (1978).
10. J. Anandan, J. Math. Phys. 20, 260 (1979).
11. L. Castillejo, M. Kugler and R. Roskies, Phys. Rev. D 19, 1782 (1979).
12. M. Carmeli and D.H. Wohl, Nuovo Cimento Lett. 25, 230 (1979).
13. M. Carmeli and M. Fischler, Phys. Rev. D 19, 3653 (1979).
14. B. Kostant and S. Rallis, Amer. J. Math. 93, 753 (1971).
15. I.M. Gelfand, M.I. Graeva and N. Ya. Vilenkin, *Generalized Functions, Vol. 5: Integral Geometry and Representation Theory*, Academic Press, New York 1966.
16. M. Carmeli, *Group Theory and General Relativity*, McGraw-Hill, New York 1977.
17. M.A. Naimark, *Linear Representations of the Lorentz Group*, Pergamon, New York 1964.
18. H. Weyl, *The Concept of a Riemann Surface*, Addison-Wesley, Mass. 1964.
19. L.V. Ahlfors and L. Sario, *Riemann Surfaces*, Princeton Univ. Press, N.J. 1960.

GAUGE ASTHENODYNAMICS (SU(2/1))*

(classical discussion)

Y. Ne'eman
Tel Aviv University
Tel Aviv, Israel
and
University of Texas
Austin, Texas

and

J. Thierry-Mieg
California Institute of Technology
Pasadena, California
and
GAR, Observatoire de Meudon, 92190
France

Table of Contents

Abstract

We present non-linear $SU(2/1)_{x\epsilon}$ exact gauge theory which represents a constrained Weinberg-Salam model with no new particles in the multiplets. We find a universal angle (including for quarks) $\theta_W = 30^o$, $\lambda = \frac{4}{3} g^2$. The linear multiplets of SU(2/1) connect physical fields to the ghosts of renormalization theory. The mass of the Higgs meson is 245 GeV. The absence of the right neutrino implies in this model that the charge of the electron be 1 and that the neutrino be neutral.

*Research supported in part by the U.S.-Israel Binational Science Foundation.

Superunitary Gauge Theory of the (Weak-Electromagnetic) Superinteraction

1. General Introduction

The unification of Weak and Electromagnetic Interactions, suggested in the fifties by Schwinger, Salam and Ward, Glashow and others[1], seems to have been attained at the phenomenological level. The precise "standard model"[2)3)] due to Weinberg and to Salam appears vindicated by experiments[4] and should be regarded as a correct description of at least the first level of the unified interaction. The SLAC e-d polarisation experiments[4] have also settled the issue raised by the previous negative results of the atomic parity-violation measurements.

There is, however, an aspect which requires a further step forward. This is the present lack of aesthetic cohesion and mathematical simplicity: a non-simple gauge group $SU(2) \otimes U(1)$, with uncorrelated weak-hypercharge quantum numbers for the left- and right-chiral fermions ($U(\nu^e_L) = U(e^-_L) = -1$; $U(e^-_R) = -2$), and with two independent couplings g, g' (or an unconstrained $\tan\theta_W = g'/g$); the ad-hoc adjunction of a Lorentz-scalar Goldstone-Higgs multiplet $\phi(x)$ with particular assignments $I_L = \frac{1}{2}$, $U = 1$; the additional independent couplings of the potentials $\lambda\phi^4$ and $-\mu^2\phi^2$ required by the spontaneous symmetry breakdown mechanism; the independent postulate of a Yukawa interaction $\Gamma\,\psi\,\phi\,\psi$ with yet another unconstrained coupling, determining the scalar neutral currents mediated by the Higgs component ϕ_H. Indeed, the original Salam version of the model[(3)] had sketched such a derivation, basing itself on SU(3) in a (e^-, ν, μ^+) mode[5], even though it would seem difficult to generate a scalar $\phi(x)$ in an ordinary gauge mechanism.

Past great syntheses have provided some of the most aesthetic and constraining elements in Physics. It was generally assumed in recent years that such restrictions would arise here through the appearance of a simple group at the next stage in the unification pyramid, e.g. when flavors and/or colours would be included in the gauge group. In the present work, we present in contradistinction a completely constraining structure at the pure weak-electromagnetic superinteraction level, due to the gauge action of a simple supergroup. The theory removes any arbitrariness in the above listed selection of multiplets, quantum numbers and couplings. It fixes

$$\theta_W = 30^\circ,\ \lambda = \frac{4}{3}g^2 = \frac{16}{3}e^2,\ \Gamma = \frac{2}{\sqrt{3}}g,\ m(\phi_H) \approx 250\ \text{GeV}.$$

In addition, the theory incorporates the key features of the renormalization process, namely, Slavnov-Taylor symmetry[6]. Both the ghost fields[7] required by that procedure, and the BRS[8] equations constraining them, are embedded together with the physical fields within the gauge and symmetry group structure itself.

We have published a brief description of this theory elsewhere[9].

2. Definition of a Supergroup

In the physics literature dealing with Supersymmetry, the discussion is often restricted to infinitesimal transformations. We would like in this and the following section to deal with the group itself[10], and derive the Cartan-Maurer equations defining the Graded Lie Algebra[11]. Readers who are less mathematically motivated may skip sections 2-3.

Let Λ be a Z(2) graded ring of parameters of countable dimension over the complex field C. Λ is the direct sum of its even (Bose) component Λ_o and odd (Fermi) component Λ_1. The multiplication respects the gradation,

$$\Lambda = \Lambda_o \oplus \Lambda_1$$

$$\Lambda_I \Lambda_J \subset \Lambda_{I+J \bmod (2)} \tag{2.1}$$

Λ is at the same time also $|Z|$ graded,

$$\Lambda = \sum_{i=o}^{\infty} \oplus \Lambda^{(i)} \quad , \quad \Lambda^{(o)} \in C \quad ,$$

$$\Lambda_o = \sum_{r=o}^{\infty} \oplus \Lambda^{(2r)} \quad , \quad \Lambda_1 = \sum_{r=o}^{\infty} \oplus \Lambda^{(2r+1)} \tag{2.2}$$

$$\Lambda^{(r)} \Lambda^{(s)} \subset \Lambda^{(r+s)}$$

Multiplication is associative and graded-abelian

$$\forall \ \lambda^{(i)} \in \Lambda^{(i)}, \ \lambda^{(j)} \in \Lambda^{(j)} \ , \ \lambda^{(i)} \lambda^{(j)} = (-1)^{ij} \lambda^{(j)} \lambda^{(i)} \tag{2.3}$$

It is further supposed that Λ is generated by the identity 1 and the lowest Fermi subspace $\Lambda^{(1)}$, i.e., given a countable basis θ^a of $\Lambda^{(1)}$, the $\Lambda^{(r)}$ are spanned by the set of $\frac{n!}{r!\,(n-r)!}$ antisymmetrized products (for finite n)

$$\theta^{a_1} \theta^{a_2} \ldots . \theta^{a_r}$$

where n is the dimension of $\Lambda^{(1)}$. Hence Λ_1 is nilpotent,

$$\Lambda^{(r)} = 0 \quad \text{for} \quad r > n \tag{2.4}$$

Any element of Λ thus admits a unique decomposition over the θ^a and their products, such that only a finite number of coefficients do not vanish:

$$\forall \vec{x} \in \Lambda \qquad \vec{x} = x^o + x^i \theta^i + x^{ij} \theta^i \theta^j + \ldots$$

$$x^o, x^i, x^{ij}, \ldots \in C \qquad\qquad \theta^i \in \Lambda^{(1)} \tag{2.5}$$

If x^o does not vanish, $\vec{x}^{-1}$ exists and is defined by its finite power expansion. Given an involution in $\Lambda^{(1)}$:

$$\bar{\bar{\theta}} = + \theta$$

We induce an involution in Λ:

$$\bar{\vec{x}} = x^{o*} + x^{i*}\,\bar{\theta}^i + x^{ij\,*}\,\bar{\theta}^j\,\bar{\theta}^i$$

$$\overline{(\vec{x}\vec{y})} = \bar{\vec{y}}\;\bar{\vec{x}} \tag{2.6}$$

$$a \in \Lambda_o \qquad \bar{\bar{a}} = a$$

$$\alpha \in \Lambda_1 \qquad \bar{\bar{\alpha}} = + \alpha \tag{2.7}$$

A Λ vector space E(m, n, v) of dimension (m, n) and valency v is a vector whose first m components belong to Λ_v, the n others to Λ_{v+1} and are therefore of opposite statistics

$$V \in E\,(m, n, v) \qquad V_i \in \Lambda_v \qquad i = 1, \dots m$$

$$V_j \in \Lambda_{v+1} \qquad j = m+1, \dots\, m+n \tag{2.8}$$

Linear mappings of a Λ vector space V onto itself (homomorphisms which respect the valency) may be represented by matrices of type (m, n) over Λ. They are matrices M of dimension $(m + n)^2$ such that the m^2 and n^2 box-diagonal elements are bosons, the off diagonal elements fermions. As a vector space, graded matrices are of fixed valency.

$$E\,(m^2 + n^2,\; 2\,mn,\; v = 0)$$

$$M = \left(\begin{array}{c|c} A & \beta \\ \hline \gamma & D \end{array}\right) \tag{2.9}$$

If A and D are regular, M is regular and one may write:

$$M = \left(\begin{array}{c|c} A & 0 \\ \hline 0 & D \end{array}\right) \quad \left(\begin{array}{c|c} 1 & \xi \\ \hline \psi & 1 \end{array}\right) \qquad \xi = A^{-1}\beta \qquad \psi = D^{-1}\gamma$$

$$M^{-1} = \left|\begin{array}{c|c} F & -F\,\beta\,D^{-1} \\ \hline -D^{-1}\,\gamma F & D^{-1}\,(\gamma\,F\,\beta + D)\,D^{-1} \end{array}\right| \qquad F = (A - \beta\,D^{-1}\,\gamma)^{-1} \tag{2.10}$$

The set of matrices M thus forms the group GL (m/n).

The graded-trace "str" and graded determinant "sdet" depend upon the valency of the carrier space

$$\mathrm{str}\ M = (-1)^V\ (\mathrm{Tr}\ A - \mathrm{Tr}\ D) \qquad \mathrm{sdet}\ M = \exp\ \mathrm{str}\ \log M. \quad (2.11)$$

Changing the valency of the carrier space defines an outer automorphism of the graded matrices. We shall denote this operation by ε.

We define the Hermitian conjugate as the transposed Λ-conjugate matrix (other definitions are possible).

$$M^+_{ij} = \overline{M}_{ji} \quad (2.12)$$

The special superunitary group[10] of type (m, n) is defined by the usual constraints:

$$\forall\ M\ \varepsilon\ SU(m/n) \qquad \begin{array}{l} M\ \varepsilon\ GL\ (m/n) \\ M\,M^+ = 1 \\ \mathrm{sdet}\ M = 1 \end{array} \quad (2.13)$$

3. The Cartan Maurer Equation and the Graded Lie Algebra

In this section, we shall deduce the graded Lie algebra structure by exterior differentiation of the graded Lie group, and comment on the integration problem. The exterior differentials of our Λ parameters are trivially defined as new objects generating the exterior algebra χ with multiplication rule[12)]

$$\forall \ x, y \in \Lambda_0 \qquad \alpha, \beta \in \Lambda$$

$$\begin{aligned} dx_\wedge dy &= - dy_\wedge dx & x_\wedge dy &= dy_\wedge x \\ dx_\wedge d\alpha &= - d\alpha_\wedge dx & x_\wedge d\alpha &= d\alpha_\wedge x \\ d\alpha_\wedge d\beta &= + d\beta_\wedge d\alpha & \alpha_\wedge d\beta &= -d\beta_\wedge \alpha \end{aligned} \tag{3.1}$$

The exterior differential is now defined recursively by:

$$\begin{aligned} \forall \ z \in \Lambda \qquad & d^2 z = 0 \\ t \in \chi \qquad & d(zt) = dz_\wedge t + z_\wedge d t \\ & d(dz\, t) = - d z_\wedge d t \end{aligned} \tag{3.2}$$

As a result, the fundamental rule $d^2 = 0$ holds and the exterior algebra χ of Λ is a $Z^2 \times Z^2$ graded countable ring. The nilpotent elements are the fermi elements θ and the bose differentials dx, whereas power series in x or $d\theta$ do not terminate:

$$d^2 = 0 \quad , \quad dx_\wedge dx = 0 \quad , \quad (d\theta)^n \neq 0 \quad .$$

We are now ready to define the Cartan left-invariant one-forms over a graded continuous group. The differential of a matrix is the matrix of the differential of its elements

$$(d\,M)_{ij} = d\,(M_{ij})$$

Let now M belong to a continuous supergroup. We define:

$$\omega = M^{-1}\, d\, M \quad , \qquad \text{by} \quad d(M^{-1} M) = d\, M^{-1} M + M^{-1}\, dM = 0 \tag{3.3}$$

We find the Cartan-Maurer equations[13)]

$$\Omega = d\omega + \omega_\wedge \omega = 0 \tag{3.4}$$

As a vector space, the set of all the ω admits a graded basis y_A of dimension at most $(m^2 + n^2, 2\,mn)$.

$$\omega = \omega^A y_A \quad , \quad y_A \in L(m', n') \quad m' \leq m^2+n^2 \quad n' \leq 2\,mn \tag{3.5}$$

The components are of the form:

$$\omega^a \;\varepsilon\; \Lambda_o \, d\Lambda_o + \Lambda_1 \, d\Lambda_1$$
$$\omega^\alpha \;\varepsilon\; \Lambda_1 \, d\Lambda_o + \Lambda_o \, d\Lambda_1 \qquad (3.6)$$

The valency of the ω is fixed, and the ω^a components corresponding to the box diagonal matrices y_a are fermions. This implies that the y_a themselves form a graded algebra. Indeed:

$$d\omega^C \, y_C = -\omega_\wedge\omega = -\tfrac{1}{2} \left(\omega^A{}_\wedge\omega^B - (-)^{ab} \, \omega^B{}_\wedge\omega^A\right) y_A \, y_B$$
$$\Rightarrow \; [y_A, \, y_B]_\pm = f_{AB}{}^C \, y_C \qquad (3.7)$$

The structure constants $f_{AB}{}^C$ are automatically of Bose type. They are skew symmetric in AB unless two fermi generators are involved. This symmetry property in turn implies the graded Jacobi identity:

$$d\omega_\wedge\omega = d\omega^A \, \omega^B \, f_{AB}{}^C \, y_C = -\omega_\wedge \, d\,\omega$$
$$\rightarrow \; \omega_\wedge\omega_\wedge\omega = \omega^A \, \omega^B \, \omega^C \, f_{AB}{}^D \, f_{DC}{}^E \, y_E = 0 \qquad (3.8)$$

The number of generators y_A defines the dimension of the group. Any matrix of the form:

$$M\,(x) = \exp\,(x^A \, y_A)$$

is a group element provided the x^A coordinates are assigned the same statistics as the y_A. Discrete subgroups may generate additional sheets not connected to the identity. In ordinary Lie theory, we know that any matrix may be defined as:

$$G(t) = G_o \; T \exp \int_o^t \omega$$

where T denotes the chronological product along a curve in the x^a space. Just because of the vanishing of the curvature:

$$d\,\omega + \tfrac{1}{2}\,(\omega_\wedge\omega) = 0 \quad \rightarrow \quad G\,(t + t') = G\,(t)\; G\,(t')$$

The choice of the curve connecting G_o to G is irrelevant. However, the generalization to the case of a graded group is not clear because we know of no appropriate definition of the integral of a Fermi form.

4. SU(2/1)

Our gauge group is SU(2/1), whose superalgebra su(2/1) is thus isomorphic to the set of supertraceless 3 x 3 matrices μ_A, A = 1...8

$$\mu_A , \mu_B = 2C^E{}_{AB}\,\mu_E \tag{4.1}$$

$$\mathrm{str}\,\mu_A = 0 \tag{4.2}$$

Using the notation

$$a,b,c = 1,2,3 \; ; \quad u,v = a,8 \; ; \quad i,j,k = 4,5,6,7$$

$$\mu_u \in L^o \quad , \quad \mu_i \in L' \tag{4.3}$$

$$C^c{}_{ab} = i\, f_{abc} \quad \text{or} \quad -i\, C^c{}_{ab} = e_{abc} \tag{4.4a}$$

$$C^c{}_{a8} = 0 \tag{4.4b}$$

$$\begin{aligned} C^j{}_{ai} &= i\, f_{aij} & C^j{}_{8i} &= \frac{i}{3}\, f_{8ij} \\ C^j{}_{ia} &= i\, f_{iaj} & C^j{}_{i8} &= \frac{i}{3}\, f_{i8j} \end{aligned} \tag{4.4c}$$

$$C^a{}_{ij} = d_{ija} \qquad C^8{}_{ij} = -\frac{\sqrt{3}}{2}\,\delta_{ij} \tag{4.4d}$$

The f_{Aij} and d_{ijA} refer to SU(3) coefficients[14].

Equations (4.4a) - (4.4b) define a u(2) subalgebra and subgroup which we denote by $U(2)_W$. Equations (4.4c) imply that L' behaves as a complex spinor under that $U(2)_W$.

$$\begin{aligned} C^j{}_{ui} &: (146)=(157)=(245)=(256)=(346)=(356)=(347)=(357)=(846)= \\ &\quad =(847)=(856)=(857) = 0 \\ -i\, C^j{}_{ui} &: (147)=-(156)=(246)=(257)=(345)=-(367) = 1/2 \\ &\quad (845)=(867)=(2\sqrt{3})^{-1} \end{aligned} \tag{4.5}$$

The $C^j{}_{ui}$ are totally antisymmetric. Thus, (see (4.14)) we define for μ_i

$$|\vec{I}| := |C^j{}_{ai}| = 1/2 \quad , \quad |U| = 2\sqrt{3} \quad |C^j{}_{8i}| = 1 \tag{4.6}$$

Equation (4.4d) describes the $\{\mu_i, \mu_j\}$ anticommutators, with

$$C^a_{\ ij}: \quad (146)=(157)=-(247)=(256)=(344)=(355)=-(366)=-(377) = 1/2 \qquad (4.7)$$

all components other than (4.7) or $C^8_{\ ij}$ of (4.4d) vanishing.

The matrices μ_A cannot be normalized directly by $str(\mu_A)^2$ as can readily be seen for the μ_i. These matrices are in the β,γ sectors of equations (2.9) and thus supertraceless. However, hermiticity ensures that $str(\mu_A)^2 = 0$ which defeats normalization.

Normalization thus proceeds in the following manner. Under L^o, the μ_A separate in two representations: L^o itself, and L' (as given by 4.3). One could thus allow at most two normalizations by $tr(\mu_A)^2$, one for the μ^u as generators of a connected $U(2)_W$, and one for the μ_i as the $|\vec{I}| = \frac{1}{2}$, $U = \pm 1$ irreps of $U(2)_W$. We pick an overall normalization in this 3-dimensional defining representation (only!)

$$Tr\ (\mu_A \mu_B) = 2\ \delta_{AB} \qquad (4.8)$$

For SU(2/1), we can then define a Killing metric[10)],

$$g_{AB} = \tfrac{1}{2}\, str\ (\mu_A \mu_B)$$

$$g_{ab} = \delta_{ab}\ , \quad g_{88} = -1/3 \qquad (4.9)$$

$$g_{45} = -g_{54} = g_{67} = -g_{76} = i$$

Denoting g^{AB} the inverse matrix and using

$$g^{AB}\mu_B = \mu^A$$

we get

$$str(\mu_A \mu^B) = str(\mu_A \mu_C g^{CB}) = g_{AC}\, g^{CB} = \delta_A^{\ B} \qquad (4.10)$$

The diagonal (Cartan subalgebra) basis is thus given by

$$\mu_3 = \left|\begin{array}{c|c|c} 1 & & \\ \hline & -1 & \\ \hline & & \end{array}\right| \qquad \mu_8 = \frac{1}{\sqrt{3}}\left|\begin{array}{c|c|c} -1 & & \\ \hline & -1 & \\ \hline & & -2 \end{array}\right| \qquad (4.11)$$

and all other μ_a and μ_i are identical with the SU(3) λ_a and λ_i.

A ninth u(2/1) base vector μ_9 is trace-orthogonal to L^o but not supertrace-orthogonal to su(2/1). It is thus not generated by closure of su(2/1) even though it does not commute with it.

$$\sqrt{\frac{2}{3}} \left|\begin{array}{c|c|c} 1 & & \\ \hline & 1 & \\ \hline & & -1 \end{array}\right| \qquad (4.12)$$

$$\mu_9$$

Note that the identity is supertrace-orthogonal to su(2/1).
We fix the scale of the charges M_A so that

$$M_A \sim \frac{1}{2} \mu_A \quad , \qquad \{M_A, M_B\} = C^E{}_{AB} M_E \qquad (4.13)$$

and identify the following physical quantum numbers (L stands for left-chiral, I_L is Weak left isospin, U_o is Weak hypercharge, E the electric charge, U_W the connected Weak U(2),

$$I_L^a = M_a \quad , \qquad U_o = 2\sqrt{3}\, M_8 \; , \qquad E = I_L^3 + \frac{1}{2} U_o \; , \qquad U_W(I_L, U_o) \qquad (4.14)$$

SU(2/1) admits a discrete Z(3) subgroup, the valency group:

$$v_o = 1$$

$$v_\pm = \text{diag.} \quad (\exp \pm 2\, \pi i/_3, \exp \pm 2\, \pi i/_3, \exp \mp 2\, \pi i/_3) \qquad (4.15)$$

$$= \exp\sqrt{\frac{2}{3}}\, \mu_9$$

Unlike the triality subgroup of SU(3) it does not belong to the center of SU(2/1) which is trivial. If an element connected to the identity is represented as:

$$M_\pm = \exp i\; (\alpha^u\, \mu_u \pm \alpha^i\, \mu_i) \qquad (4.16)$$

we may define an F parity operation as

$$\tilde{M}_\pm = M_\mp \qquad (4.17)$$

and we have:

$$v_\pm\, M = \tilde{M}\, v_\pm \qquad (4.18)$$

$$\widetilde{(M\, M')} = \tilde{M}\, \tilde{M}' \qquad (4.19)$$

SU(2/1) admits 3 conjugations: the defining hermitean conjugation, the F parity and the valency parity, which just exchanges the valency elements:

$$M^{\vee} = M \qquad (4.20)$$

$$v_{\pm}^{\vee} = v_{\mp}$$

This amounts to exchanging the statistics of the carrier space, an operation which does not affect the generators of SU(2/1) but changes, as we have seen, the sign of the supertrace. This is the operation we denoted by ε in Section 2.

5. Representations of SU(2/1)

In this section we shall show that all the known weakly-interacting particles fit into the lowest dimensional SU(2/1) irreducible representation, provided a re-interpretation principle is accepted.

All the finite-dimensional representations of SU(2/1) have been classified by M. Scheunert, W. Nahm and V. Rittenberg[15]. It was V. Rittenberg who recently pointed out to us the existence of the quark representation (which we had missed in ref. 9). In his notation the generators of SU(2/1) read

$$
\begin{aligned}
Q\pm &= \tfrac{1}{2}(\mu_1 \pm i\mu_2) = M_1 \pm iM_2 = I^L_\pm \\
Q_3 &= \tfrac{1}{2}\mu_3 = M_3 = I^L_3 \\
B &= \sqrt{3}/2 \quad \mu_8 = \tfrac{1}{2}U_o \\
V+ &= \tfrac{1}{2}(\mu_4 + i\mu_5) = M_4 + iM_5 \\
V- &= \tfrac{1}{2}(\mu_6 + i\mu_7) = M_6 + iM_7 \\
W+ &= \tfrac{1}{2}(\mu_6 - i\mu_7) = M_6 - iM_7 \\
W- &= -\tfrac{1}{2}(\mu_4 - i\mu_5) = -(M_4 - iM_5)
\end{aligned}
\tag{5.1}
$$

with (see 4.14)

$$\{W_+ , V_-\} = -E \tag{5.2}$$

To find a general irreducible representation, choose a state (highest weight) such that:

$$\tfrac{1}{2}U_o\,\phi_o = \frac{u}{2}\phi_o, \quad I^L_+\,\phi_o = V_+\,\phi_o = W_+\,\phi_o = 0 \tag{5.3}$$

The representation is then spanned by the repeated action of I^L_- onto the four states:

$$\phi_o \;,\; V_-\,\phi_o \;,\; W_-\,\phi_o \;,\; V_-\,W_-\,\phi_o \tag{5.4}$$

The representation is characterized by two quantum numbers, the $\tfrac{1}{2}U_o$ weak hyper charge (halved) and the third component of isospin i_3 of ϕ_o. It generally consists of four isospin multiplets:

$$|\tfrac{1}{2}u_o, i_3\rangle \,,\; |\tfrac{1}{2}u_o + 1, i_3 + \tfrac{1}{2}\rangle \,,\; |\tfrac{1}{2}u_o - 1, i_3 - \tfrac{1}{2}\rangle \,,\; |\tfrac{1}{2}u_o \,,\, i_3 - 1\rangle \tag{5.5}$$

with two interesting exceptions. When $\frac{1}{2}u_o = \pm i_3$, only the first two multiplets arise:

$$\text{either} \quad |i_3\,,\, i_3> \quad \text{and} \quad |i_3 + \tfrac{1}{2}\,,\, i_3 - \tfrac{1}{2}>$$
$$\text{or} \quad |-i_3\,,\, i_3> \quad \text{and} \quad |-i_3 - \tfrac{1}{2}\,,\, i_3 - \tfrac{1}{2}> \tag{5.6}$$

The lowest example, $|-\frac{1}{2}, \frac{1}{2}>$ is the defining triplet of SU(2/1). For the general case of $i_3 = 1/2$, only 3 multiplets arise

$$(\tfrac{1}{2}u_o, \tfrac{1}{2}) \quad , \quad |\tfrac{1}{2}u_o \pm 1, \quad 0> \tag{5.7}$$

The quark representation is $|\frac{1}{6}\,,\,\frac{1}{2}>$ so that it describes indeed $U_o = \text{diag.}\ (\frac{1}{3}\,,\,\frac{1}{3}\,,\,\frac{4}{3}\,,\,-\frac{2}{3})$ and $I_3^L = \text{diag.}\ (\frac{1}{2}\,,\,-\frac{1}{2}\,,\,0,0)$ fitting all the charges of $u_L^{2/3}\,,\, d_L^{-1/3}\,,\, u_R^{2/3}\,,\, d_R^{-1/3}$.

The assignment of physical particles now proceeds as follows. The valency of their representations is arbitrary, but if the graded group SU(2/1) is supposed to commute with the Lorentz group we shall soon run into open contradiction with the spin statistics correlations, as all particles of a given multiplet should have the same spin. We propose the following solution. We only consider spin 1/2 matter fields. To each representation R of valency v and helicity h, we associate another representation of opposite valency and helicity.

$$R = (f_L\,,\, b_L) \qquad\qquad \varepsilon\psi_F = \psi_B \quad , \quad \varepsilon\psi_B = \psi_F$$
$$R' = (b'_R\,,\, f'_R) \tag{5.8}$$

$$f'_R = \varepsilon\pi b_L \qquad , \qquad b'_R = \varepsilon\,\pi\, f_L \tag{5.8'}$$

Taking

$$f_L = \gamma_L\psi_F \quad , \quad b_L = \gamma_L\,\beta\psi_B \quad , \quad b'_R = \gamma_R\beta\psi'_B \quad , \quad f'_R = \gamma_R\,\psi'_F \tag{5.8''}$$

we see that π can be realized by the Dirac β matrix. Both multiplets may be coupled to the quantum theory; however, we only consider the Fermion states $(f_L\,,\, f_R)$ as classical states. SU(2/1) and the Lorentz group are thus no longer alien to each other, since, according to our definition, the components connected to the identity of both groups commute but the parity does not commute with the valency elements. Our representations are representations of the semi-direct product $SU(2/1)_{x\varepsilon\pi}$. Indeed, only R + R' can provide a representation of the Lorentz group multiplied by $SU(2/1)_{x\varepsilon\pi}$

$$V_\pm\ \pi = \pi\ V_\mp \tag{5.9}$$

In the lowest cases, interesting relations follow that are not an input of the theory. We assign the leptons to the triplet $(-\frac{1}{2},\frac{1}{2})$ consisting of a left handed isodoublet plus a right handed isosinglet

$$(-\tfrac{1}{2}, \tfrac{1}{2}): \qquad \{ (\nu_L^o \ , \ e_L^-) \qquad , \qquad e_R^- \} \tag{5.10}$$

(the anti-leptons are in $(\frac{1}{2},\frac{1}{2})$).

The charge of the electron then has to be (-1) in terms of the charge defined by (5.2) and the structure constants and the neutrino will turn out to be electrically neutral (and massless when we shall consider symmetry breaking!)

The next representation, the $\underset{\sim}{4}$, admits particles of arbitrary charge, depending on the choice of $\frac{1}{2}U_o$. It consists of a left handed isodoublet and 2 right handed singlets. Taking $\frac{1}{2}U_o = \frac{1}{6}$ we fit the quarks:

$$(\tfrac{1}{6} , \tfrac{1}{2}): \qquad \{ (u_L \ , \ d_L) \qquad , \quad u_R \ , \ d_R \ \} \tag{5.11}$$

Notice that the <u>theory predicts that integer-charge fermions occur in triplets whereas fractional charged ones occur in quartets!</u>

If we now compare the standard interaction Lagrangian of leptons:

$$g(\bar{\psi}_L \not{W}^3 \ \tau_3 \ \psi_L \ - \ g'/g \ (\bar{\psi}_L \not{B} \psi_L \ + \ 2 \ \bar{\psi}_R \ \not{B} \psi_R) \ ) \tag{5.12}$$

to the su(2/1) minimal coupling to the Bose generators:

$$g(\bar{\psi}_L, \ \bar{\psi}_R) \quad (\not{W}^3 \mu_3 \ + \not{B} \mu_8) \quad (\psi_L \ , \ \psi_R) \tag{5.13}$$

we find

$$\mathrm{tg}\ \theta_W = g'/g = 1/\sqrt{3}\ , \quad \theta_W = 30^o, \ \sin^2\theta = 1/4 \tag{5.14}$$

<u>which is close to the present experimental value</u>[4)]

We now repeat the same check for quarks,

$$g(\bar{q}_L \ \not{W}^3 \ \tau_3 \ q_L \ + \ g'/3g \ (\bar{q}_L \ \not{B} \ q_L \ + \ 4\bar{u}_R \ \not{B} \ u_R \ - \ 2\bar{d}_R \ \not{B} \ d_R)) \tag{5.15}$$

comparing that expression with the su(2/1) minimal couplings

$$g(\bar{q}_L \ , \ \bar{\ }_R) \ (\not{W}^3 \mu_3 \ + \ \not{B} \ \mu_8) \ (q_L \ , \ q_R) \tag{5.16}$$

we find again $\theta_W = 30^o$. This <u>universality of θ_W is unique to our supergroup gauge</u> and cannot occur in the conventional "grand unification" theories, where θ_W is larger for quarks and is assumed to be renormalized by the very large symmetry breaking in those theories.

In Appendix A, we give an explicit set of matrices for the representation 4. Note that these representations are "star-hermitean" 15) rather than hermitean.

We assign all sequential lepton-types to analogous 3 representations $(\nu_L^\mu, \mu_L^- ; \mu_R^-), (\nu_L^\tau, \tau_L^-; \tau_R^-)$ .... and all sequential quark-types to analogous 4 representations. Our physical picture is one in which we assume complete degeneracy of these lepton-types under SU(2/1). The different masses of the charged components should be due to a flavour interaction.

The weak hypercharge is supertraceless, which automatically excludes the BBB triangle anomaly

$$\Sigma u_o \, \phi_L = \Sigma u_o \, \phi_R \tag{5.17}$$

The WWB anomaly is excluded only in a model relating the quarks and leptons, such as the conventional SU(5) or a generalisation of the model described here.

6. The SU(2.1) gauge multiplet and the BRS Equations.

The Yang-Mills gauge multiplet is an SU(2/1) octet

$$\omega_\mu^A = W_\mu^a \, , \; W_\mu^8 \, , \; \xi_\mu^i \tag{6.1}$$

Clearly, the ξ^i with their Fermi statistics are ghost-like fields. On the other hand, the valency of the adjoint representation is fixed since it is that of the generators, with 1-3,8 as Bose and 4-7 as Fermi. We may thus not use the same method as for the matter multiplets in our reinterpretation principle. A second gauge multiplet is, however, defined in any Yang-Mills theory, i.e., the ghost multiplet: (see App. C)

$$C^A = X^a, \; X^8, \; \phi^i \tag{6.2}$$

Indeed, the Feynman De Witt Faddeev Popov ghost of the $U(2)_W$ subgroup X^a is a Fermi-like scalar field. However, for a graded group, the ϕ^i are Bose scalars and have the exact quantum numbers of the Higgs-Goldstone particles of the Weinberg-Salam model. We have shown elsewhere how the ghost fields (but not the "antighost") are defined at the classical level[16)17)] and that their existence does not depend on the choice of a particular quantum Lagrangian. The gauge and ghost fields are related by the BRS transformation:

$$s\,\omega_\mu = D_\mu C \tag{6.3}$$

We shall formally integrate this relation as:

$$\begin{aligned} \xi_\mu^i &= s^{-1} D_\mu \phi^i = \varepsilon_\mu \phi^i \\ \phi^i &= \frac{1}{N} \tilde{\varepsilon}^\mu \xi_\mu^i \end{aligned} \tag{6.4}$$

And assume the multiplication

$$\begin{aligned} \tilde{\varepsilon}^\mu \varepsilon_\nu &= N \, \delta_\nu^\mu \\ \varepsilon_\mu \varepsilon_\nu &= - \varepsilon_\nu \varepsilon_\mu \end{aligned} \tag{6.5}$$

We deal with the classical theory only.

The Lagrangian of the theory is now defined as:

$$L = - \tfrac{1}{4} \, \bar{\hat{F}}_A{}^{\mu\nu} \, \hat{F}^A_{\mu\nu} + \text{h.c.} \tag{6.6}$$

The $\bar{F}$ are not to be identified with the F, due to CPT non=invariance of a Lagrangian with ghosts. After using (6.4) and representing the ε_μ by Dirac matrices, $g_{AB} \to \delta_{AB}$.

Our group will now preserve only particles and no ghosts in the classical Lagrangian, and the symmetry we impose is thus not that of the algebraic gauge group itself. (See App. B). We denote the SU(2/1) field strengths by $\hat{F}$:

$$\hat{F}^u_{\mu\nu} = \partial_\mu W^u_\nu - \partial_\nu W^u_\mu + g f^u{}_{vw} W^v_\mu W^w_\nu - i g d^u{}_{ij} \xi^i_\mu \xi^j_\nu$$
$$\hat{F}^i_{\mu\nu} = \partial_\mu \xi^i_\nu - \partial_\nu \xi^i_\mu + g f^i{}_{uj} (W^u_\mu \xi^j_\nu - W^u_\nu \xi^j_\mu) \tag{6.7}$$

$$\hat{F}^u_{\mu\nu} = F^u_{\mu\nu} - i g d^u{}_{ij} \xi^i{}_\mu \xi^j{}_\nu \tag{6.8}$$

where the F stand for field-strengths defined over the $U(2)_W$ subgroup only. The Lagrangian (6.6) obeys $SU(2/1)_{x\epsilon\pi}$, i.e., it is invariant under the Bose action of $\mu_i x\epsilon\pi$ so as to act between physical fields only. Indeed, using (6.4) and (6.5), we get:

$$L = -\tfrac{1}{4} \{(F^u{}_{\mu\nu})^2 + 6N (D_\mu \phi^i)^2 + 12 N^2 g^2 (\phi^i \phi_i)^2\} \tag{6.9}$$

Rescaling the Higgs field by $\frac{1}{\sqrt{3N}}$ we obtain the canonical form:

$$L = -\tfrac{1}{4} (F^u{}_{\mu\nu})^2 + \tfrac{1}{2} (D_\mu \phi^i_s)^2 - \tfrac{1}{3} g^2 (\phi_s{}^i \phi_s{}^i)^2 \tag{6.10}$$

$$D_\mu \phi^i_R = \partial_\mu \phi^i + g f^i{}_{uj} W^u_\mu \phi_s{}^j \tag{6.11}$$

Or, upon going over to a complex Higgs field:

$$\phi^{(+)} = 1/\sqrt{2} (\phi^4 + i \phi^5) \tag{6.12}$$

$$\phi^{(o)} = 1/\sqrt{2} (\phi^6 + i \phi^7)$$

$$L = -\tfrac{1}{4} (F^u{}_{\mu\nu})^2 + D_\mu \phi^+ D_\mu \phi - \tfrac{4}{3} g^2 (\phi^+ \phi)^2 \tag{6.13}$$

The $SU(2/1)_{x\epsilon\pi}$ Lagrangian is thus exactly that of the standard Weinberg Salam model. All fields are gauge fields, including the Higgs-Goldstone multiplet which behaves as a $I = \tfrac{1}{2}$, $U_o = 1$ representation. The weak angle and the self couplings of the Higgs field are defined by the structure constants:

$$\theta_W = 30^o \quad , \quad \lambda = \tfrac{4}{3} g^2 \tag{6.14}$$

We do not discuss in this article the ghost states occuring in the representations of SU(2/1). As seen here (and in Section 8) they drop out of the classical theory altogether. From the identification of C^u with the FDWFP ghost we can assume that they are all related to the quantum version, which we do not treat here.

The BRS transformation (6.3) for other than the adjoint representation is (see

5.8)

$$s\psi_F = [C, \psi_F] = \psi_B$$

and using (5.8') and (5.8")

$$s^{\pi} f_L = [\beta C, \gamma_L \psi] = \gamma_R \beta\psi_B = b_R$$

$$s^{-\frac{1}{\pi}} f_R = [\beta C, \gamma_R \psi] = \gamma_L \beta\psi_B = b_L \qquad (6.15)$$

Thus, ε is realized by s, and π by β. Note also that equations of the type (6.15) can be taken to define as composite fields all the ghost fields b_R and b_L for all matter representations f_L , f_R. The only true ghosts in the theory are thus C^u and ξ^i_μ.

7. Symmetry Breaking

The symmetric Lagrangian (6.6) may possibly produce an effective quadratic term with negative μ^2 through radiative corrections[18]. However, since we have only one parameter (g), there is no room for dimensional transmutations in the usual sense. Moreover, the cases studied in ref.[18] refer to $\lambda \sim g^4$, whereas our $\lambda \sim g^2$. We thus assume that the symmetric theory is either a zero-mass theory, or a theory with logarithmically divergent masses.

To produce the conventional spontaneous symmetry breakdown, we first break $SU(2/1)_{x\varepsilon}$ down to $U(2)_W$. This is done by the explicit addition of a term

$$-3N \mu^2 \phi^i \phi^i = -\mu^2 \phi^i{}_s \phi^i{}_s \quad (7.1)$$

This term is in fact proportional to $C^8{}_{ij}$, as can be seen in (4.4d). It will trigger a spontaneous symmetry breakdown of $U(2)_W$ down to U(1) of electric charge, through

$$< o \mid \phi^6{}_s \mid o > = v \quad (7.2)$$

$$\eta^i = \phi^i{}_s - < o \mid \phi^i{}_s \mid o > \quad (7.3)$$

from

$$\frac{G}{\sqrt{2}} = \frac{1}{2v^2} \quad (7.4)$$

we get

$$v = 247 \text{ GeV}$$

$$g = \frac{e}{\sin\theta} = 2e = 0.606$$

$$M_W = \tfrac{1}{2} g v = 74.8 \text{ GeV}$$

$$M_Z = \frac{M_W}{\cos\theta} = \frac{2\text{x}74.8}{\sqrt{3}} = 86.4 \text{ GeV}$$

$$\frac{M_\eta^2}{M_W^2} = \frac{8\lambda}{g^2} = \frac{32}{3} = 10.7$$

$$M_\eta = 245 \text{ GeV} \quad (7.5)$$

We note that in SU(2/1), the vacuum degeneracy in the ϕ^6 direction indeed leaves only electric charge invariant, since $\{\mu^6, \mu^6\} \neq 0$. In linear SU(3), we would have had a U(1) x U(1) invariance, since $[\lambda^6, \lambda^6] = 0$.

8. Leptons and quarks: scalar currents and contributions to the mass.

In the symmetric theory, the universal coupling of the $\underset{\sim}{8}(\omega^u, \xi^i{}_F)$ SU(2/1) gauge multiplet to the lepton multiplets, e.g., $\underset{\sim}{3}$ $(\nu^o_{eL}, e^-_L, x^-_L)$ is induced by the covariant derivative and describes ξ^i_μ mediated currents, aside from (5.13). If we now go over to $SU(2/1)_{x\varepsilon}$, we shall get

$$g\ \bar{\nu}_{eL}\gamma^\mu(\xi^4_{\mu F} - i\xi_\mu{}^5{}_F)\ x^-_{LB} \to g\ \bar{\nu}_{eL}\ \gamma^\mu\ \varepsilon^o_\mu\ (\phi^4 - i\phi^5)\ \varepsilon^o\ (\pi e^-_R) \tag{8.1}$$

From the requirement (6.5) we may have a solution. We write

$$\varepsilon^o_\mu = \varepsilon^o\ \gamma_\mu\ \sqrt{N} \tag{8.2}$$

which yields the interactions (compare with $\frac{1}{2}g\ \mu_8$ in (5.13))

$$2\ \sqrt{2N}\ \ g\ (\bar{\nu}^o_{eL}\ \ \phi^{(+)} + \bar{e}^{+}_L\ \phi^o)\ \ \ (\pi(e^-_R)\ \ )_L$$

After spontaneous symmetry breakdown, only ϕ^6_s subsists, and the interaction it mediates will be

$$\frac{2}{\sqrt{3}}\ g\ \bar{e}_L\ \ \eta^6_s\ \ (\pi(e^-_R)\ \ )_L \tag{8.3}$$

This is a scalar current whose strength compares with electromagnetism, except that it is mediated by a meson with $M^2_\eta/M^2_W = 10.7$. <u>It should be detectable in electromagnetic experiments.</u>

For quark fields, the 4 x 4 matrices of the Appendix will replace the 3 x 3 we used in (8.1). After spontaneous breakdown, only ϕ^6_s subsists. However, the 4 x 4 matrix for μ_6 is star-hermitean (see ref.15) rather than hermitean. The requirement of hermiticity will cancel the $\bar{u}_L\ \eta^6_s\ u_R$ and contribute only to

$$\frac{2}{3}\ g\ \bar{d}_L^{\,-1/3}\ \ \eta^6_s\ d_R \tag{8.4}$$

Similarly, the $u^{2/3}$ quark gets no mass; neither does the neutrino. The $d^{-1/3}$ quark mass is $\frac{1}{\sqrt{3}}$ that of the electron (or charged lepton). Notice that (except for the size of the Higgs field contribution) we get a mass which is larger for $I_3 = -\frac{1}{2}$ than for $I_3 = +\frac{1}{2}$, i.e., an effect which gives $m_{neutron} > m_{proton}$. This would, however, conflict with the evidence in the second and third generation quarks - clearly, the entire issue of fermion masses should wait for an understanding of flavours. This should also provide the necessary cancellations, since all non-vanishing contributions of ϕ^6_s to fermion masses are of the order of 100 - 170 GeV.

Alternatively, (8.1) may be solved by using (6.4) and (6.15)

$$\rightarrow \frac{2}{\sqrt{3}} g \, \bar{e}_L \, \gamma^{\mu} \, D_{\mu} \, \phi^6 \, e_R \qquad (8.5)$$

In that case the Higgs field does not contribute directly to the Fermion masses.

Appendix A

The quark representation $(\frac{1}{6}, \frac{1}{2})$ of SU(2/1).

1. The Matrices

$$\mu_1 = \begin{pmatrix} & 1 & & \\ 1 & & & \\ & & & \\ & & & \end{pmatrix} \qquad \mu_2 = \begin{pmatrix} & -i & & \\ i & & & \\ & & & \\ & & & \end{pmatrix} \qquad \mu_3 = \begin{pmatrix} 1 & & & \\ & -1 & & \\ & & & \\ & & & \end{pmatrix} \qquad \mu_8 = \frac{1}{\sqrt{3}}\begin{pmatrix} \frac{1}{3} & & & \\ & \frac{1}{3} & & \\ & & -\frac{2}{3} & \\ & & & \frac{4}{3} \end{pmatrix}$$

$$\mu_4 = \frac{1}{\sqrt{3}}\begin{pmatrix} & & 1 & \\ & & & -\sqrt{2} \\ 1 & & & \\ & \sqrt{2} & & \end{pmatrix} \qquad \mu_5 = \frac{1}{\sqrt{3}}\begin{pmatrix} & & i & \\ & & & -i\sqrt{2} \\ i & & & \\ & -i\sqrt{2} & & \end{pmatrix} \qquad \mu_6 = \frac{1}{\sqrt{3}}\begin{pmatrix} & & & \sqrt{2} \\ & & 1 & \\ & 1 & & \\ -\sqrt{2} & & & \end{pmatrix} \qquad \mu_7 = \frac{1}{\sqrt{3}}\begin{pmatrix} & & & i\sqrt{2} \\ & & -i & \\ & i & & \\ i\sqrt{2} & & & \end{pmatrix}$$

Note: The quarks are ordered as diag (u_L, d_L, d_R, u_R) and not as in section 5.
The representation is star-hermitean: $\mu_A = (\mu_A)^{\neq}$

2. X-Conjugation

$$(\mu_{1..3,8})^{x} = \mu_{1..3,8} \qquad (\mu_4)^{x} = i\,\mu_5$$

$$(\mu_5)^{x} = -i\,\mu_4 \qquad (\mu_6)^{x} = i\,\mu_7$$

$$(\mu_7)^{x} = -i\,\mu_6$$

3. Super-transposition:

$$M = \begin{vmatrix} \alpha & \gamma \\ \delta & \beta \end{vmatrix} \quad , \quad {}^{T}M = \begin{vmatrix} \tilde{\alpha} & -\tilde{\delta} \\ \tilde{\gamma} & \tilde{\beta} \end{vmatrix}$$

where $\sim$ is ordinary transposition.

4. Star-Conjugation:

$$(\mu_A)^{\neq} = {}^{T}(\,(\mu_A{}^{x})^{*}\,)$$

The relationship to the lepton representation $\underset{\sim}{3}$ is obvious from the matrices. Note that $\underset{\sim}{3}$ is also star-hermitean.

Appendix B

The action of $SU(2/1)_{x\varepsilon}$ is defined by (e is the identity)

$$SU(2/1)_{x\varepsilon} \begin{cases} U(2)_W \otimes e \\ \\ SU(2/1)/_{U(2)_W} \otimes \varepsilon \end{cases} \tag{B.1}$$

Thus, the group transforms fields into fields and ghosts into ghosts, i.e., with no change of statistics. For example, the ν_L or e_L^- of rep. $\underset{\sim}{3}$ provide the $(\mu_{4..7}x\varepsilon)$ variations to e_R^- (belonging to $\underset{\sim}{3}'$). It is easy to check that this requires a metric (1, 1, ..., 1) as in SU(3), even though the components themselves are the SU(2/1) field-strengths.

Appendix C

1. Connections on a Principle Bundle: Gauge (Potentials) and Ghost Fields.

We introduce the concept of a Connection in a Principle Fibre Bundle $(P,M,\pi,G,\cdot)$. Previous authors used definitions in which the connection (a 1-form $\omega^{A}_{(YM)}$) was restricted to the base manifold M of dimension m=4, so that writing

$$\omega^{A}{}_{(YM)} = \omega^{A}_{\mu}\, dx^{\mu} \quad (A=1 \ldots n,\ \mu=0,\ 1,\ ..\ 3)$$

the ω^{A}_{μ} were identified with the Yang-Mills potentials. Our connection ω will have a larger dimensionality, being in P rather than in M. We denote the (vertical) projection by $\pi : P \to M$, the structure group by G and right-multiplication on P by the dot $(\cdot) : P \times G \to P$, so that

$$\forall\, p \in P,\ \forall\, g,\, g' \in G, \qquad \begin{aligned} \pi\,(p\cdot g) &= \pi\,(p) \\ (p\cdot g)\cdot g' &= p\cdot(gg') \end{aligned} \tag{C.1}$$

and for U_x a neighborhood of $x\epsilon M$, we get "local triviality" (a direct product) in P :

$$\begin{aligned} &\pi^{-1}(U_x) \to U_x \times G \\ &p \to (\pi(p),\, \tau(p)), \quad \text{where} \quad \tau(p\cdot g) = \tau(p)g \end{aligned} \tag{C.2}$$

(τ is a projection onto the fiber G).

The dot $(\cdot)$ induces a map t from the Lie algebra A of G into P_*, the tangent manifold to P. Thus,

$$\forall\, \lambda_a,\ \lambda_b,\ \lambda_e\ \epsilon\ A \qquad (a,\ b,\ e = 1 \ldots n)$$

with $\qquad\qquad$ (C.3)

$$[\lambda_a,\ \lambda_b] = C^{e}{}_{ab}\, \lambda_e$$

we have

$$t : A \to P_*,\ \lambda \to \tilde{\lambda}\ \epsilon\ P_* \tag{C.4a}$$

By differentiation of (C.1), one proves that t is an homomorphism of A, with the Lie Bracket operation realized on P_* as a Poisson Bracket

$$\widetilde{[\lambda\,,\ \lambda']}_{L.B.} = [\tilde{\lambda}\,,\ \tilde{\lambda}']_{P.B.} \tag{C.4b}$$

However, this map t has no inverse because the image of A (of dimension n) does not span P_*, of dimension $(n + m)$.

A linear mapping from P_* to A, <u>the connection ω</u>, is now chosen so as to provide the missing inverse

$$\omega : P_* \to A$$
$$\forall \lambda \in A, \quad \omega(\tilde{\lambda}) = \lambda \tag{C.5}$$

ω is Lie-algebra valued, and belongs to the cotangent manifold $\overset{*}{P}$. It is thus a one-form. If z^R are local coordinates over P, one may explicitly write

$$\forall v \in P_*, \quad v = v^R(z) \frac{\partial}{\partial z^R} \qquad (R,S = 1, 2, \ldots n+m)$$
$$\omega = \omega^a{}_S(z) \quad dz^S \lambda_a \tag{C.6}$$
$$\omega(v) = v \lrcorner \omega = \omega^a_R \; v^R \; \lambda_a \equiv \omega^a(v) \; \lambda_a \quad .$$

($\lrcorner$ denotes a contraction, $\frac{\partial}{\partial z^R} \lrcorner dz^S = \delta^S_R$)

For $\tilde{\lambda}_b$ as v, we have $\omega^a(\tilde{\lambda}_b) = \delta^a_b$.
As P_* is larger than A, there is a non-trivial kernel H of ω. In other words, to each point $p \in P$, ω associates a subspace $H_p \subset P_{*p}$. This is known as the "horizontal" tangent vector space at p, and defines an exact splitting of P_*

$$h \in H_p \iff \omega_p(h) = 0$$
$$P_{*p} = V_p + H_p$$
$$H_p = \mathrm{Ker}\,(\omega_p) \tag{C.7}$$
$$V_p = \mathrm{Im}_t\,(A) \quad , \quad (\tilde{\lambda})_p \in V_p$$

One also assumes an equivariance condition

$$H_{p\cdot g} = H_p \cdot g \tag{C.8a}$$

The equivariance condition (C.8a) can be written infinitesimally as

$$\nabla_{\tilde{\lambda}} h \lrcorner \omega = 0 \tag{C.8b}$$

i.e., the new increment $(h_p \cdot g - h_p) \sim [\tilde{\lambda}, h]_{P.B.} \sim \nabla_{\tilde{\lambda}} h$ is still horizontal. $\nabla_{\tilde{\lambda}}$ denotes a Lie derivative in the $\tilde{\lambda}^P$ direction.
Taking the Lie-derivative of (C.7), we have

$$\nabla_{\tilde{\lambda}} (h \lrcorner \omega) = \nabla_{\tilde{\lambda}} h \lrcorner \omega + h \lrcorner \nabla_{\tilde{\lambda}} \omega = 0$$

yielding by (1.8b) a statement of the verticality of $\nabla_{\tilde{\lambda}} \omega$,

$$h \lrcorner \nabla_{\tilde{\lambda}} \omega = 0$$

so that the Lie derivative of ω can be written linearly in A,

$$\nabla_{\tilde{\lambda}} \omega = f(z) \qquad [\lambda , \omega]_{L.B.} \tag{C.8c}$$

(remember ω is Lie-algebra valued). To fix f(z) we take the Lie derivative of (C.5), which vanishes since λ' is constant, (we use C.9b)

$$\nabla_{\tilde{\lambda}} \omega(\tilde{\lambda}') = [\tilde{\lambda} , \tilde{\lambda}']_{P.B.} \lrcorner \omega + \tilde{\lambda}' \lrcorner \nabla_{\tilde{\lambda}} \omega = 0$$

Replacing the second term by (C.8c) we have, using (C.4b)

$$[\lambda , \lambda']_{L.B.} + f(z) \qquad [\lambda , \lambda']_{L.B.} = 0$$

so that f(z) = -1 in (C.8c), and we have equivariance stated as

$$\nabla_{\tilde{\lambda}} \omega = - [\lambda , \omega]_{L.B.} \tag{C.8d}$$

Note that the action of the Lie derivative on functions, vector-fields and one-forms reads:

$$\nabla_v f(z) = v^R \frac{\partial}{\partial z^R} f \tag{C.9a}$$

$$\nabla_v v' = [v, v']_{P.B.} \tag{C.9b}$$

$$\nabla_v \omega = d (v \lrcorner \omega) + v \lrcorner d\omega \tag{C.9c}$$

We now define the Curvature 2-form,

$$\Omega = d\omega + \frac{1}{2} [\omega , \omega] \tag{C.10}$$

and contract it with a vertical vector field $\tilde{\lambda}$

$$\tilde{\lambda} \lrcorner \Omega = \tilde{\lambda} \lrcorner d\omega + \frac{1}{2} [\tilde{\lambda} \quad \omega , \omega] - \frac{1}{2} [\omega, \tilde{\lambda} \lrcorner \omega]$$

The first term is given by (C.9c); the last two by (C.5)

$$= \nabla_{\tilde{\lambda}} \omega + \frac{1}{2} [\lambda , \omega] - \frac{1}{2} [\omega , \lambda]$$

and using (C.8d)

$$= - [\lambda , \omega] + \frac{1}{2} [\lambda , \omega] - \frac{1}{2} [\omega , \lambda] = 0$$

The curvature 2-form is thus purely horizontal, (while (C.5) can be read to imply that ω is vertical)

$$\tilde{\lambda} \lrcorner \Omega = 0 \tag{C.11}$$

This equation is the Cartan-Maurer structural equation of a principle fiber bundle. Up to this point, we have just used textbook geometry. We can now identify the ghost fields.

Since we are in P_*, a gauge choice corresponds to defining a section, i.e., a surface Σ in P, locally diffeomorphic to the base manifold M. We fit the Z^R coordinates to Σ by lifting local x^μ coordinates from the base M, and α^i (group parameters) coordinates from G, using the maps (π^{-1}, τ^{-1}) of equation (C.2), to get the equation for Σ:

$$\Sigma: \qquad \alpha^i (x) = 0 \quad , \qquad i = 1, \dots n \tag{C.12}$$

We now express the vertical connection form ω in this basis

$$\frac{\partial}{\partial x^\mu} \lrcorner \omega = \phi_\mu \quad , \qquad \frac{\partial}{\partial \alpha^i} \lrcorner \omega = X_i$$

$$\omega = X_i \, d\alpha^i + \phi_\mu \, dx^\mu \tag{C.13}$$

It was originally suggested to identify the ghost fields C^a as

$$C^a \equiv X^a_i \, d\alpha^i$$

while ϕ^A_μ is the Yang-Mills potential. More precisely, for C^a to have the dimensions of a field, we should redefine (ℓ is a constant length)

$$\ell C^a = X^a_i \, d\alpha^i \tag{C.14}$$

According to (C.6), had we taken a topologically trivial P and a global flat section, $C^A_{(0)}$ would have coincided explicitly with the Cartan L.I. one-forms of the rigid group. It would then carry no x^μ dependence and would not be a true field. However, under a gauge transformation,

$$\delta\omega^a (x , \alpha) = D\varepsilon^a (x , \alpha) \tag{C.15}$$

so that $C^a_{(0)} = \frac{1}{\ell} (\alpha^{-1} d\alpha)^a$ receives x^μ-dependent contributions,

$$\delta C^a = \frac{1}{\ell} d\alpha^i \left[\frac{\delta}{\delta\alpha^i} \varepsilon^a (x , \alpha)\right] - \frac{1}{\ell} C^a_{\ be} C^b \varepsilon^e (x , \alpha) \tag{C.16}$$

similar to those of the Yang-Mills potential,

$$\delta\phi^a_\mu = \partial_\mu \varepsilon^a (x , \alpha) - C^a_{\ be} \phi^b_\mu \varepsilon^e (x , \alpha) \tag{C.17}$$

We now rewrite Ω of (C.10) in component form, applying what we learned from the Cartan-Maurer equation. Defining

$$df = sf + \bar{d}f \quad ; \quad sf = d\alpha^i \frac{\partial}{\partial\alpha^i} f \quad ; \quad \bar{d}f = dx^\mu \frac{\partial}{\partial x^\mu} f \tag{C.18}$$

Cohomology implies

$$\bar{d}^2 = s\bar{d} + \bar{d}s = s^2 = 0 \tag{C.19}$$

$\bar{d}$ is our "ordinary" horizontal d which depends on the section Σ, s is the exterior differential normal to the section. Ω can be broken into three pieces, i.e., terms in $d\alpha^i \wedge d\alpha^j$, in $d\alpha^i \wedge dx^\mu$ and in $dx^\mu \wedge dx^\nu$:

$$\frac{1}{2} \Omega^a_{ij} \, d\alpha^i \wedge d\alpha^j = sX^a + \frac{1}{2} [X , X]^a \tag{C.20}$$

$$\begin{aligned} \Omega^a_{i\mu} \, d\alpha^i \wedge dx^\mu &= s\phi^a + \bar{d}X^a + \frac{1}{2} ([X , \phi]^a + [\phi , X]^a) \\ &= s\phi^a + \bar{d}X^a + [\phi , X]^a \\ &= s\phi^a + \bar{D}X^a \end{aligned} \tag{C.21}$$

$$\frac{1}{2} \Omega^a_{\mu\nu} \, dx^\mu \wedge dx^\nu = \bar{d} \phi^a + \frac{1}{2} [\phi , \phi]^a \tag{C.22}$$

Applying (C.11) and identifying the field and ghost we have

$$sC^a = - \frac{\ell}{2} [C , C]^a \tag{C.23}$$

$$s\phi^a_\mu = \ell\bar{D}_\mu C^a \tag{C.24}$$

These are the BRS equations for ϕ^a_μ and C^a. $\frac{1}{\ell}$ s is thus the BRS operator.

The covariant quantization path-integral, used in summing over all configurations of the potential satisfying BRS, can be given a geometrical form. In this representation, Feynman diagrams involve non-integrated exterior forms (the ghosts) together with anticommuting Lagrange multipliers (the antighosts). One can then check that the minus sign required by ghost loops, which led to the assignment of Fermi statistics to spin-zero fields $C^a(x)$, is indeed just the sign due to self anticommutation of one-forms.

When the Lie Group G is replaced by a Lie Supergroup and the Lie Algebra A by a Graded Lie Algebra (GLA), some connection one-forms commute instead of anticommuting.

For an internal GLA, the one-forms

$$\omega^i = G^i_\mu \, dx^\mu + \phi^i \qquad \text{(C.25)}$$

commute when i represents an odd-grading (using (C.25)) and ϕ^i is thus a Lorentz-scalar physical Bose field. These fields are identified with Nambu-Goldstone (Higgs-Kibble) fields. The internal supergroup represents a Ghost-Symmetry (i.e., a symmetry between physical and ghost fields). The Higgs fields thus become in this approach the appropriate gauge fields for the odd part of the ghost symmetry.

A "global" group with Nambu-Goldstone realization through a (pseudo) scalar field multiplet is thus replaced by a local supergauge.

References

1. J. Schwinger, Ann. Phys. 2, 407 (1957); J. C. Ward and A. Salam, Nuovo Cim. 11, 568 (1959); Phys. Lett. 13, 168 (1964); Phys. Rev. 136, 763 (1964).
2. S. Weinberg, Phys. Rev. Lett. 19, 1264 (1967).
3. A. Salam, in Elementary Particle Theory, Proc. VIII Nobel Symp., N. Svartholm, ed., Almquist & Wiksell, Pub., Stockholm (1968), pp. 367-377.
4. C. Y. Prescott, et al., Phys. Lett. 77B, 347 (1978). A soft renormalization correction to θ is expected due to the symmetry breaking at 250 GeV.
5. Y. Ne'eman, Nuovo Cim. 27, 992 (1963).
6. A. A. Slavnov, Theor. Mat. Fiz. 10, 99 (1972) and 13, 174 (1972); J. C. Taylor, Nucl. Phys. B33, 436 (1971).
7. R. P. Feynman, Acta Phys. Polon. 24, 697 (1963); B. S. de Witt, Phys. Rev. 162, 1195 (1967); L. D. Faddeev and V. N. Popov, Phys. Lett. 25B, 29 (1967).
8. C. Becchi, A. Rouet and R. Stora, Comm. Math. Phys. 42, 127 (1975); I. V. Tyutin, report FIAN 39, (1975).
9. Y. Ne'eman, Phys. Lett. 81B, (1979), 190-194, (U. of Texas report ORO 3992-349, October, 1978). A model utilizing SU(2/1) has recently been independently suggested by Dr. D. Fairlie, Phys. Lett. 82B, 97 (1979).
10. V. Rittenberg, in Group Theoretical Methods in Physics, Proc. VI. Int. Conf. (Tubingen, 1977), P. Kramer and A. Rieckers, eds. Springer-Verlag Lect. Notes in Phys. 79, Berlin-Heidelberg-N.Y. 1978, pp. 3-21.
11. L. Corwin, Y. Ne'eman and S. Sternberg, Rev. Mod. Phys. 47, 573 (1975).
12. V. G. Kac, Func. Analys. and Applications, 9, 91 (1975); B. Zumino, in Proceedings of the Conf. on Gauge Theories and Modern Field Theory, Northeastern University, Boston, 1975, edited by R. Arnowitt and P. Nath (Cambridge, Mass, 1976), p. 255.
13. Y. Ne'eman and T. Regge, La Rivista del Nuovo Cim. Ser. 111, #5, pp.1-43 (1978).
14. M. Gell-Mann and Y. Ne'eman, The Eightfold Way, W. A. Benjamin, Pub., N.Y. (1964).
15. M. Scheunert, W. Nahm and V. Rittenberg, Jour. Math. Phys. 18, 155 (1977).

16. J. Thierry-Mieg, These de Doctorat d'Etat, Universite d'Orsay (1978); J. Thierry-Mieg, J. Math. Phys., to be published; J. Thierry-Mieg, Nuovo Cim A., to be published; Y. Ne'eman, Proc. 19th Int. Conf. High Energy Physics (Tokyo 1978), S. Homma et al., eds., Phys. Soc. of Jap. pub., Tokyo 1979, p.552.

17. J. Thierry-Mieg and Y. Ne'eman, to be published in Ann. of Physics.

18. S. Coleman and E. Weinberg, Phys. Rev. D73, 1888 (1973).

SPINORS ON FIBRE BUNDLES AND THEIR USE IN INVARIANT MODELS.

Richard Kerner
Département de Mécanique, Université P. et M. Curie,
4 Place Jussieu, 75005 Paris.

1. The fibre bundle formulation of the gauge theory has become a commonplace by now, and it can be found in several papers |1|,|2|,|3|. We shall just remind the notations. Let $P(V_4,G)$ be a principal fibre bundle with the base space V_4 being a 4-dimensional Riemannian space-time endowed with metric tensor $\overset{V}{g}_{i,j}$, $i,j = 0,1,2,3$; the structural group of $P(V_4,G)$ is a Lie group G which is supposed to be compact and semi-simple. Therefore, the non-degenerate Cartan-Killing metric is defined on G; we denote its metric tensor by $\overset{G}{g}_{ab}$, $a,b = 1, 2, \dots N = \dim G$. The group G acts on the points of $P(V_4,G)$ on the left effectively and transitively. The orbits are the fibres in $P(V_4,G)$, and the tangent subspaces to the fibres are called the vertical subspaces of $TP(V_4,G)$. The connection A on the principal fibre bundle $P(V_4,G)$ is a left-invariant Lie algebra valued 1-form of type ad, i.e.

$$A(gp) = ad(g^{-1})\, A(p) \qquad p \in P(V_4,G) \quad , \quad g \in G$$

We define the horizontal subspaces of $TP(V_4,G)$ at any point as the subspaces of tangent vectors X for which

$$A(X) = 0.$$

There are the following natural mappings defined :

$$A : TP(V_4,G) \longrightarrow \mathcal{A}_G$$
$$\sigma : \mathcal{A}_G \longrightarrow TP(V_4,G)$$

any element of the Lie algebra $\mathcal{A}_G$ is mapped into a left invariant vertical vector field over $P(V_4,G)$.

The differential of the canonical projection :

$$d\pi : TP(V_4,G) \rightarrow TV_4$$

mapping any tangent vector over $P(V_4,G)$ into a tangent vector over $P(V_4,G)$ into a tangent vector to the base space, and the lift τ :

$$\tau : TV_4 \longrightarrow TP(V_4,G)$$

giving the unique horizontal vector such that $d\pi \circ \tau = Id_{TV_4}$.

Also $\quad A \circ \sigma = Id_{\mathcal{A}_G} \quad , \quad d\pi \circ \sigma = 0 \quad ,$

$$A \circ \tau = 0$$

Any vector of $TP(V_4,G)$ can be decomposed into its vertical and horizontal parts :

$$X = \text{hor } X + \text{ver } X = \tau \circ d\pi(X) + \sigma \circ A(X)$$

We define the metric over $P(V_4,G)$ by putting

$$g^P(X,Y) = g^V(\text{hor } X, \text{ hor } Y) + g^G(\text{ver } X, \text{ ver } Y) .$$

By definition, in this metric the horizontal and vertical subspaces are orthogonal.

The covariant differential of an exterior p-form θ over $P(V_4,G)$ is defined as

$$D\theta(X_1, X_2, \dots X_{p+1}) = d\theta(\text{hor } X_1, \text{ hor } X_2, \dots \text{ hor } X_{p+1})$$

The structure equations of Maurer-Cartan state that for a left-invariant 1-form θ

$$d\theta = -\tfrac{1}{2}[\theta,\theta]$$

Therefore

$$DA = dA + \frac{1}{2}[A,A]$$

where the bracket means the Lie algebra skew product.

We call DA the curvature 2-form F. It is a horizontal 2-form, and of course, DF = 0.

Now we have everything in order to unify the theory of gravitation and the Yang-Mills theory. We construct the bundle of orthogonal frames over $P(V_4,G,A,g^P)$. The metric tensor g^P induces the Christoffel connection Γ_C in the bundle of frames over P, however, the parallel transport along the fibres with respect to that symmetric connection does not coincide with the left-invariant translation. In order to make these two coincide, we add the torsion term S to Γ_C. The torsion tensor S is a vertical 2-form which has its image in the Lie algebra $\mathcal{A}_G$. It is equal to half the structure constants (considered as a tensor). The new connection $\Gamma = \Gamma_C + S$ is now defined. The Riemann scalar R of this connection is equal to F.F + K + constant. Here F.F means the scalar product of form induced by g^V and g^G in Grassmann algebras, K is the Riemann scalar of V_4. We take $R\sqrt{|g^P|}$ to be the Lagrangian density over $P(V_4,G)$, and the correspondent variational principle is

$$\delta \int_{P(V_4,G)} \sqrt{|g^P|}\; R\, d^4x \;\, dG = 0$$

If our bundle is trivial, then

$$\int_{P(V_4,G)} \sqrt{|g^P|} R d^4x \quad dG = V_G \int_{V_4} \sqrt{|g^V|} \, R \, d^4x$$

Here dG means the left-invariant Haar measure over G, V_G is the volume of the group. The corresponding field equations are

$$DF^* = 0$$

$$K_{ij} - \frac{1}{2} g_{ij} K = - T_{ij}(F) \qquad i,j = 0, 1, 2, 3.$$

F^* means the 2-form dual to F in V_4; the Hodge duality being defined as usual by the metric and the volume element. T_{ij} is the energy-momentum tensor of the field F.

2. We want to unify the spinors with this theory. Following the suggestions formulated in |4|,|5|,|6|, the most natural thing to do is to define the spinors directly on $P(V_4,G;A,g^P)$. Let us remind the construction of spinors on M_4, the Minkowskian space-time. The symmetry group of M_4 is SO(3,1). We construct the principal fibre bundle $P(M_4,SO(3,1))$, which can be assumed to be a trivial one (i.e. isomorphic to $M_4 \times SO(3,1)$ globally). The associate spinor bundle is defiend as follows :

$$\frac{P(M_4,\ SO(3,1)) \times \mathbb{C}^4}{SO(3,1)} \quad .$$

In order to give some sense to this formula, we have to define the action of SO(3,1) on $\mathbb{C}^4$ via some representation. The spinor representation is obtained via the Clifford algebra. The generators of the Clifford algebra are the Dirac matrices γ_i , satisfying

$$\gamma_i \gamma_j + \gamma_j \gamma_i = 2 g_{ij} \, \mathrm{Id}$$

The lowest faithful representation of such matrices has the dimension $M = 2^{\left[\frac{n+1}{2}\right]}$, where n is the dimension of the space (here n = 4 for M_4), $[k]$ means the integer part of K. The matrices $\sigma_{ij} = \frac{1}{8}(\gamma_i \gamma_j - \gamma_j \gamma_i)$ generate the Lie algebra of SO(3,1). A spinor ψ, which has values in C^4, transforms as follows under an infinitesimal Lorentz rotation :

$$\delta\psi = \alpha^{ij} \sigma_{ij} \psi$$

Here $\alpha^{ij} = - \alpha^{ji}$ are the parameters of the infinitesimal Lorentz rotation. If we define the matrix β such that $\gamma_i^+ = \beta^{-1} \gamma_i \beta$, then the conjugate spinor $\bar{\psi}$ is defined as $\bar{\psi} = \psi^+ \beta$, and

$$\delta\bar{\psi} = - \bar{\psi} \sigma_{ij} \alpha^{ij} \quad .$$

In local coordinates the connection in the bundle of orthogonal frames is a 1-form :

$$\omega = \omega^k_\ell \, E^\ell_k = \Gamma^k_{m\ell} \, E^\ell_k \, \theta^m .$$

Here E_k^{ℓ} form the basis of the Lie algebra of orthogonal transformations of frames, and θ^m is the local basis of 1-forms. Usually, in holonomic coordinates, we put $\theta^m = dx^m$. Of course, for the orthogonal frames, $\omega_{k\ell} = -\omega_{\ell k}$. The covariant derivative of a spinor ψ is now :

$$D\psi = d\psi + \omega^{k\ell}\sigma_{k\ell}\psi = (\nabla_k\psi)\,\theta^k$$

and

$$D\bar{\psi} = d\bar{\psi} - \bar{\psi}\sigma_{k\ell}\,\omega^{k\ell}$$

Define now the density $\eta_{ijk\ell}$, a totally antisymmetric tensor on V_4, such that $\eta_{0123} = |\det g_{ij}|^{1/2}$. Define also the forms

$$\eta_{ijk} = \theta^{\ell}\,\eta_{ijk\ell}\,, \qquad \eta_{ij} = \frac{1}{2}\,\theta^k \wedge \eta_{ijk}\,,$$

$$\eta_i = \frac{1}{3}\,\theta^j \wedge \eta_{ij} \quad \text{and} \quad \eta = \frac{1}{4}\,\theta^i \wedge \eta_i$$

Their dual forms will be

$$\eta^k = g^{k\ell}\,\eta_{\ell}\,, \quad \eta^{ij} = g^{ik}\,g^{j\ell}\,\eta_{k\ell}\,, \quad \text{etc.}$$

Define the Clifford algebra valued forms, $\gamma = \gamma_k\,\theta^k$, the 1-form, and its dual $\mu = \gamma_k\,\eta^k$, the 3-form. The Dirac Lagrangian density is taken to be

$$L = \frac{i}{2}(\bar{\psi}\mu \wedge D\psi + D\bar{\psi} \wedge \mu\psi) + m\,\eta\,\bar{\psi}\,\psi$$

It gives rise to a conserved current

$$j = \bar{\psi}\,\mu\,\psi \quad , \quad dj = 0,$$

and the Dirac equation

$$\mu \wedge D\psi + D(\mu\psi) = 2\,i\,m\,\eta\,\psi$$

which, in the case $D\mu = 0$ what we suppose, reads explicitly

$$\gamma^j\,\nabla_j\,\psi - im\,\psi = 0$$

Our aim is to generalize the Dirac spinors and Dirac's equation to the whole of $P(V_4,G;A,g^P)$. We face the problem which can be concisely summarized in the following two diagrams, that can be made realized some supplementary hypothesis on the topology of the base space V_4. In the first diagram Spin(3,1) means the covering group of SO(3,1), Spin (N) is the covering group of SO(N), etc.

The same is not true for the associated spinor bundles represented in the second siagram, in which some of the homomorphisms can be defined only locally.

Diagram 1 :

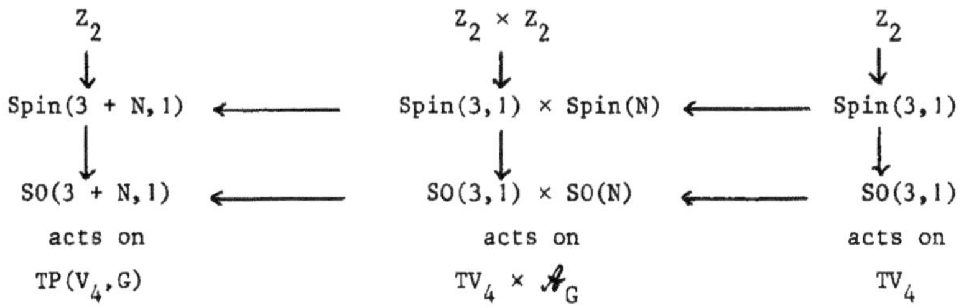

Diagram 2 :

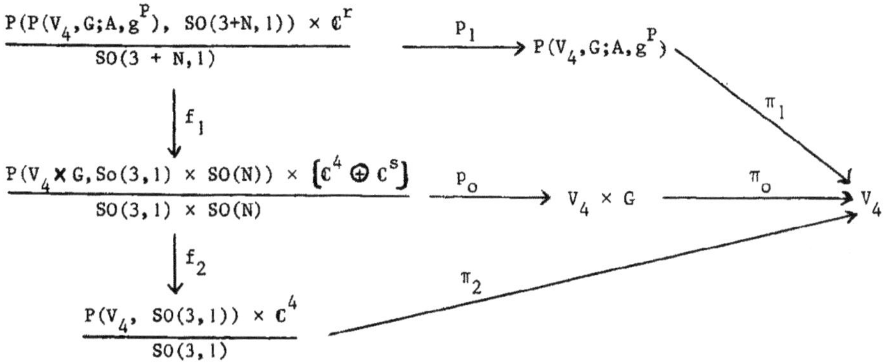

We suppose that all our fibre bundles are trivial in order to make such a diagram realizable.

The mappings π_1, π_o, π_2, p_1, p_o are the canonical projections. The mappings f_1, f_2 are the homomorphisms of fibre bundles we want to construct. The generalized Dirac spinors will be constructed on

$$\frac{P(P(V_4,G;A,g^P),\ SO(3+N,1)) \times \mathbb{C}^r}{SO(3+N,1)}$$

Here $r = 2^{\left[\frac{N+5}{2}\right]}$, $S = 2^{\left[\frac{N+1}{2}\right]}$.

The generalization is obvious. The metric is defined on each of the three principal bundles; the connections and the covariant derivatives are defined too. The Clifford algebras corresponding to any of metrics above, i.e. the metric on V_4, on $V_4 \times G$ and on $P(V_4,G)$ are very easy to define, because any Clifford algebra can be decomposed into a tensor product of the four elementary Clifford algebras :

$C(0,1) = \mathbb{C}$ = the complex numbers

$C(1,0) = R \oplus R$ = two real lines

$C(0,2) = \mathbb{H}$ = quaternions

$$C(1,1) = C(2,0) = \mathrm{Mat}_2\, R = \text{real } 2 \times 2 \text{ matrices}$$

Here C(p,q) means the Clifford algebra corresponding to the diagonal metric with the signature p +, q −. Also

$$C(p+1,\ q+1) = C(1,1) \otimes C(p,q)$$
$$C(p+2,q) = C(2,0) \otimes C(q,p)$$
$$C(p,q+2) = C(0,2) \otimes C(q,p)$$

To give an exemple, if G = SU(2), then N = 3 and we have to construct

$$C(3+N,1) = C(6,1).$$

It is easy to see that $C(6,1) = [C(3,1) \otimes H] \oplus [C(3,1) \otimes H]$.

This means that if we want to define

$$\Gamma^{\alpha} = \{\gamma^{i},\ \gamma^{a}\} \qquad i = 0,1,2,3\ ;\quad a = 1,2,3$$

then one of the possible realisations is

$$\Gamma^{i} = \gamma^{i} \otimes \tau^{1} \otimes \mathbb{1}$$
$$\Gamma^{a} = \mathbb{1} \otimes \tau^{a} \otimes \tau^{3}$$

where γ^{i} are the usual 4 × 4 Dirac matrices, τ^{a} are the 2 × 2 Pauli matrices, and $\mathbb{1}$ is the 2 × 2 identity matrix.

Then the only thing is left now is the generalization of the density tensor η. For M_4 we take η_{0123}, for the group G we take $\eta_{a_1 a_2 \ldots a_N}$, for the direct product $M_4 \times G$ we shall take the product of the two $\eta_{0123}\, \eta_{a_1 a_2 \ldots a_N}$, and for the fiber bundle $P(M_4,G)$ we shall define $\eta_{123\ldots N+4} = \eta_{0123}\, \eta_{a_1 a_2 \ldots a_N}$, other components just by antisymetrisation.

The new Lagrangian will be analogous to the Dirac Lagrangian on M_4 :

$$L = \frac{i}{2}\,(\bar{\psi}\mu \wedge D\psi + D\bar{\psi} \wedge \mu\psi) + m\,\eta\,\bar{\psi}\,\psi$$

and the corresponding Dirac equation is

$$\gamma^{\alpha}\, \nabla_{\alpha}\, \psi - im\,\psi = 0 \qquad \alpha = (i,a) = 1,2,\ \ldots\ N+4$$

When calculated explicitly, it gives the following terms :

$$\left[\ \overset{\circ}{\gamma}{}^{i}(\partial_i - A_i^a\, C_{ac}^b\, g_{bd}\, \overset{\circ}{\sigma}{}^{cd}) + \overset{\circ}{\gamma}{}^{a}\, C_{ac}^b\, g_{bd}\, \overset{\circ}{\sigma}{}^{cd} - im)\right]\psi = 0$$

Here $2C_{ab}^c$ are the structure constants of the gauge group, A_i^a is the local expression of the connection form A (the Yang-Mills potential).

The interaction with the Yang-Mills field is gauge-invariant, as it should be, and what is even more interesting, there is a non-diagonal correction to the mass term. We see that even when m = 0, the generalized spinor is not massless. It is easy to see it if we suppose A = 0, and act on the last equation with its conjugate operator; then we get

$$\{\Box + (\overset{\circ}{\gamma}{}^{a} C^{b}_{ac}\, g_{bd}\, \overset{\circ}{\sigma}{}^{cd})^{2} + m^{2}\}\,\psi = 0$$

The massless spinors can be present in this theory only if the gauge groups contains an abelian subgroup (some rows of the C^{b}_{ac} matrix vanish).

Still the correction to the mass term cannot be taken seriously in this form, because the spinot ψ does not have any direct physical meaning - it does not belong to an irreducible representation of the Lorentz group SO(3,1); neither it does belong to an irreducible representation of the gauge group G. The comparison with physics can begin only after the decomposition of the generalized spinors into the Dirac spinors (strictly speaking, some multiplets of the Dirac spinors) is performed. We shall describe briefly how such a decomposition can be done. First of all we remind that we supposed the generalized spinor fields over the bundle $P(V_4,G;A,g^{P})$ to be left invariant with respect to left translations induced on the spinorial bundle by left action of G on $P(V_4,G;A,g^{P})$. If $\tilde{L}_g$ is a finite left translation by $g \in G$, then we should have

$$\tilde{L}_g\, \psi = \psi$$

If we take the corresponding infinitesimal generator L_X, then

$$\tilde{L}_X\, \psi = 0$$

The last formula can be written explicitly in local coordinates. Let ξ^{a}, a = 1,2,..N, be some local coordinates in G, x^{i}, i = 1,2,3 be some local coordinates in V_4, so that $p^{\alpha} = (x^{i},\xi^{a})$ are the corresponding local coordinates in $P(V_4,G)$. Let us define the infinitesimal generator of the Lie algebra $\mathcal{A}_G$ acting on vertical vector fields on $P(V_4,G)$:

$$L_a\, \varphi^{b} = \frac{\partial}{\partial \xi^{a}}\varphi^{b} + C^{b}_{ac}\, \varphi^{c}$$

It is easy to verify that

$$[L_a, L_b] = 2\, C^{c}_{ab}\, L_c$$

A left-invariant vector field satisfies

$$L_a\, \varphi^{b} = 0$$

as it should be. The generalization to left-invariant spinors is immediate :

$$\tilde{L}_a\, \psi = \frac{\partial}{\partial \xi^{a}}\psi + C^{b}_{ac}\, g_{bd}\, \overset{\circ}{\sigma}{}^{cd}\, \psi = 0$$

We see that in principle we can calculate the values of ψ at any point $(x^i, g) \in P(V_4, G)$ if we know its values at (x^i, e). The Lie algebra $\mathcal{A}_G$ acts on spinors through its adjoint representation and via its imbedding into the Lie algebra of SO(N+3,1) represented in K-dimensional space (remind that $K = 2^{\left[\frac{N+5}{2}\right]}$) :

$$\mathcal{A}_G \longrightarrow \mathrm{ad}\, \mathcal{A}_G \longrightarrow SO(N+3,1) \longrightarrow GL(K)$$

The action of finite translations on ψ can be obtained by exponentiating the infinitesimal action, because

$$\exp \mathrm{Ad}(g)\, \chi = g\, (\exp \chi)\, g^{-1} ,$$

where $g \in G$, $\chi \in \mathcal{A}_G$.

So we can write, for $p \in P(V_4, G)$:

$$\psi(p) = \psi(x,g) = D(g)\, \psi\, (x,e) = D(g)\, \psi\, (x)$$

where we put for simplicity $\psi(x) = \psi(x,e)$.

The same procedure can be performed for the action of the Lorentz group on ψ; if $\alpha_{k\ell}$ are the infinitesimal parameters of the Lorentz transformation, then the formula

$$\delta \psi = \alpha_{k\ell}\, \overset{\circ}{\sigma}{}^{k\ell}\, \psi$$

defines a reducible representation of SO(3,1) in a K-dimensional space. This representation can be reduced into a sum of irreducible representations, i.e. the fibre-bundle spinor ψ can be represented as a linear combination of Lorentz spinors.

Then the corresponding terms in the Lagrangian density will represent the sum of Dirac terms with different masses : these masses will be given by integration of representation matrices over the group space. Symbolically, we can write :

$$\begin{aligned} m\, \bar{\psi}\, \psi &= m\, \overline{D(g)\, \psi\, (x)}\, D(g)\, \psi\, (x) = \\ &= m\, \bar{\psi}(x)\, \bar{D}(g)\, D(g)\, \psi\, (x) = \\ &= m\, \bar{\mathfrak{u}}_A\, \bar{D}(g)\, D(g)\, \left(\begin{smallmatrix} A \\ B \end{smallmatrix}\right) \mathfrak{u}_B \end{aligned}$$

Here by $\left(\begin{smallmatrix} A \\ B \end{smallmatrix}\right)$ we denote symbolically the 3-j Racah symbols, A, B are the generalized indices (their nature depends on the representation), and $\mathfrak{u}_A$ are the multiplets of Dirac spinors. After performing the integration of $\bar{D}(g)\, D(g)$ over the group space we get some numbers which will will contribute to the mass term.

The same things should be done with the interaction term

$$\bar{\psi}\, A^a_j\, C^c_{ab}\, g_{cd}\, \overset{\circ}{\sigma}{}^{bd}\, \psi$$

Such a program has been performed for the case G = SU(2) by G. Domokos and S. Kövesi-

Domokos |5| . For simplicity they have chosen an unfaithful representation of C(6,1) in the product $\mathbb{H}\otimes C(3,1)$, i.e. in eight dimensions.

In this representation the generators of C(6,1) are chosen to be :

$$\gamma^{\alpha} =: \{ \tau^3 \otimes \gamma^i,\ \tau^1 \otimes \mathbb{1},\ \tau^2 \otimes \mathbb{1},\ \tau^3 \otimes \gamma^5 \}$$

here $\alpha = 0,1,\ldots,6$, and $\gamma^5 = \gamma^0\gamma^1\gamma^2\gamma^3$, τ^1, τ^2 and τ^3 are the usual 2×2 Pauli matrices, γ^i are the usual Dirac matrices, and 1 means 4×4 identity matrix. Then, if we introduce the projection operators $\frac{1}{2}(1+\tau^3\otimes 1)$ and $\frac{1}{2}(1-\tau^3\otimes 1)$, and denote

$$\psi_1 = \frac{1}{2}(1+\tau^3\otimes 1)\psi$$
$$\psi_2 = \frac{1}{2}(1-\tau^3\otimes 1)\psi$$

In this case ψ_1 and ψ_2 are the Lorentz spinors embedded in 8-dimensional space. If we denote ψ_1, ψ_2 by ψ_A (A,B = 1,2), then $\psi_A(x,g)$ can be expanded into irreducible representations of the group SO(3) as follows :

$$\begin{aligned}\psi_A(x,g) &= \sum_{B,\ell,m} D^{1/2}_{AB}(g)\, D^{\ell}_{om}(g)\, \psi_B^{\ell m}(x) = \\ &= \sum_{\ell,m,B,j,M} (2j+1)\begin{pmatrix}1/2 & \ell & j\\ A & 0 & -A\end{pmatrix}\begin{pmatrix}1/2 & \ell & j\\ B & m & -M\end{pmatrix}(-1)^{M-A}\, D^j_{AM}(g)\, \psi_B^{\ell m}(x) = \\ &= \sum_{j,M,\ell} (2j+1)^{1/2}\begin{pmatrix}1/2 & j-\varepsilon/2 & j\\ A & 0 & -A\end{pmatrix}(-1)^{M-A}\, D^j_{AM}(g)\, \chi_{j\varepsilon M}(x)\end{aligned}$$

Where $\varepsilon = \pm 1, |\ell|, |m| < 2j+1$, and

$$\chi_{j\varepsilon M}(x) = (2j+1)^{1/2} \sum_{B,\ell,m} \begin{pmatrix}1/2 & \ell & j\\ B & m & -M\end{pmatrix} \delta_{\ell, j-\varepsilon/2}\, \psi_B^{\ell m}(x)$$

We shall not write down the similar expansion for other terms in the Dirac lagrangian. Instead we give the final result of averaging the term $\bar\psi\gamma^{\alpha}\nabla_{\alpha}\psi + \nabla_{\alpha}\bar\psi\gamma^{\alpha}\psi$ over the group G, in absence of mass m, and of the field A^a_i. We get the following expression :

$$\frac{1}{V_G}\int (\bar\psi\gamma^{\alpha}\nabla_{\alpha}\psi + \nabla_{\alpha}\bar\psi\gamma^{\alpha}\psi)\, dG =$$
$$= \sum_{j,\varepsilon}\left[\bar\chi_{j\varepsilon}(\gamma^i\partial_i)\chi_{j\varepsilon}\right] + \sum_{j,\varepsilon}\left[\bar\chi_{j\varepsilon}\,\sigma^3\left(\frac{3}{4i}\gamma^5 + \sigma^a C_{abc}\sigma^{bc} + \gamma^5\sigma^3 C_{3bc}\sigma^{bc}\right)\chi_{j\varepsilon}\right]$$

Here $\chi_{j\varepsilon}$ is a 4(2j+1)-column belonging to the irreducible subspace (j) of the gauge group SU(2), i.e. a column of 2j+1 Dirac spinors. Let us define new multiplets of Dirac spinors $u_{j\varepsilon}$ by putting

$$\chi_{j\varepsilon} = \frac{1}{2}\left[(1+\gamma^5) + e^{\frac{i\pi\sigma^3}{2}}(1-\gamma^5)\right] e^{\frac{i\pi\sigma^3}{2}(\varepsilon-1)\gamma^5} u_{j\varepsilon}$$

This ansatz eliminates the negative-mass (ghost) terms. As a matter of fact, we get

$$\frac{1}{V_G}\int (\bar\psi\gamma^{\alpha}\nabla_{\alpha}\psi + \nabla_{\alpha}\bar\psi\gamma^{\alpha}\psi)\, dG = \sum_{j,\varepsilon}\left[\frac{1}{2}\bar u_{j\varepsilon}\gamma^i\partial_i u_{j\varepsilon} + \left(j+\frac{1}{2}-\frac{\varepsilon}{4}\right)\bar u_{j\varepsilon}u_{j\varepsilon}\right]$$

The most important result is therefore the fact that the mass term is given by linear family $(j + \frac{1}{2} - \frac{\varepsilon}{4})$, depending on the representation index j. This is similar to the well-known Regge trajectories, whose origin is of course different (here we have the representations of the internal symmetry group instead of the kinematical group in Regge's formalism). Averaging the interaction term over the group,

$$\frac{1}{V_G} \int \bar{\psi} \gamma^i A_i^a C_{abc} \overset{\circ}{\sigma}{}^{bc} \psi \, dG$$

will also give the decomposition into irreducible representations, and the coefficients can be regarded upon as the generalized charges.

The comparison with the experimental data seems to be precarious; anyway, the true symmetry of the theory should be at least $SU(2) \times U(1)$ or $SU(3)$, in order to describe weak and electromagnetic, or the strong interactions.

References.

[1] Trautman A., Rep. Math. Phys., Vol.1, 29, (1970).
[2] Kerner R., Ann. Inst. H. Poincaré, Vol. 9, 147 (1968).
[3] Cho Y.M., J. Math. Phys., Vol. 16, 2029 (1975).
[4] Kerner R., in "Group Theoretical Methods in Physics", ed. Janner and Boon, Springer-Verlag, Phys. Ser. (1976).
[5] Domokos G., Kövesi-Domokos S., Phys. Rev. D, Vol. 16, 3060 (1977)
[6] Kerner R., to appear in Journal of Math. Phys. (1980).
[7] Kopczynski W., Acta Phys. Polon. B 10 (1979), 365.
[8] Kopczynski W., This volume.

GLUEING BROKEN SYMMETRIES TOGETHER

R.N. Sen

Department of Mathematics
Ben Gurion University, 84120 Beersheba, Israel

1. INTRODUCTION

Much of current mathematical physics - be it quantum field theory or statistical mechanics - is done in the framework of topological *-algebras, notably C^*-algebras and their representations. The reason is that when one passes from systems with finitely many degrees of freedom to systems with a countable infinity of degrees of freedom, one encounters qualitatively different mathematical problems. These problems require an altogether more sophisticated mathematical approach. One such is the so-called *algebraic approach*, which with a few important exceptions means a C^*-algebra approach. The representation theory of C^*-algebras has turned out to be particularly well adapted to a large class of physical problems. One of the important exceptions is a very concrete one: The algebraic formulation of Wightman field theory given by Borchers [1] fails to yield a C^*-algebra, and as a result one has to start practically from scratch.

Moreover, the algebraic approach has two weaknesses, both arising from its emphasis on the local (in the physical sense) as opposed to the global. One is that the very spirit of the approach has to be violated in order to include any kind of gauge invariance. The other is that the treatment of broken symmetries (this term will be defined precisely in the following) which one can give in this framework is cumbersome, and not as illuminating as it should be. Both of these weaknesses can be remedied to a large extent by the injection of a few ideas from geometry and topology. This will be sketched briefly in the following.

In section 2 we shall state the "mathematical framework for physics" in the spirit of the algebraic approach. I learnt it essentially from H.J. Borchers. As far as I know, the line of development was initiated by I.E. Segal, on the one hand, and H. Araki; R. Haag and D. Kastler; and H.J. Borchers, on the other hand. For early surveys, see the lectures of I.E. Segal [2] and D.W. Robinson [3], where the original memoirs are cited. In section 3 we shall discuss the question of "glueing broken symmetries together", and in section 4 the question of gauge invariance. We shall give only a bird's-eye view, leaving the details to the references. Moreover, the discussion in section 2 has been slightly generalized to permit the inclusion of applications which have not yet been made, but ought, perhaps, to be made, and others which are being developed.

2. A MATHEMATICAL FRAMEWORK FOR PHYSICS

One postulates that in any physical theory, one is dealing simultaneously with two sets, a *set of observables* $\mathcal{O}$ and a *set of states* Ω. These sets are not defined by axioms of specification. Rather, one imposes mathematical structures on them and studies the consequences; the spirit is exploratory. Regarding the *relationship* between the sets $\mathcal{O}$ and Ω, one tentatively demands the following:

Requirement 1: For any $\omega \in \Omega$ and any $a \in \mathcal{O}$, there exists a unique real number $\omega(a)$, called the *expectation* of the observable a in the state ω. That is, there exists a physically distinguished map $\Omega \times \mathcal{O} \to R$, which might be called the *expectation map:* $(\omega,a) = \omega(a)$.

Requirement 2: The sets $\mathcal{O}$ and Ω *separate* each other under the expectation map. That is, if $a, a' \in \mathcal{O}$, $a \neq a'$, then $\exists\, \omega \in \Omega$ such that $\omega(a) \neq \omega(a')$: If $\omega, \omega' \in \Omega$, $\omega \neq \omega'$, then $\exists\, a \in \mathcal{O}$ such that $\omega(a) \neq \omega'(a)$.

The additional structures which one imposes on $\mathcal{O}$ are generally algebraic and topological; for example one might identify $\mathcal{O}$ with some subset of symmetric elements of a C^*-algebra A. Those on Ω may be topological, linear, differentiable, or some consistent combination of these;. for example, that of a Hilbert space, or the G-structure of a symplectic manifold. Once these structures are imposed, one refers to the *algebra* A and the *state space* Ω. It is evident that A has to be mapped into a class of maps of Ω into itself, or, briefly, A has to be *represented* on Ω. We therefore arrive at a triplet; algebra A, state space Ω and continuous representation π of A on Ω. There will generally be a lot of redundancy in the description of a physical system by such a triplet. One might try to reduce this redundancy, or else one might try to exploit it to facilitate, and sometimes guide, the physical interpretation of the formalism.

The above framework has been specified *only* to the extent necessary for a discussion of symmetries and broken symmetries. One cannot evern *begin* to discuss dynamical questions without further assumptions.

Consider, to start with, the case of symmetries. First of all, we observe that the algebra A furnishes its group of automorphisms $\mathrm{Aut}(A)$. Similarly, we have a group of structure-preserving invertible maps (hereafter called *admissible maps*) $\Omega \to \Omega$, for example, unitary transformations, bundle maps, diffeomorphisms, etc. Let $\theta \in \mathrm{Aut}(A)$. Then there are two possibilities:

(a) There exists an admissible map θ_Ω such that $(\omega,a) = (\theta_\Omega\omega, \theta a)\ \forall\, \omega \in \Omega$, $a \in A$.

(b) There exists no such θ_Ω.

If (a) holds, we say that $\theta \in \mathrm{Aut}(A)$ is *implementable*. If (b) holds, we say that

θ is *not implementable*.

Now the group Aut(A) is very large. Most of its elements permute the labels of the elements of A in a way which has no useful physical interpretation. To sort out the physically significant elements, we may proceed as follows. First we choose a subgroup of Aut (A) which may be identified with some group (e.g. Galilei or Poincaré) with a secure physical interpretation. We shall call this the *distinguished subgroup* of Aut (A) and denote it by G_0. As a rule, G_0 will come furnished with the required topological and analytic structures. Then we choose the *normaliser* of G_0 in Aut (A) and denote it by G. G will have to be topologized in such a way that G_0 is a closed subgroup of G. We shall call G the group of *physical automorphisms* of A relative to G_0. Finally, we add the following to the defining conditions of the triple (A, Ω, π).

Requirement 3: Either G_0 itself, or a closed subgroup G_s of G_0 is implemented on Ω.

Clearly, the set of implementable elements θ of G is a subgroup K of G, and by definition $K \supset G_s$. We have to check whether or not this inclusion is proper. If it is, we have to check whether or not K is closed in G. We assume that it is; this question arises only when the inclusion $G \supset K$ is proper, which we assume to be true. Then we may make the following statements, which are essentially definitions.

(1) The symmetry group of the theory is G (an element of G will be called a *symmetry*).

(2) The subgroup K of G is implemented on Ω.

(3) The symmetries contained in $G-K$ are *broken* (*spontaneously broken*), or better, π-*broken*.

Observe that we cannot always require all of G_0 to be implemented on Ω. Some symmetries of the distinguished subgroup G_0 itself may be π-broken, for example the boosts in the case of a system with infinitely many massive particles.

3. GLUEING BROKEN SYMMETRIES TOGETHER

If the inclusions $G \supset K \supset G_s$ are proper, and K is a closed subgroup of G, then we may *enlarge the state space* Ω in such a way that the group G is represented continuously on the new state space, and this new description is, for some purposes, far more advantageous than the previous one. This is done as follows.

A is an algebra. *Forget the multiplication*. Then we are left with a vector space over $\mathbb{C}$. For physics, it is reasonable to assume that this vector space has a countable basis. Look upon it as an abelian group, and denote this group by A.

Then G is a group of automorphisms of A, and we may therefore define the semidirect product $G \wedge A$ (the carat symbol denoting the semidirect product), with the multiplication

$$\{g,a\}\{g',a'\} = \{gg', a+g(a')\} .$$

Here $g, g' \in G$, $a, a' \in A$, and $g(a')$ is the image of a' under the automorphism g. If K is closed in G, then $K \wedge A$ is closed in $G \wedge A$, and $G \wedge A/K \wedge A$ is naturally homeomorphic to G/K; we use left-cosets for definiteness. Denote $G/K \approx G \wedge A/K \wedge A$ by M. There exists a natural embedding

$$\nu: G \to F = G \wedge A$$

given by $\nu(g) = \{g,0\}$. Using this, we may extend any map $X \to G$ to a map $X \to G \wedge A$.

For the next step, we need to make an assumption which depends on the nature of Ω.

(1) If Ω is a differential manifold (its linear structure, *if any*, is physically irrelevant), then we need K to have a global cross-section in G. Such cases have not yet been studied.

(2) If Ω is a linear space over C, then the following weaker condition suffices: K has a Borel section in G. All cases which have been studied so far belong to this class.

Now let $\tilde{\eta}: M \to G$ be a suitable section of K in G. Then

$$\tilde{c}(g,x) = \tilde{\eta}(gx)^{-1} g\tilde{\eta}(x)$$

(where $g \in G$, $x \in M$, and gx is the left-translate of x by g), is a (G, M, K) cocycle. Since the natural action of $F = G \wedge A$ on M, namely

$$fx = \{g,a\}x = gx ,$$

where $f = \{g,a\} \in F$, $g \in G$, $a \in A$, is the same as the action of G on M, the quantity

$$c(f,x) = \eta(fx)^{-1} f\eta(x)$$

is a $(F, M, K\wedge A)$ cocycle. In the above,

$$\eta = \nu \circ \tilde{\eta} .$$

Now recall that π is a continuous representation of $K \wedge A$ on the state space Ω. Form the product bundle $M \times \Omega$. The formula

$$f(x,\omega) = (fx, \pi(c(f,x))\omega)$$

gives a continuous *bundle representation* of F on $M \times \Omega$ [4,5]. This representation has not been defined in a coordinate-free manner. One has therefore to define, first, coordinate transformations, and then *bona fide* bundle representations as equivalence classes of representations such as the above under coordinate transformations. When this is done properly, it turns out that a bundle representation is actually defined on a bundle B which is homeomorphic to $M\times\Omega$, *but there is no need to assume that this homeomorphism is natural* [6], as we had done to start with.

The time has come to recall that A was obtained from the algebra $\mathcal{A}$. The construction given above works for any continuous representation of the group $K \wedge A$. The representation π which we used was originally a representation of $\mathcal{A}$, and therefore *ipso facto* a representation of A, and moreover K was so chosen that π was a representation of $K \wedge A$ on Ω. That is, the algebra $\mathcal{A}$, and the subgroup K of G, are continuously represented on Ω. The remaining elements of G act on the base space, and on the fibres through the cocycle c. The π-broken symmetries have been "glued together" *by enlarging the state space*, from Ω to the *bundle* $B \approx M\times\Omega$.

The following special cases have been studied in some detail. (1) G is the Galilei group or its central extension, K is the Euclidean group, a crystallographic (space) group, or some other subgroup of the Euclidean group. (2) G is the Poincaré group, K is the Euclidean group. One obtains a fairly detailed theory of symmetry in the quantum mechanics of infinite systems, with *elementary excitations* appearing in place of the elementary particles which appear in the ordinary (Wigner) theory of symmetry in the quantum mechanics of systems with a finite number of degrees of freedom. For details, the reader is referred to the original memoirs [4,5,7].

It has been suggested by S.K. Bose (*private communication*) that the question of inequivalent representations of the canonical commutation (CCR) and anticommutation (CAR) relations should also be studied in the above framework. More specifically, one should take $\mathcal{A}$ to be the CCR or the CAR algebra, G_s the Poincaré group, and G a group containing, in addition, the Bogoliubov-type automorphisms of the CCR and the CAR algebra. The question is whether or not one gains some effective control over the maze of representations of the CCR and CAR [8] by enlarging the state space from a Hilbert space to a Hilbert bundle. The answer appears to be at least partially in the affirmative, but details remain to be worked out.

Finally, it should be pointed out that the theory of symmetry of infinite classical systems - either systems containing infinitely many particles, or a continuous medium with variable density - have not been studied at all. The formalism outlined above may be useful for studying the first of these problems; the second appears to be wholly open.

To sum up, the introduction of bundles as state spaces and group representations by bundle maps does indeed appear to facilitate the study of symmetry problems which are relatively difficult to approach through purely functional-analytic methods.

4. THE QUESTION OF A GAUGE PRINCIPLE

The study of gauge invariance through purely "algebraic" means presents the difficulty that all observables are supposed to be strictly gauge-invariant. Thus the approach of section 2 has to be extensively modified if one wishes to incorporate gauge invariance. Finally, one has the well-known difficulties in electrodynamics with the subsidiary condition and manifest covariance. The Gupta-Bleuler resolution of this problem [9,10] gives up requirement 2) of section 2. The set of states is enlarged to accommodate unobservable gauge transformations which are required for manifest covariance, i.e. the observables and the states no longer separate each other.

One of the main motivations for the current resurgence of interest in gauge theories is that they seem to offer some hope of leading, eventually, to a unified theory of "particles" and their interactions. A key problem is to find a principle which is considerably stronger than gauge invariance alone; it should rule out the Pauli term in electrodynamics, which is gauge invariant, but does not occur in nature.

The notion of an algebra of observables is an abstraction from a concrete algebra of quasi-local (i.e. smeared over with test functions of compact support, or of rapid decrease) observables defined on space-time. From the study of gauge theories it appears that a useful generalization of space-time is a principal bundle based upon space-time, with a gauge group as structure group and fibre. Then what one calls space-time is a particular cross-section of this principal bundle. However, this bundle has no distinguished trivialization. *It may be* that this mathematical fact has physical consequences, one of them being, possibly, the emergence of a gauge principle of the required strength. For details, we refer the reader to [11].

In this case as well, the enlargement of the geometrical infrastructure of the algebraic approach, by the introduction of some geometrical-topological ideas and techniques, seems to lead both to technical improvements and to new physical insights.

ACKNOWLEDGEMENTS

I would like to thank Professor P.L. Garcia and his colleagues for their warm hospitality in Salamanca, the organizing committee of the conference for financial support, and Professors H.D. Doebner and K. Bleuler for some important suggestions which have been incorporated in this article and in ref. [11]. I would like to thank Mrs. Yael Ahuvia for preparing this typescript with such exquisite care.

REFERENCES

1. H.J. Borchers, in: *Statistical Mechanics and Field Theory*, edited by R.N. Sen and C. Weil, Keter Publishing House (Jerusalem) and John Wiley (1972). The subsequent work of H.J. Borchers and J. Yngvason has appeared mainly in Commun. math. Phys. over the years.
2. I.E. Segal, *Mathematical Problems of Relativistic Physics*, Vol. II in Lectures in Applied Mathematics, American Mathematical Society, Providence, R.I. (1963).
3. D.W. Robinson, in: *Axiomatic Field Theory*, Gordon and Breach (1966).
4. H.J. Borchers and R.N. Sen, Commun. math. Phys. 42, 101 (1975).
5. R.N. Sen, Physica 94A, 39 (1978).
6. R.N. Sen, *The domain of definition of bundle representations*, J. Math. Phys. (in press).
7. R.N. Sen, Physica 94A, 55 (1978).
8. L. Gårding and A.S. Wightman, Proc. Natl. Acad. Sci. Washington, 40, 617; 622 (1954).
9. S.N. Gupta, Proc. Phys. Soc. London A63, 681 (1950).
10. K. Bleuler, Helv. Physica Acta 23, 567 (1950).
11. R.N. Sen, *Fibre bundles and gauge theories* (Ben Gurion University Math. preprint no. 226), Lectures given at the summer school on Differential Geometric Methods in Mathematical Physics, Clausthal-Zellerfeld (1979); to appear in the proceedings of the summer school.

DEFORMATIONS AND QUANTIZATION.

André LICHNEROWICZ
Collège de France - Paris

It is well-known that it is possible to give a complete description of Classical Mechanics in terms of symplectic geometry and Poisson bracket. It is the essential of the hamiltonian formalism. In a common program with M. Flato, D. Sternheimer, J. Vey and others, we have studied properties and applications of the deformations of the Poisson Lie algebra and of a trivial associative algebra. Such deformations give a new approach of Quantum Mechanics which corresponds to an autonomous generalization of the Weyl-Wigner quantization. The quantization of a system described by a symplectic manifold is given by the choice of a so-called $*_\nu$-product. I consider here only dynamical systems with a finite number of degrees of freedom, but the approach and a significative part of the results may be extended to physical fields. A new study of the $*_\nu$-automorphisms is given here.

I have given in my lecture of Aix (refered as [A]) the definitions and the existence and equivalence theorems concerning the $*_\nu$-products. I will use here these results for problems of quantization. The notations of the present lecture are identical to the notations of [A] .

1- Lie algebras associated to a symplectic manifold.

a) Let (W,F) be a smooth symplectic manifold of dimension $2n$ and fundamental 2-form F. We denote by $b_k(W)$ the Betti numbers of W. A symplectic infinitesimal transformation (i.t.) is defined by a vector field X such that $\mathcal{L}(X)F = 0$ (where $\mathcal{L}$ is the Lie derivative); it is an infinitesimal automorphism of the structure. We denote by L the (infinite dimensional) Lie algebra of the symplectic vector fields; X belongs to L if and only if the 1-form $\mu(X)$ is closed ($d\mu(X) = 0$). If $X,Y \in L$, we have :

$$\mu([X,Y]) = d\, i(\Lambda)\, (\mu(X) \wedge \mu(Y)) \tag{1-1}$$

Let L^* be the subspace of L defined by the converse images of the exact 1-forms ($X_u = \mu^{-1}(du)$; $u \in N = C^\infty(W; \mathbb{R})$). An element of L^* is a hamiltonian vector field. Consider the commutator ideal $[L,L]$ of L : each element of $[L,L]$ is, by definition, a finite sum of brackets of elements of L. It follows trivially from (1-1) that $[L,L] \subset L^*$. It has been proved (Arnold, Calabi, myself) that we have $[L,L] = L^*$;

L/L^* is abelian and dim. $L/L^* = b_1(W)$ (homology with compact supports).

b) Let $\bar{N}$ be the space of the classes of elements of $N = C(W; \mathbb{R})$, modulo the additive constants; $\pi : u \in N \to \bar{u} \in \bar{N}$ is the projection of N onto $\bar{N}$. The natural isomorphism between the spaces L^* and $\bar{N}$ induces on $\bar{N}$ a structure of Lie algebra defined in the following way : if $\bar{u}, \bar{v} \in \bar{N}$ it follows from (1-1) that the function :

$$w = i(\Lambda)(d\bar{u} \wedge d\bar{v})$$

defines a class $\bar{w}$ which is the bracket of $\bar{u}$ and $\bar{v}$. The function w is the Poisson bracket of $\bar{u}$, $\bar{v}$, or of two representants u, v in N. We put :

$$\text{(1-2)} \qquad \{u,v\} = i(\Lambda)(du \wedge dv) = \mathcal{L}(X_u)v = P(u, v)$$

P is a bidifferential operator of order 1 for each argument, null on the constants, which defines on N itself a structure of Lie algebra, <u>the Poisson Lie algebra</u> of the manifold; we have a homomorphism of (N, P) onto L^* since $X_{\{u,v\}} = [X_u, X_v]$.

Avez and myself [1] have determined all the derivations of the Poisson Lie algebra (N,P), that is all the endomorphisms of N which are infinitesimal automorphisms of the bracket. In the non compact case, the derivations are given by suitable first order differential operators; in the compact case, there exist non local derivations. The most useful result is the following

<u>Proposition - The derivations $\mathcal{D}$ of (N,P) which are null on the constants are given by $\mathcal{D}u = \mathcal{L}(X)u$, where X is a symplectic vector field.</u>

2 - Classical Dynamics and Symplectic Manifolds.

a) Consider a dynamical system with time independent constraints and n degrees of freedom. The corresponding configuration space is an arbitrary differential manifold M of dimension n. It is well-known that the cotangent bundle T^*M admits an exact natural symplectic structure defined by the Liouville 2-form which may be written locally in terms of classical variables

$$F = \sum_\alpha dp_\alpha \wedge dq^\alpha$$

For the hamiltonian formalism, a dynamical state of the system is nothing other than a point of $W = T^*M$ which is the usual <u>phase space</u>. The analysis of the equations of Mechanics has showed, from a long time, that it is essential to can introduce change of the variables (p_α, q^α) which does not respect the cotangent structure of W. We are thus lead to introduce as <u>phase space</u> a symplectic manifold (W,F) of dimension 2n .

On this manifold, Dynamics is determined by a function $H \in N$, the hamiltonian of the system, which defines a hamiltonian vector field X_H. A motion of the dynamical system is given, by an integral curve c(t) of the hamiltonian vector field X_H, the parameter t being the time.

Such is the geometrical meaning of the classical equations of Hamilton.

b) We can adopt another viewpoint. We have seen [A] that the space N admits the following two algebraic structures.

1) a structure of associative algebra defined by the usual product of functions (which is here commutative).

2) a structure of Lie algebra given by the Poisson bracket.

The Poisson bracket defines derivations of this product. Consider a family u_t of elements of N satisfying the differential equation

$$du_t/dt = \{H,u_t\} \tag{2-1}$$

and taking the initial value u_o at $t = 0$. We see that the evolution of $u_t \in N$ in the time processes from the integral curves which appear in the first viewpoint; (2-1) may be considered as the intrinsic equation of classical Dynamics.

We have completely described Classical Mechanics in terms of the two laws of composition defined on N. It is natural to study if it is possible to deform these two algebraic laws in a consistent way and in the sense of [A], so that we obtain a tentative model isomorphic to the conventional Quantum Mechanics. The answer is positive.

3 - The $*_\nu$-products and the main example.

a) We have seen ([A]) that a $*_\nu$-product is defined by an associative deformation of the algebra (N,.) of the form :

$$u *_\nu v = u.v + \nu P(u,v) + \sum_{r=2}^{\infty} \nu^r C_r(u,v) \qquad (u,v \in N) \tag{3-1}$$

where the C_r are differential Hochschild 2-cochains which are null on the constants and satisfy the parity assumption :

Parity assumption - C_r is even if r is even, odd if r is odd.

Introduce the space $N^c = C^\infty(W; C)$ and suppose that ν is purely imaginary. The $*_\nu$-product defines a formal associative algebra on the space $E(N^c;\nu)$. The nullity of the cochains on the constants corresponds to the fact that :

$$u *_\nu 1 = 1 *_\nu u = u \qquad (u \in N^c) \tag{3-2}$$

that is that 1 is an unit element for the $*_\nu$-product; the parity assumption is equivalent to the identity :

(3-3) $$\overline{u *_\nu v} = \bar{v} *_\nu \bar{u} \qquad (u,v \in N^c)$$

where $\bar{u}$ is the complex conjugate element of $u \in N^c$.

b) Each symplectic manifold (W,F) admits atlases of charts for which F (or Λ) has constant components (natural charts $\{x^i\}$; $i,j,.. = 1,.. 2n$).

A symplectic connection Γ is a linear connection without torsion such that $\nabla F = 0$, where ∇ is the operator of covariant differentiation defined by Γ. If $\{\Gamma^i_{jk}\}$ are the usual coefficients of a connection Γ in a natural chart $\{x^i\}$, introduce the coefficients $\Gamma_{ijk} = F_{i\ell}\,\Gamma^\ell_{jk}$. Such coefficients $\{\Gamma_{ijk}\}$ define a symplectic connection iff they are completely symmetric for every natural chart. A symplectic manifold admits infinitely many symplectic connections; the difference between two symplectic connections is given by a tensor of type (1,2) deduced from an arbitrary symmetric covariant 3-tensor.

c) Suppose that (W,Λ) admits a symplectic connection without curvature; if such is the case, the manifold (W,Λ,Γ) is sayed to be a flat symplectic manifold. The simplest example is the cotangent bundle of $\mathbb{R}^n$, that is $\mathbb{R}^n \times \mathbb{R}^n$. Introduce on a flat symplectic manifold the bidifferential operator P^r of maximum order for each argument, defined by the following expression on each domain U of an arbitrary chart $\{x^i\}$:

(3-4) $$P^r(u,v)\big|_U = P^r_\Gamma(u,v)\big|_U = \Lambda^{i_1 j_1} \ldots \Lambda^{i_r j_r}\, \nabla_{i_1\ldots i_r} u\, \nabla_{j_1\ldots j_r} v \qquad (u,v \in N)$$

We put $P^o(u,v) = uv$. For $r = 1$, we obtain the Poisson bracket operator P.

Given a formal function $f(z)$ with constant coefficients such that $f(o) = 1$, substitute P^r to z^r in the expression of $f(\nu z)$; we obtain a bilinear map $(u,v) \in N \times N \to u *_\nu v = f(\nu P)(u,v) \in E(N;\nu)$. We wish to choose f so that we define thus a deformation of $(N,.)$. The answer is given by the following :

Proposition - If (W,Λ,Γ) is a flat symplectic manifold, there is only one formal function of the Poisson bracket P (up to a constant factor and a linear change of the deformation parameter ν) that generates a formal deformation of the associative algebra $(N,.)$: it is the exponential function.

We have :

(3-5) $$u *_\nu v = \sum_{r=0}^{\infty} (\nu^r/r!)\; P^r(u,v) = \exp(\nu P)(u,v)$$

which generates the deformation of the Poisson Lie algebra (with $\lambda = \nu^2$)

(3-6) $$[u,v]_\lambda = \sum_{r=0}^{\infty} (\lambda^r/(2r+1)!)\; P^{2r+1}(u,v) = \nu^{-1} \sin h(\nu P)(u,v)$$

It is remarkable that, for $\nu = -i\hbar/2$, we deduce from (3-6) a bracket $\frac{2}{\hbar}\sin(\frac{\hbar}{2}P)$ given in 1949 by Moyal in the context of the Hermann Weyl-Wigner quantization [2] .

Consider the term P^3 of (3-6). If this Chevalley 2-cocycle were exact, it would be the coboundary of a 1-cochain, which may be supposed 3-differential, according to the proposition of [A], §1,c. But it is easy to see that such a coboundary has no term of bidifferential type (3,3); P^3 defines the cohomology 2-class β which is a generator of $H^2(N;N)$. The deformations (3-5) (3-6) are non trivial even for the order 1.

d) The Vey $*_\nu$-products and strong Vey $*_\nu$-products (see [A]) constitute natural generalizations of the notion of Moyal product (3-5). I have given in [A] existence theorems and equivalence theorems concerning the Vey $*_\nu$-products.

But it is useful to obtain effective construction processes of Vey $*_\nu$-products for classes of natural symplectic manifolds. I have given [5] such processes for large classes of cotangent bundles of classical groups or homogeneous spaces. I will limite myself to the simplest example. Consider the flat symplectic manifold defined by the cotangent bundle of $\mathbb{R}^n - \{0\}$, let $E = (\mathbb{R}^n - \{0\}) \times \mathbb{R}^n$.
The solvable group G of dimension 2 acts on E in the following way :

$$(x,y) \in E = (\mathbb{R}^n - \{0\}) \times \mathbb{R}^n \to ((x' = e^\rho x,\ y' = e^{-\rho}(y + \sigma x) \qquad (\rho, \sigma \in R)$$

G leaves invariant the natural symplectic structure of E and the flat connection; it preserves then the P^r's defined by (3-4) and the corresponding $*_\nu$-product defined on E. The space of the orbits of E by G is isomorphic to $T^* S^{n-1}$, where $S^{n-1} = SO(n) / So(n-1)$ is the sphere of dimension (n-1). We deduce from the $*_\nu$-product defined on E a natural strong Vey $*_\nu$-product on $T^* S^{n-1}$ which is invariant under SO(n). It is well-known that the regularized Kepler problem of dimension n admits as phase space the so-called Moser manifold, that is $T^*_0 S^n$ (the cotangent bundle of the sphere without the null section). We have obtained for the Moser manifold a natural invariant $*_\nu$-product.

We can deduce from this quotient process the construction of natural $*_\nu$-products for example for the cotangent bundles of the Stiefel manifolds and of the Grassmann manifolds.

4 - Introduction to a spectral theory and quantization.

a) Come back to the flat symplectic manifold $\mathbb{R}^n \times \mathbb{R}^n$. Under suitable assumptions, Hermann Weyl has defined in this case, in terms of Fourier transform, a map Ω (the Weyl map) which associates with each element u of a large class of classical functions or distributions an operator $\hat{u}$ of a Hilbert space and conversely. The conventional quantization processes in terms of these operators. But the $*_\nu$-product defined by (3-5) corresponds by Ω to the product of operators (for $\nu = - i\hbar/2$). If

$$u * v = \exp\left(\frac{\hbar}{2i} P\right)(u,v) \qquad \text{we have} \qquad \Omega(u * v) = \Omega(u) \cdot \Omega(v)$$

The Moyal bracket is (up to the factor $2\nu = -i\hbar$) the image by Ω^{-1} of the natural commutator of the corresponding operators. We note that if u or v has a compact support, we have :

(4-1) $$\int_W (u * v)\eta = \int_W uv\,\eta$$

where η is the symplectic volume element.
It appears as possible to develop directly Quantum Mechanics in terms of ordinary functions (or distributions) and $*$ - products, without reference to some Ω and to operators, in a complete and autonomous way.

b) Consider a symplectic manifold admitting a $*_\nu$-product. Let H be the classical hamiltonian of our problem. If we consider the value $\nu = -i\hbar/2$ of the parameter of deformation suggested by the Moyal product, we are led to translate the dynamical Heisenberg equation by

(4-2) $$\frac{du_t}{dt} = \frac{i}{\hbar} 2\nu [H, u_t]_{\nu^2} \qquad (u_t \in E(N^c;\nu) \times \mathbb{R})$$

If we put $\tilde{H} = iH/\hbar$, we have

(4-3) $$du_t/dt = \tilde{H} *_\nu u_t - u_t *_\nu \tilde{H}$$

Introduce the $*_\nu$-powers of $\tilde{H}$ ($\tilde{H}^{(*)p} = \tilde{H}^{(*)p-1} *_\nu \tilde{H}$). It is easy to see that it follows from the parity assumption that $\tilde{H}^{(*)p}$ depends only upon the even powers of ν. We can define the $*_\nu$-exponential of $\tilde{H}t$ in the following way :

(4-4) $$\mathrm{Exp}_*(\tilde{H}t) = \sum_{p=0}^{\infty} (t^p/p!)\, \tilde{H}^{(*)p}$$

If $u_o \in E(N^c;\nu)$, define u_t formally by :

(4-5) $$u_t = \mathrm{Exp}_*(\tilde{H}t) *_\nu u_o *_\nu \mathrm{Exp}_*(-\tilde{H}t)$$

(4-5) gives the formal solution of (4-3) taking the value u_o at $t = 0$.

c) Consider now the viewpoint of the mathematical analysis and give to ν the value $-i\hbar/2$. Assume that H is such that, for t in a complex neighborhood of the origin, the right-side of (4-4) converges to a distribution on W denoted by $\mathrm{Exp}_*(\tilde{H}t)$ again. Suppose that, for t fixed in a neighborhood of the origin, $\mathrm{Exp}_*(\tilde{H}t)$ has a unique Fourier-Dirichlet expansion

(4-6) $$\mathrm{Exp}_*(\tilde{H}t) = \sum_{\lambda \in I} e^{\frac{i\lambda t}{\hbar}}\, \Pi_\lambda$$

where I is a set of $\mathbb{C}$ and $\Pi_\lambda \in N^c$. This expansion is similar to the spectral expansion of an operator. It is easy to see that :

(4-7) $$\Sigma\, \Pi_\lambda = 1 \;, \qquad \Pi_\lambda * \Pi_{\lambda'} = \delta_{\lambda\lambda'}\, \Pi_\lambda \;, \qquad H * \Pi_\lambda = \Pi_\lambda * H = \lambda\, \Pi_\lambda$$

and $H = \Sigma\, \lambda\, \Pi_\lambda$; I gives the spectrum of H and Π_λ is the eigenprojector corresponding to $\lambda \in I$. A $\star$ - product is sayed to be <u>non degenerated</u> if, for any $u \in N^c$, $\bar{u} \star u = 0$ on a domain implies $u = 0$ on this domain; it follows from (4-1) that the Moyal $\star$ - product and the $\star$ - products deduced by quotient are non degenerated. If such is the case, <u>the spectrum of each real-valued function admitting a spectral expansion in the sense of (4-6) is real and the corresponding Π_λ are real-valued</u>.

Define N_λ by (see [3])

(4-8) $$N_\lambda = \int_W \Pi_\lambda\, \tilde{\eta}$$

where $\tilde{\eta} = \eta/(2\pi\hbar)^n$, if η is the symplectic volume element. If N_λ is finite, a normalized state ρ_λ is defined by $\rho_\lambda = \Pi_\lambda / N_\lambda$ and N_λ is the multiplicity of the state in the usual Quantum Mechanics sense. More generally, we may consider the Fourier transform in the sense of the distributions

$$\mathrm{Exp}_\star(\tilde{H}t) = \int e^{\frac{i\lambda t}{\hbar}}\, d\mu(\lambda, x)$$

and the support of $d\mu$ will be referred as the spectrum of H.

A <u>state ρ</u> is here a real (pseudo probability) distribution on the phase space normalized by the condition

$$\int_W \rho\, \tilde{\eta} = 1$$

and such that $\rho \star \rho = (1/N)\rho$. The measured value $\langle u\rangle_t$ of the observable u at time t for the state ρ is given by :

$$\langle u\rangle_t = \int_W (u_t \star \rho)\, \tilde{\eta}$$

d) The previous algorithm directly applied to the flat case gives, for the n-dimensional harmonic oscillator, the correct energy levels and multiplicities. For the Hydrogen Atom, we may consider the Moser-Kepler manifold $T_o^\star S^3$ as phase space and introduce the corresponding $\star$ - product invariant under SO(4). We obtain then the complete spectrum, that is the negative discrete spectrum and the positive continuous spectrum [3] .

5 - The group of the automorphisms of a $\star_\nu$-product [8] .

a) <u>Suppose that (W,F) admits a $\star_\nu$-product</u>. Consider an automorphism of the space $E(N;\nu)$

$$A_\nu = A_o + \sum_{s=1}^{\infty} \nu^s A_s$$

where A_o is an automorphism and the $A_s (s \geqslant 1)$ are endomorphisms of the space N; A_ν

is an automorphism of the $\star_\nu$-product if we have :

(5-1) $$A_\nu(u \star_\nu v) = A_\nu u \star_\nu A_\nu v$$

If $A_o = Id.$, we say that A_ν has a trivial main part. Let $\hat{E}(N;\nu)$ the subset of $E(N;\nu)$ defined by the elements $a_\nu = \Sigma \nu^s a_s$ such that $a_o > 0$; each element $a_\nu \in \hat{E}(N;\nu)$ admits a unique inverse in the sense of the $\star_\nu$-product. An inner automorphism of $\star_\nu$ is given by $u \to a_\nu \star_\nu u \star_\nu a_\nu^{(\star)-1}$ and its main part is trivial.

b) Let Symp.(W,F) be the group of all the symplectomorphisms of (W,F) and $Symp_c(W,F)$ its component of the identity connected by piece wise differentiable arcs. We denote by K the space of the differential operators null on the constants. Suppose that (W,F) admits a formal Lie algebra null on the constants. It is possible to show that for each $\sigma \in Symp_c(W,F)$, there is an automorphism of the formal Lie algebra of the form :

(5-2) $$A_\nu = (Id + \sum_{s=1}^{\infty} \nu^{2s} B_{2s})\sigma^\star \qquad (B_{2s} \in K;\ \lambda = \nu^2)$$

c) It is possible to prove that each automorphism of the $\star_\nu$-product has necessarily the form :

(5-3) $$A_\nu = (Id + \sum_{s=1}^{\infty} \nu^s B_s)\ \sigma^\star \qquad (B_s \in K)$$

where σ is a symplectomorphism. Conversely, using the uniqueness theorem of [A] §3, we can prove by recursion the following

Proposition - The group $Aut_t(\star_\nu)$ of the automorphisms of the $\star_\nu$-product which have a trivial main part coincides with the group of the automorphisms of the corresponding Lie algebra which have the form

$$A_\nu = Id. + \sum_{s=1}^{\infty} \nu^s A_s \qquad (A_s \in K)$$

If $b_1(W) = 0$, $Aut_t(\star_\nu) = Aut_i(\star_\nu)$, where $Aut_i(\star_\nu)$ is the group of the inner automorphisms.

It is possible to deduce from the study of the even automorphisms with respect to ν.

Theorem - The group $Aut(\star_\nu)$ of the automorphisms of the $\star_\nu$-product coincides with the group of the automorphisms of the corresponding Lie algebra which have the form (5-3).

The group $Aut(\star_\nu)/\ Aut_t(\star_\nu)$ is thus isomorphic to a subgroup H of Symp.(W,F). It follows from b that $Symp_c(W,F) \subset H$. It is possible to conjecture that H = Symp(W,F).

c) Suppose that our $\star_\nu$-product is a Vey $\star_\nu$-product : there is a unique symplectic connection Γ such that $Q^2_\nu = P^2_\Gamma + \tilde{\partial}H$, where H is a differential operator of order $\leqslant 2$ ([A] §4,c). The study of the automorphisms which are independent of ν gives the following :

Proposition - The group of the automorphisms of a Vey $*_\nu$-product which are independant of ν is a closed subgroup of the group of the symplectomorphisms which are affine for the connection Γ. It is thus a finite dimensional Lie group.
In the Moyal case, this group is the complete affine symplectic group. These results on the automorphisms clarify the study of the invariance problems of Quantum Mechanics.

References.

[1] A. Avez and A. Lichnerowicz. C.R. Acad. Sci. Paris 275 A, (1972), p 11-14.
[2] J.E. Moyal . Proc. Cambridge Phil. Soc. 45 (1949), p 99-124.
[3] F. Bayen, M. Flato, C. Fronsdal, A. Lichnerowicz, D. Sternheimer, Lett. in Math. Phys. 1 (1977), p 521-530; Ann. Of Physics 111, (1978), p 61-152.
[4] J. Vey , Comm. Math. Helv. 50, (1975), p 421-454.
[5] A. Lichnerowicz, Lett. in Math. Phys. 2, (1977), p 133-143.
[6] H. Weyl , The theory of Groups and Quantum Mechanics, Dover New-York (1931).
[7] E.P. Wigner, Phys. Rev. 40, (1932), p 749.
[8] A. Lichnerowicz, C. R. Acad. Sci. Paris 286 A, (1978), p 49-53; Ann. di Matem. pura e appl. (to appear).

Stability Theory and Quantization

I. E. Segal*
Massachusetts Institute of Technology
Invited Address, Conference on Differential Geometrical Methods in Mathematical Physics, University of Salamanca
September 10-14, 1979

1. Introduction. It is a mathematical commonplace that algebraic manifolds have defined on them canonical differential geometric structures, which have played an important part in their theory since the work of Hodge. It is a physical commonplace that symplectic structures and hermitian forms play a fundamental role in quantization. Formally, if $\varphi(x)$ is a hermitian quantum field, the commutator $[\varphi(x),\varphi(y)]$ is an antisymmetric generalized function of x and y which is appropriately construed as the kernel of a symplectic form; and the real part of the vacuum expectation value $\langle\varphi(x)\varphi(y)v,v\rangle$ (v = vacuum state vector),- the imaginary part being determined by the commutator,- is a symmetric generalized function of x and y, similarly usefully construed as the kernel of a symmetric form. Indeed, putting together these two forms in the case of the Klein-Gordon field, $\Box\varphi + m^2\varphi = 0$, one obtains the unique Lorentz-invariant hermitian structure on the solution manifold of the classical equation.

Our aim is to show that there is a natural and useful way to combine these familiar circumstances in the case of wave equations capable of representing non-trivial interactions. In so doing we make essential use of forthcoming work by and with Stephen Paneitz combining a development of invariant causality theory in Lie groups with stability theory for differential equations.

In particular, we indicate the establishment of an hermitian structure which serves formally, and rigorously in certain interesting cases, to determine a canonical functional integral over the solution manifold, which is instrumental in quantization.

---

*Research supported in part by the NSF.

2. Technical preliminaries and the symplectic structure. I shall recall briefly the development of the symplectic structure in a form which serves to establish notation.

The simplest and prototypical wave equations are of the form

$$\Box\varphi + F(\varphi) = 0$$

$$\Box\varphi + V(\vec{x},t)\varphi = 0 \;;$$

the first has the complication of nonlinearity, but the advantage of temporal (and even Lorentz) invariance; the second is linear, but its time-dependence has hitherto rendered its unique quantization elusive. Both of these equations, as well as much more general ones, such as the Yang-Mills equation, may be put in the form

$$(*) \qquad u'' + B^2u + K_t(u) = 0 \;,$$

where $u(t)$ takes its values in a real Hilbert space $\mathcal{H}$; B is a non-negative self-adjoint operator in $\mathcal{H}$; and K_t is for each t a given operator (possibly non-linear) on $\mathcal{H}$. In cases such as the Yang-Mills equations there is given in addition a constraint

$$(**) \qquad \Phi(u,u') = 0$$

on the Cauchy data, of a form that is conserved by the temporal evolution defined by the equation (*).

For generality and conciseness, we shall work with equations of the form (*), and assume, as in practice is satisfied, that the mapping $(t,y) \rightarrow K_t(u)$ is smooth from $R^1 \times H^{(n)}$ to H , where $H^{(n)}$ denotes the Sobolev-type space with norm

$$\|x\|_n^2 = \|x\|^2 + \|Bx\|^2 + \ldots + \|B^nx\|^2 \;.$$

There is then no difficulty is solving uniquely the Cauchy problem for (*) locally in time for arbitrary data in suitable Sobolev-type spaces; for the moment, the question of global existence does not intervene. (The latter is because the symplectic structure is determined locally,- unlike the hermitian structure, which is extremely non-local.)

Setting $\mathcal{M}$ for the totality of solutions of equation (*) in the more precise and cogent form of an integral equation:

$$(*') \quad u(t) = \cos(t-t_o)B\,u(t_o) + \frac{\sin(t-t_o)B}{B}u'(t_o) + \int_{t_o}^{t} \frac{\sin(t-s)B}{B} K_s(u(s))ds \quad ;$$

the map $t \to u(t)$ is required to be continuous into the Sobolev-type space in question. $\mathcal{M}$ forms a species of manifold, whose grammatical description is a useful and necessary exercise which however for brevity we shall forego, except to note that its tangent spaces are naturally definable by its first-order variational equations. Thus, for any point $u(\cdot) \in \mathcal{M}$, the tangent space consists of all solutions $V(\cdot)$ of the equation

$$v(t) = \cos(t-t_o)Bv(t_o) + \frac{\sin(t-t_o)B}{B} v'(t_o) + \int_{t_o}^{t} \frac{\sin(t-t_o)B}{B} ((\partial_{u(s)}K_s(u(s)))v(s)ds$$

where $(\partial_u K)v$ denotes the differential of K evaluated at v, i.e. $\lim \epsilon^{-1}\{K(u+\epsilon v) - K(u)\}$.

The canonical symplectic structure Ω on $\mathcal{M}$ may now be defined explicitly by the equation

$$\Omega_{n(\cdot)}(v_1(\cdot),v_2(\cdot)) = \langle v_1(t),v_2'(t)\rangle - \langle v_1'(t),v_2(t)\rangle \quad ;$$

the apparent dependence of the right-hand side on t is specious. With this definition Ω becomes a closed, non-degenerate (modulo the gauge group, in the case of the Yang-Mills equation; in fact, the gauge group may be defined by this condition, as having for its infinitesimal form precisely those vector fields on $\mathcal{M}$ in the radical R of Ω,- where $R = [x: \Omega(x,y) = 0, \forall y]$). Ω is conserved by temporal evolution, and indeed defines it, relative to an appropriate hamiltonian, which in practice can be given explicitly.

In the case of conformally invariant equations, such as $\Box \varphi + \varphi^3 = 0$, or the Yang-Mills equation, Ω is also conformally invariant; for Lorentz-invariant equations, Ω is Lorentz-invariant; and so on. It plays a fundamental role both from the standpoint of classical mechanics, being the canonical 'bilinear covariant', usually expressed as $\sum_i dp_i \wedge dq_i$ in the finite-dimensional case of a cotangent bundle; and the standpoint of quantum mechanics, in determining the form of the commutator of the corresponding quantized field (in the case of non-linear equation, the commutator at equal times,- the quantized formalism otherwise remains somewhat nebulous).

It is clear however from the consideration of linear equations that Ω does not include all fundamentally relevant information; it gives no means of describing the 'vacuum', or of designating 'creation or annihilation' operators. Relatedly, in the infinite-dimensional case Ω does not really determine an appropriate invariant measure in phase space; unlike the finite-dimensional case, there seems to be no question but that formal integrals based on Ω (although these abound in the heuristic literature) are inherently devoid of mathematical meaning, in the absence of additional structure. In its simplest form the point is that no useful meaning as a measure space can be given to an infinite product of infinite measure spaces.

In fact, in functional integration in its mathematically precise form, the Gaussian rather than euclidean measures play the fundamental role, as first shown by Wiener in connection with Brownian motion, and now doubly established by virtue of the highly developed connection with quantum field theory. To determine infinitesimally a Gaussian measure in the solution manifold one needs a Riemannian rather than a symplectic structure, and it is to the question of the canonical determination of the former that we now turn. For some other aspects of the symplectic structure see [1].

3. The hermitian structure. Before undertaking the general treatment, it should be helpful to consider a prototypical and important special case: that of the time-dependent equation

$$(\#) \qquad \Box\varphi + m^2\varphi + V(\vec{x},t)\varphi = 0 \quad ,$$

where $V(\vec{x},t)$ is a given function. The symplectic structure just treated yields in particular the 'commutator function', $D(x,x')$, so-called because in addition to serving as a kernel for the symplectic structure, it represents the quantized field commutator:

$$(CR) \qquad [\tilde{\varphi}(x),\tilde{\varphi}(y)] = iD(x,x') \quad ,$$

where $\tilde{\varphi}$ denotes a 'quantization' of the field defined by equation (#). It follows that bona fide self-adjoint operator-valued distributions φ exist that satisfy (#) and (CR); but there is no unicity about them. When $V(\vec{x},t)$ is time-independent, say $V(\vec{x})$, unicity can be attained from imposition of the further constraint of 'positive energy'. But 'energy' makes sense in this connection only when the underlying manifold admits a group of translations in time; and this is precisely what is lacking in the time-dependent case.

Despite the considerable literature which has grown up about this question,- mostly in the related form of the quantization of canonical equations (Klein-Gordon, Dirac, etc.) on convex space-times, which has been cogently developed by Lichnerowicz and his students,- the issue has been unresolved. Although from the standpoint of C*-algebra quantum field theory there is some unicity,- there is a unique one-parameter family of automorphisms defining the temporal evolution in the C*-algebra generated by the bounded functions of finitely many field variables (i.e. the space-time averaged $\tilde{\varphi}(x)$),- there is no distinguished state of this algebra, or distinguished representation of the algebra on a Hilbert space, such as appears necessary for a comprehensive physical analysis, in the literature.

It might even appear that none exists; but I want to show that a combination of scattering, stability, and group theory considerations lead to a direction of resolution that is natural and an essential step towards the treatment of non-linear equations.

Physically, the key idea is that the vacuum should be

invariant under the scattering transformation, as emphasized e.g. in [2]. But viability of this idea in the present context depends on the validity of existence and unicity results only recently developed by S. Paneitz and myself. For brevity and logical clarity, a simple illustrative, but not at all optimal, result will be quoted which is

Theorem 1. Suppose that on the real Hilbert space $\mathcal{R}$, the self-adjoint operator $V(t) \geq 0$, that $\int \|C^{-1}V(t)C^{-1}\|_2\,dt$ is sufficiently small, where $C = \sqrt{B}$ $(B > \epsilon I > 0)$ and $\|\cdot\|_2$ denotes the Hilbert-Schmidt norm; and that $V(\cdot)$ is a continuous function of compact support.

There is then a complex-Hilbert-space structure on the solution manifold of the equation

$$(*) \qquad u'' + B^2u + V(t)u = 0$$

in its earlier-given integrated form, with $u(t)$ and $u'(t)$ constrained to lie in $\mathcal{D}_C$ and $\mathcal{D}_{C^{-1}}$ respectively, such that

(1) The imaginary part of the inner product determines the symplectic structure earlier given (i.e. $\mathrm{Im}\langle u,v\rangle = \langle u,v'\rangle_{\mathcal{R}} - \langle u',v\rangle_{\mathcal{R}}$).

(2) The scattering operator $\mathbf{S}$ is unitary with respect to the complex $\langle\cdot,\cdot\rangle$; and this Hilbert space structure is essentially unique, (i.e. within the sign of its complex structure).

The proof depends partly on the exploitation of causality ideas in the symplectic group, which use the non-negativity of $V(t)$ to define an arc in the group whose forward tangent is for all t within the 'positive' cone of the symplectic Lie algebra, together with stability theory ideas generalizing some aspects of the work of M. G. Krein.

As a corollary, one obtains a unique full 'quantization' for the equation $(*)$. This includes notably:

(1) The quantum field state vector space K. This consists of all holomorphic square-integrable functionals on the complex Hilbert space, say $\mathcal{L}$, formed by the solutions of equation $(*)$.

(ii) An operator-valued quasi-distribution satisfying both equation (*) and the canonical commutation relations, in senses described in the literature. ('Quasi-' refers to the utilization of test functions in an abstract Hilbert space, rather than defined on space-time.)

(iii) A unitary temporal evolution propagator $\Gamma(t,t')$ on K, which carries the field at time t' into that at time t, and satisfies corresponding conditions.

(iv) A vacuum vector v in K, invariant under the induced scattering operator on the quantum field, and cyclic for the field operators.

3. Non-linear wave equations. The preceding section applies indirectly to non-linear wave equations; at each point of the solution manifold, the first-order variational equation is approximable by one of the form just considered. The scattering operator for this linear equation is simply the differential of the non-linear scattering transformation for the non-linear equation.

Again, for brevity and simplicity, our illustration avoids generality and significant complications which however are presently extravacant to this and we treat the equation

$$u'' + B^2 u + g\kappa(t)K(u) = 0 ,$$

where $\kappa(t)$ is a non-negative C^* function of t of compact support, and $d_{n(t)}K$ is an operator of the type $V(t)$ considered in the previous section.

Theorem 2. *If g is sufficiently small, each tangent space to the solution manifold admits a unique smooth hermitian structure, whose imaginary part is the symplectic structure for the first-order variational equation.*

In particular, a distinguished Riemannian structure in $\mathfrak{M}$ is obtained; in each tangent space there is a corresponding Gaussian measure; the complex of these measures appears to serve as a natural basis for functional integration considerations on the entire manifold, in the direction developed by L. Gross [3].

Moreover, since the hermitian structure is determined in a canonical fashion by the S-operator, it is invariant under automorphisms of the manifold which leave S fixed. This means that when the temporal cutoff $\kappa(t)$ is removed,- in practice, a formidable technical problem, but one on which much progress has been made in the past two decades,- the resulting hermitian structure (and associated structures) will be temporally invariant. Further in the case of a Lorentz-invariant wave equation, for which the S-operator is expected to be invariant,- and as confirmed by Morawetz and Strauss [4] in interesting cases,- the hermitian structure will also be Lorentz invariant. It remains however, among other matters, to show that the differentials of these S-operators have positions in the infinite symplectic group sufficiently close to the identity to admit a unique invariant hermitian structure. This involves in addition to a suitable modification and extension of the Floquet-Liapounoff-Krein stability theory (instead of periodic functions satisfying equations whose coefficients are bounded operators, one has functions satisfying boundary conditions at $\pm\infty$ whose coefficients are unbounded operators), decay estimates for nonlinear wave equations along lines developed by myself and Strauss, among others.

Detailed accounts of the work cited and of ramifications will be given by Paneitz [5], and in a joint publication.

References

1. I. Segal (1974), Symplectic structures and the quantization problem for wave equations. Symposia Mathematica vol. XIV, pp. 99-117.

2. I. Segal (1959). Foundations of the theory of dynamical systems of infinitely many degrees of freedom, I. Mat.-fys. Medd. K. Danske Vidensk. Selsk. 31, no. 12, pp. 1-38.

3. L. Gross (1963), Harmonic analysis on Hilbert space, Mem. Amer. Math. Soc. No. 46, et seq.

4. C. Morawetz and W. A. Strauss (1973). Asymptotics of a nonlinear relativistic wave equation. Proc. Symp. Pure Math. vol. 23, Amer. Math. Soc., Providence, R.I.

5. S. M. Paneitz (1980). Doctoral dissertation, M.I.T., work in progress.

PRESYMPLECTIC MANIFOLDS AND THE QUANTIZATION OF RELATIVISTIC PARTICLE SYSTEMS

by

C. Günther

Dept. of Mathematics
TU Berlin, Germany

Abstract: Dynamics of classical and relativistic particle systems is described on the evolution space, which is a presymplectic manifold. Hamiltonian formalism on presymplectic manifolds is investigated and a prequantization procedure for presymplectic manifolds is constructed. A general Pre-Klein-Gordon equation is defined and the prequantization of classical and relativistic particles discussed. The problem of quantizing presymplectic evolution spaces is for (topologically) regular systems reduced to the symplectic case.

Contents:

0. Introduction

In the Hamiltonian formulation of classical Galilean mechanics (e.g. [1],[7]) a system is represented by its phase space, which is a smooth symplectic manifold (N,ω_n).
Introducing time into the description leads to the Cartan formulation (see [1] or [7]) on the evolution space $M = \mathbb{R} \times N$, $\mathbb{R}$ being the time coordinate. The dynamics of the system is in this formulation given by a presymplectic form ω on M and the time function $t : M \longrightarrow \mathbb{R}$, t being just the projection onto the first factor of $\mathbb{R} \times N$. ω determines the direction and t the parametrization of the trajectories describing the evolution of the system. In this formalism the phase space $N \simeq \{t_0\} \times N \subset M$ represents all simultaneous states at a time t_0.
In a relativistic theory no invariant time coordinate exists and the concept of simultaneous states fails. Therefore a description of dynamics on the phase space is not possible. The system has to be described on the evolution space, which is a presymplectic manifold. The directions of the trajectories describing the evolution of the system are given by the symplectic structure. Any uniform parametrization of the trajectories is equivalent to a time structure (see below), i.e. the introduction of a global observer (see [6]), which is not a covariant concept.
The geometric quantization procedure of Kostant/Souriau [3] ,[7] is based on the phase space formulation of mechanics. In order to obtain a covariant quantization procedure for relativistic systems, the Kostant/Souriau procedure has to be extended to evolution spaces, i.e. presymplectic manifolds.
This generalization is primarily performed at the first step of quantization, the prequantization. The second step, involving polarizations remains in principle the same as in the symplectic case.
In the following we present a prequantization procedure for presymplectic manifolds and apply it to the evolution space of a particle without spin (in external field) in space-time. A general Pre-Klein-Gordon equation is formulated and we show that a time structure which induces a uniform parametrization of the classical trajectories transforms the Klein-Gordon equation into the Schrödinger equation.
Some basic facts on presymplectic geometry are given in the first section. We will use the results on prequantization in [2] and our notation will also be found in [2] .

1. Presymplectic Geometry

A closed two form $\omega \in \Omega^2 M$ on a smooth finite dimensional manifold is called a presymplectic form, iff dim ker $\omega =: k$ is constant on M. In this case $CM := \ker\omega$ is a k-dimensional smooth involutive subbundle of TM and defines a fibration of M into k-dimensional submanifolds. This fibration is called the characteristic fibration. (M,ω) is called a regular presymplectic manifold, iff $\underset{\sim}{M}$ = set of all leaves has a smooth manifold structure inducing the quotient topology on $\underset{\sim}{M}$. In this case the natural projection $p : M \to \underset{\sim}{M}$ is a surjective submersion of constant rank and we have an exact sequence of vector bundles:

$$0 \longrightarrow CM \longrightarrow TM \longrightarrow p^*T\underset{\sim}{M} \longrightarrow 0$$

In addition there exists a unique symplectic form $\underset{\sim}{\omega} \in \Omega^2\underset{\sim}{M}$ on M with $p^*\underset{\sim}{\omega} = \omega$.

As in the symplectic case it is possible to develop a canonical formalism on presymplectic manifolds:

1.1 Definition: Let (M,ω) be a presymplectic manifold.

$\mathcal{C}M := \{X \in \mathcal{X}M \mid X \lrcorner \omega = 0\}$ the characteristic vectorfields

$\mathcal{H}_\ell M := \{X \in \mathcal{X}M \mid \mathbf{L}_X\omega = 0\}$ the local Hamiltonian vectorfields

$\mathcal{H}M := \{X \in \mathcal{X}M \mid X \lrcorner \omega$ is exact$\}$ the global Hamiltonian vectorfields

$\mathcal{F}hM := \{f \in \mathcal{F}M \mid X \lrcorner df = 0$ for all $X \in \mathcal{C}M\}$ the Hamiltonian functions

$\left.\begin{array}{l} \omega^b : TM \longrightarrow T^*M,\ v_m \longmapsto \omega(v_m,\cdot) \\ .^b : \mathcal{X}M \longrightarrow \Omega^1 M,\ X \longmapsto X \lrcorner \omega \end{array}\right\}$ the musical isomorphismus

The following results are simple consequences of this definition:

1.2 Lemma

a) $\mathcal{C}M \subset \mathcal{H}M \subset \mathcal{H}_\ell M \subset \mathcal{X}M$, $\mathcal{C}M$ is $\mathcal{F}M$-submodul; $\mathcal{H}M$, $\mathcal{H}_\ell M$ are $\mathbb{R}$-subspaces } of $\mathcal{X}M$

b) $\mathcal{H}M$, $\mathcal{H}_\ell M$ are Lie-subalgebras of $\mathcal{X}M$

c) $\mathcal{C}M$, $\mathcal{H}M$ are Lie-ideals in $\mathcal{H}_\ell M$ (not in $\mathcal{X}M$).

d) $f \in \mathcal{F}hM \iff f$ constant on all characteristic fibres $\iff df \in .^b(\mathcal{X}M)$

1.3. Definition:

a) $\underset{\sim}{\mathcal{H}}M := \mathcal{H}M/\mathcal{C}M$ is because of 1.2.b) a Lie algebra and the projection $p : \mathcal{H}M \to \underset{\sim}{\mathcal{H}}M$ is a Lie algebra morphism.

b) For $f,g \in \mathcal{F}hM$ define $\{f,g\} := X \lrcorner Y \lrcorner \omega$, for any $X \in \mathcal{X}M$ with $X \lrcorner \omega = df$ and $Y \in \mathcal{X}M$ with $Y \lrcorner \omega = dg$, $\{f,g\}$ is called the Poisson brackets of f and g and $\{.,.\}$ defines on $\mathcal{F}hM$ a Lie algebra structure.

c) $\underset{\sim}{G} : \mathcal{F}hM \to \underset{\sim}{\mathcal{H}}M$, $f \longmapsto \underset{\sim}{G}_f := \{X \in \mathcal{X}M \mid X \lrcorner \omega = df\}$

$\underset{\sim}{G}$ is a Lie algebra morphism.

1.4 Theorem: Let (M,ω) be a presymplectic manifold, Then the sequence of Lie algebras

$$0 \longrightarrow \mathbb{R} \longrightarrow \mathcal{F}hM \longrightarrow \underset{\sim}{\mathcal{H}}M \longrightarrow 0$$

is exact and a central extension of $\underset{\sim}{\mathcal{H}}M$ by $\mathbb{R}$.

Presymplectic manifolds (M,ω) can at least locally be interpreted as fibered manifolds over a symplectic base. In this interpretation $\mathcal{C}M$ is the $\mathcal{F}M$-modul of all vertical vectorfields and $\mathcal{F}hM$ is the ring of smooth functions on the base manifold. The operation $\underset{\sim}{G}$ corresponds to the construction of a Hamiltonian vectorfield from its Hamiltonian function and 1.4. is the natural generalization of the exact sequence for symplectic manifolds.
In the regular case we obtain therefore Lie algebra isomorphisms $\mathcal{F}hM \longrightarrow \mathcal{F}\underset{\sim}{M}$, $\underset{\sim}{\mathcal{H}}M \longrightarrow \mathcal{H}\underset{\sim}{M}$ and the diagram of Lie algebras

$$\begin{array}{ccccccccc} 0 & \longrightarrow & \mathbb{R} & \longrightarrow & \mathcal{F}hM & \longrightarrow & \underset{\sim}{\mathcal{H}}M & \longrightarrow & 0 \\ & & \| & & \downarrow & & \downarrow & & \\ 0 & \longrightarrow & \mathbb{R} & \longrightarrow & \mathcal{F}\underset{\sim}{M} & \longrightarrow & \mathcal{H}\underset{\sim}{M} & \longrightarrow & 0 \end{array}$$

is commutative with exact lines.

A smooth map $F : (M,\omega) \longrightarrow (M',\omega')$ is said to be presymplectic, iff $F^*\omega' = \omega$. Presymplectic maps leave the characteristic fibration invariant and we have $TF(CM) \subset CM'$, $\ker TF \subset CM$, $F^*(\mathcal{F}hM') \subset \mathcal{F}hM$ for any presymplectic F.

We define:
$\mathrm{Pres}(M,\omega) := \{F \in \mathrm{Diff}(M) \mid F^*\omega = \omega\}$ the group of presymplectomorphisms
$\mathrm{Pres}^{o}(M,\omega) := \{F \in \mathrm{Diff}(M) \mid F$ leaves each char. fibre invariant$\}$
$\mathrm{Pres}^{o}(M,\omega)$ is a normal subgroup of $\mathrm{Pres}(M,\omega)$.
$\underset{\sim}{\mathrm{Pres}}(M,\omega) := \mathrm{Pres}(M,\omega)/\mathrm{Pres}^{o}(M,\omega)$.

For $X \in \mathcal{H}_e M$ the flow of X consists of local presymplectomorphisms.
For $X \in \mathcal{C}M$ the flow of X consists of local morphisms leaving each characteristic fibre invariant.
Therefore formally $\mathcal{H}_e M$ is the Lie algebra of $\mathrm{Pres}(M,\omega)$ and $\mathcal{C}M$ the Lie algebra of $\mathrm{Pres}^{o}(M,\omega)$, and the Lie algebra of $\underset{\sim}{\mathrm{Pres}}(M,\omega)$ is $\mathcal{H}_e M$.

If (M,ω) is regular, then every presymplectomorphism F induces a symplectomorphism $\underset{\sim}{F}$ on $\underset{\sim}{M}$. We have $F \in \mathrm{Pres}^{o}(M,\omega)$ iff $\underset{\sim}{F} = 1_{\underset{\sim}{M}}$.

A subbundle BM of TM is called a complementary bundle, iff $BM \oplus CM = TM$. For regular (M,ω) complementary bundles are connections of the fibered manifold $p : M \longrightarrow \underset{\sim}{M}$.
By partition of unity, on any presymplectic manifold (M,ω) a complementary bundle can be constructed.

A k-form $\alpha \in \Omega^k M$ on a presymplectic manifold (M,ω) is called a parameter form iff $\ker \alpha$ is a complementary bundle. A parameter form induces uniformly a volume on all characteristic leaves.

A presymplectic manifold (M,ω) is called an <u>almost contact manifold</u> iff $\dim \ker \omega = 1$.

<u>1.5. Definition</u>: Let (M,ω) be an almost contact manifold.
A 1-form $\eta \in \Omega^1 M$ is called a <u>time form</u> iff η is closed and $\ker \eta \oplus \ker \omega = TM$, i.e. η is an orientation form.

On an almost contact manifold $\mathcal{C}M$ is a local one dimensional $\mathfrak{F}M$-modul.
A time form η determines uniquely a vectorfield $Z \in \mathfrak{X}M$ with

$$Z \lrcorner \omega = 0 \quad \text{and} \quad Z \lrcorner \eta = 1$$

Locally a time form η can be written $\eta = dt$ with $t \in \mathfrak{F}M$, the time function. Then the condition $Z \lrcorner \eta = 1$ can be written as $\mathbb{L}_Z t = 1$, i.e. t is the parameter function for the integral curves of Z.

Therefore a time defines locally a uniform parametrization of the characteristic leaves, the time function being the parameter.
Furthermore M is locally a product of a real line and symplectic manifold:

For each $m \in M$ there exists an open neighbourhood $U \subset M$ with $U = I_\varepsilon \times V$, $I_\varepsilon \subset \mathbb{R}$, V spl.
I_ε is leaf of $\ker \omega$ and V is leaf of $\ker \eta$
and $t : I_\varepsilon \times V \longrightarrow \mathbb{R}$ is just the projection

2. <u>Classical and relativistic Particle Dynamics</u>

In this section we want to demonstrate how the dynamics of particle systems is described by presymplectic manifolds. We present the examples of Galilean mechanics and of a relativistic particle without spin.
For further examples and alternative descriptions compare [7] and [8].

a) <u>Classical time dependent mechanics</u> [1], [7] .

Let (N,ω_N) be a symplectic manifold, the <u>phase space</u> of the system.
Define $M := \mathbb{R} \times N$ with the projections $P_N : \mathbb{R} \times N \longrightarrow N$ and $t : \mathbb{R} \times N \longrightarrow \mathbb{R}$. M is called the <u>evolution space</u> of the system and t represents the <u>time</u>.
Define $\omega_o := P_N^* \omega_N \in \Omega^2 M$. Then we have

(M,ω_o, dt) is an almost contact manifold with time form

(i) The dynamics of the system is given by a <u>time dependent Hamiltonian</u> $f \in \mathfrak{F}M$.

Let $f_t := f|_{\{t\}\times N}$ and $X_t \in \mathfrak{X}N$ be defined by $X_t \lrcorner \omega_N = df_t$,

i.e. X_t is the Hamiltonian system on N at the time t.
Define $X_f \in \mathfrak{X}M$ by $X_f(t,n) = (0, X_t(n)) \in T\mathbb{R} \times TN \cong TM$
X_f is a $\mathbb{R}$-parameter family of Hamiltonian systems on M.

The law of classical dynamics can now be expressed:

The trajectories of a classical mechanical system in the evolution space M are given by the vectorfield

$$\boxed{Z_f := X_f + \frac{\partial}{\partial t}}$$

(ii) A formulation independent of the product structure $M = \mathbb{R} \times N$ can be given by using the Cartan form

Define: $\omega_f := \omega_o + df \wedge dt$

then we have the proposition:

(M, ω_f, dt) is an almost contact manifold with time form and the vectorfield Z_f is uniquely determined by the equations:

$$\boxed{\begin{aligned} Z_f \lrcorner\, \omega_f &= 0 \\ Z_f \lrcorner\, dt &= 1 \end{aligned}}$$

(iii) Finally we can describe the system in the homogeneous formalism.

Define $M_{\mathbb{R}} := \mathbb{R} \times M$ the homogeneous evolution space of the system with the projections $P_M : \mathbb{R} \times M \longrightarrow M$ and $s : \mathbb{R} \times M \longrightarrow \mathbb{R}$ and the injection $i_o : M \longrightarrow M_{\mathbb{R}}$, $m \longmapsto (0,m)$.

Define $\Omega := P_M^* \omega_f + ds \wedge d(t \circ P_M)$

Ω is a symplectic form and we have:

The vectorfields $-ds^{\#}$ and Z_f ($\#$ taken by Ω) are i_o-related.

i.e. $-s$ is the homogeneous Hamiltonian of the system and $M = s^{-1}(0)$ is an energysurface of $-s$.

b) General Relativistic Particles [5] , [6] , [7]

Let (B,<•,•>) be a 4-dimensional pseudoriemannian manifold. B represents the space-time of general relativity, <•,•> describes the gravitational field.

(i) Dynamics of relativistic particles is usually described in the homogeneous formalism:

Let $\dot{T}^*B := T^*B \setminus$ zero section be the homogeneous evolution space, Ω a symplectic form on $\dot{T}^*B$.

Remark: For pure gravitational forces is $\Omega = \Omega_o$ the natural symplectic form on $\dot{T}^*B$.
In the presence of an <u>electromagnetic field</u> $F \in \Omega^2 B$ we define $\Omega = \Omega_o + e \cdot \tau^* F$ with $\tau : \dot{T}^*B \longrightarrow B$ the natural projection.
e is the <u>charge</u> of the described particle.

The homogeneous Hamiltonian s is given by

$s(v) = |\langle v,v\rangle^*|^{1/2}$, with $\langle \cdot\cdot\rangle^*$ the pseudoriemannian structure induced on the vectorbundle T^*B.

The trajectories of the system are for $s \neq 0$ given by the vectorfield $-ds$.

(ii) By restriction on energy hypersurfaces of s we get the Cartan formulation of relativistic particles:

Define: $S^m B := s^{-1}(m)$ for $m \in \mathbb{R}$, the <u>mass-shell-bundle</u>

$S^m B$ is a bundle with 3-dimensional fibres over B. The cases $m = 0$ or $m < 0$ are possible in this formulation. This is an advantage against the homogeneous formulation.

With the restriction of Ω the $S^m B$ are almost contact manifolds and the characteristic fibres of $S^m B$ give the direction of the trajectories.
There is no special parametrization of the trajectories required, since a parametrization has no direct physical interpretation.

For $\Omega = \Omega_o$ and $m \neq 0$ the fundamental 1-form θ gives a parametrization by the postulate $Z \lrcorner\, \theta = 1$.
This gives the parametrization of the 'Eigenzeit' and coincides with the parametrization of the homogeneous formalism. However, there is no uniform parameter for all trajectories and for $m = 0$ this parametrization fails.

3. Prequantization of Presymplectic Manifolds

In order to obtain a quantization procedure for relativistic particles, quantization has to be defined for evolution spaces. In this section we construct a prequantization procedure for general presymplectic manifolds.
We will use the results on prequantization presented in [2]. For

the notations we refer to [2], too.

3.1. Definition: Let (M,ω) be a presymplectic manifold.
A principal-circle-bundle (L^c,α^c,M) over M with connection α^c is called a (presymplectic) prequantum bundle (PQB) iff:

$$\omega = -\mathrm{curv}\ \alpha^c$$

Remark: By the results of [2] we have an equivalence between:

(i) Principal-circle-bundles with connection (L^c,α^c)

(ii) Principal-$\dot{\mathbb{C}}$-bundles with connection $(\dot{L},\alpha)$ $(\dot{\mathbb{C}} := \mathbb{C}\setminus\{0\})$

(iii) Line bundles $(L,\nabla,\langle\cdot\cdot\rangle)$ with covariant derivative and ∇-affine Hermite structure.

We define:

$$\mathcal{P}L^c := \{X \in \mathfrak{X}L^c \mid X \text{ is } S^1\text{-invariant and } \mathbf{L}_X\alpha^c = 0\}$$

$\mathcal{P}\dot{L}$ and $\mathcal{P}L$ denote the isometric vectorfields, uniquely defined as extensions of elements of $\mathcal{P}L^c$.

For $X \subset \mathfrak{X}M$ let $\tilde{X}$ be the uniquely determined invariant horizontal lifting of X to L^c, resp. $\dot{L}$ or L.

3.2 Lemma Let (L^c,α^c) be a PQB over (M,ω). We define

$$\mathcal{C}L^c := \{\tilde{X} \in \mathfrak{X}^{inv}_{hor} L^c \mid X \in \mathcal{C}M\}$$

then $\mathcal{C}L^c$ is a Lie ideal in $\mathcal{P}L^c$.

Proof: Le $X \in \mathcal{C}M$, $Y \in \mathcal{P}L^c$. We have uniquely $Y = Y_h + Y_v \in \mathfrak{X}^{inv}_{hor}L^c \oplus \mathfrak{X}^{inv}_{ver}L^c$ and therefore $Y_h = \tilde{V}$ for some $V \in\ \ M$.
Since $(\pi^c)^* \mathbf{L}_V\, \omega = \mathbf{L}_{\tilde{V}}(\pi^c)^*\omega = -\mathbf{L}_Y\, d\alpha^c = 0$ we have $V \in \mathcal{H}_\ell M$ and $[\tilde{X},Y_v] = 0$. We get:

$$[\tilde{X},Y] = [\tilde{X},Y_h] = [\tilde{X},\tilde{V}] = [X,V]^{\sim} + X \lrcorner\ V \lrcorner\ \omega = [X,V]^{\sim}$$

Since $V \in \mathcal{H}M$ the lemma is proven by 1.

Now we introduce the morphismgroups, which belong to the constructed Lie algebras.

3.4 Definition: Let (L^c,α^c) be a PQB over (M,ω). Define:

$\mathrm{Pqu}(L^c,\alpha^c) := \{F : L^c \longrightarrow L^c \mid F \text{ is principal-isomorphism and } F^*\alpha^c = \alpha^c\}$

$\mathrm{Pqu}^o(L^c,\alpha^c) := \{F \in \mathrm{Pqu}(L^c,\alpha^c) \mid \hat{F} \in \mathrm{Pres}(M,\omega)\}$

$\mathrm{Pres}_L(M,\omega) := \{\hat{F} \in \mathrm{Pres}(M,\omega) \mid F \in \mathrm{Pqu}(L^c,\alpha^c)\}$

$\mathrm{Pres}^o_L(M,\omega) := \{\hat{F} \in \mathrm{Pres}(M,\omega) \mid F \in \mathrm{Pqu}^o(L^c,\alpha^c)\}$

$\underset{\sim}{\mathrm{Pqu}}(L^c,\alpha^c) :\doteq \mathrm{Pqu}(L^c,\alpha^c)/(\mathrm{Pqu}^o(L^c,\alpha^c)/S^1)$

($\hat{F}$ denotes the base map of F. see diagramm)

$$\begin{array}{ccc} L^c & \xrightarrow{F} & L^c \\ \downarrow & & \downarrow \\ M & \xrightarrow{\hat{F}} & M \end{array}$$

Since $Pqu^o(L^c,\alpha^c)/S^1$ is a normal subgroup of $Pqu(L^c,\alpha^c)$, $\underset{\sim}{Pqu}(L^c,\alpha^c)$ is a group. We define $\underset{\sim}{\mathfrak{P}}L^c := \mathfrak{P}L^c/\mathcal{C}L^c$. Then:

$\mathfrak{P}L^c$ is formally the Lie algebra of $Pqu(L^c,\alpha^c)$
$\tilde{\mathcal{C}}L^c$ is formally the Lie algebra of $Pqu^o(L^c,\alpha^c)$
$\underset{\sim}{\mathfrak{P}}L^c$ is formally the Lie algebra of $\underset{\sim}{Pqu}(L^c,\alpha^c)$.

Corresponding to 3.3 we have:

3.5 Theorem: Let (L^c,α^c) be a PQB over (M,ω).
Then the following sequences of groups are exact and central:

$$0 \longrightarrow S^1 \longrightarrow Pqu(L^c,\alpha^c) \longrightarrow Pres_L(M,\omega) \longrightarrow 0$$
$$0 \longrightarrow S^1 \longrightarrow Pqu^o(L^c,\alpha^c) \longrightarrow Pres_L^o(M,\omega) \longrightarrow 0$$
$$0 \longrightarrow S^1 \longrightarrow \underset{\sim}{Pqu}(L^c,\alpha^c) \longrightarrow \underset{\sim}{Pres}_L(M,\omega) \longrightarrow 0$$

Define for $f \in \mathcal{F}hM$: $\mathbb{G}_f := \underset{\sim}{\tilde{G}}_f := \{\tilde{X} \in \mathcal{H}^{inv}_{hor} L^c \mid X \in \underset{\sim}{G}_f\}$

$\mathbb{G}_f$ is the invariant horizontal lifting of the class of Hamiltonian Vectorfields defined by f.

Remark: Let (L^c,α^c) be a PQB over (M,ω) and $\dim \ker \omega = k$.
Then the associated Principal-$\dot{\mathbb{C}}$-bundle $\dot{L}$ is in a natural way a presymplectic manifold with the same dimension of the characteristic fibres. The elements of $\underset{\sim}{\mathfrak{P}}\dot{L}$ are invariant Hamiltonian classes i.e. $\underset{\sim}{\mathfrak{P}}\dot{L} = \underset{\sim}{\mathcal{H}}^{inv}\dot{L}$ induced by this presymplectic structure on $\dot{L}$.

3.3 Theorem: Let (L^c,α^c) be a PQB over (M,ω).
Then the map $P.$: $\mathcal{F}hM \longrightarrow \underset{\sim}{\mathfrak{P}}L^c$, $f \longmapsto \mathbb{G}_f + f\cdot\hat{e}$
($\hat{e}$ being the vectorfield induced by the unity of S^1) is a Lie algebra isomorphism and the following diagramm commutes with exact lines.

$$\begin{array}{ccccccccc} & & & & \underset{\sim}{\mathfrak{P}}L^c & & & & \\ & & & \nearrow & \uparrow & \searrow & & & \\ 0 & \longrightarrow & \mathbb{R} & & P. & & \underset{\sim}{\mathcal{H}}M & \longrightarrow & 0 \\ & & & \searrow & | & \nearrow & & & \\ & & & & \mathcal{F}hM & & & & \end{array}$$

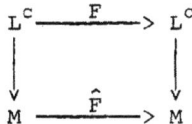

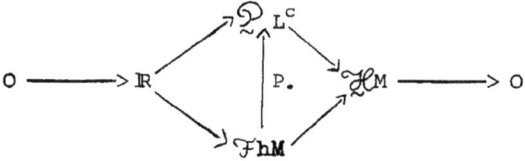

Proof: For $x \in L^c$ write $T_x L^c = Hor_x L^c \oplus Ver_x L^c$ and $Hor_x L^c = B_x L^c \oplus C_x L^c$, where $C_x L^c$ is given by all vectors π^c-related to elements of CM and $B_x L^c$ is given by a complementary bundle on M. Then $B_x L^c$ is a symplectic vector space and the theorem is proven as in the symplectic case.

Let (L^c,α^c) be a PQB over (M,ω) and $(L,\nabla,\langle\cdot,\cdot\rangle)$ the associated line bundle with covariant derivative and ∇-affine Hermitian structure $\langle\cdot\cdot\rangle$.

$Pqu(L^c,\alpha^c)$ operates on $\Gamma L := \{s : M \longrightarrow L \mid s \text{ section}\}$, the space of all sections in L naturally by $F^*s(m) = F_m^{-1}(s(F(m)))$.

Unlike the symplectic case there is no Pres-invariant volume form on M, therefore the L^2-construction for a prequantizing Hilbert space cannot be applied. For this reason the following construction has to be restricted to PQB-morphisms.

For a PQB$(L,\nabla,\langle\cdot\cdot\rangle)$ over (M,ω) and a PQB morphism $F : L \longrightarrow L$ define:

$K(L) := \Gamma^\infty L$ = space of all smooth sections in L

$K(F) := (F^*)^{-1} \in Aut(K(L))$

$\underset{\sim}{K}(L) := \{s \in K(L) \mid \nabla_X s = 0 \text{ for all } X \in \mathcal{C}M\}$

(i.e. $\underset{\sim}{K}(L)$ contains all sections being covariant constant along the characteristic fibres)

3.6 Proposition

a) $\underset{\sim}{K}(L)$ is a $\mathfrak{F}$hM-submodul of K(L)

b) $\underset{\sim}{K}(L)$ is stable under $Pres(M,\omega)$ and $Pqu(L^c,\alpha^c)$.

c) The operation of $Pres^o(M,\omega)$ and $Pqu^o(L^c,\alpha^c)/S^1$ on $\underset{\sim}{K}(L)$ are trivial.

Proof: a) $\nabla_X f\cdot s = X \lrcorner df + \nabla_X s = 0$ for $X \in \mathcal{C}M$, $f \in \mathfrak{F}hM$, $s \in \underset{\sim}{K}(L)$

b) For $X \in \mathcal{C}M$ and $s \in \underset{\sim}{K}(L)$ we have $\nabla_X F^*s = \nabla_{\hat{F}_* X} s = 0$

c) is an implication of 3.5.

As result we have an operation $\underset{\sim}{K}$

$$\underset{\sim}{K} : Pqu(\dot{L},\alpha) \longrightarrow Aut\, \underset{\sim}{K}(L),\ F \longmapsto (F^*)^{-1}|_{\underset{\sim}{K}(L)}$$

Now the infinitesimal version of this operation will be constructed. Let $(\dot{L},\alpha)$ be a PQB over (M,ω) and $f \in \mathfrak{F}hM$.

For each $X \in \underset{\sim}{G}_f$ we define a map

$$k(f)_X : \mathfrak{F}hM \longrightarrow End(K(L)),\ k(f)_X\cdot y = \nabla_{X+y} + i\hbar\cdot f,\ \hbar \in \mathbb{R}$$

$$k(f) := \{\nabla_X + i\hbar f \mid X \in \underset{\sim}{G}_f\} \subset End\, K(L)$$

k(f) has to be interpreted as a $\mathcal{C}M$-parameter family of operators. Since elements of $\underset{\sim}{G}_f$ have the form $X + \mathcal{C}M$, $X \in \underset{\sim}{G}_f$, all operators of

this family are the same on $\underset{\sim}{K}(L)$.
Moreover $\underset{\sim}{K}(L)$ is the largest subspace of $K(L)$, on which $k(f)$ induces one operator. By the Kommutator $\mathrm{End}(K(L))$ and $\mathrm{End}(\underset{\sim}{K}(L))$ are in a natural way Lie algebras and we obtain:

3.7. Theorem: For each $f \in \mathfrak{F}hM$ $k(f)$ induces uniquely an operator $\underset{\sim}{k}(f)$ on $\underset{\sim}{K}(L)$ by restriction of k. The map

$$\underset{\sim}{k} : \mathfrak{F}hM \longrightarrow \mathrm{End}(\underset{\sim}{K}(L))$$

is a Lie algebra homomorphism, called the prequantization map.
$\underset{\sim}{k}$ is the infinitesimaloperation of $\underset{\sim}{K}$, (cf.[2], 7.6.).

Remark: For $X \in \underset{\sim}{G}_f$ we have: $\langle \underset{\sim}{k}(f)\phi,\psi\rangle = \langle\phi,\underset{\sim}{k}(f)\psi\rangle + \mathbf{L}_X\langle\phi,\psi\rangle$.
Since there is in general no Pres-invariant volume on M, we cannot integrate over M in order to obtain a skalar product on $\underset{\sim}{K}(L)$.
But for regular (M,ω) the natural invariant volume on the quotient manifold M makes such a L^2-construction possible. This leads to the Pre-Klein-Gordon equation:

4. The Pre-Klein-Gordon Equation

4.1. Definition: Let (M,ω) be a regular presymplectic manifold with symplectic quotient $(\underset{\sim}{M},\underset{\sim}{\omega})$ and the natural projection p. Let $\underset{\sim}{L}$ be a vector bundle over M and $L := p^*\underset{\sim}{L}$ the pull back of $\underset{\sim}{L}$ via p over M and let $p_L : L \longrightarrow \underset{\sim}{L}$ be the natural projection:

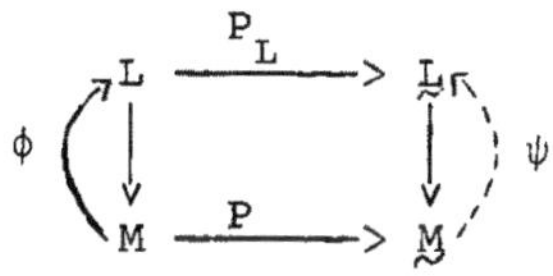

A section $\phi \in \Gamma L$ is called projectable, iff there exists a section $\psi \in \Gamma\underset{\sim}{L}$ with $p_L \circ \phi = \psi \circ p$.

The set of all projectable sections in L is a $\mathfrak{F}hM$-modul and isomorphic to $\Gamma\underset{\sim}{L}$, the set of all projections in $\underset{\sim}{L}$.

The following simple lemmata give a local description of projectable sections.

4.2 Lemma: ϕ is projectable iff

$$T(p_L \circ \phi)(CM) = 0, \text{ i.e. } CM = \ker(Tp_L \circ T\phi)$$

Proof: a) ϕ projectable $\Longrightarrow p_L \circ \phi = \psi \circ p \Longrightarrow T(p_L \circ \phi) = T\psi \circ Tp$

$\Longrightarrow \ker(Tp_L \circ T\phi) = \ker(T\psi \circ Tp) = \ker Tp = \mathcal{C}M$

since ψ is an immersion.

b) $T(p_L \circ \phi)(\mathcal{C}M) = 0 \Longrightarrow p_L \circ \phi$ is constant on characteristic fibres Define ψ by: $\psi(p(m)) = p_L \circ \phi(m)$

4.3. Lemma: Let $\underset{\sim}{\nabla}$ be a covariant derivative on $\underset{\sim}{L}$ and ∇ the induced covariant derivative on L. Then:
ϕ is projectable, iff $\nabla_X \phi = 0$ for all $X \in \mathcal{C}M$, i.e. iff ϕ is covariant constant along the characteristic fibres of (M,ω).

Proof: Let C and $\underset{\sim}{C}$ be the connection maps of ∇ and $\underset{\sim}{\nabla}$.

Then $\nabla_X \phi = C \circ T\phi \circ X = 0$ iff $X \in \ker(C \circ T\phi) = \ker(\underset{\sim}{C} \circ Tp_L \circ T\phi)$ since $C = \underset{\sim}{C} \circ Tp_L$.

Therefore $X \in \ker(C \circ T\phi) \Longleftrightarrow Tp_L \circ T\phi \circ X(m) \in \mathrm{Hor}_{p(m)}\underset{\sim}{L} \Longleftrightarrow Tp_L \circ T\phi \circ X(m) = 0$

Since $Tp_L \circ T\phi \circ X(m)$ can only be $\underset{\sim}{\pi}^c$-related to 0 for $X \in \mathcal{C}M$.

4.4. Corollary: Let $(L,\nabla,\langle\cdot,\cdot\rangle)$ be a PQB (see remark after 3.1.) over the regular presymplectic manifold (M,ω) induced by a PQB $(\underset{\sim}{L},\underset{\sim}{\nabla},\langle\cdot,\cdot\rangle_{\sim})$ over $\underset{\sim}{M}$.
Then $\underset{\sim}{K}(L)$ is the space of all smooth projectable sections in L and with the definition $K(\underset{\sim}{L}) := \Gamma^\infty \underset{\sim}{L}$ we have:

$$\underset{\sim}{K}(L) \cong K(\underset{\sim}{L})$$

Remark: In general there are more PQB's over (M,ω) than over $(\underset{\sim}{M},\underset{\sim}{\omega})$, i.e. not every PQB over (M,ω) is induced by one over $(\underset{\sim}{M},\underset{\sim}{\omega})$.

In case of a PQB over M induced by a PQB over $\underset{\sim}{M}$ we can shift all operations from M to $\underset{\sim}{M}$ and we obtain the following result:

4.5 Theorem: Let (M,ω) be a regular presymplectic manifold with symplectic quotient $(\underset{\sim}{M},\underset{\sim}{\omega})$ and the natural projection $p : M \longrightarrow \underset{\sim}{M}$, and let $(\underset{\sim}{L}^c,\underset{\sim}{\alpha}^c)$ be a PQB over $(\underset{\sim}{M},\underset{\sim}{\omega})$ and (L^c,α^c) the pull back of $(\underset{\sim}{L}^c,\underset{\sim}{\alpha}^c)$ with induced connection.

$$\begin{array}{ccc} L^c & \xrightarrow{\;p_L\;} & \underset{\sim}{L}^c \\ \big\downarrow & & \big\downarrow \\ M & \xrightarrow{\;p\;} & \underset{\sim}{M} \end{array}$$

Then we have:

1. (L^c,α^c) is a PQB over (M,ω).
2. For the associated PQB's $(L,\nabla,\langle\cdot,\cdot\rangle)$ and $(L,\nabla,\langle\cdot,\cdot\rangle_{\sim})$ we have

$$\underset{\sim}{K}(L) \cong K(\underset{\sim}{L})$$

3. $\underset{\sim}{Pqu}(L^c,\alpha^c) \cong Pqu(\underset{\sim}{L}^c,\underset{\sim}{\alpha}^c)$ and the operations

$$\underset{\sim}{K} : \underset{\sim}{Pqu}(L^c,\alpha^c) \longrightarrow \mathrm{Aut}\ \underset{\sim}{K}(L) \text{ and } \underset{\sim}{K}' : Pqu(\underset{\sim}{L}^c,\underset{\sim}{\alpha}^c) \longrightarrow \mathrm{Aut}\ K(\underset{\sim}{L})$$

can be identified by this isomorphism (see diagramms below)

4. $\mathfrak{F}\underset{\sim}{M} \cong \mathfrak{F}hM$ and $\mathfrak{P}\underset{\sim}{L}^c \cong \underset{\sim}{\mathfrak{P}}L^c$ as Lie algebras.
The prequantization maps

$$\underset{\sim}{k} : \mathfrak{F}hM \longrightarrow \mathrm{End}\ \underset{\sim}{K}(L) \text{ and } \underset{\sim}{k}' : \mathfrak{F}\underset{\sim}{M} \longrightarrow \mathrm{End}\ k(\underset{\sim}{L})$$

can be identified by this isomorphy (see diagramm below).

$$\begin{array}{ccc} \underset{\sim}{Pqu}(L^c,\alpha^c) & \xrightarrow{\cong} & Pqu(\underset{\sim}{L}^c,\underset{\sim}{\alpha}^c) \\ \downarrow \underset{\sim}{K} & & \downarrow K' \\ \mathrm{Aut}\ \underset{\sim}{K}(L) & \xrightarrow{\cong} & \mathrm{Aut}\ K(\underset{\sim}{L}) \end{array} \qquad \begin{array}{ccc} \mathfrak{F}hM & \xrightarrow{\cong} & \mathfrak{F}\underset{\sim}{M} \\ \downarrow \underset{\sim}{k} & & \downarrow \underset{\sim}{k}' \\ \mathrm{End}\ \underset{\sim}{K}(L) & \xrightarrow{\cong} & \mathrm{End}\ K(\underset{\sim}{L}) \end{array}$$

Remark: If we represent $\underset{\sim}{k}$ by $\underset{\sim}{k}'$, $\underset{\sim}{k}$ has the form:

$$\boxed{\underset{\sim}{k}'(f)\psi = \nabla_{df^{\#}} + i\hbar\cdot f}$$

For the proof of 4.5. we only have to note that k is infinitesimal-operation of K.

Now we are able to construct a Pre-Klein-Gordon operator:

4.6. Definition: Let (M,ω) be a regular presymplectic manifold with symplectic quotient $(\underset{\sim}{M},\underset{\sim}{\omega})$ and let $(L,\nabla,\langle\cdot,\cdot\rangle)$ be a PQB induced by a PQB $(\underset{\sim}{L},\underset{\sim}{\nabla},\langle\cdot,\cdot\rangle_{\sim})$ over $(\underset{\sim}{M},\underset{\sim}{\omega})$.
Then we define the Hilbert space

$$\mathbb{H} := \left\{\phi \in \underset{\sim}{\Gamma L} \;\middle|\; \int_{\underset{\sim}{M}} \langle\psi(m), \psi(m)\rangle \cdot \underset{\sim}{\omega}^n(\underset{\sim}{m}) < \infty, \text{for } p_L \circ \phi = \psi \circ p\right\}$$

of all projectable over $\underset{\sim}{M}$ square integrable sections in L.
For $f \in \mathfrak{F}hM$ $i\cdot\underset{\sim}{k}(f) \in \mathrm{End}\ \underset{\sim}{K}(L)$ has a natural continuation to an antisymmetric, in general unbounded operator $\kappa: \mathbb{H} \longrightarrow \mathbb{H}$.
$i\cdot\kappa(f)$ is called the Pre-Klein-Gordon operator and the equation

$$\kappa(f)\psi = 0$$

is called the Pre-Klein-Gordon equation.

Remark: Sometimes an extended Pre-Klein-Gordon equation on K(L)

$$k(f)\phi = 0$$

has to be considered. Since $k(f) = \nabla_{\underset{\sim}{G}_f} + i\hbar\cdot f$ this is a $\mathcal{C}M$-parameter family of equations.

For $f \equiv 0$ the Pre-Klein-Gordon equation is trivial: Allelements of $\mathbb{H}$ are solutions.
If (M,ω) is an almost contact manifold with time form η any Pre-Klein-Gordon equation for a PQB $(L,\nabla,<\cdot,\cdot>)$ over (M,ω) can be transformed into the case $f \equiv 0$:

4.7. Theorem: Let (M,ω) be an almost contact manifold with time form η, and $Z \in \mathcal{C}M$ the unique characteristic vectorfield with $Z \lrcorner \eta = 1$.
Let $(L,\nabla,<\cdot\cdot>)$ be a PQB over (M,ω) and $f \in \mathfrak{F}hM$.
Then there exists a covariant derivative $\tilde{\nabla}$ on L and an almost contact form $\tilde{\omega}$ on M with:

(i) $(L,\tilde{\nabla},<\cdot,\cdot>)$ is a PQB over $(M,\tilde{\omega})$

(ii) $\tilde{\omega} = \omega + df \wedge \eta$ and (M,ω,η) is almost contact manifold with time form

(iii) For all $\phi = \Gamma^{\infty}L$ we have:

$$\tilde{\nabla}_Z\phi = 0 \quad \text{iff} \quad \nabla_Z\phi + i\hbar\cdot f\cdot\phi = 0$$

Proof: Let (L^c,α^c) correspond to $(L,\nabla,<\cdot\cdot>)$.
Define on L^c the connection form

$$\tilde{\alpha}^c := \alpha^c - (f\circ\pi^c)\cdot(\pi^c)^*\eta$$

Denote by Z^h the invariant horizontal lift of Z on $(L,\nabla,<\cdot,\cdot>)$ and by $Z^{\tilde{h}}$ the invariant horizontal lift of Z on $(L,\tilde{\nabla},<\cdot,\cdot>)$. Then

$$Z^{\tilde{h}} = Z^h - i\hbar(f\circ\pi)\cdot\hat{e} \quad (\hat{e} \text{ the fundamental field on } L^c)$$

Since L is a vector bundle associated to L^c we have the natural projection $q : L^c \times \mathbb{C} \longrightarrow L$. Then we get:

$$q^*\tilde{\nabla}_Z\phi = \mathbb{L}_{Z^{\tilde{h}}}(q^*\phi) + \mathbb{L}_{i\hbar(f\circ\pi)\hat{e}}q^*\phi = q^*(\nabla_Z\phi + i\hbar f\cdot\phi)$$

This proves the theorem.

5. Prequantization of Classical and Relativistic Systems

In this section we apply the results of the last two sections to the examples from section 2. We assume generally, that all presymplectic manifolds considered in this section are quantizable, i.e. all PQB's exist.

a) Prequantization of classical mechanics

(i) Let (L_N^c,α_N^c) be a PQB over the (symplectic) phase space (N,ω_N).
For each $t \in \mathbb{R}$ we get a Pre-Schrödinger operator $k_N(f_t) \in \mathrm{End}\,\Gamma^{\infty}L_N$

$(L_N, \nabla^N, \langle\cdot\cdot\rangle^N)$ being the PQB corresponding to (L_N^c, α_N^c).
This induces a time dependent unbounded vectorfield (or a $\mathbb{R}$-parameter family of vectorfields) on $\mathbb{H}_N$, the Hilbertspace of square integrable sections in L_N.
We obtain the time dependent Pre-Schrödinger equation (PSE) for curves $t \longrightarrow \phi_t \in \mathbb{H}_N$ in $\mathbb{H}_N$:

$$\boxed{-k_N(f_t)\phi_t = \frac{\partial}{\partial t}\phi_t} \quad \text{or} \quad \boxed{\nabla^N_{X_t}\phi_t + i\hbar f\cdot\phi_t = -\frac{\partial}{\partial t}\phi_t}$$

Let $(p^*L_N, \nabla^o, \langle\cdot,\cdot\rangle)$ the induced PQB over the evolution space (M,ω_o).
Notice: $p^*L \simeq \mathbb{R}\times\mathbb{L}_N$ as manifold.

$$\begin{array}{ccc} p^*L_N & \longrightarrow & L_N \\ \downarrow & & \downarrow \\ M=\mathbb{R}\times N & \xrightarrow{\;p_N\;} & N \end{array}$$

The sections in p^*L_N can be identified with curves $\mathbb{R} \longrightarrow \Gamma L_N$ by $\psi(t,n) := \phi_t(n)$.
The vectorfield $\frac{\partial}{\partial t} \in \mathfrak{X}(\mathbb{R}\times L_N)$ is S^1-invariant and horizontal in $(p^*L_N, \nabla^o, \langle\cdot\ \cdot\rangle)$. Therefore by identifying sections in p^*L_N with curves in $\Gamma L_N, \frac{\partial}{\partial t}$ is identified with $\nabla^o_{\frac{\partial}{\partial t}}$.

Therefore we can write the PSE:

$$\nabla^o_{X_f}\phi_t + i\hbar f\cdot\phi_t = \nabla^o_{\frac{\partial}{\partial t}}\phi_t \quad \text{or} \quad \boxed{\nabla^o_{Z_f}\psi + i\hbar f\cdot\psi = 0}$$

where ψ means sections in p^*L_N and ϕ_t curves in ΓL_N .

(ii) Now we can apply theorem 4.7. and get:
On $L := p_N^*L_N$ there exists a covariant derivative ∇^f, such that we can write the PSE for $\psi \in \Gamma^\infty L$:

$$\boxed{\nabla^f_{Z_f}\psi = 0}$$

By 4.7. $(L,\nabla^f,\langle\cdot,\cdot\rangle)$ is a PQB over (M,ω_f) with $\omega_f = \omega_o + df\wedge dt$.
But this is just the Cartan form. So we have:

<u>Theorem:</u> The time dependent Pre-Schrödinger equation can be written as the Pre-Klein-Gordon equation on the evolution space for the zero function: $k(\psi) = 0$

(k the prequantization map of $(L,\nabla^f,<\cdot\cdot>)$)

For the special case of (M,ω_f) being regular we get the result of Kostant [4] :

<u>Corollary:</u> Assume (M,ω_f) to be regular and (L^c,α^f) to be induced by $(\underset{\sim}{L}^c,\underset{\sim}{\alpha}^f)$. Then the solutions of the time dependent Pre-Schrödinger equation are the projectable sections in $(L,\nabla^f,<\cdot\cdot>)$.

(iii) Finally we discuss the method of prequantizing the homogeneous evolution space:
Let $(L^c_{\mathbb{R}},\alpha^c_{\mathbb{R}})$ be a PQB over $(M_{\mathbb{R}},\Omega)$ and (L^c,α^f) the restriction of $L^c_{\mathbb{R}}$ on M. Then (L^c,α^c) is a PQB over (M,ω_f) and $L^c_{\mathbb{R}} \cong \mathbb{R}\times L^c$ as manifold.

$$\begin{array}{ccc}
(L^c_{\mathbb{R}},\alpha^c_{\mathbb{R}}) & \hookleftarrow & (L^c,\alpha^c) \\
\downarrow & & \downarrow \\
(M_{\mathbb{R}},\Omega) & \hookleftarrow & (M,\omega_f)
\end{array}$$

Sections in L are given by restriction of sections in $L_{\mathbb{R}}$ and the prequantization map in L is the restriction of the prequantization map in $L_{\mathbb{R}}$.
The Pre-Klein-Gordon equation for $-s \in \mathfrak{F}M_{\mathbb{R}}$ is:

$$k(-s) = 0 \quad \text{or} \quad \nabla_{-ds^\#}\,\phi - i\hbar\, s\cdot\phi = 0$$

Restriction to $s^{-1}(0)$ gives again the PSE : $\nabla_{Z_f} = 0$

Generally we have

> The prequantization of the evolution space is the restriction of the prequantization of the homogeneous evolution space.

b) <u>Prequantization of Relativistic Particles</u>

Let (L^c,α^c) be a PQB over $(\dot{T}^*B,\Omega)$ with prequantization map k. Then we have the Pre-Klein-Gordon equation (PKGE)

$$k(s) = 0 \ .$$

If we restrict to S^mB and put $\hbar = 1$, we obtain the extended PKGE for sections in L:

$$\nabla_Z\phi + im\phi\ , \quad Z \in \mathcal{C}S^mB.$$

i.e. the PKGE on the mass-shell-bundles is given by prequantizing the constant function m.

Remark: If we construct on S^mB a time form (this is always possible) and if S^mB is regular, then we can represent the solutions of the PKGE as sections in a PQB over the 6-dimensional quotient of S^mB. But, since the time form is not covariant, this form of the PKGE is not covariant.

Finally we give a local expression for the PKGE of a relativistic particle:

Let $(q_o, q_1, q_2, q_3, p_o, p_1, p_2, p_3)$ be canonical coordinates of $\dot{T}^*B$.

$$s(q, p_o, p_1, p_2, p_3) = \left(\sum_1^3 p_i^2 - p_o^2\right)^{1/2}, \quad \Omega = \sum_0^3 dq_i \wedge dp_i$$

Then $ds = -\frac{1}{s}\left(\sum_1^3 p_i dp_i - p_o dp_o\right)$ and therefore

$$-s \cdot X_s = \sum_1^3 p_i \frac{\partial}{\partial q_i} - p_o \frac{\partial}{\partial p_o} .$$ So the PKG operator is:

$$k(m) = \left(\nabla_{\left(\sum_1^3 p_i \frac{\partial}{\partial q_i} - p_o \frac{\partial}{\partial q_o}\right)} + im\right) \quad \text{on } S^mB$$

6. Remark on Quantization of Evolution spaces

For regular evolution spaces theorem 4.5. reduces the quantization problem, to the symplectic case, if we restrict to those PQB's being induced by PQB's on the quotient.
Therefore the quantization of relativistic systems gives for regular systems no principally new problems. But the identifications of theorem 4.5. make the physical interpretation of the procedure more complicated, since we get equivalence classes instead of coordinates.

Some standard notations:

The notation is although slightly modified based on the book [1] and the same as in [2]. In particular we use the following standard symbols:
TM the tangent bundle of a manifold M, Tf the tangent map of a smooth map f, $\mathfrak{F}M$ the ring of smooth real valued functions on M, $\mathfrak{X}M$ the smooth vector fields, Ω^pM the alternating p-forms.
$\mathbb{L}_X$ the Lie derivative, d the exterieur derivative, $\lrcorner$ the inn. product.

References

[1] Abraham,R./Marsden,J.E.: Foundations of Mechanics (2. ed)
[Benjamin/Cummings] Reading, Mass. 1978

[2] Günther, C.: Prequantum Bundles and Projective Hilbert Geometries
Int. Jour. Theor. Phys. 16, 447 - 464 (1977)

[3] Kostant, B.: Quantization and Unitary Representations
in: Lectures in Modern Analysis and Applications III
Taam, C.T. ed., Springer Lecture Notes in Math. 170

[4] Kostant, B.: Line Bundles and the Prequantized Schrödinger Equation
Symp. Math. (1973)

[5] Sachs, R.K/Wu, H.: General Relativity for Mathematicians
[Springer] New York 1977

[6] Sniatycki, J./Tulczyjew, W.M: Canocical Relativistic Charged Particles
Ann. Inst. H. Poincaré, Sec A, XV, 177-1 (1971)

[7] Souriau, J.M.: Structure des Systemes Dynamiques
[Dunod] Paris 1970

[8] Sternberg, S./Ungar, Th.: Classical and Prequantized Mechanics without Lagrangians or Hamiltonians
(1978)

GEOMETRIC QUANTISATION FOR SINGULAR LAGRANGIANS

D. J. Simms
Mathematical Institute, Bonn[1]

In this talk I want to indicate how the procedures of geometric quantisation, in the sense of Kostant [3] and Souriau [5] can be adapted to the case of a dynamical system with a singular Lagrangian.

Let X be a smooth manifold of finite dimension n, let TX denote the tangent bundle of X and let L be a smooth real valued function on $TX \times R$. We regard L as the time-dependent Lagrangian function of a dynamical system based on the configuration space X. We introduce the corresponding homogeneous Lagrangian Λ in the usual way, as follows. Let $T_+(X \times R)$ denote the subset of $T(X \times R) = T(X) \times R^2$ of vectors of the form (v, t, λ) with $\lambda > 0$. Let $TX \times R$ be identified with a subset of $T_+(X \times R)$ by the map $(v, t) \mapsto (v, t, 1)$. Then Λ is defined to be the real valued function on $T_+(X \times R)$ which is homogeneous of degree 1 on each fibre and which equals L on $TX \times R$. Thus

$$\Lambda(v, t, \lambda) = \lambda\Lambda(\lambda^{-1}v, t, 1) = \lambda L(\lambda^{-1}v, t).$$

For each smooth path $\gamma : [\tau_1, \tau_2] \to X \times R$ with strictly increasing second (time-) component we denote by $\dot{\gamma}$ the corresponding path in $T_+(X \times R)$. Then the action of the path γ is defined to be

$$S[\gamma] = \int_{\tau_1}^{\tau_2} \Lambda(\dot{\gamma}(\tau))d\tau.$$

Hamilton's principle states that the classical paths from b to c in $X \times R$ are critical points of the action functional S on the space of paths from b to c.

Associated with Λ we have the corresponding Legendre transformation

$$\Lambda : T_+(X \times R) \to T^*(X \times R)$$

(1) On leave of absence from Trinity College Dublin, supported by the Sonderforschungsbereich Theoretische Mathematik.

into the cotangent bundle of $X \times R$. This is defined by

$$\langle \Lambda'(v), w \rangle = \frac{d}{ds} \Lambda(v + sw)\Big|_{s=0}.$$

Since Λ is homogeneous of degree 1 it follows that Λ' is homogeneous of degree 0 and that $\langle \Lambda'(v), v \rangle = \Lambda(v)$.

If α denotes the canonical 1-form on the cotangent bundle $T^*(X \times R)$ and if $\tilde{\gamma}$ denotes the path in $T^*(X \times R)$ obtained from $\dot{\gamma}$ by composing it with the Legendre transformation Λ', then we have

$$S[\gamma] = \int_{\tilde{\gamma}} \alpha .$$

To see this we let $\xi_\tau = \frac{d}{d\tau}\tilde{\gamma}(\tau)$ and $v_\tau = \dot{\gamma}(\tau)$ denote the velocity vectors at τ to $\tilde{\gamma}$ and γ respectively. Then

$$\langle \alpha, \xi_\tau \rangle = \langle \tilde{\gamma}(\tau), v_\tau \rangle = \langle \Lambda'(v_\tau), v_\tau \rangle = \Lambda(v_\tau)$$

as required.

Let M denote the range of the Legendre transformation Λ', and suppose that M is a smooth submanifold of $T^*(X \times R)$. Let $\omega = d\alpha$ be the canonical symplectic form and let $|M$ denote the pull back of ω to M. If γ is a classical path in $X \times R$ then we call $\tilde{\gamma}$ a classical path in M. It follows from Hamilton's principle that the velocity vector of $\tilde{\gamma}$ lies in the kernel of $\omega|M$ at each point.

Now Λ' is homogeneous of degree 0 and hence its image M has codimension at least 1 in $T^*(X \times R)$. When M has codimension 1 we call the Lagrangian L non-singular. In this case $\omega|M$ has a 1-dimensional kernel and the classical paths in M are the leaves of a 1-dimensional foliation. This is the case considered in [4].

In the singular case M has codimension greater than 1 and here the various possibilities were analysed by Dirac; see for example his Yeshiva lectures [1]. His analysis may be described as follows. M is called the primary constraint submanifold. If $\tilde{\gamma}$ is a classical path in M then $\tilde{\gamma}$ must lie in the closed subset M_1 where $\omega|M$ is degenerate. If M_1 is empty, $\omega|M$ everywhere non-degenerate, M a symplectic submanifold, then no curve in M can be a classical path,

path, and the Lagrangian equations of motion are inconsistent in this case.

If M_1 is a non-empty submanifold then any classical path $\tilde{\gamma}$ must lie in the closed subset $M_2 \subset M_1$ where the kernel of $\omega|M$ has a non-zero intersection with TM_1. Similarly $\tilde{\gamma}$ must lie in the closed subset $M_3 \subset M_2$ where the kernel of $\omega|M$ has a non-zero intersection with TM_2. Continuing in this way, suppose we get a sequence of submanifolds $M \supset M_1 \supset M_2$ ...... terminating in a submanifold N, say. N is called the secondary constraint submanifold. If N is empty then the equations of motion are inconsistent.

Let K denote the intersection of the kernel of $\omega|M$ with TN and suppose K has constant rank. Let D denote the foliation of N generated by K under the Lie bracket. In Dirac's terminology K corresponds to the generators of the first class primary constraints. The tangent vectors to classical paths all lie in K, and we suppose that K is equal to the set of all such tangent vectors. Two classical paths, in N which intersect are regarded as representing the history of the same physical state. From this we are led to regard two classical paths which lie in the same leaf of D as representing the history of the same physical state. In the case when D is the kernel of $\omega|N$, the space of leaves of D is called the phase space of the dynamical system.

We now further assume that N is a coisotropic submanifold of $T^*(X \times R)$, in the sense that the orthogonal complement of TN under the bilinear form ω is contained in TN. This orthogonal complement is then equal to the kernel of $\omega|N$ and we assume it is also equal to D. This corresponds to the case where the constraints are all first class in the sense of Dirac. This is also the case considered by Faddeev [2] .

For $x = (\underline{x}, t) \in X \times R$ let F_x denote the union of the leaves of D which intersect the cotangent fibre over x. If N has codimension k in $T^*(X \times R)$ then the leaves of D are k-dimensional and the intersection of N with a general cotangent fibre is $(n + 1 - k)$-dimensional. We suppose that $F^t = \{F_{(\underline{x},\, t)} | \underline{x} \in X\}$ is an $(n + 1)$-dimensional foliation of N. The leaves of F^t are then Lagrangian submanifolds of $T^*(X \times R)$.

To apply the geometric quantisation procedure in this context we take the product line bundle $\mathcal{L} = N \times \mathbb{C}$ over N with covariant derivative ∇ defined on the space of smooth sections $\Gamma(\mathcal{L}) = C^\infty(N, \mathbb{C})$ by

$$\nabla_\zeta f = \zeta f - 2\pi i h^{-1} \langle \alpha, \zeta \rangle f$$

for all $\zeta \in \Gamma TN$ and $f \in C^\infty(N, \mathbb{C})$. Here h is Planck's constant. Thus $(\mathcal{L}, \nabla)$ is a line bundle with connection having $-h^{-1}\omega|N$ as curvature form.

Let W^t be the space of $\mathcal{L}$-valued half-forms on N normal to F^t and covariant constant along F^t. We now proceed as in [4] to use the Blattner-Kostant-Sternberg pairing of half-forms to give a formal integral transform

$$T^{s,t} : W^t \to W^s .$$

The following is proposed as an interpretation of these constructions. The space W^t represents the quantum mechanical wave functions at time t. The transform $T^{s,t}$ represents, in the limit $s \to t$, the quantum mechanical time evolution from time t to time s. The evolution from time t to time s is to be obtained as a limit of composite operators of the form

$$T^{s_N, s_{N-1}} \circ \ldots \circ T^{s_1, s_0}$$

where $s_r = N^{-1} [rs + (N - r)t]$ and $N \to \infty$. This may be viewed formally as a functional integral of the type considered by Faddeev [2].

REFERENCES

1. P. A. M. Dirac. Lectures on Quantum Mechanics. Yeshiva University, New York 1964.

2. L. D. Faddeev. The Feyman integral for singular Lagrangians. Theoretical and Mathematical Physics 1, 1 - 13, 1969.

3. B. Kostant. Quantization and unitary representations. Lecture Notes in Mathematics 170, Springer, Berlin 1970.

4. D. J. Simms. Geometric quantization and the Feynman integral. Lecture Notes in Physics 106, Springer, Berlin 1979.

5. J. M. Souriau. Structure des systèmes dynamiques. Dunod, Paris 1970.

ELECTRON SCATTERING ON MAGNETIC MONOPOLES

Herbert-Rainer Petry
Institut für Theoretische Kernphysik der Universität Bonn

Introduction: It has been recognized within the framework of geometric quantization[1),2)], that it is very convenient to interpret the wave--function of a particle as a section in a complex line-bundle. For the quantum-mechanical description of an electron in the field of a magnetic monopole, this interpretation is in fact essential[3),4),5)] because it yields in the most natural way the mathematical explanation of the quantization of magnetic charge, first discovered by Dirac[6)]. However, as we shall see, this new interpretation forbids to apply the standard methods of scattering theory. It is the purpose of this note to demonstrate how these methods have to be changed in order to obtain a consistent theory of electron scattering on magnetic monopoles.

1. Standard scattering theory

In order to see what changes we have to perform, it is useful to recall the two basic approaches of scattering theory[7)]. To be as close as possible to the physical situation which we finally want to describe, let us first consider an electron moving in a time-independent magnetic field which is represented by a two-form B. Assume that

$$B = dA \qquad (A = \sum_{j=1}^{3} A_j \cdot dx^j) \ , \tag{1}$$

and consider the Schrödinger-equation

$$E\psi = -\frac{1}{2m} \sum_{j=1}^{3} \left(\frac{\partial}{\partial x^j} + iqA_j\right)^2 \psi \ , \tag{2}$$

which is valid for a particle with mass m and charge q.

a) The prescription of the time-independent scattering theory reads then as follows: Set

$$E = k^2/2m$$

and look for a solution ψ, of equation (2) which asymptotically $(|x|\to\infty)$ behaves as

$$\psi \simeq e^{ikx} + A(\Omega)e^{i|k||x|}/|x| \tag{3}$$

where A is function of the angle variables Ω alone. The differential cross-section σ is then given by

$$\sigma = |A|^2 .$$

b) In contrast to this approach, the time-dependent scattering theory[7)] compares the time-evolution governed by the Hamiltonian

$$H = -\frac{1}{2m}\sum_{j=1}^{3}\left(\frac{\partial}{\partial x^j} + iqA_j\right)^2 \tag{4}$$

with the evolution governed by the free Hamiltonian

$$H_o = -\frac{1}{2m}\Delta . \tag{5}$$

To this end one considers the Möller-operators $\Omega_\pm$, which are given by the strong limits

$$\Omega_\pm = \lim_{t\to\pm\infty} e^{iHt}e^{-iH_ot} , \tag{6}$$

and studies the so-called scattering matrix S, formally defined by

$$S = \Omega_+^*\Omega_- . \tag{7}$$

The physical importance of these operators lies in the following facts:
Choose $\psi \in L^2(R^3)$ arbitrarily and consider the time-evolution of $\Omega_-\psi$, i.e.

$$\psi(t) = e^{-iHt}\Omega_-\psi .$$

Then, by construction of $\Omega_\pm$, it follows that

$$\lim_{t\to-\infty} \| \psi(t) - e^{-iH_ot}\psi \| = 0$$

which shows that $\psi(t)$ behaves as $e^{-iH_ot}\psi$ when $t\to-\infty$. Hence, at large negative times, the time-evolution of $\Omega_-\psi$ equals the free evolution of ψ. On the other hand we find that

$$\lim_{t\to+\infty} \| e^{-iH_o t} S\psi - \psi(t) \| = 0 \tag{8}$$

which shows that, at large positive times, $\psi(t)$ behaves as $e^{-iH_o t} S\psi$, i.e. we observe the free evolution of $S\psi$. If there is no magnetic field, then

$$\psi(t) = e^{-iH_o t} \psi$$

will hold for all times, (and not just, when $t\to-\infty$). Hence we write

$$\psi(t) = e^{-iH_o t} \psi + \psi_{sc}(t) \tag{9}$$

in the presence of a nonvanishing field and interpret $\psi_{sc}(t)$ as the scattering wave due to the interaction with the external field. Consequently, the probability that the particle is scattered into a cone C with apex at the origin, is given by the formula

$$P(C,\psi) = \lim_{t\to+\infty} \int_C d^3x |\psi_{sc}(t)(x)|^2 . \tag{10}$$

Using equation (8) and some mathematical properties of the free evolution operator one finds[7]:

$$P(C,\psi) = \int_C d^3p |(S-1)\hat{\psi}(p)|^2, \tag{11}$$

where $\hat{\psi}$ denotes the Fourier transform of ψ. Equation (11) is known as the "scattering into cones formula"; it is shown in standard text--books, how the differential cross-section can be extracted out of it.

2. Electron-monopole scattering

Let us now consider electron scattering in the field of a magnetic monopole with magnetic charge μ fixed at the origin; i.e. the electron moves in a magnetic field described by the two-form

$$B = \mu B_o$$

$$B_o = (x^1 dx^2 \wedge dx^3 + x^3 dx^1 \wedge dx^2 + x^2 dx^3 \wedge dx^1)/|x|^3 . \tag{12}$$

B is closed, but not exact; hence there is no vector-potential A in the domain $R^3 - O =: \dot{R}^3$, where B is well-defined. Therefore, equation (1) and, as a consequence, equation (2) seem to be mathematically meaningless. The way out of this difficulty was shown by Dirac. His arguments become, however, more transparent, when we use the language of modern differential geometry. In this language the quantum-mechanical description of a charged particle in a magnetic field B reads as follows[4)]:

The particle is described by a section in complex line-bundle $\xi \overset{\pi}{\to} D(B)$; ($D(B) \subset R^3$ denotes the domain, where B is well-defined). The line bundle ξ must have the following properties:

a) There is a fibre metric $<,>$ and a covariant derivative ∇ in ξ which are compatible; i.e. for arbitrary sections σ_1, σ_2 and vector fields X the equation

$$X(<\sigma_1,\sigma_2>) = <\nabla_X\sigma_1,\sigma_2> + <\sigma_1,\nabla_X\sigma_2> \tag{13}$$

holds.

b) The curvature $\omega(\nabla)$ and the external magnetic field B are related by the formula

$$\omega(\nabla) = iqB \; . \tag{14}$$

It has been shown in ref. 4) that this new formulation fulfills all the requirements of quantum theory. In particular, if the particle is described by a section σ, then $<\sigma,\sigma>(x)$ represents the probability to find the particle at the point x. It follows that our quantum mechanical Hilbert-space H consists of all square-integrable sections σ:

$$H = :\{\sigma; \int_{D(B)} d^3x <\sigma,\sigma>(x) < \infty\} \; , \tag{15}$$

with scalar product

$$<\sigma_1,\sigma_2> = \int_{D(B)} d^3x <\sigma_1,\sigma_2>(x) \; .$$

Moreover, the Schrödinger equation (2) reads now as follows:

$$E\sigma = -\frac{1}{2m}\sum_{j=1}^{3} \nabla_j^2\sigma \; . \tag{16}$$

(In this equation ∇_j denotes the covariant derivative in the direction of the vector field $\frac{\partial}{\partial x_j}$).

The two requirements a) and b) determine ξ (together with $<,>$ and ∇) up to strong bundle isomorphisms, provided $D(B)$ is simply connected. Physically, such isomorphisms correspond to gauge transformations and do not lead to observable consequences. More important is, therefore, the question, under which conditions on B a suitable line bundle ξ exists. The answer is extremely simple: ξ exists if and only if the cohomology class $[qB/2\pi]$ defined by B is integer[2]. If, in particular, B is given by formula (12) the last statement requires

$$2q \cdot \mu = n \in \mathbb{Z}\ , \tag{17}$$

which is precisely Dirac's quantization of the monopole charge μ.

Hence we will assume that (17) holds indeed with $n \neq 0$; so that the general requirements of quantum theory are fulfilled. We must expect, of course, that the fact, that from now on we have to deal with sections in a nontrivial bundle, will present us new kinds of difficulties not known in ordinary Schrödinger theory. In particular, let us investigate what happens to the two standard approaches of scattering theory. Clearly, the time-independent approach a) cannot be used, because there is no way to formulate a boundary condition like equation (3) for sections in a nontrivial line-bundle. This has been realized by several authors who have changed this condition[8),9),10)]. We will come to this point later and consider first the time-dependent approach b). Also this approach is, in our case, no longer useful. The reason lies in the fact that we have to compare the time evolution governed by

$$H = -\frac{1}{2m} \sum_{j=1}^{3} \nabla_j^2\ , \tag{18}$$

(which acts on sections in ξ), with the time evolution given by

$$H_o = -\frac{1}{2m}\Delta\ ,$$

(which acts on ordinary functions). Hence the Möller-operators can no longer be defined as in equation (6), because H and H_o act on completely different spaces.

At this point it is useful to recall that the definition (6) of the Möller operators is known to be no longer valid, when long range forces are present. In particular, equation (6) does not hold for the Coulomb problem[7]. In this case H_o has to be replaced by a "free" Hamiltonian H_o' which differs from the expression (5). Note, however, that in the Coulomb problem, the particular choice of H_o' is not predicted by the general theory, but has to be guessed from the particular form of the corresponding Schrödinger-equation and its solutions[11]. This is now the procedure which we want to adopt for our problem, too; i.e. we want to find an operator H_o' replacing H_o in equation (6) which yields the Möller operators. We hope that H_o' will be somehow naturally determined by H itself, and consequently, the next sections deal with a more detailed investigation of H.

However, we have to mention before, that there is another subtle difficulty concealed in our problem, which arises from the "scattering into cones formula" (equ. (11)). This formula contains the Fourier transform, or, in physical terms, the momentum distribution of the wave-function ψ. Obviously, sections in a nontrivial line bundle cannot be Fourier transformed like ordinary functions. Hence we expect for our particular scattering problem a "scattering into cones formula" where the Fourier transformation is replaced by a suitable unitary operator F. If σ is a section, we may, by obvious analogy, interpret $F\sigma$ as the "momentum space wave-function" corresponding to σ. We shall see in the following, how F is determined by the appropriate choice of H_o'.

3. The Hamiltonian of electron-monopole scattering

According to the last section, we have to construct a line bundle $\xi_n \xrightarrow{\pi} \dot{R}^3$ with a fibre metric $<,>$ and a compatible covariant derivative ∇, such that

$$\omega(\nabla) = -i\,\frac{n}{2}\,B_o \qquad \text{(compare equ. (12))}$$

with

$$2q\mu = n \in Z\ .$$

This construction has been shown elsewhere[4]; at this place we just give the result without proofs.

Identify $U(1)$ with the group of unimodular complex numbers and let

$d_n: U(1) \to GL(\mathbb{C})$ denote the representation $d_n(z) = z^n$; $(z \in U(1))$.

It turns out[4] that there is a fixed principal U(1)-bundle $P \overset{\tilde{\pi}}{\to} \dot{R}^3$ such that ξ_n is associated to P via the representation d_n for all n ; i.e.

$$\xi_n = P \times_{d_n} \mathbb{C} . \tag{19}$$

Moreover, there is a connection form α in P , such that

$$d\alpha = - \tilde{\pi}^* B_o / 2 . \tag{20}$$

Let $Sec(n)$ and $H(n)$ denote the vector spaces of all sections in ξ_n and the Hilbert space of all square-integrable sections, respectively (compare equ. (15)); let $F(n)$ denote the space of n-equivariant functions on P , i.e.

$$F(n) = \{\psi : P \to \mathbb{C};\ \psi(q \cdot z) = z^{-n} \psi(q), \quad \text{for all } q \in P,\ z \in U(1)\} \tag{21}$$

Let X be any vector field on $\dot{R}^3$ and let $H(X)$ its horizontal lift to P (induced by the connection form α).

By a standard theorem in differential geometry[12], there is a linear isomorphism

$$d_n^{\#} : F(n) \to Sec(n) ;$$

moreover, a covariant derivative ∇ in ξ_n is given by the formula, $(\sigma \in Sec(n))$:

$$\nabla_X \sigma = d_n^{\#} H(X) (d_n^{\#-1} \sigma) . \tag{22}$$

∇ is compatible with the fibre metric

$$\langle \sigma_1, \sigma_2 \rangle (x) = (d_n^{\#-1} \sigma_1)(q) \overline{(d_n^{\#-1} \sigma_2)(q)} , \tag{23}$$

where $q \in \tilde{\pi}^{-1}(x)\ P$ is any point in the fibre over $x \in \dot{R}^3$. It is now convenient to define a suitable volume form ω_4 on P :

$$\omega_4 = \frac{1}{2\pi} \tilde{\pi}^* (\omega_3) \wedge \alpha , \qquad (\omega_3 = d^3x) , \tag{24}$$

and to introduce the Hilbert space $\tilde{H}(n) \subset F(n)$:

$$\tilde{H}(n) = \{\psi \in F(n) \ ; \quad \int_P \bar{\psi}\psi\omega_4 < \infty\} \ , \tag{25}$$

with scalar product given by the formula

$$<\psi_1,\psi_2> = \int_P \bar{\psi}_1\psi_2\omega_4 \ . \tag{26}$$

Then $d_n^{\#}$ restricts to a Hilbert space isomorphism

$$d_n^{\#}:\tilde{H}(n) \to H(n) \ .$$

According to equation (18), the Hamiltonian of electron-monopole scattering is given by

$$H = -\frac{1}{2m}\sum_{j=1}^{3} \nabla_j$$

In view of the definition of ∇ (compare equ. (22)) it is more convenient to study not H , but

$$\tilde{H} = d_n^{\#-1} H d_n^{\#} = -\frac{1}{2m}\sum_{j=1} H(e_j)^2 \tag{27}$$

which acts on $\tilde{H}(n)$. (e_j denotes a unit vector in the direction of the j-th coordinate axis). Since $d_n^{\#}$ is a Hilbert space isomorphism we loose nothing in doing so, but have the advantage of dealing with an ordinary partial differential operator.

It remains to specify $P \overset{\tilde{\pi}}{\to} \dot{R}^3$ together with α . Let Q denote the algebra of quaternions, with orthonormal basis e_o,e_1,e_2,e_3 . (e_o is the unit element, and the elements e_1,e_2,e_3 generate the usual multiplication table). Set

$$P = \dot{Q} = Q - 0 \tag{28}$$

and identify $\dot{R}^3$ with the elements $q \in \dot{Q}$ orthogonal to e_o . If $z \in U(1)$ has the form $z=\alpha+i\beta$, define, for all $q \in \dot{Q}$

$$q \cdot z = q(\alpha+e_3\beta) . \tag{29}$$

Let $\tilde{\pi}:P \to \dot{R}^3$ denote the map given by

$$\tilde{\pi}(q) = q e_3 \bar{q} \ ; \tag{30}$$

by (28), (29) and (30) we have then indeed defined the U(1)-principal bundle P. The connection form α is given by

$$\alpha_q(h_q) = \langle qe_3, h_q \rangle / |q|^2 , \tag{31}$$

(for arbitrary tangent vectors h_q at q).
By a straightforward calculation, one finds

$$\omega_4 = \frac{4}{\pi} |q|^2 d^4 q \tag{32}$$

and

$$\tilde{H} = - \frac{1}{8m|q|^2} (\Delta_4 - n^2/4|q|^2) . \tag{33}$$

In the derivation of (33), it has been used that $\tilde{H}$ acts on n-equivariant function; therefore, n appears in this formula. (Δ_4 denotes the Laplacian in four dimensions).

According to section 2, we have to look for an operator H_o' which replaces H_o in equation (6). H_o' has to fulfill the following physical requirements:

a) There should be no charge-dependence in H_o .

b) At large distances from the origin H and H_o should look more or less equal.

Now observe that $|q|^2 = |\tilde{\pi}(q)|$. Hence, at large distances from the origin, we find from equation (33):

$$\tilde{H} \rightarrow \tilde{H}_o = - \frac{1}{8m|q|^2} \Delta_4 \tag{34}$$

Moreover, $\tilde{H}_o$ shows no explicit dependence on n , that is on the charge q . There is, of course, some hidden n-dependence, in $\tilde{H}_o$, because it is assumed that it acts on n-equivariant functions; but note that, for all n , $\tilde{H}_o$ has the same form. The fact that $\tilde{H}_o$ acts on n-equivariant functions has, however, a certain effect on the corresponding angular momentum spectrum (compare ref.[4),8)]): the allowed angular momenta have the quantized values $j = |\frac{n}{2}|, |\frac{n}{2}|+1, \ldots$ etc., i.e. they are modified by the field of the monopole. This phenomenon is already well-known from the classical theory[13),14)]; moreover, in the classical theory, one shows that this modification persists even at large, (and in fact infinite) distances from the origin. This physical effect will, therefore, be taken into account if we make the following choice for H_o' :

$$H_o' = d_n^{\#}\tilde{H}_o d_n^{\#-1} \quad . \tag{35}$$

Besides the physical arguments leading to (35), it should, however, be stressed, that (35) represents from a formal point of view, the only choice for H_o' which appears to be mathematically natural.

4. The scattering into cones formula

After having defined H and H_o', we are now able to study the Möller operators $\Omega_\pm$ given by equation (6). From the last section it is clearly more convenient to work within the Hilbert space $\tilde{H}(n)$, of n-equivariant functions on P, and to investigate the operators

$$\tilde{\Omega}_\pm = d_n^{\#-1}\Omega_\pm d_n^{\#} = \lim_{t\to\pm\infty} e^{i\tilde{H}t}e^{-i\tilde{H}_o t} \tag{36}$$

To this end we need some technical preparations. First recall that the space of unit quaternions forms the group $SU(2)$. The Wigner coefficients[15] $D^j_{m,m'}(q_o)$, $(q_o\in SU(2))$, of the standard irreducible representations of $SU(2)$ in $\mathbb{C}^{2j+1}$, yield (by the Peter-Weil theorem)[16], a complete basis for the Hilbert space of square-integrable functions on S^3 (with standard volume form). Consider the function $\psi: P\to\mathbb{C}$

$$\psi(q) = \varphi(r)D^j_{m,m'}(q/|q|) \tag{37}$$

where $r=|q|^2$ and φ has compact support on the positive real line. ψ is n-equivariant, if and only if

$$m' = -n/2 \tag{38}$$

(This follows from the well-known properties of the Wigner coefficients[15]). (38) implies[15] that $j=|m'|, |m'|+1, |m'|+2,\ldots,$ and that m has the values $m=-j, -j+1,\ldots,j$.

Assume now that (38) holds. It follows from the completeness of the Wigner coefficients, that finite sums of functions which have the form (37) generate a dense subspace $\hat{H}(n)$ of $\tilde{H}(n)$. Let now $p\in R$ be greater than $-1/2$ and define

$$h_p(\varphi)(r) = \int_0^\infty dr' J_p(rr')\sqrt{rr'}\varphi(r') \quad . \tag{39}$$

J_p denotes a Besselfunction and h_p is known as the Hankel transformation[17] of order p. Define T and T_o by the formulas

$$(T\psi)(q) = (t_j\varphi)(r) \cdot D^j_{m,m'}(q/|q|)$$

$$(T_o\psi)(q) = (t^o_j\varphi)(r) \cdot D^j_{m,m'}(q/|q|) \tag{40}$$

where t_j and t^o_j are given by the equations

$$t_j\varphi = \frac{1}{r}\, h_{\lambda(j)}(r\varphi) \ ,$$

$$t^o_j\varphi = \frac{1}{r}\, h_{\lambda_o(j)}(r\varphi) \ ,$$

with

$$\lambda(j) = ((j+\tfrac{1}{2})^2 - m'^2)^{1/2} \ ,$$

$$\lambda_o(j) = j + \frac{1}{2} \ .$$

It is well-known[17] that h_p extends to a unitary transformation of the Hilbert space of all square-integrable functions on the positive real line. Using this fact one can easily prove that the operators T and T_o extend to unitary transormations of $\tilde{H}(n)$.
Moreover, one can show that T and T_o are hermitean and idempotent, i.e.

$$T^2 = T_o^2 = \mathrm{id} \ . \tag{41}$$

By partial integration, one proves furthermore that the equations

$$(T\tilde{H}\psi)(q) = \frac{1}{2m}\, r^2 (T\psi)(q) \ ,$$

$$(T_o\tilde{H}_o\psi)(q) = \frac{1}{2m}\, r^2 (T_o\psi)(q) \ , \tag{42}$$

hold for any function of the form (37) and hence for any $\psi\in\hat{H}(n)$.
Since $\hat{H}(n)$ is dense in $\tilde{H}(n)$, equation (42) shows that the differential operators H and $\tilde{H}_o$ have self-adjoint extensions which are unitarily equivalent to the operator of multiplication with $r^2/2m$. Hence both operators have a continuous spectrum, (the positive real line), and no bound states.

In addition to the operators T and T_o we define for functions ψ of the form (37), (compare equ. (40)),

$$U_{\pm}\psi = \exp\,[\pm i\pi(\lambda(j)-\lambda_o(j))/2]\cdot\psi.$$

$$U_o\psi = \exp\,[-i(\lambda_o(j)+1))/2]\cdot\psi \tag{43}$$

The operators $U_{\pm}$ and U_o extend trivially to isometries of $\hat{H}(n)$, and hence of $\tilde{H}(n)$.

After these preparations we can present the result for the limits (36):

$$\tilde{\Omega}_{\pm} = U_{\pm}TT_o \tag{44}$$

(44) yields the scattering matrix $\tilde{S}$, (defined in the space of equivariant functions),

$$\tilde{S} = \tilde{\Omega}_{+}^{*}\tilde{\Omega}_{-} = U_{-}^{2}\ . \tag{45}$$

With the help of $d_n^{\#}$ we obtain the S-matrix in the space $H(n)$, of sections in ξ_n:

$$S = d_n^{\#}\tilde{S}d_n^{\#-1} \tag{46}$$

With the help of (36) and (46) one can then establish the "scattering into cones formula" which holds for our problem. The derivation is completely analogous to the one of section 2; the only difference consists in the replacement of the ordinary wave-function ψ by a section σ in the bundle ξ_n. One finds that the probability of a particle being scattered into the cone C is given by the formula:

$$P(C,\sigma) = \int_C d^3x\ \langle (S-1)F\sigma,\ (S-1)F\sigma\rangle(x) \tag{47}$$

with

$$F\sigma = d_n^{\#}U_oT_od_n^{\#-1} \tag{48}$$

The operator F replaces the ordinary Fourier transformation, as was anticipated in section 2.

The differential cross section of electron-monopole scattering can be derived from (47). For completeness we state the result here.

Let θ denote the scattering angle and P the momentum of the charged particle; the differential cross section $d\sigma/d\Omega$ is given by the equation:

$$\frac{d\sigma}{d\Omega} = |2f(\theta)/P| \tag{49}$$

with

$$f(\theta) = [(\sin\theta)^{-\frac{1}{2}} \frac{d^2}{d\theta^2} (\sin\theta)^{\frac{1}{2}} + \frac{1}{16(\sin\frac{\theta}{2})^2} + \frac{1-4n^2}{16(\cos\frac{\theta}{2})^2}] f_o(\theta)$$

and

$$f_o(\theta) = \sum_{k=0}^{\infty} \frac{T(k)}{2k+|n|+1} \cdot P_k^{(0,|n|)}(\cos\theta) \cdot (\cos\frac{\theta}{2})^{|n|} .$$

$P_k^{(0,|n|)}$ denotes a Jacobi polynomial and $T(k)$ is given by

$$T(k) = -1 + \exp -i\pi([(|n|/2+k+1/2)^2 - (|n|/2)^2]^{\frac{1}{2}} -k-|n|/2-1/2) .$$

The proofs of the equations (44)-(49) are lengthy and have to be omitted here. They can be found in ref.[18].

Discussion:

In the last section we have shown, that after some reasonable changes in the definition of the Möller operators, the time-dependent approach to scattering theory can be successfully applied to the electron--monopole scattering problem, despite of the fact, that wave-functions have to be replaced by sections in a complex line bundle. Actually all the physically relevant quantities can even be computed in closed form.

We have already mentioned in section 2 that our problem was already treated within a modified time-independent approach by several authors[8),9),10)]. Their result can be more easily compared to ours if we consider phase shifts, which can be read off from equation (45). As a function of angular momentum j, $j=|n|/2, |n|/2+1, \ldots$ etc., the phase shifts $\delta(j)$ are of the form (compare also equation (49)):

$$\delta(j) = -\pi([(j+\tfrac{1}{2})^2 - (|n|/2)^2]^{\frac{1}{2}} - j - \tfrac{1}{2}) \tag{50}$$

In contrast to this, one finds in the references mentioned before

$$\delta(j) = -\pi[(j+\tfrac{1}{2})^2 - (|n|/2)^2]^{-\frac{1}{2}} \tag{51}$$

The expression (50) has the (physically reasonable) property that $\delta(j)$ vanishes for large angular momenta j, whereas (51) diverges. As a consequence the corresponding expression for the differential cross section is a divergent series which has to be regularized ad hoc. Our expression for the cross section behaves perfectly well in this respect.

We have already mentioned that the time-independent approach uses a modified boundary condition at infinity. It might be that this new boundary condition is mathematically not adequate, but that the numerical regularization (which is in fact done on the computer) compensates for this defect. Our result could then be compared only numerically with the time-independent approach presented so far in the literature; no effort in this direction has yet been done.

References

1) J.M. Souriau, "Structure des systèmes dynamiques" (Dunod, Paris, 1970).
2) B. Kostant, "Quantization andunitary representations", Springer Lecture Notes in Mathematics (Springer, Heidelberg, 1970).
3) J. Sniatycki, J. Math. Phys. 15, 619 (1974)
4) W. Greub, H.R. Petry, J. Math. Phys. 16, 1347 (1975)
5) T.T. Wu, C.N. Yang, Phys. Rev. D 12, 3845 (1975)
6) P.A.M. Dirac, Proc. Roy. Soc., A 133, 60 (1931)
7) W.O. Amrein, J.M. Jauch, K.B. Sinha, "Scattering theory in Quantum Mechanics", (W.A. Benjamin, London, 1977)
8) P.P. Banderet, Helv. Phys. Acta 19, 503 (1946)
9) K. Ford, J.A. Wheeler, Ann. Phys. (N.Y.) 7, 287 (1959)
10) J. Schwinger, K.A. Milton, Wu-Yang Tsai, L.L. de Raad, D.C. Clark, Ann. Phys. (N.Y.) 101, 451 (1976)
11) J.D. Dollard, J. Math. Phys. 5, 729 (1964)
12) W. Greub, S. Halperin, R. Vanstone, "Connections, Curvature and Cohomology", (Academic Press, New York, 1973)
13) H. Poincaré, Compt. Rend. 123, 530 (1896)
14) M. Fierz, Helv. Phys. Acta, 17, 27 (1944)
15) A.R. Edmonds, "Angular momentum in Quantum Mechanics", (Princeton University Press, Princeton, N.J., 1957)
16) C. Chevalley, "Theory of Lie Groups", (Princeton University Press, Princeton, N.J., 1946)
17) N.F. Achieser, I.M. Glasman, "Theorie der linearen Operatoren im Hilbertraum", (Akademie Verlag, Berlin, 1960)
18) H.R. Petry, "Zur Streutheorie geladener Teilchen an magnetischen Monopolen", (thesis, Bonn 1979)

The Metaplectic Representation, Weyl Operators and Spectral Theory

V. Guillemin
Massachusetts Institute of Technology
Cambridge, MA 02139, USA

S. Sternberg
Harvard University
Cambridge, MA 02138, USA

Let X be a compact manifold and $P: C^\infty(X) \to C^\infty(X)$ a positive self-adjoint elliptic pseudodifferential operator of order $m > 0$. The symbol of P, $\sigma(P)$, is a smooth everywhere positive function on $T^*X - 0$. Let $p = \sigma(P)^{1/m}$ and let H_p be the Hamilton vector field associated with p. Since X is compact, this vector field generates a global flow, $\exp T H_p$, on $T^*X - 0$. If one thinks of $\exp t H_p$ as representing a classical dynamical system and the operator P as representing its quantum mechanical counterpart, one is not surprised to find a "correspondence principle" governing the behavior of the large eigenvalues of P. We will describe three results which can, in some sense, be regarded as manifestations of such a principle:

I. The Weyl theorem. Let $N(\lambda)$ be the number of eigenvalues of P which are less than λ and let $V(\lambda)$ be the symplectic volume of the set $\sigma(P) = p^m \leq \lambda$. Then

1.1 $$N(\lambda) \sim (1/2\pi)^n V(\lambda) \quad \text{for } \lambda >> 0 .$$

(In the generality stated here this theorem is due to Hormander. See [7].)

II. The trace theorem. By definition the period spectrum of

$\exp tH_p$ is the set of periods of its periodic trajectories, i.e. T is in the period spectrum $\Leftrightarrow \exists\, z \in T^*X - 0$ such that $(\exp TH_p)(z) = z$. The trace theorem asserts that

$$\text{(1.2)} \qquad \operatorname{trace} \exp \sqrt{-1}\, t\, P^{1/m} \ \underline{\underline{\text{def.}}} \sum_{\lambda^m \in \operatorname{spec} P} e^{\sqrt{-1}\,\lambda t}$$

is well-defined as a distribution on the real line, (the sum on the right converges in $\mathcal{D}'(\mathbb{R})$) and that its singular support is contained in the period spectrum of $\exp tH_p$. (See [2], [3], and [5].)

III. The <u>clustering theorem</u>. Let

$$\text{(1.3)} \qquad 0 \leq \lambda_1 \leq \lambda_2 \leq \cdots$$

be the points in the spectrum of $P^{1/m}$ and let Σ be the cluster set of $\{\lambda_i - \lambda_j\}$. The clustering theorem asserts that $\Sigma \neq \mathbb{R}$ if and only if all the trajectories of $\exp tH_p$ are periodic. This theorem is due to Helton, [6]. For refinesments of it, see [4], [18], and [19].

The purpose of this paper is to discuss analogues of the theorems above for a class of operators which are of considerable interest in quantum mechanics. These operators are obtained from Hamiltonian systems on classical phase space by the process of "Weyl quantization". Explicitly let $p(x,\xi)$ be a smooth function on $\mathbb{R}^{2n}$. For simplicity we will assume that, for $|x|^2 + |\xi|^2 \geq 1$, p is homogeneous of degree m in (x,ξ), i.e.

$$\text{(1.4)} \qquad p(\lambda x, \lambda\xi) = \lambda^m p(x,\xi)$$

for $\lambda \geq 1$ and $|x|^2 + |\xi|^2 \geq 1$. (This assumption will be weakened later on.) Let $S(\mathbb{R}^n)$ be the space of Schwartz functions on $\mathbb{R}^n$. Associated with p is an operator, $P: S(\mathbb{R}^n) \to S(\mathbb{R}^n)$, defined by

$$(1.5) \qquad (Pu)(x) = (1/2\pi)^n \int e^{\sqrt{-1}(x-z,\xi)} p((x+z/2),\xi)u(z)dzd\xi \quad .$$

If p is real valued, P is formally self-adjoint. Moreover, if p is everywhere positive, the spectrum of P is discrete and bounded from below. Therefore, Theorems I, II, and III make sense for P. Are they true? It has been known for a long time that the Weyl theorem is true for P. Indeed it is true with much less restrictive assumptions on P than (1.4). For instance, for symbols of the form

$$p(x,\xi) = (\xi^2/2) + q(x) \quad .$$

Titchmarsh proves that the Weyl theorem is true providing $q(x) \to +\infty$ when $x \to \infty$. (See [15], page 174.) Tulovskii and Shubin have recently established the validity of the Weyl theorem for a very large class of symbols in [14].

For symbols of the type (1.4) it is easy to obtain the Weyl theorem from the results on pseudodifferential operators mentioned earlier. The idea is to "compactify" $\mathbb{R}^n$ by means of the Stone-Von Neumann representation. This idea turns out to be very close, in spirit, to the underlying idea of this article; so we will sketch some of the details: Let $\mathcal{H}_{2n+1}$ be the 2n+1-dimensional Heisenberg group. The underlying manifold of this group is

$$\mathbb{R}^{2n} \times S^1$$

and the group law is

(1.6) $(a,e^{i\alpha}) \circ (b,e^{i\beta}) = (a+b,\ e^{i\{(\alpha+\beta)+2\pi\omega(a,b)\}})$

where ω is the standard symplectic form on $\mathbb{R}^{2n}$. $\mathcal{N}_{2n+1}$ has an irreducible representation, unique up to unitary equivalence, with the property that on the subgroup, S^1, of $\mathcal{N}_{2n+1}$ the representation restricts to

(1.7) $$e^{\sqrt{-1}\,\theta} \rightarrow e^{\sqrt{-1}\,\theta}\ \text{Identity}.$$

In the usual construction of this representation, the underlying Hilbert space is $L^2(\mathbb{R}^n)$; however, an alternative construction is the following: Let Z^{2n} be the integer lattice in $\mathbb{R}^{2n}$. If $a,b \in Z^{2n}$, $\omega(a,b) \in Z$; so the set

$$\Gamma = Z^{2n} \times \{1\}$$

is a subgroup of $\mathcal{N}_{2n+1}$. It is discrete and cocompact; so $\mathcal{N}_{2n+1}/\Gamma$ is a compact manifold on which $\mathcal{N}_{2n+1}$ acts on the left. This action induces, in turn, an action of $\mathcal{N}_{2n+1}$ on $L^2(\mathcal{N}_{2n+1}/\Gamma)$. Let H be the Hilbert subspace of $L^2(\mathcal{N}_{2n+1}/\Gamma)$ consisting of those functions which satisfy

(1.8) $$f(e^{\sqrt{-1}\,\alpha}m) = e^{\sqrt{-1}\,\alpha} f(m)$$

for $e^{\sqrt{-1}\,\alpha} \in S^1$ and $m \in \mathcal{N}_{2n+1}/\Gamma$. Since S^1 is the center of $\mathcal{N}_{2n+1}$, this subspace is invariant; and one can show that it is irreducible. (See [16].) Comparing (1.7) and (1.8), we conclude that the representation of $\mathcal{N}_{2n+1}$ on this subspace is the unique irreducible representation of $\mathcal{N}_{2n+1}$ with property (1.7).

Let $X = S^1\backslash\mathcal{N}_{2n+1}/\Gamma$. As a manifold X is the $2n$-torus, T^{2n}, and $\mathcal{N}_{2n+1}/\Gamma$ is a circle bundle over X. If Θ is the attached line bundle we can identify H with the space of L^2 sections of Θ ; i.e., in a canonical way,

$$H = L^2(\Theta) .$$

Let Ω_n be the universal enveloping algebra of $\mathcal{N}_{2n+1}$. Given a unitary representation of $\mathcal{N}_{2n+1}$ on a Hilbert space, H , we will denote by H^∞ the space of C^∞ vectors for the representation. For instance, for the Stone-Von Neumann representation on $L^2(\mathbb{R}^n)$, $H^\infty = S(\mathbb{R}^n)$ and, for the alternative form of the Stone-Von Neumann representation described above, $H^\infty = C^\infty(\Theta)$. The representation of $\mathcal{N}_{2n+1}$ on H induces a representation of Ω_n on H^∞. Moreover, if the representation of $\mathcal{N}_{2n+1}$ has the property (1.7), then in the induced representation on H^∞ , the generator, Z , of S^1 gets represented as $\sqrt{-1}\,I$. Therefore, one gets a representation of the quotient algebra

$$\mathcal{A}_n = \Omega_n / (z - \sqrt{-1})\Omega_n$$

on H^∞ . In particular, let us consider the two examples above. Let P be an element of $\mathcal{A}_n$. If one takes as a model of the Stone-Von Neumann representation, the usual model with $H = L^2(\mathbb{R}^n)$, then P gets transformed into an operator, P_1 , on $S(\mathbb{R}^n)$. If one takes the model with $H = L^2(\Theta)$, P gets transformed into an operator, P_2 , on $C^\infty(\Theta)$. Both P_1 and P_2 are differential operators, and they turn out to be related to each other, in a very simple way, on the symbolic level. To begin with, P_1 is a differential operator with polynomial coefficients; so it has the form

$$\sum_{|\alpha|+|\beta| \leq m} a_{\alpha\beta} x^\alpha D^\beta .$$

Let

$$p(x,\xi) = \sum_{|\alpha|+|\beta|=m} a_{\alpha\beta} x^{\alpha} \xi^{\beta} .$$

The symbol of P_2 is a function on T^*X. Since $X = T^{2n}$, $T^*X = X \times \mathbb{R}^{2n}$, so there is a canonical map $\pi: T^*X \to \mathbb{R}^{2n}$. The symbol of P_2 turns out to be just π^*p.

Now let us apply the Weyl theorem of Hormander to P_2. Since P_1 and P_2 are unitarily equivalent, we get a corresponding Weyl theorem for P_1. Moreover, because of the formula for $\sigma(P_2)$ described above, the symplectic volume of the set, $\sigma(P_2) < \lambda$, is the same as that of the set, $\sigma(P_1) < \lambda$; so in fact we get the Weyl theorem for P_1 in its usual form. This, of course, only proves the Weyl theorem for polynomial differential operators; however, if we replace Ω_n by its pseudodifferential completion, the argument we have just sketched extends to symbols of type (1.4).

What about the other theorems described above, i.e. the trace theorem and the clustering theorem? For these theorems, the compactification trick we have just outlined gives rather discouraging results. For instance if P is the m-th order pseudodifferential operator associated with the symbol (1.4), then the singularities of the generating function

$$\sum_{\lambda^m \in \operatorname{spec} P} e^{\sqrt{-1}\lambda t} \tag{1.9}$$

are contained in the period spectrum of the flow, $\exp tH_q$, on $T^*X - 0$. Here $q = q(\eta)$ with $\eta = (x,\xi)$ and $q(x,\xi) = p(x,\xi)^{1/m}$. If $(y_1,\ldots,y_{2n})$ is the set of linear coordinates on $T^{2n} = \mathbb{R}^{2n}/Z^{2n}$ dual to the coordinates $\eta_1,\ldots,\eta_{2n}$, the flow, $\exp tH_q$, is the linear flow

$$(y,\eta) \to (y + t(\partial q/\partial\eta)(\eta),\eta) . \tag{1.10}$$

This flow is completely unrelated to the classical flow on $\mathbb{R}^{2n} - \{0\}$ associated with the Hamiltonian

$$(1.11) \qquad H_q = \sum (\partial q/\partial \xi_i)(\partial/\partial x_i) - (\partial q/\partial x_i)(\partial/\partial \xi_i) \quad .$$

In particular, there is no relation between the period spectrum of (1.11) and the period spectrum of (1.10). Notice also that (1.10) always has non-periodic trajectories; so the clustering theorem has no non-trivial applications.

The negative implications of these results will be puzzling to those of our readers who are familiar with the beautiful work of Balian-Block, Voros and others on quasi-modes; for their results seem to suggest that analogues of the trace theorem and clustering theorem <u>are</u> true. It turns out that they are not quite true as stated above, but that they become true if one makes a minor adjustment in the usual convention of assigning an "order" to the operator (1.5). Explicitly let p in (1.4) be an <u>even</u> function, i.e. $p(-x,-\xi) = p(x,\xi)$, and let $d = n/2$. It turns out that if one considers in place of 1.9 the generating function

$$(1.12) \qquad \sum_{\lambda^d \in \operatorname{spec} P} e^{\sqrt{-1}\lambda t}$$

its singularities <u>do</u> lie on the period spectrum of the flow (1.11). Also the clustering formula is true, with the λ_i's in (1.3) replaced by their squares, and is non-vacuous. (The harmonic oscillator provides a non-trivial example of clustering.) In addition the various refinements and extensions of the trace theorem and clustering theorem alluded to above are true, (for example, the trace formula of Duistermaat-Guillemin for generic flows, [5], and the Weinstein-Widom theorem on the asymptotic behavior of clusters,

[18].) As above these results can be proved by a compactification trick involving an explicit unitary representation of an explicit Lie group. This time, however, the representation involved will not be the Stone-Von Neumann representation but instead the <u>metaplectic</u> representation of $Mp(n)$.

We will briefly review some facts about the metaplectic representation referring the reader to [13] or [17] for details. Let $Sp(n)$ be the group of symplectic linear mappings of $\mathbb{R}^{2n}$ and $Mp(n)$ its double cover. $Sp(n)$ can be thought of as the centralizer in $Aut(\mathcal{N}_{2n+1})$ of the subgroup, S^1 , of $\mathcal{N}_{2n+1}$; so there is a homomorphism

$$\tau: Mp(n) \to Aut(\mathcal{N}_{2n+1}) \quad .$$

Let $H = L^2(\mathbb{R}^n)$. The metaplectic representation is a representation of $Mp(n)$ on H having the property that

$$(1.3) \qquad (\tau(a)b)v = a(bv)$$

for all $v \in H$, $b \in \mathcal{N}_{2n+1}$ and $a \in Mp(n)$. Indeed, (1.13) determines the representation of $Mp(n)$ on H up to unitary equivalence. For the existence of a representation with the property (1.13), see [13].

As above we will make the identification

$$\mathcal{N}_{2n+1} \cong \mathbb{R}^{2n} \times S^1 \quad .$$

Let ω be the canonical symplectic two-form on $\mathbb{R}^{2n}$, and let Λ be a Lagrangian subspace of $\mathbb{R}^{2n}$. By (1.6), $\Lambda \times \{1\}$ is an abelian subgroup of $\mathcal{N}_{2n+1}$. Let H^∞ be the space of C^∞ vectors in H and $H^{-\infty}$ its topological dual. (If we identify H with $L^2(\mathbb{R}^n)$, $H^\infty = S(\mathbb{R}^n)$ and $H^{-\infty} = S^*(\mathbb{R}^n)$ = the space of tempered distributions.) Let Θ_Λ^* be the set of all vectors in $H^{-\infty}$ left fixed by $\Lambda \times \{1\}$. It turns out that Θ_Λ^* is a one-dimensional subspace of $H^{-\infty}$ and can be canonically identified with the space of $\frac{1}{2}$-forms on Λ. (See [10].) Let X be the manifold of all Lagrangian subspaces of $\mathbb{R}^{2n}$, and let Θ be the line bundle over X whose fiber at Λ is the dual space to Θ_Λ^*, i.e. the space of negative half-forms on Λ. Given an element, f, of $S(\mathbb{R}^n)$, we can define a section $f^\#$ of Θ as follows. A point of X is a Lagrangian subspace, Λ, of $\mathbb{R}^{2n}$. The space, Θ_Λ^*, is a subspace of $H^{-\infty}$, and f is in H^∞; so because of the canonical pairing of H^∞ and $H^{-\infty}$, f determines an element in the dual space to Θ_Λ^*, in other words an element of Θ_Λ. This is, by definition, the value of $f^\#$ at Λ. One can show that Θ is a smooth line bundle over X and that $f^\#$ is a smooth section of Θ. Thus we get a transformation

$$M\colon S(\mathbb{R}^n) \to C^\infty(\Theta) \quad . \tag{1.14}$$

Unfortunately, M is not injective; its kernel is $S_{odd}(\mathbb{R}^n)$. Also its range is not all of $C^\infty(\Theta)$; in fact its range is quite tricky to describe. (See below.) Finally, M is not a unitary mapping, at least not in any obvious sense. (One can show, however, that it satisfies a "Plancherel formula" which is relatively simple for n even and more complicated for n odd.) One thing which is clear, however, is that (1.14) is an $Mp(n)$-morphism. In particular if P is an element of the universal enveloping algebra of $Mp(n)$ and P_1 and P_2 are the corresponding operators on $S(\mathbb{R}^n)$ and $C^\infty(\Theta)$ then

$$P_2 M = M P_1 \quad . \tag{1.15}$$

It turns out, as above, that both P_1 and P_2 are differential operators; however, it is not true as above that they have the same degree. In fact, the degree of P_2 is half the degree of P_1. This accounts for the curious convention about degrees in (1.12).

From (1.15) we get an identification of the spectrum of P_1 on $S_{even}(\mathbb{R}^n)$ with the spectrum of P_2 on the range of M. To get control of the latter we need to interpret the restriction of P_2 to the range of M as a Fourier integral operator. We will briefly describe how this is done. The cotangent space to X at Λ can be canonically identified with $S^2(\Lambda)$ and the Lie algebra, $sp(n)$, of $Mp(n)$ can be canonically identified with $S^2(\mathbb{R}^{2n})$. Let Σ be the fiber bundle over X whose fiber at Λ is the set $\{v \otimes v, v \in V\}$. Modulo the identification we have just described, Σ is a fiber subbundle of $T^*X - 0$. Let $\mu: \Sigma \to sp(n)$ be the map whose restriction to the fiber above Λ is the inclusion map $v \otimes v \in S^2(\Lambda) \to v \otimes v \in S^2(\mathbb{R}^{2n})$. The image, O, of μ is the set

$$\{v \otimes v, v \in S^2(\mathbb{R}^{2n})\} = \mathbb{R}^{2n} - \{0\} / v \sim -v \quad .$$

It inherits from $\mathbb{R}^{2n}$ a symplectic form which we will denote by ω_0. One can show that if ω is the cotangent symplectic form on T^*X then $\omega \mid \Sigma = \mu^*\omega_0$, i.e. μ is a canonical relation. Moreover, if P_1 and P_2 are as in (1.15) then $\sigma(P_2) = \mu^*\sigma(P_1)$. Roughly speaking this says that M is an $Mp(n)$-invariant "quantization" of the canonical relation, μ. To describe the range of M microlocally let K be the maximal compact subgroup of $Mp(n)$ and let $L^2(\Theta)$ be the L^2-completion of $C^\infty(\Theta)$ with respect to a K-invariant Hilbert structure. Let π be orthogonal projection of $L^2(\Theta)$ onto the range of M. Then there exists a K-invariant elliptic pseudodifferential operator P such that $P\pi = MM^*$. In particular π is a Fourier Integral operator associated with the canonical relation $\mu \circ \mu^{-1}$. (The proofs of all of these assertions will appear in a forth-coming paper.)

We conclude by pointing out that the results of this paper are really part of a much larger picture. Let G be a Lie group and $\underline{g}$ its Lie algebra. G acts on $\underline{g}^*$ by its co-adjoint action and each of the G-orbits possesses a canonical symplectic structure. (See [9].) If G is a non-compact simple Lie group it turns out that there is a unique orbit O in $\underline{g}^* - o$ with the following properties:

a) $z \in O \Rightarrow \lambda z \in O$ for all $\lambda \in \mathbb{R}^+$,

b) $O / \mathbb{R}^+$ is compact.

What we do in this paper can be viewed as a special case of the problem of "quantizing" these so-called minimal nilpotent orbits. We hope to take up this problem, in full generality, in a later article. For certain groups, notably $SO(n,2)$ and $U(n,n)$, the solution of this problem appears to have interesting physical ramifications (see [1], [11], and [12].)

Acknowledgements: Our compactification trick was implicitly already in Kostant's paper on "symplectic spinors". Also, a local version of the transform (1.14), occurs in the paper [8].

References

1. H. Bacry, "The de Sitter group and the bound states of the hydrogen atom", Nuovo Cimento, Vol. 41 (1966), 222-234.

2. J. Chazarain, "Formule de Poisson pour les variétés riemanniennes", Invent. Math. 24 (1974), 65-82.

3. Y. Colin de Verdiére, "Spectre du laplacien et longueurs des geodesiques periodiques", Comp. Math. 27 (1974), 159-184.

4. Y. Colin de Verdiére, "Sur le spectre des operateurs elliptiques a bicharacteristiques toutes periodiques", Commentarii Math. Helvetici 54 (1979), 508-522.

5. J. J. Duistermaat and V. Guillemin, "The spectrum of positive elliptic operators and periodic geodesics", Invent. Math. 29 (1975), 184-269.

6. W. Helton, "An operator-theoretic approach to partial differential equations, propagation of singularities and spectral theory", Indiana U. Math. Jour. Vol. 26, no. 6 (1977), 997-1018.

7. L. Hörmander, "The spectral function of an elliptic operator", Acta Math. 121 (1968), 193-218.

8. M. Kashiwara and M. Vergne, "Functions on the Shilov boundary of the generalized half-plane", in Non-commutative Harmonic Analysis, Lecture Notes in Math., No. 728, Springer, New York (1979).

9. B. Kostant, "Quantization and unitary representations", in Lectures in Modern Analysis and Applications, Springer Lecture Notes No. 170 (1970), 87-208.

10. B. Kostant, "Symplectic Spinors", Conv. di Geom. Simp. e Fis. Math., INDAM Rome (1973).

11. J. Moser, "Regularization of Kepler's problem and the averaging method on a manifold", Comm. Pure Appl. Math. 23 (1970),609-636.

12. E. Onofri, "Dynamical quantization of the Kepler manifold", Istituto di Fisica, Parma (1965).

13. D. Shale, "Linear symmetries of free boson fields", Trams. Amer. Math. Soc. 103 (1962), 149-167.

14. M. A. Shubin and V. N. Tulovskii, "On asymptotic distribution of eigenvalues of pseudodifferential operators on $\mathbb{R}^n$", Math. USSR Sbornik, 21 (1973), 565-583.

15. E. C. Titchmarsh, Eigenfunction Expansions Associated with Second-order Differential Equations Part II, Clarendon Press, Oxford (1958).

16. N. Wallach, Symplectic Geometry and Fourier Analysis, Math. Sci. Press, 53 Jordan Road, Brookline, Mass. (1977).

17. A. Weil, "Sur certaines groupes d'operateurs unitaires", Acta Math. 111 (1964), 143-211.

18. A. Weinstein, "Asymptotics of eigenvalue clusters for the Laplacian plus a potential", Duke Math. Journal 44 (1977), 883-892.

19. H. Widom, "Eigenvalue distribution theorems in certain homogeneous spaces", Journal of Functional Analysis, (to appear).

SUPERGRAVITY: A UNIQUE SELF-INTERACTING THEORY

S. Deser*

Department of Physics
Brandeis University
Waltham, Massachusetts 02254

Starting from a free flat space spin 2 plus 3/2 system, it is shown that its self-interacting generalization is supergravity. No local gauge principles are invoked, and the theory is unique.

INTRODUCTION

This is a report on a recently completed program to derive from first principles the structure of supergravity as a self-interacting system in a way entirely analogous to that by which the usual non abelian bosonic gauge theories may be understood. The work is the result of a collaboration with D.G. Boulware and J.H. Kay and a detailed version will soon appear [1]. My aim will be to give a physicist's discussion of the theory, starting from its humble origins as the sum of two free fields (with simple abelian invariances) which if they are to couple at all must do so in a unique way. The whole discussion will be carried out in flat space - curvature emerges as a consequence. To be sure, the fact that we are dealing here with graded algebras emerges as a property of the system but all the steps in the process are dictated by simple considerations of Noether currents associated with the global invariances inherent in the initial system, and grading does not alter the procedure in any way. I shall therefore deliberately be working with unsophisticated notation, avoiding superspace and superfields, although it should be instructive to have a parallel development in that language.

Let me begin with the main ideas underlying free gauge theories, which represent massless higher spin (>½) fields. In a Lorentz invariant description of higher spin particles, the number of components outruns what we know to be the number of degrees of freedom, which is always two; consequently there are restrictions which ensure that the spin points in the only possible directions - along or opposite the propagation vector $\vec{k}$ (there being no rest frame). The statement that the polarization is orthogonal to $\vec{k}$ translates, in a field description, to local abelian gauge invariance. This in turn implies that only conserved sources can be coupled. For the Maxwell case, the Bianchi identity (which is what the conservation requirement is) is simply $\partial_\mu(\partial_\nu F^{\mu\nu}) \equiv 0$. If there were several free gauge fields, say a set of photons A^a_μ, $a = 1-n$, then all sources J^a_μ must be conserved. So far there is not apparent difference between the single field and the global multiplet case. However, the difference is profound: one cannot retain the global invariance with dynamical sources:

$\partial_\mu J^\mu_a$ will not vanish, precisely because of the coupling $A^a_\mu J^\mu_a$. This is well-understood in both the Yang-Mills and gravity systems: the source of the latter is the matter stress tensor, which cannot be conserved since gravitational coupling means the matter system is no longer isolated, and gravitational energy must be included in the total stress tensor if it is to be conserved. The corresponding quantity for Yang-Mills is the isotopic charge: that carried by the gauge field itself must be included. This is the basis of nonlinear, nonabelian gauge theories: coupling requires the total-Noether-current to be included, and the scale of the gauge field's contribution is fixed with the same coupling constant as used for matter. These are only necessary conditions, and in general, when the spin exceeds 2, no consistent theory exists. When it does, there automatically emerges a local gauge invariance which is a combination of the original abelian helicity one and the global symmetry. Supergravity is apparently different from the vector and tensor cases because there are two separate initial systems (spins 2, 3/2) and two separate global invariances with associated Noether currents, one space-time the other graded. The first "miracle" about supergravity then is that there is nevertheless only a single coupling constant, the gravitational one.

GENERAL RELATIVITY

Let us begin with the more familiar case of pure Einstein theory, but in a way which has not been done before, namely in the vierbein rather than metric [2] formulation. As a reminder, we first write the full Einstein action

$$I^E = \frac{1}{2\kappa^2}\int e_{\mu a} e_{\nu b} {}^*R^{*\mu\nu\alpha\beta}(\omega)$$

$${}^*R^{*\mu\nu\alpha\beta} \equiv \frac{1}{4}\varepsilon^{\mu\nu\lambda\rho}\varepsilon^{abcd}R_{\lambda\rho cd} \quad , \quad R_{\mu\nu ab} \equiv \partial_\mu \omega_{\nu ab} - \omega_{\mu ba}\omega_{\nu ad} - (\mu\leftrightarrow\nu) \qquad (1)$$

$$\equiv R_L + R_Q$$

Note that the action in this form is polynomial in the basic variables $(e_{\mu a}, \omega_{\mu ab})$ of the first order formulation. What is the humble origin of this geometric theory? It is just the quadratic part of I^E in the variables $\kappa h_{\mu a} \equiv e_{\mu a} - \eta_{\mu a}$ and ω, where $\eta_{\mu a}$ is the Minkowski metric:

$$I_2^E(h,\omega) = \int [\eta_{\mu a} h_{\nu b} {}^*R_L^{*\mu\nu ab}(\omega) + \frac{1}{2}\eta_{\mu a}\eta_{\nu b} {}^*R_Q^{*\mu\nu ab}(\omega)] \ . \qquad (2)$$

The local abelian invariance which ensures that there are only pure helicity ±2 excitations is just

$$\delta h_{\mu a} = \partial_\mu \xi_a(x) \quad , \quad \delta\omega_{\mu ab} = 0 \qquad (3a)$$

while the global invariance we will use is given by

$$\delta h_{\mu a} = -\omega_{\mu ab}\rho^b \quad , \quad \rho^b = \text{const.}$$

$$\delta\omega_{\rho cd} = (\eta_{\rho b}\eta_{\mu c}\eta_{\nu d} - \frac{1}{2}\eta_{\rho d}\eta_{\mu c}\eta_{\nu b} + \frac{1}{2}\eta_{\rho c}\eta_{\mu d}\eta_{\nu b})\rho_a {}^*R_L^{*\mu\nu ab} \qquad (3b)$$

The Bianchi identities corresponding to (3a) are just $\eta_{\nu b}\partial_\mu {}^*R_L^{*\mu\nu ab} \equiv 0$, while the conserved Noether current corresponding to (3b) is the stress tensor

$$T_2^{\mu a} = h_{\nu b} {}^*R_L^{*\mu\nu ab} + \eta_{\nu b} {}^*R_Q^{*\mu\nu ab} \sim h\partial\omega + \omega^2 \ . \qquad (4)$$

Thus, if any coupling to matter is to be permitted, $T_2^{\mu a}$ must be included as part of the source, that is to say the field equation must read $\eta_{\mu a} {}^*R_L^{*\mu\nu ab} = T_2^{\nu b}$. This is achieved by adding to the linear action (2) the cubic term

$$I_3^E = \kappa\int [\frac{1}{2}h_{\mu a}h_{\nu b} {}^*R_L^{*\mu\nu ab} + h_{\mu a}\eta_{\nu b}\varepsilon^{\mu\nu\lambda\rho}\varepsilon^{abcd}\omega_{\lambda cf}\omega_{\rho df}]. \qquad (5)$$

But this added action has a stress tensor contribution of its own, namely

$$T_3^{\mu a} = \frac{1}{2}h_{\nu b}\varepsilon^{\mu\nu\lambda\rho}\varepsilon^{abcd}\omega_{\lambda cf}\omega_{\rho df} \qquad (6)$$

and inclusion of it is ensured by the quartic term

$$I_4^E = -\frac{\kappa}{4}\int h_{\mu a}h_{\nu b}\varepsilon^{\mu\nu\lambda\rho}\varepsilon^{abcd}\omega_{\lambda cf}\omega_{\rho df} \ . \qquad (7)$$

Here the process stops: $T_4^{\mu a}$ vanishes identically, as is clear because Noether currents arise only when there are explicit derivatives in the action. But this final model, with action $I_2 + I_3 + I_4$ is nothing but the Einstein action (1) in terms of the notation $\eta + \kappa h \equiv e$. Note that geometry has entered because the Minkowski metric $\eta_{\mu a}$ no longer enters explicitly, being replaced everywhere by the vierbein $e_{\mu a}$. We note parenthetically that there was also a separate local abelian invariance associated with (2), namely that of Lorentz rotations, which reverses the roles of δe, $\delta \omega$ with respect to the initial abelian invariance. The full action is invariant under

$$\delta e_{\mu a} = e_{\mu b} \Lambda_{ab}(x) \; , \quad \delta \omega_{\mu ab} = D_\mu \Lambda_{ab} \equiv (\partial + \omega)\Lambda . \qquad (8)$$
$$(\Lambda_{ab} \equiv -\Lambda_{ba})$$

This invariance alone is not sufficient to "bootstrap", roughly because it is one on the torsion rather than the true metric sector. This is clear because the associated Bianchi identity is the one linking the antisymmetric part of the Ricci tensor to the torsion:

$$^{*}R^{*cd}(\omega) + D_\rho {}^{*}C^{\nu}{}_{a}{}^{\rho} \varepsilon^{abcd} e_{\nu b} \equiv 0 \quad , \quad ^{*}R^{*cd} \equiv e_{\nu}{}^{c} e_{\mu b} {}^{*}R^{*\mu\nu bd} \qquad (8a)$$

rather than the important one on the Einstein tensor. On the other hand, had we used both invariances (3) and the linearized part of (8) together, we would still have ended up with the same final action and only one self-coupling constant κ despite having two ostensibly separate gauges. Here $C_{\mu\nu a} \equiv D_\mu(\omega)e_{\nu a} - D_\nu(\omega)e_{\mu a}$ is the torsion tensor.

I will not present here the rather complicated proof that I^E is unique. Essentially, it shows that had we started from a different initial description of the free system in terms of different variables, or used different Noether currents (remember that these are only unique up to identically conserved "superpotentials"), the end result would be the same. The basic idea uses the Ward identities associated with the final gauge invariance. One shows that all theories satisfying these identities are equivalent, and in particular equivalent to that form of the action which we obtained here.

We may therefore conclude that <u>any</u> theory which represents pure $s = 2$ massless particles is, at low frequencies at least, identical with general relativity. Also, all (minimal) coupling to matter is uniquely fixed, and this coupling includes as a corollary the equivalence principle on which Einstein originally based his considerations. Finally we mention that the cosmological constant version of gravity could also have been obtained in this way, although the physics is clearer here since we have stayed in flat space throughout where the familiar particle representations of the Lorentz group are the starting point. Armed with this example, we now proceed to supergravity, one component of which is of course the graviton, the other a fermionic spin 3/2 particle.

SUPERGRAVITY

Once again we begin with the end-result, namely the full supergravity [3,4] action I^{SG} which we will then derive from first principles.

$$I^{SG} = \kappa^{-2}\int \frac{1}{2} e_{\mu a} e_{\nu b} {}^{*}R^{*\mu\nu ab}(\omega) + \frac{i}{2}\int \bar{\psi}_\mu \gamma^a \gamma_5 e_{\nu a} {}^{*}f^{\mu\nu}$$
$${}^{*}f^{\mu\nu} \equiv \frac{1}{2}\epsilon^{\mu\nu\lambda\rho} f_{\lambda\rho} \quad , \quad f_{\lambda\rho} \equiv (\partial_\lambda - \frac{1}{2}\omega_{\lambda cd}\sigma^{cd})\psi_\rho - (\lambda \leftrightarrow \rho) \tag{9}$$

Here ψ_μ is an anticommuting vector-spinor which describes spin 3/2. We shall see that (9) is unique, consistent and describes a geometry with torsion whose source is ψ_μ. The full supergravity action (9) is invariant under three separate local transformations. The first is ordinary coordinate invariance with a local translation function $\xi_a(x)$,

$$\delta e_{\mu a} = D_\mu(\omega)\xi_a(x) \quad , \quad \delta\psi_\mu = e^{\nu a}\xi_a f_{\mu\nu}$$
$$\delta\omega_{\rho cd} = -|e|^{-1}\xi_a B^{\mu\nu ab}(e_{\rho b}e_{\mu c}e_{\nu d} - \frac{1}{2}e_{\mu c}e_{\rho d}e_{\nu b} + \frac{1}{2}e_{\mu d}e_{\rho c}e_{\nu b}) \tag{10}$$
$$B^{\mu\nu ab} \equiv {}^{*}R^{*\mu\nu ab} + \frac{i}{8}\epsilon^{\mu\nu\lambda\rho}\bar{f}_{\sigma\lambda}e^{\sigma a}\gamma_5\gamma^b\psi_\rho$$

where we have made use of the gauge invariance of ψ_μ to recast $\delta\psi_\mu$ in this form. The additional factor in $\delta\omega$ is due to the derivative coupling nature of the theory. The $\delta\psi_\mu$ is precisely the same as δA_μ for the Maxwell action, once a gauge transformation is also performed there. The new invariance under local supersymmetry transformations is of course the special aspect in which fermions and bosons are interchanged in terms of local parameters $\alpha(x)$ which are Grassmann elements:

$$\delta\psi_\mu = 2D_\mu(\omega)\alpha(x) \quad , \quad \delta e_{\mu a} = i\bar{\alpha}\gamma_a\psi_\mu$$
$$\delta\omega_{\mu ab} = B_{\mu ab} - \frac{1}{2}e_{\mu b}B_{ca}{}^{c} + \frac{1}{2}e_{\mu a}B_{cb}{}^{c} \quad , \quad \frac{1}{2}B_a{}^{\lambda\mu} \equiv i\bar{\alpha}\gamma_5\gamma_a {}^{*}f^{\lambda\mu} \tag{11}$$

The third invariance is again under local Lorentz transformations, but is omitted because it is neither sufficient to give the correct theory, nor necessary to include explicitly. Instead, it is automatically respected at each stage of the construction below.

To begin the construction, one takes the abelian free field system I_2^E of the spin 2 plus spin 3/2, I_2^{RS}. We don't bother to write the latter action explicitly; it is simply the flat space limit of the expression in (10). Now the combined system has two obvious independent local abelian invariances each of whose jobs it is to ensure pure helicity content for the respective free fields:

$$\delta h_{\mu a} = \partial_\mu \xi_a(x) \quad , \quad \delta\omega_{\mu ab} = 0 \quad , \quad \delta\psi_\mu = 0 \tag{12a}$$

$$\delta h_{\mu a} = 0 \quad , \quad \delta\omega_{\mu ab} = 0 \quad , \quad \delta\psi_\mu = 2\partial_\mu\alpha(x) \tag{12b}$$

The role of supersymmetry enters as the hidden global invariance which rotates the two fields into each other with a Grassmann parameter β (totally independent of $\alpha(x)$); there are therefore two global invariances when we include the global constant translations:

$$\delta e_{\mu a} = -\omega_\mu{}^{ab}\rho_b \quad , \quad \delta\psi_\mu = \rho^\nu f_{\mu\nu} \tag{13a}$$

$$\delta e_{\mu a} = i\bar{\beta}\gamma_a\psi_\mu \quad , \quad \delta\psi_\mu = -\omega_{\mu ab}\sigma^{ab}\beta \quad . \tag{13b}$$

We are now forced (if there is to be any interaction at all) to a double bootstrap on (as before) $T^{\mu a}$ and on J^μ, the spinorial Noether current as the source of the ψ_μ-field (we omit the explicit spin index throughout). Of course $T^{\mu a}$ is now to be augmented by the 3/2 contribution, namely

$$T_2^{\mu a} = h_{\nu b}{}^*R_L^{*\mu\nu ab} + \eta_{\nu b}{}^*R_Q^{*\mu\nu ab} + \frac{i}{2}\bar{\psi}_\nu\gamma_5\gamma^{a*}f_L^{\mu\nu} \quad , \quad f_{\mu\nu}^L \equiv \partial_\mu\psi_\nu - \partial_\nu\psi_\mu \tag{14a}$$

while the fermionic Noether current has the form

$$\bar{J}_2^\lambda = \frac{i}{2}\varepsilon^{\mu\nu\lambda\rho}\bar{\psi}_\mu\gamma_s\eta_\nu{}^a\{\gamma_a,\sigma_{cd}\}\omega_{\rho cd} + i{}^*\bar{f}_L^{\lambda\nu}h_\nu{}^a\gamma_5\gamma^a \quad . \tag{14b}$$

In obtaining the cubic action I_3^{SG} which will simultaneously accomplish both objectives of having (T_2,J_2) appear as sources, one would expect a combination $I_3 \sim \kappa_1\int\bar{J}_2^\mu\psi_\mu + \kappa_2\int h_{\mu a}T_2^{\mu a}$. In fact, however, there is an overlap between the terms required in the κ_1 and κ_2 parts; one is common to both and as a result, κ_1 must equal κ_2, which is one way of seeing that there is only one coupling constant (at another level, this is due to the fact that we are building up a single graded algebra in which (T,J) transform in a rigid way). So we are forced to

$$I_3^{SG} = \frac{\kappa}{2}\int[h_{\mu a}h_{\nu b}{}^*R_L^{*\mu\nu ab} + h_{\mu a}\eta_{\nu b}{}^*R_Q^{*\mu\nu ab} - \frac{i}{2}\bar{\psi}_\mu\gamma_5\gamma_a h_\nu{}^{a*}f_L^{\mu\nu}$$
$$+ \frac{i}{2}\varepsilon^{\mu\nu\lambda\rho}\bar{\psi}_\mu\gamma_5\gamma_a\eta_{\nu a}\omega_{\lambda cd}\sigma^{cd}\psi_\rho] \quad ; \tag{15}$$

again this generates both a $T_3^{\mu a}$ and a J_3^μ according to

$$T_3^{\mu a} = \frac{1}{2}h_{\nu b}\varepsilon^{\lambda\nu\mu\rho}\varepsilon^{abcd}\omega_{\nu cf}\omega_{\rho df} - \frac{i}{4}\varepsilon^{\mu\nu\lambda\rho}\bar{\psi}_\lambda\gamma_5\gamma^a\omega_{\nu cd}\sigma^{cd}\psi_\rho \tag{16b}$$

$$\bar{J}_3^\mu = \frac{i}{2}\varepsilon^{\mu\nu\lambda\rho}\bar{\psi}_\mu\gamma_5 h_\nu{}^a\omega_{\rho cd}\{\gamma_a,\sigma^{cd}\} \quad . \tag{16b}$$

The required I_4^{SG} which might have been of the form $I_4^{SG} \sim \kappa' T_3 h + \kappa'' J_3\psi$ again has unique κ because of the same mechanism as occurred at the previous step. Furthermore, this is again the end of the line in that this final form

$$I_4^{SG} = \frac{\kappa^2}{4}\int[-h_{\nu b}h_{\lambda a}\varepsilon^{\mu\nu\lambda\rho}\varepsilon^{abcd}\omega_{\mu cf}\omega_{\rho df} + i\varepsilon^{\mu\nu\lambda\rho}\bar{\psi}_\mu\gamma_5\gamma_a h_\nu{}^a\omega_{\lambda cd}\sigma^{cd}\psi_\rho] \quad . \tag{17}$$

has no explicity derivatives and consequently no quartic Noether currents. The sum $I_2 + I_3 + I_4$ is again precisely the full supergravity action (9), (with $e \equiv \eta + \kappa h$) which is thereby (in this first order form) seen to be purely polynomial - quartic in fact - in the basic variables (e,ω,ψ) . It is worth mentioning, however, that polynomiality is lost as soon as one goes beyond the simple supergravity model and considers any of the extended O(N>1) theories which include lower spins than the basic multiplet (e,ψ), for example, inclusion of spin 1 in the O(2) case brings in the contravariant metric in the Maxwell action, and it is nonpolynomial in $e_{\mu a}$.

Note how our original invariances (12,13) have been amalgamated into the full non-abelian graded symmetry (10,11). In particular, the anticommutator of two local supersymmetry transformations just reproduces the coordinate one. It may then seem strange that we could not have achieved our end result purely from the graded aspect alone, but this may be understood from the discussion of pure gravity in the previous section where it is clear that all energy must also be included as the source of the spin 2 field quite apart from any other invariances.

We leave as an open problem the derivation of extended supergravities in the same way from a free multiplet of spins (2,3/2,1,1/2,0). It is particularly impressive that in these models the non-gauge fields (1/2,0) are part of a genuine gauge multiplet.

The uniqueness of supergravity may be demonstrated as for pure gravity via Ward identities, (again it is a rather complicated derivation); the only freedom is the trivial one of field redefinitions. In particular, there are no fermionic contact self-coupling terms $\sim\psi^4$ in this formulation [4], although the explicit torsion contribution may of course be recast into this form when one goes over to second order formulation [3]. [It should be mentioned that field redefinitions, while trivial in principle, can complicate the proof immensely in practice. Thus, had we used second order form, we would have needed an infinite series in the vierbein deviations to obtain the final action!]

CONCLUSIONS

We have seen that both for pure gravity and its graded extension there is one and only one consistent theory beyond the free field level. Everything is determined by the original free fields and their pure helicity ±2, ±3/2 content, which are physical kinematical requirements. These in turn require conserved sources and therefore self-coupling. No use is made of the ultimate local gauge invariances possessed by the theory, which emerge as consequences of the required form of the coupling. Finally, we note that there is no a priori guarantee that this bootstrap process always works for any initial free gauge theory. Indeed, were one to try the same procedure starting from an initial system of free spin 2 and 5/2, no consistent theory could be obtained [5], despite the fact that everything looks the same at the free level. That is, we have the analogs of both (12',13) to start with, but the bootstrap fails because the Noether current remains a vector-spinor, while the (necessarily conserved) sources of spin 5/2 would have to be tensor-spinors. Thus supergravity appears (unfortunately for some purposes) to be unique in this way as well.

Although we do not yet have any direct evidence that supergravity is physically necessary to establish the unification of matter and geometry so dear to Einstein, it provides at the very least an elegant and unique model of how such unifications might come about. In particular it shows how (purely quantically definable) fermions can be unified with geometry into a rigid supermultiplet in exactly the same way as all other gauge theories can be derived.

REFERENCES

* Work supported in part by NSF Grant PHY 78 - 09644 .

1. D.G. Boulware, S. Deser, and J.H. Kay, Physica 96A, 141 (1979).
2. S. Deser, Gen. Relativ. Gravit. 1, 9 (1970)
3. D.Z. Freedman, P. van Nieuwenhuizen, and S. Ferrara, Phys. Rev. D13, 3214 (1976).
4. S. Deser and B. Zumino, Phys. Lett. 62B, 335 (1976).
5. C. Aragone and S. Deser, Phys. Lett. 86B, 161 (1979), and forthcoming papers.

GENERAL RELATIVITY AS A GAUGE THEORY

A. Pérez-Rendón
Universidad de Salamanca

Introduction

Given two classical fields, whose Lagrangians are known, a capital problem in field physics is to obtain their interaction equations.

It is well known that this problem was partially solved by Yang-Mills and Utiyama for invariant fields under a group of internal symmetries, in two papers ([16] and [17]) — that have become classics — and can be summarized as follows:

Let a classical field be given by means of a vector bundle $\pi_1 : E \longrightarrow \mathbb{R}^4$, with the Minkowskian space-time $(\mathbb{R}^4, T_2)$ as base space, satisfying the following conditions:

a) $\pi_1 : E \longrightarrow \mathbb{R}^4$ is a vector bundle with standard fibre F, associated to a principal bundle $p : P \longrightarrow \mathbb{R}^4$ with structure group G (group of internal symmetries).

b) Its Lagrangian density $\mathcal{L}\omega$ ($\mathcal{L}$ is a function on the bundle J^1E of 1-jets of local sections of E and ω is the volume form on $\mathbb{R}^4$ canonically associated to T_2) is G-invariant.

That is, there exists a representation of the Lie algebra $\mathfrak{g}$ of G in the module of vertical vector fields on E, whose 1-jets extensions, denoted by $\overline{\eta(e)}$, leave the Lagrangian density invariant.

In a mathematical language this condition states as:

The Lie derivative $\overline{\eta(e)}^L(\mathcal{L}\omega)$ vanishes for every $e \in \mathfrak{g}$.

Let us now consider the group of all automorphisms of the principal bundle $p : P \longrightarrow \mathbb{R}^4$ inducing the identity map on $\mathbb{R}^4$; we shall call it gauge group and denote it by Gau(P).

The Lie algebra of Gau(P) is the infinite-dimensional real Lie algebra (gauge algebra) $A \otimes_{\mathbb{R}} \mathfrak{g}$ endowed with the bracket operation $[f \otimes e, f' \otimes e'] = ff' \otimes [e, e']$, $(A = C^\infty(\mathbb{R}^4))$.

If for every $f_i \otimes e_i \in A \otimes_{\mathbb{R}} \mathfrak{g}$ we have $\overline{f_i \eta(e_i)}^L(\mathcal{L}\omega) = 0$, we shall say that the classical field we are dealing with is gauge invariant.

Now, we cannot infer at all the gauge invariance of a classical field from the second hypothesis stated above (invariance under the group G of internal symmetries). Moreover, checking that a G-invariant field is not gauge invariant is precisely the starting point of the Yang-Mills-Utiyama theory.

Before going on with thus, we should point out that most classical fields studied in Physics admit a third hypothesis:

c) $p : P \longrightarrow \mathbb{R}^4$ is a trivial principal bundle, i.e., $P = G \times \mathbb{R}^4$; in such case G is a subgroup of Gau(P).

Communication presented at the Conference on "Differential Geometrical Methods in Mathematical Physics", Salamanca, September 10th– 14th, 1979.

The Yang-Mills-Utiyama theory can now be summarized as follows:

In order to obtain the gauge invariance of the given field one must:

1) Add a second field, defined on the vector bundle $\pi_2 : K \longrightarrow \mathbb{R}^4$, associated to the affine bundle of connections on $p:P \longrightarrow \mathbb{R}^4$, by means of any possible Lagrangian V (which is not unique) given by Utiyama's theorem ([16] and [2]).

Such fields are called Yang-Mills fields.

2) Modify the Lagrangian of the first field, by adding to it a term $\mathcal{L}_{int}$ called interaction Lagrangian (which is certainly unique!)

Thus, the variational problem defined by the total Lagrangian density

$$\mathcal{L}_T \omega = (\mathcal{L} + V + \mathcal{L}_{int})\omega$$

on the fibred product $E \times_{\mathbb{R}^4} K$ is gauge invariant and its Euler-Lagrange equations are the interaction equations of the fields respectively defined on $\pi_1 : E \longrightarrow \mathbb{R}^4$ and $\pi_2 : K \longrightarrow \mathbb{R}^4$ by the Lagrangians $\mathcal{L}$ and V.

From a physical point of view two important restrictions appear in this theory of field interaction:

1. It cannot be used to study interaction between arbitrary fields, but just between a field defined on a vector bundle associated to $p:P \longrightarrow \mathbb{R}^4$ and its corresponding Yang-Mills field.

2. Both fields are invariant under a group of internal symmetries (Utiyama could not use his theory in General Relativity).

However, a purely mathematical viewpoint leads to more important criticisms:

1. The characterization of the field $\pi_2 : K \longrightarrow \mathbb{R}^4$ and the determination of the functions V and $\mathcal{L}_{int}$ are obtained by simple calculation and such calculation is based upon a quite unnatural requirement: that a variational problem invariant under the subgroup $G \subset Gau(P)$ be invariant under the whole group.

2. The entire theory looks like the mere consequence of the invariance of the first field $\pi_1 : E \longrightarrow \mathbb{R}^4$ with respect to the structure group of the principal bundle $p:P \longrightarrow \mathbb{R}^4$; this makes difficult its generalization to classical fields invariant under space-time symmetries.

In works [3] and [4], together with P.L. García, about intrinsic and global formulation of a variational problem (of first order) we proved that in order to formulate and solve a variational problem on the bundle $\pi_1 : E \longrightarrow \mathbb{R}^4$, one must build the bundle J^1E and give on it, besides the Lagrangian density $\mathcal{L}\omega$, the structure 1-form θ with values in the module of sections of VE (bundle of vertical vector fields on E).

As a matter of fact, J^1E is an affine bundle over E whose associated vector bundle is

$$\bar{E} = \mathrm{Hom}(\pi^* T(\mathbb{R}^4), VE)$$

$\pi^*T(\mathbb{R}^4)$ being the pull-back of the tangent bundle $T(\mathbb{R}^4)$ on E, and, so, the following question arises:

How about formulating the variational problem on $\bar{E}$ instead of doing it on J^1E ? .

We have a very simple relation between both bundles: Given a connection σ on $\pi_1 : E \longrightarrow \mathbb{R}^4$, we can define the following isomorphism of affine bundles on E :

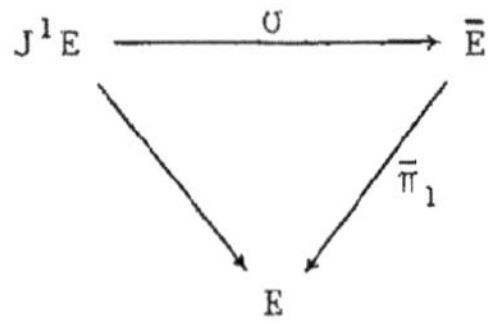

We associate to every point $j^1_x s$ of J^1E the point $(ds)_x - \sigma_{s(x)}$ of $\bar{E}$, s belonging to the coset $j^1 s$. If θ^σ denotes the image of the structure form θ of J^1E by such isomorphism, then we obtain as many structure forms θ^σ on $\bar{E}$ as connections σ defined on $\pi_1 : E \longrightarrow \mathbb{R}^4$, i.e., as many as connections on the principal bundle $p : P \longrightarrow \mathbb{R}^4$.

However, as was shown in [12] and [13] — both of them will be summarized in section 2.1 of this paper — , the structure forms can be defined directly on $\bar{E}$, without making use of the isomorphism $\sigma : J^1E \longrightarrow \bar{E}$; moreover, if θ^{σ_1} and θ^{σ_2} are the structure forms associated to the connections σ_1 , σ_2 on $\pi_1 : E \longrightarrow \mathbb{R}^4$, we have:

$$\theta^{\sigma_1}_{\bar{y}}(\bar{D}) = \theta^{\sigma_2}_{\bar{y}}(\bar{D}) + (\sigma_2 - \sigma_1)_x D'$$

where $\bar{y}$ is an arbitrary point of $\bar{E}$, $\bar{D}$ a vector of $T_{\bar{y}}(\bar{E})$ and D'_x its projection on $\mathbb{R}^4$. Summing up, there is no unique structure form on $\bar{E}$ and the set of all such forms can be parametrized by cross sections of the bundle $\pi_2 : K \longrightarrow \mathbb{R}^4$.

Let now a classical field with an arbitrary Lagrangian $\mathcal{L}$ be given on $\pi_1 : E \longrightarrow \mathbb{R}^4$ (no G-invariance is required here); we can define as many different variational problems on the vector bundle $\bar{\pi}_1 : \bar{E} \longrightarrow E$ as pairs $(\mathcal{L}\omega, \theta^\sigma)$, i.e., there are as many variational problems associated to a classical field as cross sections of the Yang-Mills field $\pi_2 : K \longrightarrow \mathbb{R}^4$, even if the Lagrangian density $\mathcal{L}\omega$ is unique and non-conditioned.

Nevertheless, if we restrict ourselves to a coordinate neighborhood and choose suitable local coordinates, there exists, for every pair $(\mathcal{L}\omega, \theta^\sigma)$, a local system (x_i, z_j, p_{ij}) carrying the Euler-Lagrange equations to the form

$$\frac{\partial}{\partial x_i}\left(\frac{\partial \mathcal{L}}{\partial p_{ij}}\right) - \frac{\partial \mathcal{L}}{\partial z_j} = 0$$

we expected a unique system of field equations and this is what has been found locally.

Could the uniqueness of the variational problem be recovered at a global level?.

In [13] we proved also the following theorem:

There exists a 1-form $\bar{\theta}$ defined on the fibred product $\bar{E}\times_{\mathbb{R}^4}\bar{K}$, with values in the module of sections of $V(E\times_{\mathbb{R}^4}K)$ and a family of immersions $i_\sigma:\bar{E}\longrightarrow\bar{E}\times_{\mathbb{R}^4}\bar{K}$ such that $\bar{\theta}|_{Im(i_\sigma)} = \theta^\sigma$.

In other words:

The non-uniqueness of the structure form θ^σ on $\bar{E}$ is avoided by enlarging the initial bundle to $\bar{E}\times_{\mathbb{R}^4}\bar{K}$ and by taking $\bar{\theta}$ as structure form on it.

Futhermore, since $V(E\times_{\mathbb{R}^4}K)$ can be identified with the Whitney sum $VE\oplus VK$ on $E\times_{\mathbb{R}^4}K$, $\bar{\theta}$ splits into the sum of two terms $\bar{\theta}_1$ and $\bar{\theta}_2$, with values in VE and VK, respectively. Taking a local coordinate system $(x_i,z_j,p_{ij},A_{ij},B_{\ell ij})$ in $\bar{E}\times_{\mathbb{R}^4}\bar{K}$ we can write (see [4]):

$$\bar{\theta} = \bar{\theta}_1 + \bar{\theta}_2 = \left[dz_j - (p_{ij} + A_{i\ell}\, a^{\ell}_{jm}\, z_m)dx_i\right] \frac{\partial}{\partial z_j} + (dA_{hk} - B_{ihk}\, dx_i) \frac{\partial}{\partial A_{hk}}$$

(the constants a^{ℓ}_{jm} depend only on the representation ρ of the structure group G in the standard fibre F of E).

Finally, we change our local coordinate system by taking

$$p'_{ij} = p_{ij} + A_{i\ell}\, a^{\ell}_{jm}\, z_m$$

Then, we can write $\bar{\theta}$ as

$$\bar{\theta} = (dz_j - p'_{ij}\, dx_i) \frac{\partial}{\partial z_j} + (dA_{hk} - B_{ihk}\, dx_i) \frac{\partial}{\partial A_{hk}}$$

and the Lagrangian $\mathcal{L}(x_i,z_j,p_{ij})$ becomes $\mathcal{L}(x_i,z_j,p'_{ij} - A_{i\ell}\, a^{\ell}_{jm}\, z_m)$.

Let us take now a cross section s of $\bar{E}\times_{\mathbb{R}^4}\bar{K}\longrightarrow\mathbb{R}^4$ such that $\bar{\theta}|_{S(\mathbb{R}^4)} = 0$; the restriction of $\mathcal{L}$ to $S(\mathbb{R}^4)$ is

$$\mathcal{L}(x_i,z_j,\frac{\partial z_j}{\partial x_i} - A_{i\ell}\, a^{\ell}_{jm}\, z_m)$$

which, according to Utiyama, is $\mathcal{L}+\mathcal{L}_{int}$.

That is, in a field interaction we can distinguish two parts which are utterly different:

a) Adjunction of the corresponding Yang-Mills field to the field E and modification of the initial Lagrangian by means of a interaction term $\Longleftrightarrow$ Non-canonical choice of a structure form on $\bar{E}$.

b) Calculation of possible Lagrangians V of the Yang-Mills field — assuming the initial field to be G-invariant — and requiring gauge invariance of the variational problem defined on $\bar{E}\times_{\mathbb{R}^4}\bar{K}$ by $(\bar{\theta},(\mathcal{L}+V)\omega)$.

Therefore, invariance under G or under Gau(P) is involved only in the second part and, since the required condition is very weak, the number of possible

Lagrangians is infinite. (For a formalization of the second part of Yang-Mills-Utiyama theory, see [2]).

This paper starts a series planned to generalize the Yang-Mills-Utiyama theory and the ideas developed so far, in order to give the theory of interaction between a classical field and a gravitational one.

As we pointed out above, Utiyama himself attempted to give a gauge theory of gravity, regarding Einstein's General Relativity as a special case of his own theory. (The communication by Y. Ne'eman at the Bonn Conference (1977) on Differential Geometrical Methods in Mathematical Physics is an excellent review of Utiyama's attempt, of the contributions by Sciama [14] and Kible [8] — which led to the rediscovery of the forgotten Cartan's theory — and of later contributions by Hehl et al. ([6], [7])).

Starting form some ideas due to Ne'eman and Regge ([11], [18]), we obtain the main result in this paper:

The Poincaré gauge field theory by Hehl and Von der Heyde when formulated in a bundle language is equivalent to solving a variational problem with a non-canonical choice of the structure form and a volume form that must be non-canonical too, since one cannot use the Minkowskian space-time as base manifold.

1. Automorphisms of the principal bundle $P = L \times \mathbb{R}^4$.

1.1. We shall denote by L the Lorentz group and by $\mathbb{R}^4$ the group of translations in the affine space $\mathbb{R}^4$. It is well known that the Poincaré group P is the semidirect product of both groups. That is, P is the set of all pairs (ϕ,a), where $\phi \in L$ and $\phi \in \mathbb{R}^4$, satisfying the following composition law:

$$(\phi,a)(\psi,b) = (\phi\psi,\phi(b) + a)$$

Moreover, since L and $\mathbb{R}^4$ are Lie groups, P is also the product manifold $L \times \mathbb{R}^4$ and $p:P \longrightarrow \mathbb{R}^4$ is now a regular projection.

In other words:

(1) $p:P \longrightarrow \mathbb{R}^4$ is a trivial principal bundle with structure group L.

We are going to study its automorphisms. An automorphism of (1) is a diffeomorphism α of P onto P commuting with the action of the structure group:

$$\alpha(\xi\phi) = \alpha(\xi)\phi$$

for all $\xi \in P$ and $\phi \in L$.

Let Aut(P) be the group of all automorphisms of (1) and Gau(P) the subgroup of automorphisms inducing the identity map on $\mathbb{R}^4$, i.e., such that $p(\alpha(\xi)) = p(\xi)$. This group is the gauge group of the principal bundle $p:P \longrightarrow \mathbb{R}^4$ (see [2] and [15]) and this fact suggests the notation Gau(P).

Proposition 1: We have the following inclusions between the groups P, L, Aut(P) and Gau(P):

$$P \subset \mathrm{Aut}(P) \quad ; \quad L \subset \mathrm{Gau}(P)$$

Proof: Every element in P acting on the left defines obviously an automorphism of $p:P \longrightarrow \mathbb{R}^4$.

On the other hand, since P is the product manifold $L \times \mathbb{R}^4$, there exists an isomorphism between Gau(P) and the group of differentiable functions on $\mathbb{R}^4$ with values in L.

Indeed, we can associate to every automorphism $\beta(1,x) = (\phi,x)$, for any $x \in \mathbb{R}^4$, the mapping $\tilde{\beta} : \mathbb{R}^4 \longrightarrow L$ given by $\tilde{\beta}(x) = \phi$ and, conversely, we can associate to every mapping $\tilde{\beta}$ the automorphism $\beta(\psi,a) = (\tilde{\beta}(a) \cdot \psi, a)$. Finally, L is isomorphic to the subgroup of differential mappings of $\mathbb{R}^4$ into L which are constant along $\mathbb{R}^4$.

1.2. Let us consider the exact sequence of vector bundles

(2) $$0 \longrightarrow \mathrm{Ad}\, P \longrightarrow T(P)_L \longrightarrow T(\mathbb{R}^4) \longrightarrow 0$$

where $T(P)_L$ is the bundle of L-invariant vector fields on P, Ad P (adjoint bundle of P) is the subbundle of vector fields tangential to the fibres and $T(\mathbb{R}^4)$ is the tangent bundle of $\mathbb{R}^4$.

The corresponding sequence of global sections over $\mathbb{R}^4$ is:

$$0 \longrightarrow \Gamma(\mathrm{Ad}\, P) \longrightarrow \Gamma(T(P_L)) \longrightarrow \Gamma(T(\mathbb{R}^4)) \longrightarrow 0$$

where $\Gamma(\mathrm{Ad}\, P)$ and $(T(P_L))$ are, respectively, the Lie algebras of Gau(P) and of Aut(P).

From Proposition 1 we infer that the Lie algebra of the Poincaré group, $\mathfrak{g}_P$, is a subalgebra of the Lie algebra of infinitesimal automorphisms of $p: P \longrightarrow \mathbb{R}^4$ and that $\mathfrak{g}_L$ (Lie algebra of the Lorentz group) is a subalgebra of the gauge algebra of P.

Using once more the fact $P = L \times \mathbb{R}^4$, we have:

The gauge algebra $\Gamma(\mathrm{Ad}\, P)$ can be identified with the tensor product $A \otimes_{\mathbb{R}} \mathfrak{g}_L$ (denoting by A the ring $C^\infty(\mathbb{R}^4)$) if we define

$$[f \otimes e, f' \otimes e'] = ff' \otimes [e, e']$$

However, the algebra of infinitesimal automorphisms of P, $\Gamma(T(P_L))$ and $A \otimes_{\mathbb{R}} \mathfrak{g}_P$ are isomorphic only as A-modules.

1.3. Let F be a vector space of dimension n, ρ a linear representation of the Lorentz group in F and $\pi_1: E \longrightarrow \mathbb{R}^4$ the vector bundle associated to $p: P \longrightarrow \mathbb{R}^4$ with standard fibre F. Of course, $\pi_1: E \longrightarrow \mathbb{R}^4$ is also a trivial bundle $E = F \times \mathbb{R}^4$.

Let e be a vector in F and q_e the differential mapping of P into E such that

$$q_e(\xi) = (\overline{\xi, e}) \in E$$

If ξ is the point of P given by (ϕ, x), then $((\phi, x), e)$ and $((I, x), \rho(\phi^{-1})e)$ belong to the same coset $(\overline{\xi, e})$, and $(\overline{\xi, e})$ can be identified with the point $(\rho(\phi^{-1})e, x)$ in $F \times \mathbb{R}^4$. The mapping q_e can also be written as follows:

$$q_e(\phi, x) = (\rho(\phi^{-1})e, x)$$

Hence:

a) The diagram

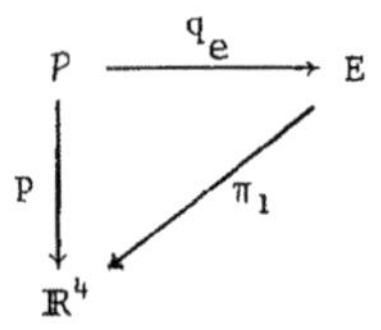

is commutative.

b) We can associate to every infinitesimal automorphism D of P a vector field on E, $\tilde{D}$, such that

$$\tilde{D}_{q_e(\xi)} = (q_e)_* D_\xi$$

Thus, if $D \in \mathrm{Ad}\, P$, then $(q_e)_*(D)$ is a vector field tangential to the fibres.

c) There exists a representation η of $\mathfrak{g}_P$ in the module of vector fields on E.

In order to prove this assert, let $(\tau_{\alpha\beta})$ be a basis of the Lie algebra $\mathfrak{g}_L$, $D_{\alpha\beta}$ the fundamental vector field on P corresponding to $\tau_{\alpha\beta}$ and $f_{\alpha\beta} = (q_e)_* D_{\alpha\beta}$. We choose a basis $(\tau_{\alpha\beta}, \varepsilon_\gamma)$ in $\mathfrak{g}_P$, with $\alpha < \beta = 1,\ldots,4$ and $\gamma = 1,\ldots,4$. The six infinitesimal rotations and the four infinitesimal translations satisfy the well-known relations:

$$[\varepsilon_\gamma, \varepsilon_\delta] = 0$$

$$[\tau_{\alpha\beta}, \varepsilon_\gamma] = g_{\gamma\alpha}\varepsilon_\beta - g_{\gamma\beta}\varepsilon_\alpha = g_{\gamma[\alpha}\varepsilon_{\beta]}$$

$$[\tau_{\alpha\beta}, \tau_{\gamma\delta}] = g_{\gamma[\alpha}\tau_{\beta]\delta} - g_{\delta[\alpha}\tau_{\beta]\gamma}$$

$$\tau_{\alpha\beta} = -\tau_{\beta\alpha}$$

Finally, if (x_i, z_j), with $i = 1,\ldots,4$ and $j = 1,\ldots,n$, is a coordinate system in E, z_j being linear functions on F, then the representation is given by

$$\eta(\tau_{\alpha\beta}) = x_\alpha \frac{\partial}{\partial x_\beta} - x_\beta \frac{\partial}{\partial x_\alpha} + f_{\alpha\beta} = x_\alpha \frac{\partial}{\partial x_\beta} - x_\beta \frac{\partial}{\partial x_\alpha} + a^{\alpha\beta}_{jk}\, z_j \frac{\partial}{\partial z_k}$$

$$\eta(\varepsilon_\gamma) = \frac{\partial}{\partial x_\gamma}$$

2. Variational calculus in $\pi_1 : E \longrightarrow \mathbb{R}^4$.

2.1. Let us build on the vector bundle $\pi_1 : E \longrightarrow \mathbb{R}^4$ its vertical bundle $\bar{\pi}_1 : \bar{E} \longrightarrow E$, where $\bar{E} = \mathrm{Hom}(\pi_1^* T(\mathbb{R}^4), VE)$ (see [3]), $\pi_1^* T(\mathbb{R}^4)$ is the pull-back of $T(\mathbb{R}^4)$ on E and $VE \subset T(E)$ is the subbundle of the vector fields tangential to the fibres of E.

We proved in [3] that, in order to define a variational problem in $\pi : E \longrightarrow \mathbb{R}^4$, a structure 1-form θ with values in the module of sections of VE and a density form with Lagrangian $\mathcal{L}\omega$ must be given on $\bar{E}$. However, as we said above, the difficulty lies in the non-uniqueness of the form θ.

Let us now summarize the results obtained in [12] and [13]:

Given a connection σ on $\pi_1 : E \longrightarrow \mathbb{R}^4$, we can define on $\bar{E}$ a structure form θ^σ in the following way:

$$\theta^\sigma_{\bar{y}}(\bar{D}) = D_y - \sigma_x(D') + \psi(D'_y) \tag{3}$$

where $\bar{D}_{\bar{y}}$ is any tangent vector at $\bar{y} = (y, \psi)$, $\psi \in \mathrm{Hom}(\pi_1^* T_x(\mathbb{R}^4), VE_y)$, and D'_x, D_y are the projections of $\bar{D}_{\bar{y}}$ on $\mathbb{R}^4$ and E, respectively, i.e., $D_y = (\bar{\pi}_1)_* \bar{D}_{\bar{y}}$, $D'_x = (\pi_1)_* D_y$, with $x = \pi_1(y)$.

Furthemore, if we choose the flat connection σ_0 on $\pi_1 : E \longrightarrow \mathbb{R}^4$ and if θ^{σ_0} is the associated structure form, then we have:

$$\theta^{\sigma}_{\bar{y}}(\bar{D}) = \theta^{\sigma_0}_{\bar{y}}(\bar{D}) + (\sigma_0 - \sigma)_x D' \tag{4}$$

Consequently, one may define as many structure forms on $\bar{E}$ as substractions $(\sigma - \sigma_0)$. Now, a 1-form on $\mathbb{R}^4$ with values in $\mathfrak{g}_L$ can be uniquely assigned to every pair (σ,σ_0), since $K = T^*(\mathbb{R}^4) \otimes \mathrm{Ad}\, P$ is the vector bundle associated to the affine bundle of connections on $p : P \longrightarrow \mathbb{R}^4$ ([2]). This leads to the following important result:

<u>The set of structure forms on $\bar{E}$ can be parametrized by means of the cross sections of the bundle</u> K.

We choose in $\bar{E}$ the following coordinate system: (x_i, z_j, p_{ij}), where (x_i, z_j) are the coordinates of E defined in 1.3 and p_{ij} are the 4n functions given by

$$p_{ij}(\bar{y}) = -\left[\theta^{\sigma_0}_{\bar{y}}\left(\frac{\partial}{\partial x_i}\right)\right] z_j$$

for every point $\bar{y}$ of $\bar{E}$.

The form θ^{σ} can be written in terms of the first system as:

$$\theta^{\sigma} = \left[dz_j - (p_{ij} + \Gamma^{\alpha\beta}_i\, a^{\alpha\beta}_{kj}\, z_k)\, dx_i\right] \frac{\partial}{\partial z_j} \tag{5}$$

where $\Gamma^{\alpha\beta}_i\, dx_i\, f_{\alpha\beta}$ is the coordinate expression of the 1-form on $\mathbb{R}^4$, with values in the bundle VE, associated to the pair of connections (σ,σ_0). In particular, θ^{σ_0} can be written as

$$\theta^{\sigma_0} = (dz_j - p_{ij}\, dx_i)\frac{\partial}{\partial z_j}$$

which is the well-known structure form of the bundle of 1-jets $j^1\pi_1 : J^1E \longrightarrow \mathbb{R}^4$ (see [3] and [4]).

Finally, we shall see how θ^{σ_0} yields all structure forms θ^{σ}.

Let s be the cross section of the bundle K associated to (σ,σ_0) and β_s the $\bar{\pi}_1$-vertical automorphism of the bundle $\pi_1 : \bar{E} \longrightarrow E$ such that

$$\beta_s(\bar{y}) = (y, \psi + \sigma_0 - \sigma)$$

where $\bar{y} = (y,\psi)$.

<u>All structure forms θ^{σ} are the image of the form θ^{σ_0} by β^*_s</u>.

Indeed, the automorphism β_s acts on the functions (x_i, z_j, p_{ij}) by carrying p_{ij} into $p_{ij} + \Gamma^{\alpha\beta}_i\, a^{\alpha\beta}_{kj}\, z_k$ and, so:

$$\theta^{\sigma} = (\beta_s)^* \, \theta^{\sigma_0}$$

2.2. In order to develop a gauge theory of gravity, the former results must be generalized as follows:

Let $Aut(T(\mathbb{R}^4))$ be the trivial bundle on $\mathbb{R}^4$ whose fibre at any point x is $Aut_{\mathbb{R}}(T_x(\mathbb{R}^4))$. A cross section ϕ of $Aut(T(\mathbb{R}^4))$ is characterized when sixteen functions $e^i{}_\ell$ are given on $\mathbb{R}^4$ such that the square matrix $(e^i{}_\ell(x))$ has non-zero determinant for all x.

Given a connection σ on $\pi_1 : E \longrightarrow \mathbb{R}^4$ and a cross section ϕ of the bundle $Aut(T(\mathbb{R}^4))$, we can define the following structure form on $\bar{E}$:

$$\theta^{\sigma,\phi}_{\bar{y}}(\bar{D}) = D_y - \sigma_x(D') + \psi \cdot \phi^{-1}(D'_x) \tag{3'}$$

In particular, if we choose the flat connection σ_0 on the bundle $\pi_1 : E \longrightarrow \mathbb{R}^4$ and the identity section of $Aut(T(\mathbb{R}^4))$, we obtain:

$$\theta^{\sigma_0,I}_{\bar{y}}(\bar{D}) = D_y - \sigma^0_x(D') + \psi(D'_x) = \theta^{\sigma_0}_{\bar{y}}(\bar{D})$$

and

$$\theta^{\sigma,\phi}_{\bar{y}}(\bar{D}) = \theta^{\sigma_0,\phi}_{\bar{y}}(\bar{D}) + (\sigma_0 - \sigma)_x D' \tag{4'}$$

Therefore: <u>The set of structure forms on $\bar{E}$ can be parametrized by cross sections</u> of the bundle $Y = K \times_{\mathbb{R}^4} Aut(T(\mathbb{R}^4))$ (fibred product of K and $Aut(T(\mathbb{R}^4))$ over $\mathbb{R}^4$). In terms of the system (x_i, z_j, p_{ij}) the form $\theta^{\sigma,\phi}$ is:

$$\theta^{\sigma,\phi} = \left[dz_j - (e_i{}^\ell p_{\ell j} + \Gamma^{\alpha\beta}_i a^{\alpha\beta}_{kj} z_k)dx_i\right] \frac{\partial}{\partial z_j} \tag{5'}$$

if $\phi^{-1}(\frac{\partial}{\partial x_i}) = e_i{}^\ell \frac{\partial}{\partial x_\ell}$; hence, the functions $e_i{}^\ell$ satisfy the following conditions:

$$e^j{}_\ell \, e_i{}^\ell = \delta_{ij}$$

Finally, let $\bar{s}$ be the cross section of Y given by $(\sigma - \sigma_0, \phi)$ and $\beta_{\bar{s}}$ the $\bar{\pi}_1$-vertical automorphism of $\bar{E}$ such that

$$\beta_{\bar{s}}(\bar{y}) = (y, \psi \cdot \phi^{-1} + \sigma_0 - \sigma)$$

where $\bar{y} = (y,\psi)$. <u>All structure forms $\theta^{\sigma,\phi}$ on $\bar{E}$ are the image of θ^{σ_0} by β^*_s</u>.

2.3. A last question arises when defining a variational problem in E: <u>which Lagrangian density should be chosen?</u>

In classical field theory the base space $\mathbb{R}^4$ is usually assumed to be <u>the Minkowskian space-time</u> $(\mathbb{R}^4, T_2)$. Thus, all possible Lagrangian densities are $\mathcal{L}\omega$, where $\mathcal{L}$ is an arbitrary function on $\bar{E}$ and ω is the volume form on $\mathbb{R}^4$ canonically associated to T_2.

Obviously, the former hypothesis cannot be used in a gravitatory theory. Nevertheless, all metrics $\bar{T}_2$ that may be defined on $\mathbb{R}^4$ obey the following law: $(\bar{T}_2)_x$ and

$(T_2)_x$ are linearly equivalent at every point x. Consequently, all possible volume forms can be parametrized by cross sections of the bundle $\mathrm{Aut}(T(\mathbb{R}^4))$ and ω^ϕ is the 4-form on $\mathbb{R}^4$ given by

$$\omega^\phi(D_1,\ldots,D_4) = \omega(\phi^{-1}(D_1),\ldots,\phi^{-1}(D_4))$$

where D_i are four arbitrary vector fields.

If $(x_1,\ldots,x_4)$ is an inertial system of the space-time $(\mathbb{R}^4, T_2)$, then

$$\omega^\phi = e\, dx_1 \wedge \ldots \wedge dx_4$$

where e(x) is the determinat of the matrix associated to $\phi(x)^{-1}$.

In short, if $\pi_1 : E \longrightarrow \mathbb{R}^4$ is a classical field with Lagrangian $\mathcal{L}$, then its stationary sections cannot be characterized globally by means of a unique variational problem. Each pair of forms $(\theta^{\sigma,\phi}, \mathcal{L}\omega^\phi)$, defined on $\bar{E}$, stands for a different variational problem. However, the Euler-Lagrange equations are all

$$\frac{\partial}{\partial x_i}\left(e \frac{\partial \mathcal{L}}{\partial p_{ij}}\right) - e \frac{\partial \mathcal{L}}{\partial z_j} = 0$$

when choosing a coordinate system in $\bar{E}$ consisting of the functions (x_i, z_j, p_{ij}) such that

$$e = \omega^\phi\left(\frac{\partial}{\partial x_1},\ldots,\frac{\partial}{\partial x_4}\right) \ ; \quad \theta^{\sigma,\phi} = (dz_j - p_{ij}dx_i)\frac{\partial}{\partial z_j}$$

3. Minimal interaction with gravitational fields.

3.1. Let $\bar{\pi}_2 : \bar{Y} \longrightarrow Y$ be the vertical bundle on $\pi_2 : K \times_{\mathbb{R}^4} \mathrm{Aut}(T(\mathbb{R}^4)) \longrightarrow \mathbb{R}^4$, where $\bar{Y} = \mathrm{Hom}(\pi_2^* T(\mathbb{R}^4), VY)$. Obviously, $\bar{Y} = \bar{K} \times_{\mathbb{R}^4} \overline{\mathrm{Aut}(T(\mathbb{R}^4))}$ if $\bar{K}$ and $\overline{\mathrm{Aut}(T(\mathbb{R}^4))}$ are the corresponding vertical bundles on K and $\mathrm{Aut}(T(\mathbb{R}^4))$.

Definition 1. By a gravitational field we shall mean a classical field associated to the bundle $\pi_2 : Y \longrightarrow \mathbb{R}^4$ whose Lagrantian is, for the time being, an arbitrary function V on $\bar{Y}$.

Let us consider now the fibre product $\bar{E} \times_{\mathbb{R}^4} \bar{Y}$.

Proposition. Every cross section $s = (\sigma - \sigma_0, \phi)$ of $\pi_2 : Y \longrightarrow \mathbb{R}^4$ allows us to define an injection i_s of $\bar{E}$ into $\bar{E} \times_{\mathbb{R}^4} \bar{Y}$.

If $\bar{y} = (y_1, \psi_1)$ is a point of $\bar{E}$, then $\bar{z} = i_s(\bar{y}) = (y_1, (\sigma-\sigma_0)_x, \phi(x), \psi_1, \psi_2)$, where $x = \pi_1(y_1)$ and $\psi_2 = \bar{s}(x)$.

$\bar{s}$ is the canonical lift of s to $\bar{Y}$, i.e., [3] , the cross section of $\pi_2 \circ \bar{\pi}_2 : \bar{Y}_2 \longrightarrow \mathbb{R}^4$ projecting on s through $\bar{\pi}_2$ and such that the restriction to $\bar{s}$ of the structure form on $\bar{Y}$—associated to the flat connection on Y—is zero.

The main result obtained in this section is the following

Theorem. We can build on $\bar{E}\times_{\mathbb{R}^4}\bar{Y}$ a structure form $\bar{\theta}$, with values in the bundle $V(E\times_{\mathbb{R}^4}Y)$, and a volume form $\bar{\omega}$, whose respective restrictions to $\mathrm{Im}(i_s)$ are $\theta^{\sigma,\phi}$ and ω^{ϕ}.

In other words:

The non-uniqueness, in $\bar{E}$, of the variational problem associated to a classical field $\pi_1: E \longrightarrow \mathbb{R}^4$, with Lagrangian $\mathcal{L}$, is avoided by enlarging the first vector bundle to $E\times_{\mathbb{R}^4}Y$ and considering the variational problem given by $(\bar{\theta},(\mathcal{L}+V)\bar{\omega})$ on $\bar{E}\times_{\mathbb{R}^4}\bar{Y}$.

Remark. Throughout this section, $\mathcal{L}$ and V are both arbitrary functions on $\bar{E}$ and $\bar{Y}$.

Let us define the forms θ and $\bar{\omega}$.

Definition 2. We shall denote by $\bar{\theta}$ the 1-form on $\bar{E}\times_{\mathbb{R}^4}\bar{Y}$, with values in the bundle $V(E\times_{\mathbb{R}^4}Y)$, such that

$$\bar{\theta}_{\bar{z}}(\bar{D}) = D_z - \sigma_x(D') + \psi_1(\phi_x^{-1}(D')_{y_1}) - (\sigma_0)_x D' + \psi_2(D'_{y_2})$$

where $\bar{z} = (z,\psi_1,\psi_2) = (y_1,y_2,\psi_1,\psi_2)$, $\bar{D}$ is a tangent vector at $\bar{z}\in\bar{E}\times_{\mathbb{R}^4}\bar{Y}$ and D_z, D_x are its corresponding projections on $E\times_{\mathbb{R}^4}Y$ and $\mathbb{R}^4$.

The bundle $V(E\times_{\mathbb{R}^4}Y)$ may be identified with the Whitney sum $VE\oplus VK\oplus V\,\mathrm{Aut}(T(\mathbb{R}^4))$ on $E\times_{\mathbb{R}^4}Y$ and, so, $\bar{\theta}$ splits into the sum of three terms with values in VE, VK and $V\,\mathrm{Aut}(T(\mathbb{R}^4))$, respectively.

We shall compute it in coordinates:

We take the functions $\Gamma_i^{\alpha\beta}$ $(i=1,\dots,4\,;\alpha<\beta=1,\dots,4)$ on K such that their value at a point $(\sigma-\sigma_0)_x$ is $\tau^*_{\alpha\beta}((\sigma-\sigma_0)_x\frac{\partial}{\partial x_i})$, where $\tau^*_{\alpha\beta}$ is a basis of $\mathfrak{g}^*_L$ (the dual space of $\mathfrak{g}_L$). We choose the functions e_i^{ℓ} $(i=\ell=1,\dots,4)$ on $\mathrm{Aut}(T(\mathbb{R}^4))$ such that $e_i^{\ell}(\phi_x)$ is the i,ℓ-th component of the 4×4 square matrix associated to the automorphism ϕ^{-1} of $T_x(\mathbb{R}^4)$ and to the basis $(\frac{\partial}{\partial x_1},\dots,\frac{\partial}{\partial x_4})$. The functions $(x_i,z_j,\Gamma_i^{\alpha\beta},e_i^{\ell})$ are a coordinate system in $E\times_{\mathbb{R}^4}Y$.

Finally, taking the functions $B_{ih}^{\alpha\beta}$ $(i=h=1,\dots,4\,;\alpha<\beta=1,\dots,4)$ on $\bar{K}$, such that their value at $((\sigma-\sigma_0)_x,\psi_2')$ is $-[\psi_2'(\frac{\partial}{\partial x_h})]\Gamma_i^{\alpha\beta}$, and the functions $H_{i\,k}^{\ell}$ on $\overline{\mathrm{Aut}(T(\mathbb{R}^4))}$ such that their value at (ϕ_x,ψ_2'') is $-[\psi_2''(\frac{\partial}{\partial x_h})]e_i^{\ell}$, one has a coordinate systém $(x_i,z_j,\Gamma_i^{\alpha\beta},e_i^{\ell},p_{ij},B_{ih}^{\alpha\beta},H_{i\,k}^{\ell})$ in $\bar{E}\times_{\mathbb{R}^4}\bar{Y}$, where $p_{ij}(\bar{y}) = -[\theta_{\bar{y}}^{\sigma_0}(\frac{\partial}{\partial x_i})]z_j$.

In this system the form $\bar{\theta}$ is:

$$(6)\qquad \bar{\theta} = \left[dz_j - (e_i^{\ell}p_{\ell j} + \Gamma_i^{\alpha\beta}\,a_{kj}^{\alpha\beta}\,z_k)dx_i\right]\frac{\partial}{\partial z_j} + (d\Gamma_h^{\alpha\beta} - B_{ih}^{\alpha\beta}\,dx_i)\frac{\partial}{\partial\Gamma_h^{\alpha\beta}} + (de_h^{\ell} - H_{h\,i}^{\ell}\,dx_i)\frac{\partial}{\partial e_h^{\ell}} = \bar{\theta}_1 + \bar{\theta}_2 + \bar{\theta}_3$$

Notice the essential difference between $\bar{\theta}_1$ and equation (5') obtained in the previous section: In (5') e_i^{ℓ} and $\Gamma_i^{\alpha\beta}$ are functions on $\mathbb{R}^4$ while in $\bar{\theta}_1$ they are variables, not depending on x_i and defined on Aut(T ($\mathbb{R}^4$)) and K, respectively.

From the definition of canonical lift of a cross section $s = (\sigma-\sigma_0, \phi)$ of $\pi_2 : Y \longrightarrow \mathbb{R}^4$ to $\bar{Y}$ — which was given at the beginning of this paragraph —, together with equations (5') and (6), one infers immediately that the restriction of $\bar{\theta}$ to $\mathrm{Im}(i_s)$ is $\theta^{\sigma,\phi}$.

Definition 3. $\bar{\omega}$ is the 4-form defined on $\bar{E} \times_{\mathbb{R}^4} \bar{Y}$ such that

$$\bar{\omega}_{\bar{z}}\left(\frac{\partial}{\partial x_1}, \ldots, \frac{\partial}{\partial x_4}\right) = \omega_x\left(\phi^{-1}\left(\frac{\partial}{\partial x_1}\right), \ldots, \phi^{-1}\left(\frac{\partial}{\partial x_4}\right)\right)$$

if $\bar{z} = (y_1, y_2, \psi_1, \psi_2)$, where $y_2 = ((\sigma-\sigma_0)_x, \phi_x)$ and ω is the volume element on $\mathbb{R}^4$ canonically associated to the Minkowskian metric T_2. Clearly, ω^{ϕ} becomes now the restriction of $\bar{\omega}$ to $\mathrm{Im}(i_s)$.

3.2. We conclude this section by writing, in the coordinate system $(x_i, z_j, \Gamma_i^{\alpha\beta}, e_i^{\ell}, p_{ij}, B_{ih}^{\alpha\beta}, H_i^{\ell}{}_k)$, the Euler-Lagrange equations of the variational principle defined on $\bar{E} \times_{\mathbb{R}^4} \bar{Y}$ by $(\bar{\theta}, (\mathcal{L} + V)\bar{\omega})$.

The stationary sections of a variational problem are characterized as follows (see [3], and also [5] and [9]).

A cross section s of $E \times_{\mathbb{R}^4} Y \longrightarrow \mathbb{R}^4$ is a stationary section of the variational problem stated above if and only if for every vector field $\bar{D}$ on $\bar{E} \times_{\mathbb{R}^4} \bar{Y}$ we have

(7)
$$(i\, D \cdot d\bar{\Theta})_{\bar{s}(\mathbb{R}^4)} = 0$$

where $\bar{s}$ is the canonical lift of the section s to $\bar{E} \times_{\mathbb{R}^4} \bar{Y}$ — carried out by making use of the structure form $\bar{\theta}$ —, $\bar{\Theta} = \bar{\theta} \wedge \bar{\Omega} - (\mathcal{L} + V)\bar{\omega}$ is the Poincaré-Cartan form associated to the given variational problem and $\bar{\Omega}$ is the Legendre form. In coordinates we have:

(8)
$$\bar{\Omega} = (-1)^i e^i_{\ell} \frac{\partial \mathcal{L}}{\partial p_{\ell j}}\, e\omega_i \circ dz_j + (-1)^i \frac{\partial V}{\partial B_{ih}^{\alpha\beta}}\, e\omega_i \circ d\Gamma_h^{\alpha\beta} + (-1)^i \frac{\partial V}{\partial H_h{}^{\ell}{}_i}\, e\omega_i \circ de_h^{\ell}$$

where $\omega_i = dx_1 \wedge \ldots \wedge \widehat{dx_i} \wedge \ldots \wedge dx_4$.

(6) and (8) yield

$$\bar{\Theta} = (-1)^i e^i_{\ell} \frac{\partial \mathcal{L}}{\partial p_{\ell j}}\, e\, dz_j \wedge \omega_i + (-1)^i \frac{\partial V}{\partial B_{ih}^{\alpha\beta}}\, e\, d\Gamma_h^{\alpha\beta} \wedge \omega_i + (-1)^i \frac{\partial V}{\partial H_h{}^{\ell}{}_i}\, e\, de_h^{\ell} \wedge \omega_i +$$

$$+ \left[(e_i^{\ell}\, p_{\ell j} + \Gamma_i^{\alpha\beta}\, a_{kj}^{\alpha\beta}\, z_k) e^i_{\ell} \frac{\partial \mathcal{L}}{\partial p_{\ell j}} + B_{ih}^{\alpha\beta} \frac{\partial V}{\partial B_{ih}^{\alpha\beta}} + H_h{}^{\ell}{}_i \frac{\partial V}{\partial H_h{}^{\ell}{}_i} - (\mathcal{L} + V)\right] \cdot e\omega$$

Therefore, the Euler-Lagrange equations are:

(8) $$\frac{\partial}{\partial x_i}\left(e^i_\ell \frac{\partial \mathcal{L}}{\partial p_{\ell k}}\right) + \Gamma^{\alpha\beta}_i a^{\alpha\beta}_{kj} e^i_\ell \frac{\partial \mathcal{L}}{\partial p_{\ell j}} - \frac{\partial \mathcal{L}}{\partial z_k} + \frac{1}{e}\frac{\partial e}{\partial x_i} e^i_\ell \frac{\partial \mathcal{L}}{\partial p_{\ell k}} = 0$$

(9) $$\frac{\partial}{\partial x_i}\left(\frac{\partial V}{\partial H_h{}^\ell{}_i}\right) - \left(e^h{}_\ell V + \frac{\partial V}{\partial e_h{}^\ell}\right) = e^h{}_\ell \mathcal{L} - p_{\ell j}\frac{\partial \mathcal{L}}{\partial p_{hj}} = e^h{}_\ell \mathcal{L} - \frac{\partial \mathcal{L}}{\partial p_{hj}} D_\ell z_j$$

(since $p_{\ell j} = e^i{}_\ell\left(\frac{\partial}{\partial x_i} - \Gamma^{\alpha\beta}_i f_{\alpha\beta}\right) z_j = D_\ell z_j$ on account of $\bar{\theta}_{\bar{s}}(\mathbb{R}^4) = 0$), and

(10) $$\frac{\partial}{\partial x_i}\left(\frac{\partial V}{\partial B^{\alpha\beta}_{ih}}\right) - \left(\frac{\partial V}{\partial \Gamma^{\alpha\beta}_h} - \frac{\partial V}{\partial \Gamma^{\beta\alpha}_h}\right) = - \frac{\partial \mathcal{L}}{\partial p_{hj}} f_{\alpha\beta} z_j$$

Equations (9) and (10) generalize, as we shall see in the next section, the equations given by Hehl, Nitsch and Von der Heyde in [7].

4. Hehl - Nitsh - Von der Heyde equations.

We shall develop the intrinsic characterization of all possible gauge field Lagrangians V in a further paper. Here we just check by means of a local calculation that equations (9) and (10) coincide with Hehl-Nitsh-Von der Heyde equations, if we impose the following condition on V :

Let X be the vertical vector field on Y given by

$$X = \left(-D_i\varepsilon^\ell + \omega_h{}^\ell e_i{}^h - \varepsilon^j F_{ji}{}^\ell\right)\frac{\partial}{\partial e_i{}^\ell} - \left(D_i\,\omega^{\alpha\beta} + \varepsilon^j F_{ji}{}^{\alpha\beta}\right)\frac{\partial}{\partial \Gamma_i{}^{\alpha\beta}}$$

where ε^ℓ and $\omega^{\alpha\beta}$ are arbitrary functions on $\mathbb{R}^4$, $D_i = \frac{\partial}{\partial x_i} - \Gamma_i{}^{\alpha\beta} f_{\alpha\beta}$ and $F_{ji}{}^\ell$, $F_{ji}{}^{\alpha\beta}$ are the components of the torsion and the curvature tensors associated to σ .

And let $\bar{X}$ be its canonical lift to $\bar{Y}$ through the structure form $\bar{\theta}_2 + \bar{\theta}_3$, i.e. [3], $\bar{X}$ is the only field on $\bar{Y}$ projecting on X and leaving the structure form $\bar{\theta}_2 + \bar{\theta}_3$ invariant in the following sense:

The Lie derivative $\bar{X}^L(\bar{\theta}_2 + \bar{\theta}_3) = \Phi \circ (\bar{\theta}_2 + \bar{\theta}_3)$.

If $\bar{X}V = 0$, for any functions ε^ℓ , $\omega^{\alpha\beta}$, then we have

$$\frac{\partial}{\partial x_i}\left(\frac{\partial V}{\partial H_h{}^\ell{}_i}\right) - \left(e^h{}_\ell V + \frac{\partial V}{\partial e_h{}^\ell}\right) = D_i\left(\frac{\partial V}{\partial H_h{}^\ell{}_i}\right) - \left(e^h{}_\ell V - \right.$$

$$\left. - F_{\ell j}{}^\gamma \frac{\partial V}{\partial H_j{}^\gamma{}_h} - F_{\ell j}{}^{\gamma\delta}\frac{\partial V}{\partial B^{\gamma\delta}_{jh}}\right) = D_i\, H_\ell^{hi} - \varepsilon_\ell{}^h$$

and

$$\frac{\partial}{\partial x_i}\left(\frac{\partial V}{\partial B^{\alpha\beta}_{ih}}\right) - \left(\frac{\partial V}{\partial \Gamma^{\alpha\beta}_h} - \frac{\partial V}{\partial \Gamma^{\beta\alpha}_h}\right) = D_i\, H_{\alpha\beta}{}^{hi} - \varepsilon_{\alpha\beta}{}^h$$

(with the Hehl's notation [7]).

REFERENCES

[1] Cartan, E. - Sur les variétés à connexion affine et la théorie de la relativité généralisée, I partie, Ann. Ec. Norm. 40 (1923), 325.

[2] García, P.L. - Gauge algebras, curvature and symplectic structure, Journ. Diff. Geo., 12 (1977), 209-227.

[3] García, P.L. and A. Pérez-Rendón - Symplectic Approach to the Theory of Quantized Fields I. Comm. Math. Phys., 13 (1969), 24-44.

[4] García, P.L. and A. Pérez-Rendón - Reductibility of the symplectic structure of minimal interactions. In "Differential Geometrical Methods in Mathematical Physics", Bonn, 1977. Springer Pub., 1978.

[5] Goldschmidt, H. and S. Sternberg - The Hamilton Cartan formalism in the calculus of variations. Ann. Inst. Four., 23 (1973), 203-267.

[6] Hehl, F.W., P.v.d. Heyde, G.D. Kerlick and J.M. Nester - General Relativity with spin and torsion: Foundations and prospects. Rev. Mod. Phys., 48 , (1976), 393-416.

[7] Hehl, F.W., J. Nitsch and P.v.d. Heyde. In "Einstein Commemorative Volume, Plenum Press, 1979/80.

[8] Kible, T.W.B. - Lorentz Invariance and Gravitational Field, J. Math. Phys., 2 (1961), 212-221.

[9] Kijowski, J. and W. Szczyrba - A canonical Structure for Classical Field Theories. Comm. Math. Phys., 46 (1976), 183-206.

[10] Ne'eman, Y. - Gravity is a gauge theory of parallel-transport. Modification of the Poincaré group. In "Differential Geometrical Methods in Mathematical Physics", Bonn, 1977, Springer Pub., 1978.

[11] Ne'eman, Y. and T. Regge - Gauge theory of gravity and supergravity on a group manifold. Rivista del Nuovo Cim. Vol., 1, n° 5 (1978).

[12] Pérez-Rendón, A. - A minimal interaction principle for classical fields. Sym. Math., 14 (1974), 293-321.

[13] Pérez-Rendón, A. - Yang-Mills interactions: a problem not depending on the Gauge-invariance. In "3rd Int. Coll. on Group Theoret. Method in Physics. Marseille, (1974).

[14] Sciama, D.W. - In "Recent Developments in General Relativity". Pergamon Press (1962), 415-440.

[15] Sternberg, S. - On the role of field theories in our physical conception of geometry. In "Differential Geometrical Methods in Mathematical Physics". Bonn, 1977, Springer Pub., 1978.

[16] Utiyama, R. - Invariant theoretical interpretation of interaction. Phys. Rev. 101 (1956), 1597-1607.

[17] Yang, C.N. and R.L. Mills - Conservation of Isotopic Spin and Isotopic Gauge invariance. Phys. Rev. 96 (1954), 191-

[18] Ne'eman - Gravity, Groups and Gauges. In "Contribution to the Einstein centenary GRG Volume". Plenum Press. 1979/80.

ON A PURELY AFFINE FORMULATION OF GENERAL RELATIVITY

by

Jerzy Kijowski
Institute of Mathematical Methods in Physics,
University of Warsaw, ul. Hoża 74;
00-682 Warszawa, Poland

Recent discovery of a new symplectic structure of the classical field theory /see [4] [5] [3]/ has important implications for General Relativity. It turns out that the symmetric affine connection $\Gamma^{\lambda}_{\mu\nu}$ in space-time M /and not a metric tensor $g_{\mu\nu}$/ plays the role of field potentials. Einstein equations can be derived from a first order variational principle

$$\delta \int L = 0 \qquad (1)$$

where the Lagrangian density L depends on a connection $\Gamma^{\lambda}_{\mu\nu}$, a matter field φ^A /if there is any/ and their first derivatives $\Gamma^{\lambda}_{\mu\nu\varkappa} = \partial_{\varkappa}\Gamma^{\lambda}_{\mu\nu}$ $\varphi^A_{\mu} = \partial_{\mu}\varphi^A$. There is no metric tensor in the Lagrangian. The metric appears in the theory as a component of a momentum canonically conjugate to the connection:

$$\pi_{\lambda}{}^{\mu\nu\varkappa} = \frac{\partial L}{\partial \Gamma^{\lambda}_{\mu\nu\varkappa}} \qquad (2)$$

/see [1]/. The General Relativity theory is a special case of such a theory. It is distinguished among other such theories by the following assumption: the Lagrangian L depends on derivatives of $\Gamma^{\lambda}_{\mu\nu}$ only via the symmetric part $K_{\mu\nu}$ of a Ricci tensor $R_{\mu\nu}$:

$$K_{\mu\nu} = R_{(\mu\nu)} = \frac{1}{2}\left(R^{\lambda}{}_{\mu\lambda\nu} + R^{\lambda}{}_{\nu\lambda\mu}\right) \qquad (3)$$

where

$$R^{\lambda}{}_{\mu\nu\varkappa} = \Gamma^{\lambda}_{\mu\varkappa\nu} - \Gamma^{\lambda}_{\mu\nu\varkappa} + \Gamma^{\lambda}_{\sigma\nu}\Gamma^{\sigma}_{\mu\varkappa} - \Gamma^{\lambda}_{\sigma\varkappa}\Gamma^{\sigma}_{\mu\nu} \qquad (4)$$

Thus

$$K_{\alpha\beta} = \left(\Gamma^{\lambda}_{\mu\nu\varkappa} + \Gamma^{\lambda}_{\sigma\varkappa}\Gamma^{\sigma}_{\mu\nu}\right)\left(\delta^{\varkappa}_{\lambda}\,\delta^{\mu}_{(\alpha}\,\delta^{\nu}_{\beta)} - \delta^{\mu}_{\lambda}\,\delta^{\nu}_{(\alpha}\,\delta^{\varkappa}_{\beta)}\right). \tag{5}$$

We do not know a priori wheather $\Gamma^{\lambda}_{\mu\nu}$ is the metric connection for a metric g. Therefore, we do not know a priori that $R_{\mu\nu}$ is symmetric. However, the metricity of the connection /and thus the symmetry of $R_{\mu\nu}$ / will be a consequence of field equations as we will show in the sequel.

For the sake of conceptual symplicity we assume in the present paper that L depends on $\Gamma^{\lambda}_{\mu\nu}$ also only via $K_{\mu\nu}$ /for the discussion of a general case see [1] [3]/. Thus

$$L = L(K_{\mu\nu}, \varphi^{A}, \varphi^{A}_{\mu}). \tag{6}$$

This happens in the case of a free gravitational field and also for matter which is described by a scalar field, Maxwell field, Proca field and Hydrodynamics. The assumption (6) implies the reduction of the momentum $\pi_{\lambda}^{\mu\nu\varkappa}$ according to the formula (2):

$$\pi_{\lambda}^{\mu\nu\varkappa} = \frac{\partial L}{\partial\Gamma^{\lambda}_{\mu\nu\varkappa}} = \frac{1}{2}\,\frac{\partial L}{\partial K_{\alpha\beta}}\cdot\frac{\partial K_{\alpha\beta}}{\partial\Gamma^{\lambda}_{\mu\nu\varkappa}} = \pi^{\alpha\beta}\left(\delta^{\varkappa}_{\lambda}\,\delta^{(\mu}_{\alpha}\,\delta^{\nu)}_{\beta} - \delta^{(\nu}_{\lambda}\,\delta^{\mu)}_{\alpha}\,\delta^{\varkappa}_{\beta}\right) \tag{7}$$

where

$$\pi^{\alpha\beta} = \frac{\partial L}{\partial K_{\alpha\beta}} \tag{8}$$

Finally

$$\pi_{\lambda}^{\mu\nu\varkappa} = \delta^{\varkappa}_{\lambda}\,\pi^{\mu\nu} - \delta^{(\mu}_{\lambda}\,\pi^{\nu)\varkappa} \tag{9}$$

It turns out that our theory is exactly Einstein's General Relativity if we interprete the tensor density $\pi^{\mu\nu}$ as a contravariant density of metric tensor:

$$\pi^{\mu\nu} = -\frac{1}{k}\sqrt{-g}\;g^{\mu\nu} \tag{10}$$

where $g = \det g_{\alpha\beta}$ and k is the gravitational constant. Equation (10) is a definition of the metric tensor $g_{\mu\nu}$ in terms of the canonical momentum π . The Euler-Lagrange equations of our theory are:

$$\partial_{\varkappa}\,\frac{\partial L}{\partial\Gamma^{\lambda}_{\mu\nu\varkappa}} = \frac{\partial L}{\partial\Gamma^{\lambda}_{\mu\nu}} \tag{11}$$

and

$$\partial_\mu \frac{\partial L}{\partial \varphi^A_\mu} = \frac{\partial L}{\partial \varphi^A} \tag{12}$$

We introduce the momentum canonically conjugate to the matter field

$$p^\mu_A = \frac{\partial L}{\partial \varphi^A_\mu} \tag{13}$$

Euler-Lagrange equations can thus be rewritten in the following form:

$$\partial_\varkappa \pi_\lambda{}^{\mu\nu\varkappa} = \frac{\partial L}{\partial \Gamma^\lambda_{\mu\nu}} \tag{14}$$

$$\partial_\mu p^\mu_A = \frac{\partial L}{\partial \varphi^A} \tag{15}$$

But

$$\frac{\partial L}{\partial \Gamma^\lambda_{\mu\nu}} = \frac{1}{2} \frac{\partial L}{\partial K_{\alpha\beta}} \frac{\partial K_{\alpha\beta}}{\partial \Gamma^\lambda_{\mu\nu}} = \pi^{\alpha\beta} \cdot \frac{1}{2} \frac{\partial K_{\alpha\beta}}{\partial \Gamma^\lambda_{\mu\nu}} \tag{16}$$

The last term is linear in terms of $\Gamma^\lambda_{\mu\nu}$. It turns out that the value of (16) is exactly what we need in order to replace the partial derivative on the left hand side of (14) by the covariant derivative. The equation (14) reads

$$\nabla_\varkappa \pi_\lambda{}^{\mu\nu\varkappa} = 0 \tag{17}$$

Using (9) we obtain

$$\nabla_\lambda \pi^{\mu\nu} = 0 \tag{18}$$

Finally using the equation (10) we obtain the metricity condition

$$\nabla_\lambda g_{\mu\nu} = 0 \tag{19}$$

equivalent to $\Gamma^\lambda_{\mu\nu}$ being the Levi-Civitta connection for the metric g

$$\Gamma^\lambda_{\mu\nu} = \left\{ {\lambda \atop \mu\nu} \right\} \tag{20}$$

We see that varying the Lagrangian with respect to $\Gamma^\lambda_{\mu\nu}$ we obtain the equation (11) equivalent to metricity equation (20). Varying L with respect to the matter field φ^A we obtain the matter equation (12). But where are Einstein equations?

It turns out that Einstein equations are equivatent to the equation (2) or , equivalently, to the equation (8). It gives us the relation between the momentum π and the field variables. It is analogous to the equation $p=m\dot{q}$ in particle mechanics /here the relation between the momentum $\pi^{\mu\nu}$ and the velocity $K_{\mu\nu}$ is , in general , non-linear/. The dynamical equation $\dot{p}=F$ in mechanics is analogous to the metricity equation (17) for gravitational field and to equation (15) for matter field.

In order to prove the equivalence between (8) and Einstein equations we introduce the following notation

$$J_\lambda^{\mu\nu} = \nabla_\varkappa \pi_\lambda{}^{\mu\nu\varkappa}, \qquad j_A = \partial_\mu p_A^\mu . \tag{21}$$

The complete set of field equations ce be written in the following way:

$$dL = \frac{1}{2}\pi^{\mu\nu}\, dK_{\mu\nu} + \frac{1}{2} J_\lambda^{\mu\nu}\, d\Gamma^\lambda_{\mu\nu} + j_A\, d\varphi^A + p_A^\mu\, d\varphi^A_\mu .$$

Now we perform a Legendre transformation in gravitational variables replacing "velocities" $K_{\mu\nu}$ by "momenta" $\pi^{\mu\nu}$:

$$d\left(L - \frac{1}{2}\pi^{\mu\nu}K_{\mu\nu}\right) = -\frac{1}{2}K_{\mu\nu}\, d\pi^{\mu\nu} + \frac{1}{2} J_\lambda^{\mu\nu}\, d\Gamma^\lambda_{\mu\nu} + j_A\, d\varphi^A + p_A^\mu\, d\varphi^A_\mu . \tag{22}$$

The scalar density

$$L_{mat} = L - \frac{1}{2}\pi^{\mu\nu}K_{\mu\nu} = L + \frac{1}{2k}\sqrt{-g}\cdot g^{\mu\nu}R_{\mu\nu} = L + \frac{1}{2k}\sqrt{-g}\, R \tag{23}$$

will be called the matter Lagrangian. After the Legendre transformation it becomes a function of "hamiltonian variables" for gravitational field and "lagrangian variables" for matter field:

$$L_{mat} = L_{mat}\left(\pi^{\mu\nu}, \varphi^A, \varphi^A_\mu\right) = L_{mat}\left(g_{\mu\nu} ; \varphi^A, \varphi^A_\mu\right) \tag{24}$$

Equation (22) reads

$$\frac{\partial L_{mat}}{\partial \pi^{\mu\nu}} = -K_{\mu\nu} , \qquad \frac{\partial L_{mat}}{\partial \Gamma^\lambda_{\mu\nu}} = J_\lambda^{\mu\nu} = \nabla_\varkappa \pi_\lambda{}^{\mu\nu\varkappa} \tag{25}$$

and

$$\frac{\partial L_{mat}}{\partial \varphi^A_\mu} = p_A^\mu , \qquad \frac{\partial L_{mat}}{\partial \varphi^A} = j_A = \partial_\mu p_A^\mu \tag{26}$$

But

$$\frac{\partial}{\partial \pi^{\mu\nu}} = -\frac{k}{\sqrt{-g}}\left(\frac{\partial}{\partial g^{\mu\nu}} - \frac{1}{2} g_{\mu\nu}\, g^{\alpha\beta}\frac{\partial}{\partial g^{\alpha\beta}}\right) \tag{27}$$

because of equation (10). Putting (27) into (25) we obtain finally

$$k\,\frac{1}{\sqrt{-g}}\,\frac{\partial L_{mat}}{\partial g^{\mu\nu}} = K_{\mu\nu} - \frac{1}{2} R\cdot g_{\mu\nu} \tag{28}$$

which is exactly the Einstein equation if we interprete (24) as a matter Lagrangian. Matter Lagrangian plays role of a Lagrangian with respect to matter field φ^A /equations (26)/ and plays role of a Hamiltonian for the gravitational field $\Gamma^{\lambda}_{\mu\nu}$. The quantity

$$\frac{1}{2}\pi^{\mu\nu} K_{\mu\nu} = -\frac{1}{2k}\sqrt{-g}\cdot R \qquad (29)$$

in the equation

$$L = L_{mat} - \frac{1}{2k}\sqrt{-g}\cdot R \qquad (30)$$

is not a "gravitational lagrangian" which has to be added to the matter Lagrangian L_{mat}, but is merely a Legendre transformation term analoguous to "$p\dot{q}$" in classical mechanics. Equation (30) means that the numerical value of our Lagrangian is equal to the value of the Lagrangian in the standard approach. However, the metric tensor $g_{\mu\nu}$ has to be elliminated. The situation is analogous to classical mechanics where the equation

$$L(q,\dot{q}) = p\dot{q} - H(q,p) \qquad (31)$$

gives us the Lagrangian provided we are able to elliminate p from the right-hand side.

Examples:

1^0 The Lagrangian for the Klein-Gordon-Einstein field /scalar field interacting with the gravitational field/ is equal

$$L = \frac{1}{2}\sqrt{-g}\left(g^{\mu\nu}\varphi_\mu\varphi_\nu - m^2\varphi^2\right) - \frac{1}{2k}\sqrt{-g}\; R =$$

$$= -\frac{1}{2}\left(k\pi^{\mu\nu}\varphi_\mu\varphi_\nu + k^2\sqrt{-\det\pi^{\mu\nu}}\;\; m^2\varphi^2\right) + \frac{1}{2}\pi^{\mu\nu}K_{\mu\nu} \qquad (32)$$

Einstein equations are:

$$K_{\mu\nu} = k\left(\varphi_\mu\varphi_\nu - \frac{1}{2}g_{\mu\nu}m^2\varphi^2\right) \qquad (33)$$

or, equivalently

$$g_{\mu\nu} = \left(\frac{1}{2}m^2\varphi^2\right)^{-1}\left[\varphi_\mu\varphi_\nu - \frac{1}{k}K_{\mu\nu}\right] \qquad (34)$$

Inserting (34) into (32) we obtain

$$L = \left(\frac{1}{2}m^2\varphi^2\right)^{-1}\sqrt{-\det\left(\varphi_\mu\varphi_\nu - \frac{1}{k}K_{\mu\nu}\right)} \qquad (35)$$

The reader may easily check that the above Lagrangian depending on $\Gamma^{\lambda}_{\mu\nu}$, φ^{A} and their first derivatives produces the complete Klein--Gordon-Einstein theory.

2° A configuration of a barotropic fluid in hydrodynamics is described by the mapping

$$\Psi : M \longrightarrow Z \tag{36}$$

where M is a space-time and Z is a 3-dimensional material space equipped with the volume-structure, i.e. the 3-form

$$\underline{r} = r(z)\ dz^{1} \wedge dz^{2} \wedge dz^{3} \tag{37}$$

which measures the number of moles of a fluid. The matter Lagrangian depends on field potentials (z^{a}) , a=1,2,3, and their first derivatives

$$z^{a}_{\mu} = \partial_{\mu} z^{a} \tag{38}$$

It turns out that the numerical value of L_{mat} is equal to $-\sqrt{-g}\,\mathcal{E}(\varrho)$ where $\mathcal{E}(\varrho)$ is the rest-mass /rest energy/ density of the fluid depending on the matter density:

$$= \frac{1}{\sqrt{-g}} \sqrt{j^{\mu} j^{\nu} g_{\mu\nu}} \tag{39}$$

and j^{μ} is a matter current defined by the equation

$$\Psi^{*}\underline{r} = j^{\mu}\frac{\partial}{\partial x^{\mu}} \rfloor \left(dx^{o} \wedge \dots \wedge dx^{3}\right). \tag{40}$$

The equation defines j^{μ} as an algebraic function of (z^{a}) and their first derivatives (z^{a}_{μ}) /see [2],[3]/. The gravitating fluid will thus be described by the lagrangian

$$L = L_{mat} - \frac{1}{2k} \sqrt{-g}\ R\ . \tag{41}$$

The ellimination of the metric tensor $g_{\mu\nu}$ from the above formula gives the following result

$$L = -\sqrt{\det K_{\alpha\beta}} \cdot f\left(\frac{K_{\mu\nu}\, j^{\mu} j^{\nu}}{\det K_{\alpha\beta}}\right)$$

where f is a function of a sinle real variable. The function is uniquely defined by the state equation $\mathcal{E} = \mathcal{E}(\varrho)$ /see [2]/. The quantity

$$\frac{K_{\mu\nu}\, j^{\mu} j^{\nu}}{\det K_{\alpha\beta}} \tag{43}$$

is a scalar because both numerator and denominator are scalar densities of weigt 2. The signature of $K_{\alpha\beta}$ has to be strictly positive in order to have the correct signature of $g_{\mu\nu}$. The reader may easily check that the above Lagrangian d pending on $\Gamma^{\lambda}_{\mu\nu}$, z^a and their first derivatives produces /when varying with respect to $\Gamma^{\lambda}_{\mu\nu}$ and z^a/ the complete Navier-Stokes-Einstein theory.

References

[1] Kijowski J. : G.R.G. Journal, 9 /1978/ p. 857

[2] Kijowski J., Pawlik B., Tulczyjew W.M. : A variational formulation of non-graviting and graviting hydrodynamics ; Bull. Acad. Polon. Sci. /Math., phys., astr./ in print.

[3] Kijowski J., Tulczyjew W.M. : A symplectic framework for field theories ; Springer Lecture Notes in Physics, vol. 107

[4] Tulczyjew W.M. : Symposia Math. 14 /1974/ p. 247

[5] Tulczyjew W.M. : Ann. Inst. H. Poincaré 27 A /1977/ p. 101

A FIBRE BUNDLE DESCRIPTION OF COUPLED GRAVITATIONAL AND GAUGE FIELDS

Wojciech Kopczyński[1]

Instytut Fizyki Teoretycznej, Uniwersytet Warszawski
Hoża 69, 00-681 Warszawa, Poland
and
Institut für Theoretische Physik, Universität zu Köln
Zülpicher Str. 77, 5000 Köln 41, West Germany

Abstract

This article can be understood as a preamble to consider principal fibre bundles associated to gauge fields as a true arena for physical fields and particles. The bundle approach to gauge fields and the generalized Kaluza-Klein theory are reviewed. Physically plausible conditions imposed on the metric tensor on the bundle imply restrictions on admissible gauge groups, formulated here as necessary and sufficient conditions. Geodesics on the bundle are shown to describe motion of classical particles with non-Abelian gauge charge. An attempt to give for a linear connection on the bundle a role indenendent on the metric structure leads to averting the principal difficulty of the generalized Kaluza-Klein theory: presence of an enormous cosmological constant.

Based on a lecture delivered at the Conference on Differential Geometrical Methods in Mathematical Physics, Salamanca 1979.

---

[1]Alexander von Humboldt fellow

1. INTRODUCTION

It is known from several years that principal fibre bundles provide an adequate frame for description of gauge fields. The fibre bundle approach ensures us a reasonable understanding of the role played by the gauge field potential and the gauge field strenght; it gives us moreover a transparent geometrical interpretation of gauge transformations.

The discussion on an appropriate gauge approach to gravitation, starting from fundamental Utiyama's paper (1), has a long history and is still continuated (2, 3, 4). The discussion got a new impulse in 1972 due to the proof of renormalizability of gauge theories (5). A gravitational field theory, treated as a gauge theory, has indeed certain distinguished features. The gravitational connection is a connection on the bundle of linear frames over space-time. This bundle is soldered to space-time, whereas bundles of proper gauge fields are attached to space-time in a quite arbitrary manner.

The succes of the bundle description of gauge fields stimulates us to treat the bundle manifolds as a true arena for physical fields and particles and not as a mathematical tool only. It means that one hopes that the bundle manifolds could play in future a role similar to that played by the space-time manifold today. "Similar" here does not mean identical, although one wishes to extent the similarity as much as possible. Then the following questions, concerning the role of the gravitational field, arise. Should we treat the gravitational bundle in the same manner as the bundles corresponding to electromagnetic or gluon fields? Should we attach to it the meaning of an arena for physics? In my opinion, the answer to this question is negative, because the bundle of linear frames. which is given simultaneously with space-time itself, is here superfluous. After all we do not wish to multiply the number of "beings" more times than necessary. The approach presented here is consistent with this point of view.

There exist at least two examples of the concrete realization of the idea exposed above. I do not take here into account the numerous attempts to extent the idea to supergravity. The supergravity idea, although similar in spirit to that expressed here, is beyond the scope of these notes. The first example is a simple generalization of the Kaluza-Klein theory to non-Abelian gauge groups (6, 7, 8). The other example is a modification of the latter approach, consisting in assent to the independent role of the linear connection (9). The relation between these two approaches is similar to the relation between the Einstein and the Einstein-Cartan theories of gravitation. The metric tensor is the only entity which describes gravitation in the Einstein theory, while in the Einstein-Cartan theory (10, 11, 12) (and in its extension with a nonmetric connection (13)) we have two such entities: the metric tensor and the linear connection.

However the Einstein-Cartan theory revived after 40 years just because of the development of gauge theories. The linear connection, which is the analogue of the gauge potential in gauge theories, in the Einstein theory is a secondary quantity only, derivable from the metric tensor. Therefore it is quite hard to say that gravitational field in the Einstein theory is a gauge field. In the Einstein-Cartan theory the linear connection is to large extent independent on the metric tensor (in its metric affine extension it is fully independent) and so this theory can be better understood as a gauge theory. There is indeed the metric tensor, which has no direct analogue in other gauge theories, but it can be viewed as a generalized Higgs field (3). In the generalized Kaluza-Klein theories a metric tensor on a bundle manifold determines gravitational and a gauge field, thus the described above "gauge argument" cannot be used in order to give for the linear connection on the bundle manifold an independent dynamical meaning. The linear connection constructed finally in this paper (Section 4.2) is completely determined by the metric tensor on the bundle manifold. which corresponds to gravitational and the gauge field. Gravitational field is described here by $g_{\mu\nu}$ and not by $g_{\mu\nu}$ plus $\Gamma^{\mu}{}_{\nu\rho}$. It makes this approach quite different than that of (9). I present here this new, third approach together with the basic simple generalization of the Kaluza-Klein theory.

The Kaluza-Klein theory provides us with a joint description of gravitational and electromagnetic fields. Its generalization to non-Abelian gauge groups (as well as its present modification) provides us with a joint description of gravitational and an arbitrary proper gauge field. For the reason of simplicity we say sometimes the "unified description" (9) instead of the "joint description", though it seems to be not the proper word.

As I mentioned above, the Kaluza-Klein theory and its generalizations are a preamble to treat the bundle manifolds as an arena for physics. This cannot be taken too literally. because I do not attempt to treat the vertical coordinates of the bundle manifolds as observable ones. The reason for that is that I am not going to construct here an essentially new physical theory, but rather to reformulate on old one.

The plan of this paper is following. In Part 2, I want to present a short description of the fibre bundle point of view on gauge fields. Then, in Part 3, I start with a naked bundle manifold with an arbitrary gauge group and imposing on it a structure similar to that of the general relativity, I arrive to the joint description of gravitational and gauge fields. We shall see there, how our approach leads to a restriction on physically admissible gauge groups. This restriction was already formulated by Glashow and Gell-Mann (14) in an early stage of the development of gauge theories. In the present paper condition leading to this restriction and its outcome are precisely formulated in geometrical language. The proof presented in Section 3.2 helps, I hope, to clarify argumentation of (14). Part 3 will contain also an examination of the role played by geodesics on the bundle manifold. In Part 4, I start once again with a naked bundle manifold and attempt to give it a structure similar to that of the Einstein-Cartan theory. It appears however that, following the pattern of the previous section, I obtain a structure which seems to be too rich for physical applications. Imposing an additional constraint on the linear connection on the principal fibre bundle, I use this connection in order to construct a correct Lagrangian for coupled gravitational and gauge fields.

The description of interaction between gauge and gravitational fields on one hand and matter fields on the other in the spirit of the present approach is beyond the scope of this article. The readers interested in this problem find in this volume the article by Kerner (15), who discusses the aspect of fermion fields. Compare also (16) and references therein.

2. GAUGE FIELD AS CONNECTION ON PRINCIPAL FIBRE BUNDLE

Principal fibre bundles constitute a framework for gauge fields (17, 18).

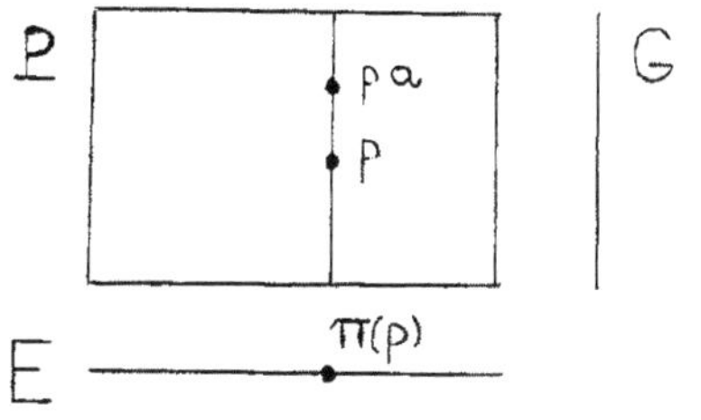

Let (P, E, π, G) be a principal fibre bundle. P is the bundle manifold. E is the base manifold, which we identify with the space-time throughout the paper. $\pi: P \to E$ is the projection. G is the structure or the gauge (Lie) group of the bundle, which acts to the right on P freely and transitively on its fibres. If $p \in P$ and $a \in G$, then the action of a on p is denoted by pa.

The generators of the action of G on P are called fundamental vector fields. They are in one-to-one correspondence with the elements of the Lie algebra G' of the group G. So if $\xi \in G'$, then $\xi^{\times}$ is a fundamental vector field on P corresponding to the element ξ. The operation $\times$ is linear and moreover it preserves the commutation relations, i. e. $[\xi^{\times}, \eta^{\times}] = [\xi, \eta]^{\times}$. All fundamental vector fields $\xi^{\times}_p$ taken at a given point $p \in P$ span the vertical subspace V_p at this point, i. e. the subspace of the tangent space $T_p(P)$ consisting of all vectors tangent to the fibre at p.

2.1. Gauge field

A gauge field is geometrically a connection on a principal fibre bundle over space-time.

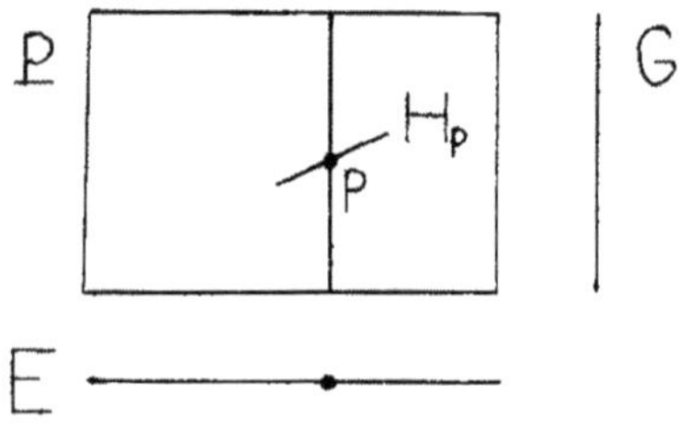

A connection is a smooth attachment of the horizontal subspace $H_p \subset T_p(P)$ for each point $p \in P$. It should satisfy the following conditions:

$$T_p(P) = V_p \oplus H_p$$

$$H_{pa} = H_p a$$

It follows from the first condition that $\pi'_{|H_p} : H_p \rightarrow T_{\pi(p)}(E)$ is a vector space isomorphism and that $\dim H_p = \dim E = 4$. The second condition is a requirement of invariance of the connection with respect to the action of the group G.

Let us define the connection 1-form A with values in the Lie algebra G', $A_p : T_p(P) \rightarrow G'$. To this aim let us decompose a vector $X_p \in T_p(P)$ on its vertical and horizontal components, $X_p = \text{ver } X_p + \text{hor } X_p$. Then, there exists a unique element $\xi \in G'$ such that $\text{ver } X_p = \xi^{\times}_p$. Subsequently, let $A_p(X_p) = \xi$. So defined the connection 1-form A satisfies the following constitutive relations

$$A(\xi^{\times}) = \xi \tag{1}$$

$$A(Xa) = \mathrm{ad}_{a^{-1}} A(X) \tag{2}$$

A smooth 1-form A on P with values in G', which satisfies conditions (1) and (2), provides us an equivalent definition of the connection. Namely, the horizontal subspaces can be defined by the formula

$$H_p = \{X_p \in T_p(P) : A(X_p) = 0\}$$

2.2. Gauge potential and gauge strenght

The connection 1-form A is the gauge-independent image of a gauge field. If we specify a gauge s, then it reduces to a 1-form of gauge potential ${}_sA$ on a subset of E. Similarly, the gauge-dependent (and defined on a subset of E rather than on P) counterpart of the curvature 2-form F of the connection A

$$F = dA + \frac{1}{2}[A, A] \tag{3}$$

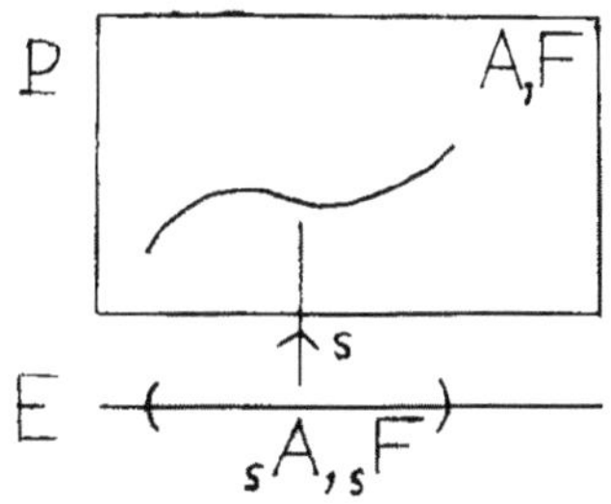

is the field strenght ${}_sF$. The gauge s is precisely defined as a local cross-section of the bundle P, $\pi(s(e)) = e$. Then ${}_sA$ and ${}_sF$ are defined as pull-backs of the connection and its curvature respectively,

$${}_sA = s^* A, \quad {}_sF = s^* F$$

Consider now two gauges s_1 and s_2. For each $e \in E$ belonging to the intersection of s_1 and s_2 there exists an element $a(e) \in G$ such that $s_2(e) = s_1(e)a(e)$. Conditions (1) and (2) for the connection 1-form A imply then the well known transformation formulas for potential and strenght of the gauge field:

$${}_{s_2}A = ad_{a^{-1}}\, {}_{s_1}A + a^{-1}da$$

$${}_{s_2}F = ad_{a^{-1}}\, {}_{s_1}F$$

2.3. A bundle frame

For our further considerations it will be useful to choose certain frame on P. Let (θ^μ) be a (local) frame of 1-forms on E; $\mu, \nu, \ldots = 1, 2, 3, 4$. For instance, if (e^μ) is a coordinate system on E, we can put $\theta^\mu = de^\mu$. Pull-backs of these frame 1-forms $\omega^\mu = \pi^*\theta^\mu$ constitute a frame of horizontal 1-forms on P. Let us choose now a certain basis (ξ_i) of the Lie algebra G'; i, j, ... = 5, ..., 4 + n, where n = dim G. Then the components A^i $(= \omega^i)$ of the connection 1-form A with respect to this basis constitute a frame of vertical 1-forms on P. Taking them altogether, we have the frame $(\omega^\alpha) = (\omega^\mu, \omega^i)$ of 1-forms on P; α, β, ... = 1, ..., 4 + n. Let us note down the differential formulas for the frame 1-forms:

$$d\omega^\mu = \pi^* d\theta^\mu$$

$$d\omega^i = \frac{1}{2} F^i{}_{\mu\nu}\, \omega^\mu \wedge \omega^\nu - \frac{1}{2} c^i{}_{jk}\, \omega^j \wedge \omega^k$$

The second formula is just the explicitly written definition (3) of the curvature $F = (F^i)$, $F^i = \frac{1}{2} F^i{}_{\mu\nu}\, \omega^\mu \wedge \omega^\nu$. $c^i{}_{jk}$ are the structure constants of the group G with respect to the basis (ξ_i).

If x is a (local) vector field on E, $\hat{x}$ denotes its horizontal lift to the bundle P, i.e. a vector field on P which is horizontal and such that $\pi'\hat{x} = x$. Let (ξ_μ) be a frame of vector fields on E dual with respect to (θ^μ). The horizontal lifts $\hat{\xi}_\mu$ of these basic fields constitute together with basic fundamental fields $\overset{\times}{\xi}_i$ a frame $(\hat{\xi}_\mu, \overset{\times}{\xi}_i)$ on P dual with respect to (ω^μ, ω^i). The commutation relations for this frame are following:

$$[\hat{\xi}_\mu, \hat{\xi}_\nu] = [\xi_\mu, \xi_\nu]^{\wedge} - F^i{}_{\mu\nu}\, \overset{\times}{\xi}_i$$

$$[\hat{\xi}_\mu, \overset{\times}{\xi}_i] = 0$$

$$[\overset{\times}{\xi}_i, \overset{\times}{\xi}_j] = c^k{}_{ij}\, \overset{\times}{\xi}_k$$

2.4. Coupling constants

Physicist usually assume that the Lagrangian for a free gauge field is quadratic in field strenght and invariant with respect to the gauge transformations. If the group G is simple (or semi-simple), we can consider the expression $k_{ij}\, {}_sF^i{}_{\mu\nu}\, {}_sF^{j\mu\nu}$, where $k_{ij} = c^k{}_{i\ell}\, c^\ell{}_{jk}$ are coefficients of the Killing metric on G with respect to the basis (ξ_i). As required, it is invariant with respect to gauge transformations (so we can omit the subscript s), it has not however the correct physical dimensions. To improve that, we postulate the free gauge field Lagrangian to be of the form

$$L_{gauge} = \frac{1}{16\pi} \frac{\hbar c}{\varepsilon^2} k_{ij} F^i{}_{\mu\nu} F^{j\mu\nu} \tag{4}$$

The dimensionless gauge coupling constant ε, defined formally by this formula, acquires the physical meaning only if we consider interaction of the gauge field with matter fields or with the gravitational field. For a semi-simple group G does not exist a single coupling constant , but for each of its simple normal subgroups we have a distinct coupling constant. In the case of the group $\mathbb{U}(1)$, i.e. in the case of electromagnetism, there is a natural metric tensor on its Lie algebra, since it is canonically isomorphic to $\mathbb{R}^1$. In this case we insert in formula (4) -1 instead of k_{ij}. Let us notice that $\frac{\sqrt{\hbar c}}{\varepsilon}\, {}_sF^i{}_{\mu\nu}$ has dimension of electromagnetic field and in a more classical approach is identified with the gauge strenght. Similarly $\frac{\sqrt{\hbar c}}{\varepsilon} A^i{}_{\mu}$ can be identified with the classical gauge potential. For electromagnetism $\varepsilon = \frac{1}{\sqrt{137}}$.

3. EINSTEIN-LIKE THEORY IN 4 + n DIMENSIONS

I attempt now to construct a theory as close as possible to general relativity, but such that the underlying manifold is a principal fibre bundle over space-time instead of the space-time itself. The theory should describe two fields: gravitational and a gauge field. The bundle manifold does not inherit the connection structure à priori. G is an arbitrary Lie group.

3.1. The metric tensor

The basic field in general relativity is the metric tensor g. Therefore, I assume that in our case the basic field will be a certain non-degenerate metric tensor $\tilde{g}$. At first, $\tilde{g}$ should be invariant with respect to the action of the gauge group G on P,

$$\tilde{g}(Xa,\ Ya) = \tilde{g}(X,\ Y) \tag{5}$$

for each vectors $X, Y \in T(P)$ and each $a \in G$. Equivalently

$$\mathcal{L}_{\xi^x} \tilde{g} = 0$$

for each fundamental vector field $\xi^{\times}$.

Further on, the orthogonality relations between each two fundamental vector fields should be constant throughout the bundle,

$$\tilde{g}(\xi^{\times}, \eta^{\times}) = \text{const} = h(\xi, \eta) \tag{6}$$

In the case of a non-Abelian gauge group conditions (5) and (6) are not independent. We have

$$\tilde{g}(\xi^{\times}a, \eta^{\times}a) = \tilde{g}(\xi^{\times}, \eta^{\times})$$

Taking into account that $\xi^{\times}a = (\mathrm{ad}_{a^{-1}}\,\xi)^{\times}$, we find that the group metric h, defined by condition (6), is subjected to the requirement

$$h(\mathrm{ad}_{a^{-1}}\,\xi, \mathrm{ad}_{a^{-1}}\,\eta) = h(\xi, \eta) \tag{7}$$

In other words, h should be ad-invariant.

The third and the last condition sounds: the metric tensor $\tilde{g}$ has the proper hiperbolic signature $(-, +, \ldots, +)$ and moreover lenght of each vertical vector is positive.

This requirement can be motivated by the fact that in the opposite case it would be difficult to attach to the Klein-Gordon equation associated to $\tilde{g}$ the meaning of an evolution equation or to formulate for it the Cauchy problem. The requirement implies that the group metric h is positively defined.

Now we are able to define the horizontal subspace H_p at p as that consisting of all vectors at p orthogonal to V_p. We see immediately that the horizontal subspaces are invariant with respect to the action of the group G due to condition (5). So we have a connection on P and are able to define the gauge field A. By the formula

$$g_{\mu\nu}(\pi(p)) = \tilde{g}(\hat{\xi}_{\mu}(p), \hat{\xi}_{\nu}(p))$$

we define the metric tensor $g = (g_{\mu\nu})$ on E, i.e. the gravitational field. Again due to requirement (5) this $g_{\mu\nu}$ is correctly defined. Requirement (6) instead ensures us that the tensor $\tilde{g}$ does not contain more information than the gravitational and the gauge fields together. This can be easily seen from the decomposition formula

$$\tilde{g} = g_{\mu\nu}\, \omega^{\mu} \otimes \omega^{\nu} + h_{ij}\, A^{i} \otimes A^{j}$$

By means of $g_{\mu\nu}$ and its inverse $g^{\mu\nu}$ we shall raise and lower the horizontal indices. h_{ij} and its inverse h^{ij} will be used for the same purpose in the case of the vertical indices.

3.2. Admissible gauge groups and their metrics

The conditions hitherto imposed on the metric $\tilde{g}$ imply restrictions on the group metric h and on the group G itself. I define an admissible (gauge) group to be a Lie group assuming an ad-invariant positively defined metric h on its Lie algebra; the Lie algebra of an admissible group is called admissible too.

Proposition

A simple Lie algebra G' is admissible if and only if it is compact. The ad-invariant positively defined metric h on the simple Lie algebra G' is proportional, $h = \lambda k$, to the Killing metric k and $\lambda < 0$.

Proof: The Killing metric

$$k(\xi, \eta) = \mathrm{Tr}(\mathrm{Ad}_\xi \circ \mathrm{Ad}_\eta)$$

where $\mathrm{Ad}_\xi(\zeta) = [\xi, \zeta]$, is ad-invariant and, for a compact group, is negatively defined. Thus, $h = \lambda k$ is ad-invariant and, if $\lambda < 0$, positively defined. The sufficiency is proved.

Condition (7) by differentiation with respect to a implies that

$$h([\zeta, \xi], \eta) + h(\xi, [\zeta, \eta]) = 0 \tag{8}$$

for every $\xi, \eta, \zeta \in G'$.

We can perform simultaneous diagonalization of h and k, choosing a basis $\Xi = (\xi_i)$ such that $h(\xi_i, \xi_j) = \delta_{ij}$ and $k(\xi_i, \xi_j) = \mu_i \delta_{ij}$. We have then

$$k(\xi_i, [\eta, \zeta]) = \mu_i\, h(\xi_i, [\eta, \zeta])$$

Putting $\eta = \xi_j$ and permutting ξ_i with ξ_j, as it is allowed for both metrics h and k due to formula (8), we get

$$k(\xi_i, [\xi_j, \zeta]) = \mu_j\, h(\xi_i, [\xi_j, \zeta])$$

The above two equalities give by subtraction

$$(\mu_i - \mu_j)\ h(\xi_i,\ [\xi_j,\ \zeta]) = 0 \qquad (9)$$

Define the subbasis $\Xi_i = (\xi_j)$ of the basis Ξ by the following requirements: 1° if the element ξ_k of Ξ does not commute with at least one element of Ξ_i then ξ_k belongs to Ξ_i; 2° Ξ_i contains the element ξ_i; 3° Ξ_i is minimal. Equation (9) implies then that all μ_j corresponding to elements of Ξ_i are the same, $\mu_j = \mu_i$. Assume that (ad absurdum) there exists an element ξ_ℓ of Ξ which does not belong to Ξ_i. Then, for each $\xi_j,\ \xi_k \in \Xi_i$ and each such ξ_ℓ

$$h([\xi_j,\ \xi_k],\ \xi_\ell) = h([\xi_\ell,\ \xi_j],\ \xi_k) = 0$$

so $[\xi_j,\ \xi_k]$ is a linear combination of the elements of Ξ_i. In effect the subspace spanned by Ξ_i is a non-trivial ideal of G'. It is however inconsistent with simplicity of G'. Therefore all μ_i are the same, $\mu_i = \frac{1}{\lambda}$.

We have $h = \lambda k$. Because there does not exist a Lie algebra with positively defined Killing metric, so $\lambda < 0$ and the Lie algebra G' is compact. q.e.d.

<u>Theorem</u>

The group G is admissible if and only if its Lie algebra G' is a (vectorial) direct sum

$$G' = J \oplus C_1 \oplus \ldots \oplus C_m \qquad (10)$$

of an Abelian ideal J and compact simple ideals $C_1, \ldots, C_m$.

<u>Proof</u>: According to E.Cartan - Levy - Malcev's Theorem each Lie algebra G' can be decomposed in the form $G' = J \oplus C$, where C is a semi-simple subalgebra of G', whereas J is the radical of G', i.e. the maximal (containing each other) solvable ideal of G'. The index of solvability of J is the number p such that $J^p \neq \{0\}$ and $J^{p+1} = \{0\}$ (J^k is defined by the induction: $J^1 = J$, $J^{k+1} = [J^k,\ J^k]$).

For $\zeta,\ \xi \in J^p$, $\eta \in G'$ condition (8) takes on the form

$$h(\xi,\ [\zeta,\ \eta]) = 0 \qquad (11)$$

therefore $[J^p,\ G'] \perp J^p$.

Assume that (ad absurdum) $p > 1$. Then $[J^p,\ J^{p-1}] \subset J^p$ and so $[J^p,\ J^{p-1}]$ is a subspace of G', whose vectors have zero lenght. Thus $[J^p,\ J^{p-1}] = \{0\}$, because the metric h is positively defined. For

$\xi \in J^p$, $\zeta \in J^{p-1}$, $\eta \in G'$ condition (8) takes on now the form (11) and so $[J^{p-1}, G'] \perp J^p$. Because $J^p \subset [J^{p-1}, G']$, it means that $J^p \perp J^p$. It is inconsistent however with the condition $J^p \neq \{0\}$. We have shown therefore that $p \leq 1$, i.e. J is an Abelian ideal of G'.

Due to the condition $[J, G'] \subset J$, the condition $[J, G'] \perp J$ means that $[J, G'] = [J, C] = \{0\}$, so the semi-simple subalgebra C is an ideal of G'.

The semi-simple Lie algebra C is a direct sum of simple ideals, $C = C_1 \oplus \ldots \oplus C_m$. For $\zeta, \eta \in C_a$, $\xi \in C_b$, where $a \neq b$, condition (8) takes on the form (11), therefore $C_a = [C_a, C_a] \perp C_b$. Thus, the Lie algebra G' is the direct sum (10) of the orthogonal ideals $J, C_1, \ldots, C_m$. In other words, h has a cellular form,

$$h = \begin{bmatrix} h_0 & & & 0 \\ & h_1 & & \\ & & \ddots & \\ 0 & & & h_m \end{bmatrix}$$

where $h_0, h_1, \ldots, h_m$ are ad-invariant positively defined metrics on $J, C_1, \ldots, C_m$ respectively. We have to investigate now, under what conditions such metrics exist.

It follows from condition (7) that h_0 can be an arbitrary positively defined metric on J. So the Abelian component of G' is arbitrary. The Proposition implies now that each C_a is compact, $h_a = \lambda_a k_a$ and $\lambda_a < 0$. q.e.d.

The large gauge group G consists of an Abelian subgroup and several simple compact normal subgroups. The mere vertical part of the metric $\tilde{g}$ is determined by the metric h_0 on the Abelian subgroup of G and by the sequence of numbers: $\lambda_1, \ldots, \lambda_m$. Let us notice that these numbers have dimension cm^2 and provide the vertical (non-Abelian part of) coordinates with the space dimensions. For an abstract Abelian group does not exist a distinguished invariant metric tensor - an analogue of k. For this reason we are not able to define in a natural manner corresponding dimensional numbers. It can be done however in the case of the group $\mathbb{U}(1)$ - the principal example for such a group from a physical point of view - since its Lie algebra is canonically isomorphic to $\mathbb{R}^1$.

3.3 Geodesics and gauge charge

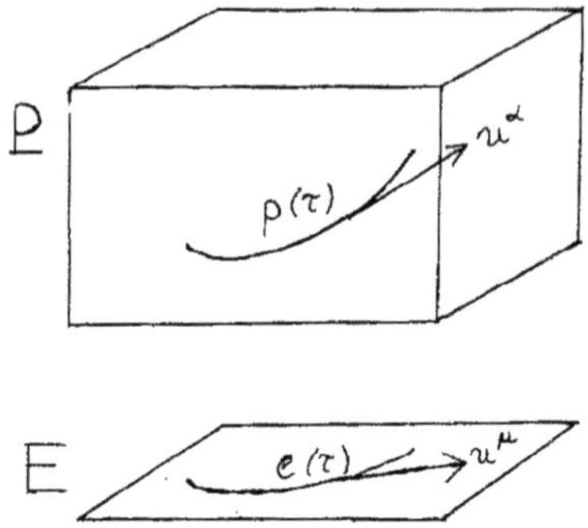

Remembering the role played by geodesics in general relativity, let us examine geodesics of the metric $\tilde{g}$. Let $(u^\alpha) = (u^\mu, u^i)$ be the vector tangent to the geodesic $\tau \mapsto p(\tau)$. Decompose then the geodesic equation

$$\frac{\tilde{D}}{d\tau} u^\alpha = 0$$

on the horizontal and vertical constituents:

$$\frac{Du^\mu}{d\tau} = h_{ij} \; u^i \; F^{j\mu}{}_\nu \; u^\nu \tag{12}$$

$$\frac{du^i}{d\tau} = 0 \tag{13}$$

$\frac{\tilde{D}}{d\tau}$ denotes here the absolute derivative with respect to the Levi-Civita connection associated to $\tilde{g}$, whereas $\frac{D}{d\tau}$ is the absolute derivative in the space-time; $\frac{D}{d\tau} u^\mu$ is well defined, since (u^μ) is the tangent vector to the projected curve $\tau \mapsto e(\tau) = \pi(p(\tau))$.

In order to examine the physical content of the geodesic equation, I reformulate it in the gauge dependent manner. Let us consider a gauge $e \mapsto s(e)$. Then one can define also a gauge dependent vertical projection $\tau \mapsto a(\tau) \in G$ of the curve $\tau \mapsto p(\tau)$ using the formula

$$p(\tau) = s(e(\tau))a(\tau)$$

Replace now F in formula (12) by the gauge dependent quantity ${}_sF$ according to the equation

$$F^{j\mu}{}_\nu(p(\tau)) = \alpha^j{}_k(a(\tau)^{-1}) \; {}_sF^{k\mu}{}_\nu(e(\tau))$$

where $(\alpha^j{}_k(a)) = ad_a$ is the matrix of the adjoint representation. Define the gauge charge ${}_sq_k$, which is the gauge dependent counterpart of the vertical component (u^i) of the tangent vector (u^α),

$${}_sq_k(\tau) = h_{ij} \; u^i \; \alpha^j{}_k(a(\tau)^{-1}) \quad M$$

Then the geodesic equations (12) and (13) become equivalent to

$$M \frac{Du^{\mu}}{D\tau} = {}_{s}q_{k}\ {}_{s}F^{k\mu}{}_{\nu}\ u^{\nu}$$

$$\frac{D_{s}q_{k}}{d\tau} \equiv \frac{d_{s}q_{k}}{d\tau} + c^{i}{}_{kj}\ {}_{s}q_{i}\ {}_{s}A^{j}{}_{\mu}\ u^{\mu} = 0$$

Similarity of these equations to the Lorentz equations of motion is apparent and so provides a justification of the name "charge" for the quantity ${}_{s}q_{k}$. The gauge charge ${}_{s}q_{k}$ is covariantly constant and in general not constant in non-Abelian gauge theories in contrast to the case of electromagnetism. The gauge independent "charge" u^{i} is instead constant in both Abelian and non-Abelian cases.

3.4. Dynamics

We shall derive now field equations for coupled gauge and gravitational fields from a principle of stationary action,

$$\delta \int_{V} L \sqrt{\det \tilde{g}}\ d_{4+n}p = 0$$

There is sometimes expressed the opinion that in order to do it, we have to take as V a tube over certain region U of E, $V = \pi^{-1}(U)$, and then, if the Lagrangian L is indenendent on the group coordinates, to integrate the Lagrangian over the group and in effect to reduce this action principle to a usual 4-dimensional one. It is not however this case because, if L is vertically constant, then the vertical constituents of derived Euler-Lagrange equations are identically satisfied and we do not have to specify the region V in advance.

According to our programme, we should use the curvature scalar $R(\tilde{g})$ as the Lagrangian. The calculation of this quantity was published by many authors (6, 7, 8). Cho as the first obtained the correct result. $R(\tilde{g})$ is the sum of the three constituents: the curvature scalar of g, the constant curvature scalar of h, and the classical Lagrangian of the gauge field,

$$R(\tilde{g}) = R(g) - \frac{1}{4} h^{ij} k_{ij} - \frac{1}{4} h_{ij} F^{i\mu\nu} F^{j}{}_{\mu\nu} \qquad (14)$$

Thus, from the principle of stationary action, we get the Einstein equations with a cosmological constant

$$G_{\mu\nu} + \Lambda\, g_{\mu\nu} = T_{\mu\nu}$$

and the Yang-Mills equations

$$\nabla_{\mu}\, F^{i\mu\nu} = 0$$

where ∇_{μ} is covariant derivative in both space-time and gauge senses. $T_{\mu\nu}$ is energy-momentum tensor (with accuracy up to a dimensional constant) of the gauge field:

$$T_{\mu\nu} = \frac{1}{2}\, h_{ij}\, (F^{i}{}_{\mu\rho}\, F^{j}{}_{\nu}{}^{\rho} - \frac{1}{4}\, g_{\mu\nu}\, F^{i}{}_{\rho\sigma}\, F^{j\rho\sigma})$$

If we introduce $\vec{E}^{i}$, $\vec{B}^{i}$ following the electromagnetic notation, $(\vec{E}^{i}, \vec{B}^{i}) = (F^{i}{}_{\mu\nu})$, then

$$T_{44} = \frac{1}{4}\, h_{ij}\, (\vec{E}^{i} \cdot \vec{E}^{j} + \vec{B}^{i} \cdot \vec{B}^{j})$$

We see that the requirement of positive definitness of the metric h is equivalent to the energy positivity condition, $T_{44} \geqslant 0$. We can appreciate now and, at the same time, to recognize necessity of the requirement that the sign of the vertical vectors lenght is the same as that of space-like vectors - i.e. the requirement of hiperbolicity of $\tilde{g}$.

It ought to be underlined that the presented approach clarifies - in a sense - an important ambiguity of classical field theory: Why the Einstein-Hilbert Lagrangian of gravitational field is linear in curvature, whereas the Lagrangians of proper gauge fields are quadratic?

Comparing formula (14) with the form of the gravitational Lagrangian $L_{grav} = (c^{4}/16\pi G)\, R(g)$ and with the form of the gauge field Lagrangian (formula (4)), we obtain

$$\lambda_{a}\, \varepsilon_{a}{}^{2} = -4\ell^{2}$$

where $\ell = \sqrt{(G\hbar/c^{3})}$ is the Planck lenght. Due to this equality, the coupling constant ε_{a} of the normal subgroup G_{a} is related to the coefficient of proportionality λ_{a} between the metrics h_{a} and k_{a}, hence it acquires a geometrical meaning.

Further on, we get the explicit expression for the upseting cosmological constant

$$\Lambda = -\frac{1}{32\ell^{2}}\, (n_{1}\, \varepsilon_{1}{}^{2} + \ldots + n_{m}\, \varepsilon_{m}{}^{2})$$

where $n_a = \dim G_a$, $a = 1, \ldots, m$. We see that each simple normal subgroup of G attributes to Λ, whereas the Abelian normal subgroup does not. We see also that the cosmological constant is never positive (I use here the conventions of Misner-Thorne-Wheeler's "Gravitation"). Numerically the above result is a nonsense: MTW's estimation of the cosmological constant, $|\Lambda| < 10^{-57}\ \mathrm{cm}^{-2}$, implies $\sum n_a \varepsilon_a^2 < 10^{-121}$.

4. METRIC-AFFINE MODIFICATION

4.1. Independent linear connection

Following the pattern of the Einstein-Cartan theory of gravitation and its metric-affine extension, I attempt to construct on the principal fibre bundle P a linear connection (i.e. a connection on the bundle of the linear frames over the bundle P) independent on the metric tensor already introduced. I postulate existence on P of the connection $\tilde{\Gamma}$ (or the covariant derivative $\tilde{\nabla}$) and impose on it two constraints analogous to conditions (5) and (6) for the metric tensor $\tilde{g}$. At first, the connection $\tilde{\Gamma}$ should be invariant with respect to the action of the group G:

$$\tilde{\nabla}_{X_p a} (Ya) = (\tilde{\nabla}_{X_p} Y) a \tag{15}$$

for each vector $X_p \in T_p(P)$, each point $p \in P$, each vector field Y on P and each $a \in G$. Equivalently

$$\mathcal{L}_{\xi^\times} \tilde{\Gamma} = 0$$

The second condition is that any fundamental vector field $\xi^\times$ on P is covariantly constant:

$$\tilde{\nabla}_{X_p} \xi^\times = 0$$

for each $X_p \in T(P)$. Conditions (15) and (16) are independent, because

$$\tilde{\nabla}_{X_p a} (\xi^\times a) = \tilde{\nabla}_{X_p a} (\mathrm{ad}_{a^{-1}} \xi)^\times = 0$$

From now we assume that the bundle manifold is provided with the metric tensor $\tilde{g}$ and with the linear connection $\tilde{\Gamma}$. Let us notice that the Levi-Civita connection $\Gamma(\tilde{g})$ associated to $\tilde{g}$ satisfies condition (15) but, if the group G is non-Abelian, it does not satisfy condition (16), since

$$\overset{R}{\nabla}_{\eta^{\times}} \xi^{\times} = \frac{1}{2} \left[\eta^{\times}, \xi^{\times} \right] \neq 0$$

where $\overset{R}{\nabla}$ is corresponding to $\Gamma(\tilde{g})$ covariant differentiation. Moreover

$$\overset{R}{\nabla}_{\hat{\xi}_{\mu}} \xi_i^{\times} = \frac{1}{2} F_{i\mu}{}^{\nu} \hat{\xi}_{\nu} \neq 0$$

if the connection A is not flat.

I define now a covariant differentiation ∇ (or a linear connection Γ) on the space-time manifold E. If $x_e \in T_e(E)$ and y is a vector field on E, I put

$$\nabla_{x_e} y = \pi' (\tilde{\nabla}_{(\hat{x}_e)_p} \hat{y})$$

where $(\hat{x}_e)_p$ denotes the horizontal lift of the vector x_e to the point $p \in \pi^{-1}(e)$. Due to invariance of $\tilde{\Gamma}$, the value of the right-hand side of the above equation is independent on the choice of the point p in the fibre over e. It can be easily shown also that ∇ satisfies the axioms of covariant differentiation, so the definition is correct.

Having the linear connection Γ on E, we can construct a linear connection $\hat{\Gamma}$ on P by means of the formulas:

$$\hat{\nabla}_{X_p} \hat{x} = \nabla_{\pi' X_p} x$$

$$\hat{\nabla}_{X_p} \xi^{\times} = 0$$

The definition can be easily extended to arbitrary vector fields on P, since horizontally lifted fields and fundamental vector fields span their module. Invariance of $\hat{\Gamma}$ is obvious, so $\hat{\Gamma}$ satisfies conditions (15) and (16) imposed on $\tilde{\Gamma}$. The difference of these two linear connections

$$\Phi(X, Y) = \tilde{\nabla}_X Y - \hat{\nabla}_X Y$$

determines a tensor field Φ on P twice covariant once contravariant.

Let us decompose Φ on its horizontal and vertical components. There are three cases in which $\Phi(X, Y)$ may not vanish: hor $\Phi(\xi^\times, \hat{y})$, ver $\Phi(\xi^\times, \hat{y})$, ver $\Phi(\hat{x}, \hat{y}) \neq 0$. I define:

$$S(\xi^\times, \hat{y}) = \pi'(\Phi(\xi^\times, \hat{y}))$$

$$T(\xi^\times, \hat{y}) = A(\Phi(\xi^\times, \hat{y}))$$

$$U(\hat{x}, \hat{y}) = A(\Phi(\hat{x}, \hat{y}))$$

The fields S, T, U together with the linear connection Γ fully describe the linear connection $\tilde{\Gamma}$. They have the equivariance properties:

$$S((ad_{a^{-1}}\,\xi)^\times_{pa}, \hat{y}_{pa}) = S(\xi^\times_p, \hat{y}_p)$$

$$T((ad_{a^{-1}}\,\xi)^\times_{pa}, \hat{y}_{pa}) = ad_{a^{-1}}\, T(\xi^\times_p, \hat{y}_p)$$

$$U(\hat{x}_{pa}, \hat{y}_{pa}) = ad_{a^{-1}}\, U(\hat{x}_p, \hat{y}_p)$$

For their components we write $S^\mu{}_{i\nu}$, $T^i{}_{j\mu}$, $U^i{}_{\mu\nu}$. These components are defined according to

$$S^\mu{}_{i\nu}(p)\ \xi_\mu(\pi(p)) = S(\xi^\times_i(p), \hat{\xi}_\nu(p))$$

and so on.

The non-vanishing coefficients of the connection $\tilde{\Gamma}$ with respect to the frame (ω^μ, ω^i) are following:

$$\tilde{\Gamma}^\mu{}_{\nu\rho} = \Gamma^\mu{}_{\nu\rho} \qquad \tilde{\Gamma}^i{}_{\nu k} = T^i{}_{k\nu}$$

$$\tilde{\Gamma}^\mu{}_{\nu k} = S^\mu{}_{k\nu} \qquad \tilde{\Gamma}^i{}_{\nu\rho} = U^i{}_{\rho\nu}$$

In order to examine the relation between the metric tensor $\tilde{g}$ and the linear connection $\tilde{\Gamma}$, we calculate the covariant derivative of the metric tensor:

$$\tilde{\nabla}_\rho\, \tilde{g}_{\mu\nu} = \nabla_\rho\, g_{\mu\nu} \qquad \tilde{\nabla}_\rho\, \tilde{g}_{i\mu} = -\,U_{i\rho\mu}$$

$$\tilde{\nabla}_j\, \tilde{g}_{\mu\nu} = -\,S_{\mu j\nu} - S_{\nu j\mu} \qquad \tilde{\nabla}_j\, \tilde{g}_{i\mu} = -\,T_{ij\mu}$$

We observe that the tensors T, U and the horizontally symmetric part of S are responsible for non-metricity of the linear connection $\tilde{\Gamma}$.

I list the non-vanishing components of the torsion tensor:

$$\tilde{Q}^{i}{}_{\mu\nu} = F^{i}{}_{\mu\nu} + 2U^{i}{}_{\mu\nu} \qquad \tilde{Q}^{\mu}{}_{\nu\rho} = Q^{\mu}{}_{\nu\rho}$$

$$\tilde{Q}^{i}{}_{\mu j} = - T^{i}{}_{j\mu} \qquad \tilde{Q}^{\mu}{}_{\nu i} = - S^{\mu}{}_{i\nu}$$

$$\tilde{Q}^{i}{}_{jk} = - c^{i}{}_{jk}$$

and of the curvature tensor:

$$\tilde{R}^{\mu}{}_{\nu\rho\sigma} = R^{\mu}{}_{\nu\rho\sigma} + S^{\mu}{}_{i\nu}\, F^{i}{}_{\rho\sigma}$$

$$\tilde{R}^{\mu}{}_{\nu\rho i} = \nabla_{\rho}\, S^{\mu}{}_{i\nu}$$

$$\tilde{R}^{\mu}{}_{\nu ij} = c^{k}{}_{ij}\, S^{\mu}{}_{k\nu} + S^{\mu}{}_{i\rho}\, S^{\rho}{}_{j\nu} - S^{\mu}{}_{j\rho}\, S^{\rho}{}_{i\nu}$$

$$\tilde{R}^{i}{}_{\mu\nu\rho} = T^{i}{}_{j\mu}\, F^{j}{}_{\nu\rho} + U^{i}{}_{\sigma\mu}\, Q^{\sigma}{}_{\nu\rho} + \nabla_{\nu}\, U^{i}{}_{\rho\mu} - \nabla_{\rho}\, U^{i}{}_{\nu\mu}$$

$$\tilde{R}^{i}{}_{\mu\nu j} = \nabla_{\nu}\, T^{i}{}_{j\mu} - c^{i}{}_{kj}\, U^{k}{}_{\nu\mu} + U^{i}{}_{\nu\rho}\, S^{\rho}{}_{j\mu}$$

$$\tilde{R}^{i}{}_{\mu jk} = c^{\ell}{}_{jk}\, T^{i}{}_{\ell\mu} + c^{i}{}_{\ell j}\, T^{\ell}{}_{k\mu} + c^{i}{}_{k\ell}\, T^{\ell}{}_{j\mu} + T^{i}{}_{j\nu}\, S^{\nu}{}_{k\mu} - T^{i}{}_{k\nu}\, S^{\nu}{}_{j\mu}$$

The derived Ricci scalar $\tilde{R} = \tilde{g}^{\beta\gamma}\, \tilde{R}^{\alpha}{}_{\beta\alpha\gamma}$ is

$$\tilde{R} = R - S^{\nu}{}_{i}{}^{\mu}\, (F^{i}{}_{\mu\nu} + U^{i}{}_{\mu\nu}) - \nabla^{\mu}\, T^{i}{}_{i\mu} \tag{17}$$

The tensors S, T and U do not appear in usual space-time descriptions of gravity and gauge theories. Their eventual physical meaning is not clear. They could serve perhaps as generalized Higgs fields, but then there arise the difficult problem of finding a reasonable Lagrangian with an appropriate ground state leading to spontaneous breakdown of the gauge symmetry. Let us notice that even the metricity condition, $\tilde{\nabla}\tilde{g} = 0$, does not kill all additional degrees of freedom contained in $\tilde{\Gamma}$; namely the horizontally skew-symmetric part of S remains.

In paper (9) I gave the general linear connection on P, which can be constructed by means of the linear connection Γ on E and the connection A. That linear connection, which contains two arbitrary parameters α and β, satisfies condition (15). Condition (16) is satisfied if and only if the parameter α vanish. The assumption $\alpha = 0$ is necessary here since I wish to have a linear connection on P constructed independently on the connection A. The linear connection of paper (9) becomes then a special case of the linear connection considered here, for which $S = 0$, $T = 0$ and $U = \beta F$.

4.2. Matching condition and its consequence

I reject now the two-pattern metric-affine approach of the previous section. The fundamental field is the metric tensor $\tilde{g}$ only. I restrict the freedom contained in $\tilde{\Gamma}$ matching it in a sense to the Levi-Civita linear connection $\Gamma(\tilde{g})$. The matching condition can be formulated as follows: the difference tensor $\tilde{\Gamma} - \Gamma(\tilde{g})$ has to be "minimal" with respect to $\tilde{\Gamma}$. More precisely speaking: decompose the difference tensor on its horizontal and vertical components and then put equal to zero those components which depend on Γ, S, T and U.

The components of the difference tensor are following:

$$\tilde{\Gamma}^{\mu}{}_{\nu\rho} - \Gamma(\tilde{g})^{\mu}{}_{\nu\rho} = \Gamma^{\mu}{}_{\nu\rho} - \Gamma(g)^{\mu}{}_{\nu\rho}$$

$$\tilde{\Gamma}^{\mu}{}_{\nu k} - \Gamma(\tilde{g})^{\mu}{}_{\nu k} = S^{\mu}{}_{k\nu} - \frac{1}{2} F_{k\nu}{}^{\mu}$$

$$\tilde{\Gamma}^{\mu}{}_{j\rho} - \Gamma(\tilde{g})^{\mu}{}_{j\rho} = \frac{1}{2} F_{j}{}^{\mu}{}_{\rho}$$

$$\tilde{\Gamma}^{\mu}{}_{jk} - \Gamma(\tilde{g})^{\mu}{}_{jk} = 0$$

$$\tilde{\Gamma}^{i}{}_{\nu\rho} - \Gamma(\tilde{g})^{i}{}_{\nu\rho} = U^{i}{}_{\rho\nu} - \frac{1}{2} F^{i}{}_{\nu\rho}$$

$$\tilde{\Gamma}^{i}{}_{\nu k} - \Gamma(\tilde{g})^{i}{}_{\nu k} = T^{i}{}_{k\nu}$$

$$\tilde{\Gamma}^{i}{}_{j\rho} - \Gamma(\tilde{g})^{i}{}_{j\rho} = 0$$

$$\tilde{\Gamma}^{i}{}_{jk} - \Gamma(\tilde{g})^{i}{}_{jk} = -\frac{1}{2} C^{i}{}_{kj}$$

So, the matching condition implies:

$$\Gamma^{\mu}{}_{\nu\rho} = \Gamma(g)^{\mu}{}_{\nu\rho} \qquad U^{i}{}_{\rho\nu} = \frac{1}{2} F^{i}{}_{\nu\rho}$$

$$S^{\mu}{}_{k\nu} = \frac{1}{2} F_{k\nu}{}^{\mu} \qquad T^{i}{}_{k\nu} = 0$$

Let us notice that the matching condition can be formulated as a variational principle. For this purpose we can take the square of the difference tensor $\tilde{\Gamma} - \Gamma(\tilde{g})$ as a Lagrangian density and vary the corresponding action integral with respect to the components of $\tilde{\Gamma}$ - i.e. with respect to $\Gamma^{\mu}{}_{\nu\rho}$, $S^{\mu}{}_{i\nu}$, $T^{i}{}_{j\mu}$ and $U^{i}{}_{\mu\nu}$. One might also try to vary it with respect to the components of $\tilde{g}$. Although the variation with respect to $A^{i}{}_{\mu}$ gives us additionally the Yang-Mills equations, it is not however reasonable, because the variation with respect to $g_{\mu\nu}$ implies vanishing of the energy-momentum $T^{\mu\nu}$ of the gauge field.

The connection $\tilde{\Gamma}$ is now fully and uniquely determined by the metric tensor $\tilde{g}$ (i.e. by the gravitational field g and the gauge field A). Construction of such a connection (determined by a metric tensor but rather different from its Levi-Civita connection) is possible due to the fact that a principal fibre bundle has a richer structure than an ordinary manifold.

As the Lagrangian density for coupled gauge and gravitational fields, I propose, as the most straightforward possibility, the Ricci scalar $\tilde{R}$ of the connection $\tilde{\Gamma}$. This Ricci scalar, derived from eq. (17),

$$\tilde{R} = R(g) - \frac{1}{4} F_{i\mu\nu} F^{i\mu\nu}$$

opposite to $R(\tilde{g})$ given by formula (14), does not contain the cosmological constant term and this is the only difference between $\tilde{R}$ and $R(\tilde{g})$. Thus the present approach allows us to maintain advantages of the previous approach and at the same time to avoid its principal disadvantage.

ACKNOWLEDGEMENTS

This is a good occasion, to express my gratitude to Andrzej Trautman, from whom I have learned bundles and gauge theories, and due to whom the present text is much better than its preliminary version. I am grateful to Jörg Hennig, who found an essential error in a preliminary version of the Theorem's proof and helped me to remove it. I wish also to thank Friedrich W. Hehl and Jürgen Nitsch for warm hospitality extended to me during first months of my stay in Cologne. Awardness of the fellowship by the Humboldt Foundation is gratefully acknowledged.

References

(1) R. Utiyama; Phys. Rev. 101 (1956) 1597.

(2) Y. Ne'eman: Gravity is the gauge theory of the parallel-transport modification of the Poincaré group. In: Differential Geometrical Methods in Mathematical Physics II, Bonn 1977; Springer-Verlag, Berlin - Heidelberg - New York.

(3) A. Trautman: Fibre bundles, gauge fields, and gravitation. In: Albert Einstein Commemorative Volume, ed. by A. Held, Plenum Press (in press).

(4) F.W. Hehl, Y. Ne'eman, J. Nitsch and P. von der Heyde; Phys. Lett. 78B (1978) 102.

(5) G. 't Hooft and M. Veltman; Nucl. Phys. B50 (1972) 318.

(6) R. Kerner; Ann. Inst. H. Poincaré 9 (1968) 143.

(7) Y.M. Cho; J. Math. Phys. 16 (1975) 2029.

(8) L.N. Chang, K.I. Macrae and F. Mansouri; Phys. Rev. D13 (1976) 235.

(9) W. Kopczyński; Acta Phys. Polon. B10 (1979) 365.

(10) A. Trautman; Bull. Polon. Acad. Sci. (Sér. Sci. Math. Astr. Phys.) 20 (1972) 185, 503.

(11) F.W. Hehl, P. von der Heyde, G. Kerlick and J. Nester; Rev. Mod. Phys. 78B (1978) 102.

(12) W. Kopczyński; Scripta Fac. Sci. Nat. UJEP Brunensis, Physica 3-4 (1975) 255.

(13) F.W. Hehl, G. Kerlick and P. von der Heyde; Z. Naturf. 31a (1976) 111, 524, 823.

(14) S.L. Glashow and M. Gell-Mann; Ann. Phys. 15 (1961) 437.

(15) R. Kerner: Spinor on fiber bundles and their use in invariant models. This volume.

(16) W. Mecklenburg: Aspects of seven dimensional relativity. Trieste preprint IC/79/87.

(17) A. Trautman; Rep. Math. Phys. 1 (1970) 29.

(18) T.T. Wu and C.N. Yang; Phys. Rev. D12 (1975) 3845.

HOMOGENOUS SYMPLECTIC FORMULATION OF FIELD DYNAMICS AND THE POINCARÉ-CARTAN FORM

W.M. Tulczyjew

Istituto di Fisica Matematica
Università di Torino
Via Carlo Alberto, 10
I-10123 Torino

1. Introduction.

The aim of this note is to establish a relation between the symplectic approach to field dynamics and the Poincaré-Cartan form associated with the multisymplectic structure. This relation is expressed within the homogenous framework. The presentation begins with the homogenous formulation of nonrelativistic particle dynamics in Section 2. A simple example of a field theory is given in Section 7.

Limitations of space imposed restrictions on the presentation. Only the infinitesimal version of dynamics is treated. No attempt is made to justify the choice of geometric concepts. Much of the material is only stated rather than derived.

The reader is expected to be familiar with basic concepts of symplectic geometry such as reduction of presymplectic structures and with the use of Lagrangian submanifolds to represent dynamics.

The literature on the Poincaré-Cartan form, known also as the Hamilton-Cartan form, is vast. The references are by no means complete.

2. Homogenous formulation of nonrelativistic particle dynamics and the Poincaré-Cartan form.

Let Q be a differential manifold representing the *configuration space* of a nonrelativistic particle system. The space-time $\mathrm{R} \times Q$ will be denoted by X. Local coordinates (t,x^A) will be used in X. The coordinate t is the canonical projection $t:\mathrm{R} \times Q \to \mathrm{R}$.

The *energy-momentum bundle* is the cotangent bundle $P = T^*(X)$ isomorphic to $\mathrm{R}^2 \times T^*(Q)$. Coordinates (t,x^A,e,p_B) will be used in P. The coordinates t and e are canonical projections $t:\mathrm{R}^2 \times T^*(Q) \to \mathrm{R}$ and $e:\mathrm{R}^2 \times T^*(Q) \to \mathrm{R}$ further characterized by the local expressions

$$\theta_X = - e dt + p_A dx^A \tag{2.1}$$

and

$$\omega_X = - de \wedge dt + dp_A \wedge dx^A \tag{2.2}$$

of the canonical 1-form θ_X on P and its differential $\omega_X = d\theta_X$.

The tangent bundle $T(X)$ is isomorphic to $\mathrm{R}^2 \times T(Q)$. The representation

$$u = t'\frac{\partial}{\partial t} + x'^A \frac{\partial}{\partial x^A} \tag{2.3}$$

of vectors $u \in T(X)$ leads to a coordinate system (t,x^A,t',x'^B). A second coordinate system $(t,x^A,\tau,\dot{x}^B)$ related to the first system by $t' = \tau$, $x'^B = \tau\dot{x}^B$ will be used in the submanifold $\mathbf{X}$ of $T(X)$ characterized by $t' \neq 0$. The submanifold $\mathbf{X}$ consists of vectors tangent to sections of the trivial bundle $\mathrm{R} \times Q$ and will be called the *infinitesimal configuration fibration.*

The tangent bundle $T(P)$ is isomorphic to $\mathrm{R}^4 \times T(T^*(Q))$. Coordinates $(t,x^A,e,p_B,t',x'^C,e',p'_D)$ will be used in $T(P)$. The subbundle $\mathbf{P} \subset T(P)$ characterized by $t' = 0$ will be called the *infinitesimal energy-momentum bundle.* Coordinates $(t,x^A,e,p_B,\tau,\dot{x}^C,\dot{e},\dot{p}_D)$ related to $(t,x^A,e,p_B,t',x'^C,e',\dot{p}_D)$ by $t' = \tau$, $x'^C = \tau\dot{x}^C$, $e' = \tau\dot{e}$ and $p'_D = \tau\dot{p}_D$ will be used in $\mathbf{P}$.

It is known that $T(P) = T(T^*(X))$ is a symplectic manifold canonically isomorphic to $T^*(T(X))$ and also to $T^*(T^*(X))$ [1],[10],[13-16]. We define a function g, a 1-form $\underline{\chi}$, a 1-form $\underline{\theta}$ and a 2-form $\underline{\omega}$ on $\mathbf{P} \subset T(P)$ by

$$g(u) = \langle u,\theta_X\rangle, \tag{2.4}$$

$$\langle w,\underline{\chi}\rangle = \langle \tau_{T(P)}(w) \wedge T\tau_P(w),\omega_X\rangle, \tag{2.5}$$

$$\underline{\theta} = \underline{\chi} + dg \tag{2.6}$$

and

$$\underline{\omega} = d\underline{\chi} = d\underline{\theta}. \tag{2.7}$$

In (2.5) we used the tangent bundle projection

$$\tau_{T(P)}:T(T(P)) \to T(P) \tag{2.8}$$

and the tangent mapping

$$T\tau_P:T(T(P)) \to T(P) \tag{2.9}$$

of the tangent bundle projection

$$\tau_P:T(P) \to P. \tag{2.10}$$

In terms of local coordinates we have

$$g = -et' + p_A x'^A = \tau(-e + p_A\dot{x}^A), \tag{2.11}$$

$$\begin{aligned}\underline{\chi} &= -e'dt + p'_A dx^A + t'de - x'^A dp_A \\ &= \tau(-\dot{e}dt + \dot{p}_A dx^A + de - \dot{x}^A dp_A),\end{aligned} \tag{2.12}$$

$$\begin{aligned}\underline{\theta} &= -e'dt + p'_A dx^A - edt' + p_A dx'^A \\ &= \tau(-\dot{e}dt + \dot{p}_A dx^A + p_A d\dot{x}^A) + (-e + p_A\dot{x}^A)d\tau\end{aligned} \tag{2.13}$$

and

$$\begin{aligned}\underline{\omega} &= -de' \wedge dt + dp'_A \wedge dx^A - de \wedge dt' + dp_A \wedge dx'^A \\ &= d\tau \wedge (-\dot{e}dt + \dot{p}_A dx^A + de - \dot{x}^A dp_A)\end{aligned} \tag{2.14}$$

$$+ \tau(- d\dot{e} \wedge dt + d\dot{p}_A \wedge dx^A + dp_A \wedge d\dot{x}^A).$$

Diffeomorphisms

$$\alpha : P \to T^*(X) \quad (2.15)$$

and

$$\beta : P \to T^*(P) \quad (2.16)$$

are characterized by

$$\underline{\theta} = \alpha^*\theta_X \quad (2.17)$$

and

$$\underline{\chi} = \beta^*\theta_P, \quad (2.18)$$

where θ_X and θ_P are the canonical 1-forms of $T^*(X)$ and $T^*(P)$ respectively. Relations

$$\underline{\pi} = \pi_X \circ \alpha \quad (2.19)$$

and

$$\underline{\rho} = \pi_P \circ \beta \quad (2.20)$$

hold. Here π_X and π_P are the cotangent bundle projections

$$\pi_X : T^*(X) \to X, \quad (2.21)$$

$$\pi_P : T^*(P) \to P, \quad (2.22)$$

the projection

$$\underline{\pi} : P \to X \quad (2.23)$$

is the mapping induced by

$$T\pi_X : T(P) \to T(X) \quad (2.24)$$

and

$$\underline{\rho} : P \to P \quad (2.25)$$

is the restriction to P of the tangent bundle projection

$$\tau_P : T(P) \to P. \quad (2.26)$$

The mapping β can be defined directly by

$$\beta(p) = p \lrcorner \omega_X. \quad (2.27)$$

Dynamics of a nonrelativistic particle system is stated in the form of a Lagrangian submanifold D of P. Elements of D are vectors tangent to energy-momentum space trajectories of the system compatible with the laws of dynamics. A characteristic feature of the homogenous formulation is that if $p \in D$ then also $ap \in D$ for any real number $a \neq 0$. Symplectomorphisms α and β make it possible to describe the dynamics D in terms of a generating function in two different ways. Thus we have the Lagrangian description and the Hamiltonian description. Related to the Hamiltonian description is the Poincaré-Cartan description. The three descriptions are formulated below.

A. Lagrangian formulation of dynamics.

In the Lagrangian formulation the dynamics D is a Lagrangian submanifold generated by a function $L:X \to \mathrm{R}$ in the sense that D is the image of the mapping

$$\alpha^{-1} \cdot dL : X \to \mathrm{P}. \tag{2.28}$$

Alternately D is given by

$$\begin{aligned} D = \{p \in \mathrm{P};\ \langle w, \underline{\theta} \rangle = \langle u, dL \rangle \text{ for each } u \in T(X) \text{ and} \\ \text{each } w \in T(\mathrm{P}) \text{ such that } T\underline{\pi}(w) = u\}. \end{aligned} \tag{2.29}$$

The function L is homogenous:

$$L(ax) = aL(x) \tag{2.30}$$

for each real number $a \neq 0$. In terms of local coordinates D is described by the variational equation

$$\tau(-\dot{e}dt + \dot{p}_A dx^A + p_A d\dot{x}^A) + (-e + p_A \dot{x}^A)d\tau = dL(t, x^A, \tau, \dot{x}^B). \tag{2.31}$$

If $L:\mathrm{R} \times T(Q) \to \mathrm{R}$ denotes the *Lagrangian* defined as the derivative $\frac{\partial}{\partial \tau} L$ at $\tau = 1$ then

$$L(t, x^A, \tau, \dot{x}^B) = \tau L(t, x^A, \dot{x}^B). \tag{2.32}$$

Equation (2.31) leads to the well known *Lagrange equations*

$$p_A = \frac{\partial L}{\partial \dot{x}^A}, \tag{2.33}$$

$$\dot{p}_A = \frac{\partial L}{\partial x^A} \tag{2.34}$$

and additional expressions

$$e = p_A \dot{x}^A - L(t, x^A, \dot{x}^B), \tag{2.35}$$

$$\dot{e} = -\frac{\partial L}{\partial t} \tag{2.36}$$

for energy and its rate of change.

B. Hamiltonian formulation of dynamics.

In the homogenous Hamiltonian formulation the dynamics D is generated by the zero function on a constraint submanifold $K \subset P$ in the sense that D is the inverse image by β of the Lagrangian submanifold

$$\begin{aligned} \{r \in T^*(P);\ p = \pi_P(r) \in K,\ \langle w, r \rangle = 0 \text{ for} \\ \text{each } w \in T_p(K) \subset T_p(P)\}. \end{aligned} \tag{2.37}$$

An alternate description of D is provided by

$$\begin{aligned} D = \{p \in \mathrm{P};\ \underline{\rho}(p) \in K,\ \langle z, \underline{\chi} \rangle = 0 \text{ for each} \\ z \in T_p(\mathrm{P}) \text{ such that } T\underline{\rho}(z) \in T(K) \subset T(P)\} \end{aligned} \tag{2.38}$$

and from (2.5) it follows that D is the characteristic distribution of $\omega_X|K$ [11-12]:

$$D = \{p \in \mathrm{P};\ \underline{\rho}(p) \in K,\ p \lrcorner (\omega_X|K) = 0\ \}. \tag{2.39}$$

In terms of local coordinates the submanifold K is described by an equation

$$e = H(t,x^A,p_B). \tag{2.40}$$

The function

$$H:\mathrm{R} \times T^*(Q) \to \mathrm{R} \tag{2.41}$$

is the *Hamiltonian* of the system. Formula (2.39) leads to the familiar *Hamilton equations*

$$\dot{x}^A = \frac{\partial H}{\partial p_A}, \tag{2.42}$$

$$\dot{p}_A = -\frac{\partial H}{\partial x^A} \tag{2.43}$$

in addition to (2.40) and

$$\dot{e} = \frac{\partial H}{\partial t}. \tag{2.44}$$

C. The Poincaré-Cartan form.

Equation (2.40) implies that K is the image of a section

$$\kappa:\mathrm{R} \times T^*(Q) \to P = \mathrm{R} \times \mathrm{R} \times T^*(Q):(t,s) \mapsto (H(t,s),t,s). \tag{2.45}$$

Let $\tilde{D}$ be the characteristic distribution of the form $\Omega = \kappa^*\omega_X$:

$$\tilde{D} = \{u \in T(\mathrm{R} \times T^*(Q)); \ u \lrcorner \Omega = 0\}. \tag{2.46}$$

Then D is obtained by lifting $\tilde{D}$ to K. The form $\Theta = \kappa^*\theta_X$ is called the *Poincare-Cartan form*. We have

$$\Omega = d\Theta. \tag{2.47}$$

Local expressions of Θ and Ω in terms of coordinates (t,x^A,p_B) of $\mathrm{R} \times T^*(Q)$ are

$$\Theta = - H(t,x^A,p_B)dt + p_A dx^A \tag{2.48}$$

and

$$\Omega = - dH(t,x^A,p_B) \wedge dt + dp_A \wedge dx^A. \tag{2.49}$$

The distribution $\tilde{D}$ is described by the Hamilton equations (2.42), (2.43) in the coordinate system $(t,x^A,p_B,\tau,\dot{x}^C,\dot{p}_D)$ of $T(\mathrm{R} \times T^*(Q))$. Lifting $\tilde{D}$ to K means supplementing these equations with equations (2.40) and (2.43).

3. Homogenous symplectic framework for field theories.

Field configurations are sections of a *configuration fibration* $\xi:X \to M$ of rank k. The base M of the configuration fibration is a manifold of dimension m interpreted as time, space or space-time depending on the particular field theory. For local considerations coordinates (t^κ); $\kappa = 1,\ldots,m$ will be used in M and adapted coordinates (t^κ,x^A); $\kappa = 1,\ldots,m$; $A = 1,\ldots,k$ in X.

We denote by P the subbundle of the m-cotangent bundle $\wedge^m T^*(X)$ defined by

$$P = \{p \in \wedge^m T^*(X); \ \langle v_1 \wedge v_2 \wedge \ldots \wedge v_m,p\rangle = 0 \text{ if two}$$

of the vectors $v_1,\dots,v_m$ are vertical}. (3.1)

We denote by

$$\pi : P \to X \tag{3.2}$$

the restriction of the m-cotangent bundle projection to P. The subbundle P will be called the *energy-momentum bundle*. An element $p \in P$ at a point $x \in X$ with coordinates $(t\ ,x^A)$ can be represented by

$$p = -\,edt^1 \wedge\dots\wedge dt^m - p_A{}^\kappa \frac{\partial}{\partial t^\kappa} \lrcorner\ (dx^A \wedge dt^1 \wedge\dots\wedge dt^m). \tag{3.3}$$

We will use in P coordinates $(t^\kappa, x^A, e, p_B{}^\lambda)$ induced by this representation. The energy-momentum bundle P has a canonical *special m-symplectic structure* which consists of the projection π and the m-form

$$\theta = \theta^m_X | P, \tag{3.4}$$

where θ^m_X is the canonical m-form on $\wedge^m T^*(X)$ defined by

$$\langle w, \theta^m_X\rangle = \langle \wedge^m T\pi^m_X(w), \tau^m_{\wedge^m T^*(X)}(w)\rangle \tag{3.5}$$

for each $w \in \wedge^m T(\,{}^m T^*(X))$. Here

$$\tau^m_{\wedge^m T^*(X)} : \wedge^m T(\wedge^m T^*(X)) \to \wedge^m T^*(X) \tag{3.6}$$

denotes the m-tangent bundle projection,

$$\pi^m_X : \wedge^m T^*(X) \to X \tag{3.7}$$

is the m-cotangent bundle projection and

$$\wedge^m T\pi^m_X : \wedge^m T(\wedge^m T^*(X)) \to \wedge^m T(X) \tag{3.8}$$

is the m-tangent mapping of π^m_X. We denote by ω the $(m+1)$-form $d\theta$. The local expression of θ is

$$\theta = -\,edt^1 \wedge\dots\wedge dt^m - p_A{}^\kappa \frac{\partial}{\partial t^\kappa} \lrcorner\ (dx^A \wedge dt^1 \wedge\dots\wedge dt^m). \tag{3.9}$$

The subfibration $\mathbf{X} \subset \wedge^m T(X)$ which consists of non-zero simple vectors tangent to sections of ξ will be called the *infinitesimal configuration fibration*. Although elements of $\mathbf{X}$ are not jets of sections the infinitesimal configuration fibration can be identified with the fibre product $J^1(X) \times_M \underline{\wedge^m T(M)}$, where $J^1(X)$ is the 1-jet bundle and $\underline{\wedge^m T(M)}$ denotes the m-tangent bundle with the zero section removed. In terms of the coordinates (t^κ, x^A) the jet of a section $x^A = \sigma^A(t^\kappa)$ at (t^κ) is represented by $(t^\kappa, x^A, x^B{}_\lambda)$, where $x^B{}_\lambda = \partial_\lambda \sigma^B(t^\kappa)$, and a m-vector $\mathbf{t}$ at (t^κ) can be written in the form

$$\mathbf{t} = \tau \frac{\partial}{\partial t^1} \wedge\dots\wedge \frac{\partial}{\partial t^m},\ \tau \neq 0 \tag{3.10}$$

and represented by (t^κ, τ). If $\mathbf{x}$ is a simple m-vector tangent to the section $x^A = \sigma^A(t^\kappa)$ and $\wedge^m T\xi(\mathbf{x}) = \mathbf{t}$ then

$$\mathbf{x} = \tau\Big(\frac{\partial}{\partial t^1} + x^A{}_1 \frac{\partial}{\partial x^A}\Big) \wedge\dots\wedge \Big(\frac{\partial}{\partial t^m} + x^B{}_m \frac{\partial}{\partial x^B}\Big). \tag{3.11}$$

We will use in X coordinates $(t^{\kappa}, x^{A}, \tau, x^{B}{}_{\lambda})$ constructed in this way.

We consider the subfibration $P' \subset \wedge^{m}T(P)$ which consists of non-zero simple m-vectors tangent to sections of $\xi\circ\pi : P \to M$. The fibration P' can be identified with the fibre product $J^{1}(P) \times_{M} \wedge^{m}T(M)$, where $J^{1}(P)$ is the bundle of 1-jets of sections of $\xi\circ\pi$. In analogy with the coordinates $(t^{\kappa}, x^{A}, \tau, x^{B}{}_{\lambda})$ in X we have coordinates $(t^{\kappa}, x^{A}, e, p_{B}{}^{\lambda}, \tau, x^{C}{}_{\mu}, e_{\nu}, p_{D}{}^{\omega}{}_{\pi})$ in P'. A projection

$$\underline{\pi}' : P' \to X \tag{3.12}$$

is induced by

$$\wedge^{m}T\pi : \wedge^{m}T(P) \to \wedge^{m}T(X), \tag{3.13}$$

or equivalently, by

$$J^{1}\pi : J^{1}(P) \to J^{1}(X). \tag{3.14}$$

A second projection

$$\underline{\rho}' : P' \to P \tag{3.15}$$

is the restriction to P' of the m-tangent bundle projection

$$\tau^{m}_{P} : \wedge^{m}T(P) \to P. \tag{3.16}$$

If μ is a $(m+l)$-form on P we define a l-form $i_{T}\mu$ on $\wedge^{m}T(P)$ by

$$\langle w, i_{T}\mu\rangle = \langle \tau^{l}_{\wedge^{m}T(P)}(w) \wedge \wedge^{l}T\tau^{m}_{P}(w), \mu\rangle \tag{3.17}$$

for each $w \in \wedge^{l}T(\wedge^{m}T(P))$. We denote by $d_{T}\mu$ the $(l+1)$-form $i_{T}d\mu - (-1)^{m} d i_{T}\mu$.

We introduce forms

$$\underline{\theta}' = (-1)^{m+1} d_{T}\theta | P' \tag{3.18}$$

and

$$\underline{\chi}' = (-1)^{m+1} i_{T}\omega | P' \tag{3.19}$$

related by

$$\underline{\theta}' = \underline{\chi}' + dg', \tag{3.20}$$

where

$$g' = i_{T}\theta | P'. \tag{3.21}$$

The corresponding local expressions are

$$\begin{aligned} \underline{\theta}' = {} & \tau(- e_{\kappa} + p_{A}{}^{\lambda} x^{A}{}_{\kappa\lambda} - p_{A}{}^{\lambda}{}_{\lambda} x^{A}{}_{\kappa}) dt^{\kappa} + \tau p_{A}{}^{\lambda}{}_{\lambda} dx^{A} \\ & + (- e + p_{A}{}^{\lambda} x^{A}{}_{\lambda}) d\tau + \tau p_{A}{}^{\lambda} dx^{A}{}_{\lambda}, \end{aligned} \tag{3.22}$$

$$\begin{aligned} \underline{\chi}' = {} & \tau[(- e_{\kappa} + p_{A}{}^{\lambda} x^{A}{}_{\kappa\lambda} - p_{A}{}^{\lambda}{}_{\lambda} x^{A}{}_{\kappa}) dt^{\kappa} \\ & + p_{A}{}^{\lambda}{}_{\lambda} dx^{A} + de - x^{A}{}_{\lambda} dp_{A}{}^{\lambda}] \end{aligned} \tag{3.23}$$

and

$$g' = \tau(- e + p_{A}{}^{\lambda} x^{A}{}_{\lambda}). \tag{3.24}$$

We note that the 2-form

$$\underline{\omega}' = d\underline{\theta}' = d\underline{\chi}' = d_T\omega|P' \tag{3.25}$$

defines a presymplectic structure on P'. We also observe that forms $\underline{\theta}'$ and $\underline{\chi}'$ are vertical with respect to projections $\underline{\pi}'$ and $\underline{\rho}'$. Consequently we construct mappings

$$\alpha' : P' \to T^*(X) \tag{3.26}$$

and

$$\beta' : P' \to T^*(P) \tag{3.27}$$

such that

$$\underline{\theta}' = \alpha'^*\theta_X \tag{3.28}$$

and

$$\underline{\chi}' = \beta'^*\theta_P, \tag{3.29}$$

where θ_X and θ_P are the canonical 1-forms on $T^*(X)$ and $T^*(P)$ respectively. The mapping α' is defined by

$$\langle v, \alpha'(p')\rangle = \langle w, \underline{\theta}'\rangle, \tag{3.30}$$

where $v \in T(X)$ and $w \in T(P')$ satisfy $\tau_{P'}(w) = p'$ and $v = T\underline{\pi}'(w)$. Similarly β' is defined by

$$\langle u, \beta'(p')\rangle = \langle w, \underline{\chi}'\rangle, \tag{3.31}$$

where $u \in T(P)$ and $w \in T(P')$ satisfy $\tau_{P'}(w) = p'$ and $u = T\underline{\rho}'(w)$. Relations

$$\pi_X \circ \alpha' = \underline{\pi}' \tag{3.32}$$

and

$$\pi_P \circ \beta' = \underline{\rho}' \tag{3.33}$$

follow directly from the definitions of α' and β'. Formula (3.28) is proved by

$$\begin{aligned} \langle w, \alpha'^*\theta_X\rangle &= \langle T\alpha'(w), \theta_X\rangle \\ &= \langle T\pi_X(T\alpha'(w)), \tau_{T^*(X)}(T\alpha'(w))\rangle \\ &= \langle T\underline{\pi}'(w), \alpha'(\tau_{P'}(w))\rangle \\ &= \langle w, \underline{\theta}'\rangle \end{aligned} \tag{3.34}$$

and (3.29) follows from

$$\begin{aligned} \langle w, \beta'^*\theta_P\rangle &= \langle T\beta'(w), \theta_P\rangle \\ &= \langle T\pi_P(T\beta'(w)), \tau_{T^*(P)}(T\beta'(w))\rangle \\ &= \langle T\underline{\rho}'(w), \beta'(\tau_{P'}(w))\rangle \\ &= \langle w, \underline{\chi}'\rangle. \end{aligned} \tag{3.35}$$

The mapping β' can also be defined by

$$\beta'(p') = (-1)^{m+1} p' \lrcorner\, \omega . \tag{3.36}$$

The *infinitesimal energy-momentum bundle* is the symplectic manifold $(P, \underline{\omega})$

obtained by reducing the presymplectic manifold $(P', \underline{\omega}')$. Since

$$\underline{\omega}' = \alpha'^{*}\omega_X \tag{3.37}$$

and

$$\underline{\omega}' = \beta'^{*}\omega_P \tag{3.38}$$

the characteristic foliation of $\underline{\omega}'$ is both the foliation by fibres of α' and the foliation by fibres of β'. The quotient manifold P has a canonical symplectic structure $\underline{\omega}$. Mappings α', β', $\underline{\pi}'$ and $\underline{\rho}'$ induce diffeomorphisms

$$\alpha : P \to T^{*}(X), \tag{3.39}$$

$$\beta : P \to T^{*}(P) \tag{3.40}$$

and fibrations

$$\underline{\pi} : P \to X, \tag{3.41}$$

$$\underline{\rho} : P \to P \tag{3.42}$$

satisfying relations

$$\underline{\omega} = \alpha^{*}\omega_X, \tag{3.43}$$

$$\underline{\omega} = \beta^{*}\omega_P, \tag{3.44}$$

$$\pi_X \circ \alpha = \underline{\pi} \tag{3.45}$$

and

$$\pi_P \circ \beta = \underline{\rho} \; . \tag{3.46}$$

Forms $\underline{\theta}'$ and $\underline{\chi}'$ induce forms $\underline{\theta}$ and $\underline{\chi}$ on P such that

$$\underline{\theta} = \alpha^{*}\theta_X, \tag{3.47}$$

$$\underline{\chi} = \beta^{*}\theta_P \tag{3.48}$$

and

$$\underline{\omega} = d\underline{\theta} = d\underline{\chi}. \tag{3.49}$$

Also the function g' induces a function g on P and

$$\underline{\theta} = \underline{\chi} + dg. \tag{3.50}$$

We conclude that the infinitesimal energy-momentum bundle carries two canonical special symplectic structures isomorphic to the canonical structures of the cotangent bundles $T^{*}(X)$ and $T^{*}(P)$.

We use in P coordinates $(t^{\kappa}, x^{A}, e, p_{B}{}^{\lambda}, \tau, x^{C}{}_{\mu}, e_{\nu}, p_{D})$ related to the coordinates $(t^{\kappa}, x^{A}, e, p_{B}{}^{\lambda}, \tau, x^{C}{}_{\mu}, e_{\nu}, p_{D}{}^{\rho}{}_{\sigma})$ of P' by

$$e_{\nu} = e_{\nu} - p_{A}{}^{\lambda}{}_{\nu} x^{A}{}_{\lambda} + p_{A}{}^{\lambda}{}_{\lambda} x^{A}{}_{\nu} \tag{3.51}$$

and

$$p_{D} = p_{D}{}^{\lambda}{}_{\lambda}. \tag{3.52}$$

In terms of these coordinates we have the local expressions

$$\underline{\theta} = - \tau e_{\kappa} dt^{\kappa} + \tau p_{A} dx^{A}$$

$$+ (-e + p_A{}^\lambda x^A{}_\lambda)d\tau + \tau p_A{}^\lambda dx^A{}_\lambda, \tag{3.53}$$

$$\chi = \tau(-e_\kappa dt^\kappa + p_A dx^A + de - x^A{}_\lambda dp_A{}^\lambda) \tag{3.54}$$

and

$$g = \tau(-e + p_A{}^\lambda x^A{}_\lambda). \tag{3.55}$$

Dynamics of a physical system can frequently be stated in form of a Lagrangian submanifold of a suitable symplectic manifold. In its homogenous infinitesimal form field dynamics is a Lagrangian submanifold D of the infinitesimal energy-momentum bundle. Multiplication by non-zero numbers is well defined in $P' \subset \wedge^m T(P)$ and induces a corresponding operation in P. A characteristic feature of the homogenous description is that if $p \in D$ then $ap \in D$ for each real number $a \neq 0$. Symplectomorphisms α and β make it possible to describe the dynamics in Lagrangian terms and Hamiltonian terms.

The submanifold $D \subset P$ consists of equivalence classes of simple m-vectors tangent to sections of $\xi\circ\pi$ compatible with the dynamical laws of the field. These vectors form a submanifold $D' \subset P'$. Submanifolds D and D' provide equivalent descriptions of dynamics. Description in terms of D has the advantage of being symplectic. The submanifold D' on the other hand is more directly characterized by the Poincaré--Cartan form.

4. Lagrangian formulation of field dynamics.

In the Lagrangian formulation the dynamics D is generated by a homogenous generating function $L:X \to \mathrm{R}$ in the sense used in Section 2. Also D' is generated by L. It is the image of the mapping

$$(\alpha')^{-1}\circ dL:X \to P' \tag{4.1}$$

and is also given by

$$D' = \{p' \in P' ;\ \langle w,\underline{\theta}'\rangle = \langle u,dL\rangle \text{ for each } u \in T(X) \text{ and each } w \in T(P') \text{ such that } T\underline{\pi}'(w) = u\}. \tag{4.2}$$

In terms of local coordinates we have variational equations

$$\begin{aligned} -\tau e_\kappa dt^\kappa + \tau p_A dx^A + (-e + p_A{}^\lambda x^A{}_\lambda)d\tau + \tau p_A{}^\lambda dx^A{}_\lambda \\ = dL(t^\kappa, x^A, \tau, x^B{}_\lambda) \end{aligned} \tag{4.3}$$

and

$$\begin{aligned} \tau(-e_\kappa + p_A{}^\lambda{}_\kappa x^A{}_\lambda - p_A{}^\lambda x^A{}_{\lambda\kappa})dt^\kappa + \tau p_A{}^\lambda{}_\lambda dx^A{}_\kappa + (-e + p_A{}^\lambda x^A{}_\lambda)d\tau \\ + \tau p_A{}^\lambda dx^A{}_\lambda = dL(t^\kappa, x^A, \tau, x^B{}_\lambda) \end{aligned} \tag{4.4}$$

for D and D' respectively. If the *Lagrangian density* $L(t^\kappa, x^A, x^B{}_\lambda)$ is defined by

$$L(t^\kappa, x^A, \tau, x^B{}_\lambda) = \tau L(t^\kappa, x^A, x^B{}_\lambda) \tag{4.5}$$

then (4.3) leads to equations

$$p_A{}^\lambda = \frac{\partial L}{\partial x^A{}_\lambda}\ , \tag{4.6}$$

$$p_A = \frac{\partial L}{\partial x^A}, \tag{4.7}$$

$$e = p_A{}^\lambda x^A{}_\lambda - L(t^\kappa, x^A, x^B{}_\lambda), \tag{4.8}$$

$$e_\kappa = -\frac{\partial L}{\partial t^\kappa} \tag{4.9}$$

and (4.4) leads to

$$p_A{}^\lambda = \frac{\partial L}{\partial x^A{}_\lambda}, \tag{4.10}$$

$$p_A{}^\lambda{}_\lambda = \frac{\partial L}{\partial x^A}, \tag{4.11}$$

$$e = p_A{}^\lambda x^A{}_\lambda - L(t^\kappa, x^A, x^B{}_\lambda) \tag{4.12}$$

and

$$e_\kappa - p_A{}^\lambda{}_\kappa x^A{}_\lambda + p_A{}^\lambda{}_\lambda x^A{}_\kappa = -\frac{\partial L}{\partial t^\kappa}. \tag{4.13}$$

5. Hamiltonian formulation of field dynamics.

In the Hamiltonian formulation of field dynamics there is a constraint $K \subset P$ and the Lagrangian submanifold of $T^*(P)$ defined by (2.37). The dynamics D is the inverse image of this submanifold by β and D' is the inverse image of this submanifold by β'. Alternate expressions for D and D' are (2.38) and

$$D' = \{p' \in P';\ \underline{\rho}'(p') \in K,\ \langle z, \underline{\chi}' \rangle = 0 \text{ for each } z \in T_{p'}(P') \\ \text{such that } T\underline{\rho}'(z) \in T(K) \subset T(P)\}. \tag{5.1}$$

From (3.19) it follows that D' can be characterized in a manner analogous to (2.39):

$$D' = \{p' \in P';\ \underline{\rho}'(p') \in K,\ p' \lrcorner (\omega|K) = 0\}. \tag{5.2}$$

No characterization of this type exists for D.

In terms of local coordinates K is described by an equation

$$e = H(t^\kappa, x^A, p_B{}^\lambda). \tag{5.3}$$

The function H can be given a coordinate invariant meaning of a density if a connection is given in X. Local equations for D are

$$x^A{}_\lambda = \frac{\partial H}{\partial p_A{}^\lambda}, \tag{5.4}$$

$$p_A = -\frac{\partial H}{\partial x^A}, \tag{5.5}$$

$$e = H(t^\kappa, x^A, p_B{}^\lambda), \tag{5.6}$$

$$e_\kappa = \frac{\partial H}{\partial t^\kappa} \tag{5.7}$$

and D' is described by

$$x^A{}_\lambda = \frac{\partial H}{\partial p_A{}^\lambda}, \tag{5.8}$$

$$p_{A}{}^{\lambda}{}_{\lambda} = -\frac{\partial H}{\partial x^{A}{}_{\lambda}}, \tag{5.9}$$

$$e = H(t^{\kappa}, x^{A}, p_{B}{}^{\lambda}), \tag{5.10}$$

$$e_{\kappa} - p_{A}{}^{\lambda}{}_{\kappa} x^{A}{}_{\lambda} + p_{A}{}^{\lambda}{}_{\lambda} x^{A}{}_{\kappa} = \frac{\partial H}{\partial t^{\kappa}}. \tag{5.11}$$

6. The Poincaré-Cartan form.

Let $E \subset P$ be the subbundle defined by

$$E = \{e \in \wedge^{m} T^{*}(X);\ v \lrcorner\, p = 0 \text{ if } v \text{ is vertical}\}, \tag{6.1}$$

and let P' denote the quotient bundle P/E. Let

$$\pi' : P' \to X \tag{6.2}$$

be the bundle projection. Coordinates $(t^{\kappa}, x^{A}, e, P_{B}{}^{\lambda})$ of P induce coordinates (t^{κ}, x^{A}, e) in E and coordinates $(t^{\kappa}, x^{A}, p_{B}{}^{\lambda})$ in P'.

Equation (5.3) implies that K is the image of a section

$$\kappa : P' \to P \tag{6.3}$$

of the canonical projection

$$\varepsilon : P \to P'. \tag{6.4}$$

Let Ω denote the $(m+1)$-form $\kappa^{*}\omega$ and let $\tilde{D} \subset \wedge^{m} T(P')$ be defined as the set of simple non-zero m-vectors $\tilde{p}$ tangent to sections of $\xi \circ \pi' : P' \to M$ and satisfying

$$\tilde{p} \lrcorner\, \Omega = 0. \tag{6.5}$$

Then D' is obtained by lifting $\tilde{D}$ to K. The m-form $\Theta = \kappa^{*}\theta$ is called the*Poincaré-Cartan form.*

The local expression of the Poincaré-Cartan form is

$$\Theta = -H(t^{\kappa}, x^{A}, p_{B}{}^{\lambda}) dt^{1} \wedge \ldots \wedge dt^{m} - p_{A}{}^{\kappa} \frac{\partial}{\partial t^{\kappa}} \lrcorner\, (dx^{A} \wedge dt^{1} \wedge \ldots \wedge dt^{m}). \tag{6.6}$$

In terms of coordinates $(t^{\kappa}, x^{A}, p_{B}{}^{\lambda}, \tau, x^{C}{}_{\mu}, p_{D}{}^{\nu}{}_{\rho})$ of a simple non-zero m-vector tangent to a section of $\xi \circ \pi'$ the set $\tilde{D}$ is described by equations (5.8) and (5.9).

7. An example: the real scalar field.

Let M be the flat space-time of special relativity with coordinates (t^{κ}); $\kappa = 0, \ldots, 3$. We denote by $g_{\kappa\lambda}$ and $g^{\kappa\lambda}$ the components of the covariant metric tensor and the contravariant metric tensor respectively and by g the determinant of the covariant tensor. The configuration bundle X of the real scalar field is the trivial bundle $M \times \mathrm{R}$ with coordinates (t^{κ}, x).

In terms of coordinates $(t^{\kappa}, x, e, p^{\lambda}, \tau, x_{\mu}, e_{\nu}, p)$ of P the dynamics D of the field is described by equations

$$p^{\lambda} = \sqrt{-g}\, g^{\lambda\mu} x_{\mu}, \tag{7.1}$$

$$p = -\sqrt{-g}\; m^2 x, \tag{7.2}$$

$$e = \tfrac{1}{2}\sqrt{-g}\;(g_{\kappa\lambda}p^\kappa p^\lambda + m^2 x^2), \tag{7.3}$$

$$e_\kappa = 0. \tag{7.4}$$

Here m is a constant and we assumed for simplicity that the metric tensor is constant. Equations (7.1) and (7.2) imply the Klein-Gordon equation

$$\frac{\partial}{\partial t^\kappa}\left(\sqrt{-g}\; g^{\kappa\lambda}\frac{\partial\phi}{\partial t^\lambda}\right) + \; -g\; m^2\phi = 0 \tag{7.5}$$

for a section

$$\phi : M \to X : (t^\kappa) \mapsto (t^\kappa, \phi(t^\lambda)) \tag{7.6}$$

if $x = \phi(t^\lambda)$ and $x_\kappa = \dfrac{\partial\phi}{\partial t^\kappa}$. The submanifold $D' \subset P'$ is described by

$$p^\lambda = \sqrt{-g}\; g^{\lambda\mu}x_\mu, \tag{7.7}$$

$$p^\lambda{}_\lambda = -\sqrt{-g}\; m^2 x, \tag{7.8}$$

$$e = \tfrac{1}{2}\sqrt{-g}(g_{\kappa\lambda}p^\kappa p^\lambda + m^2 x^2), \tag{7.9}$$

$$e_\kappa - p^\lambda{}_\kappa x_\lambda + p^\lambda{}_\lambda x_\kappa = 0, \tag{7.10}$$

in terms of coordinates $(t^\kappa, x, e, p^\lambda, \tau, x_\mu, e_\nu, p^\rho{}_\sigma)$.

The Lagrangian description of dynamics is obtained with the Lagrangian

$$L(t^\kappa, x, x_\lambda) = \tfrac{1}{2}\sqrt{-g}(g^{\kappa\lambda}x_\kappa x_\lambda - m^2 x^2) \tag{7.11}$$

and the Hamiltonian for the field is

$$H(t^\kappa, x, p^\lambda) = \tfrac{1}{2}\sqrt{-g}(g_{\kappa\lambda}p^\kappa p^\lambda + m^2 x^2). \tag{7.12}$$

The Poincaré-Cartan form is expressed locally by

$$\begin{aligned}\Theta = &-\tfrac{1}{2}\sqrt{-g}(g_{\kappa\lambda}p^\kappa p^\lambda + m^2 x^2)\,dt^0 \wedge dt^1 \wedge dt^2 \wedge dt^3\\ &+ p^0 dx \wedge dt^1 \wedge dt^2 \wedge dt^3\\ &- p^1 dx \wedge dt^2 \wedge dt^3 \wedge dt^0\\ &+ p^2 dx \wedge dt^3 \wedge dt^0 \wedge dt^1\\ &- p^3 dx \wedge dt^0 \wedge dt^1 \wedge dt^2\end{aligned} \tag{7.13}$$

in terms of coordinates (t^κ, x, p^λ) of P'. The local expression of Ω is

$$\begin{aligned}\Omega = &-\sqrt{-g}(g_{\kappa\lambda}p^\kappa dp^\lambda + m^2 x\,dx) \wedge dt^0 \wedge dt^1 \wedge dt^2 \wedge dt^3\\ &+ dp^0 \wedge dx \wedge dt^1 \wedge dt^2 \wedge dt^3\\ &- dp^1 \wedge dx \wedge dt^2 \wedge dt^3 \wedge dt^0\\ &+ dp^2 \wedge dx \wedge dt^3 \wedge dt^0 \wedge dt^1\\ &- dp^3 \wedge dx \wedge dt^0 \wedge dt^1 \wedge dt^2\end{aligned} \tag{7.14}$$

If a simple non-zero 4-vector p' tangent to a section of $\xi \circ \pi'$ is represented by

$$p' = \tau\left(\frac{\partial}{\partial t^0} + x_0\frac{\partial}{\partial x} + p^\lambda{}_0\frac{\partial}{\partial p^\lambda}\right) \wedge \left(\frac{\partial}{\partial t^1} + x_1\frac{\partial}{\partial x} + p^\lambda{}_1\frac{\partial}{\partial p^\lambda}\right)$$
$$\wedge\left(\frac{\partial}{\partial t^2} + x_2\frac{\partial}{\partial x} + p^\lambda{}_2\frac{\partial}{\partial p^\lambda}\right) \wedge \left(\frac{\partial}{\partial t^3} + x_3\frac{\partial}{\partial x} + p^\lambda{}_3\frac{\partial}{\partial p^\lambda}\right) \tag{7.15}$$

then the equation

$$p' \lrcorner \Omega = - \sqrt{-g}\,\tau g_{\kappa\lambda}p^\kappa dp^\lambda - \sqrt{-g}\,\tau m^2 x dx + \tau x_\lambda dp^\lambda$$
$$- \tau p^\lambda{}_\lambda dx + \sqrt{-g}\,\tau g_{\mu\lambda}p^\mu p^\lambda{}_\kappa dt^\kappa + \sqrt{-g}\,\tau m^2 x x_\kappa dt^\kappa \tag{7.16}$$
$$+ \tau p^\lambda{}_\lambda x_\kappa dt^\kappa - \tau p^\lambda{}_\kappa x_\lambda dt^\kappa$$

describing $\tilde{D}$ is equivalent to (7.7) and (7.8).

References.

[1] R. Abraham and J. Marsden, *Foundations of Mechanics*, Addison-Wesley, 1978.

[2] P. Dedecker, Coll. Intern. du C.N.R.S., Strassbourg, 1953.

[3] P. Dedecker, Lecture Notes in Mathematics, 570, p. 395-456, Springer-Verlag, 1975.

[4] P.L. Garcia, Symposia Math., 14, p. 219-246, 1974.

[5] P.L. Garcia and A. Pérez-Rendón, Comm. Math. Phys., 13, p. 22-44, 1969.

[6] K. Gawedzki, Rep. Math. Phys., 3, p. 307-326, 1972.

[7] H. Goldschmidt and S. Sternberg, Ann. Inst. Fourier, 23, p. 203-267, 1973.

[8] J. Kijowski, Comm. Math. Phys., 30, p. 99-128, 1973.

[9] J. Kijowski and W. Szczyrba, Comm. Math. Phys., 46, p. 183-203, 1976.

[10] J. Kijowski and W.M. Tulczyjew, *A Symplectic Framework for Field Theories*, Lecture Notes in Physics, 107, Springer-Verlag, 1979.

[11] M.R. Menzio and W.M. Tulczyjew, Ann. Inst. H. Poincaré, A 28, p. 349-367, 1978.

[12] J. Sniatycki and W.M. Tulczyjew, Ann. Inst. H. Poincaré, 14, p. 177-187, 1971.

[13] W.M. Tulczyjew, Symposia Math., 14, p. 247-258, 1974.

[14] W.M. Tulczyjew, C.R. Acad. Sc. Paris, 283, p. 15-18, 1975.

[15] W.M. Tulczyjew, C.R. Acad. Sc. Paris, 283, p. 675-678, 1975.

[16] W.M. Tulczyjew, Ann. Inst. H. Poincaré, 27, p. 101-114, 1977.

SPECTRAL SEQUENCES AND THE INVERSE PROBLEM OF THE CALCULUS OF VARIATIONS

P. Dedecker

and

W. M. Tulczyjew

Institut de Mathématique
Université Catholique de Louvain
Chemin du Cyclotron 2
B-1348 Louvain-la-Neuve

The inverse problem of the calculus of variations attacked recently by different methods [3],[7],[8] has found an elegant solution based on the "triviality" of a spectral sequence. Different versions of the technique of spectral sequences are found in references [1],[4],[5],[9],[10],[11]. Other applications of spectral sequences in the calculus of variations were presented in [4]. We construct the spectral sequence and discuss its exactness in the simplest setting of coordinate geometry.

We consider the space R^{p+q} with coordinates (x^α, x^i); $\alpha = 1,\dots,p$; $i = 1,\dots,q$, and the space $J^\infty(R^p,R^q)$ of infinite jets with coordinates

$$(x^\alpha, x^i, x^i_{\alpha_1}, x^i_{\alpha_1\alpha_2}, \dots, x^i_{\alpha_1\dots\alpha_k}, \dots);\ \alpha_1 \leq \alpha_2 \leq \dots \leq \alpha_k. \tag{1}$$

It is convenient to consider functions $x^i_{\alpha_1\dots\alpha_k}$ defined for arbitrary sequences $\alpha_1,\dots,\alpha_k$ by symmetry: $x^i_{\alpha'_1\dots\alpha'_k} = x^i_{\alpha_1\dots\alpha_k}$ if $\alpha'_1,\dots,\alpha'_k$ is a permutation of $\alpha_1,\dots,\alpha_k$. Functions on $J^\infty(R^p,R^q)$ are locally functions of finite number of coordinates (1) [2]. The space of differential forms on $J^\infty(R^p,R^q)$ will be denoted by Ω. An element of Ω is locally expressible as a combination of exterior products of a finite number of differentials

$$dx^\alpha, dx^i, dx^i_{\alpha_1}, \dots, dx^i_{\alpha_1\dots\alpha_k}, \dots \tag{2}$$

with coefficients in the space of functions.

Instead of the differentials (2) we will use the systems of 1-forms

$$dx^\alpha \tag{3}$$

and

$$\omega^i, \omega^i_{\alpha_1}, \dots, \omega^i_{\alpha_1\dots\alpha_k}, \dots \tag{4}$$

where

$$\omega^i_{\alpha_1\dots\alpha_k} = dx^i_{\alpha_1\dots\alpha_k} - \Sigma_\alpha\, x^i_{\alpha\alpha_1\dots\alpha_k} dx^\alpha. \tag{5}$$

We denote by $\Omega^{r,s}$ the space of (r+s)-forms which are of degree s with respect to the system (3) and of degree r with respect to (4). The exterior differential d is decomposed into a sum

$$d = \partial + \delta \tag{6}$$

of differentials defined by the requirement that if $\mu \in \Omega^{r,s}$ then $\partial\mu \in \Omega^{r,s+1}$ and $\delta\mu \in \Omega^{r+1,s}$. From

$$dx^i_{\alpha_1 \ldots \alpha_k} = \omega^i_{\alpha_1 \ldots \alpha_k} + \Sigma_\alpha \, x^i_{\alpha\alpha_1 \ldots \alpha_k} dx^\alpha \tag{7}$$

and

$$d\omega^i_{\alpha_1 \ldots \alpha_k} = \Sigma_\alpha \, dx^\alpha \wedge \omega^i_{\alpha\alpha_1 \ldots \alpha_k} \tag{8}$$

we derive the explicit expressions

$$\begin{aligned} \partial x^i_{\alpha_1 \ldots \alpha_k} &= \Sigma_\alpha \, x^i_{\alpha\alpha_1 \ldots \alpha_k} dx^\alpha, \\ \delta x^i_{\alpha_1 \ldots \alpha_k} &= \omega^i_{\alpha_1 \ldots \alpha_k}, \\ \partial \omega^i_{\alpha_1 \ldots \alpha_k} &= \Sigma_\alpha \, dx^\alpha \wedge \omega^i_{\alpha\alpha_1 \ldots \alpha_k}, \\ \delta \omega^i_{\alpha_1 \ldots \alpha_k} &= 0. \end{aligned} \tag{9}$$

Also

$$\delta x^\alpha = 0, \quad \partial x^\alpha = dx^\alpha, \quad \delta dx^\alpha = 0, \quad \partial dx^\alpha = 0. \tag{10}$$

Relations

$$\partial\partial = 0, \quad \partial\delta + \delta\partial = 0, \quad \delta\delta = 0 \tag{11}$$

follow from dd = 0. Hence, we have a bicomplex

$$\begin{array}{ccccccccccc}
\Omega^{0,0} & \xrightarrow{\delta} & \Omega^{1,0} & \xrightarrow{\delta} & \Omega^{2,0} & \xrightarrow{\delta} & \cdots & \xrightarrow{\delta} & \Omega^{r,0} & \xrightarrow{\delta} & \cdots \\
\downarrow{\scriptstyle\partial} & & \downarrow{\scriptstyle\partial} & & \downarrow{\scriptstyle\partial} & & & & \downarrow{\scriptstyle\partial} & & \\
\Omega^{0,1} & \xrightarrow{\delta} & \Omega^{1,1} & \xrightarrow{\delta} & \Omega^{2,1} & \xrightarrow{\delta} & \cdots & \xrightarrow{\delta} & \Omega^{r,1} & \xrightarrow{\delta} & \cdots \\
\downarrow{\scriptstyle\partial} & & \downarrow{\scriptstyle\partial} & & \downarrow{\scriptstyle\partial} & & & & \downarrow{\scriptstyle\partial} & & \\
\vdots & & \vdots & & \vdots & & & & \vdots & & \\
\downarrow{\scriptstyle\partial} & & \downarrow{\scriptstyle\partial} & & \downarrow{\scriptstyle\partial} & & & & \downarrow{\scriptstyle\partial} & & \\
\Omega^{0,p} & \xrightarrow{\delta} & \Omega^{1,p} & \xrightarrow{\delta} & \Omega^{2,p} & \xrightarrow{\delta} & \cdots & \xrightarrow{\delta} & \Omega^{r,p} & \xrightarrow{\delta} & \cdots
\end{array} \tag{12}$$

The augmented cochain complex

$$0 \longrightarrow K \xrightarrow{\kappa} \Omega^{0,p} \xrightarrow{\gamma} E_1^{1,p} \xrightarrow{d^1} \cdots \xrightarrow{d^{r-1}} E_1^{r,p} \xrightarrow{d^r} \cdots \tag{13}$$

related to the term E_1 of the spectral sequence

$$E_1^{0,p} \xrightarrow{d^0} E_1^{1,p} \xrightarrow{d^1} E_1^{2,p} \xrightarrow{d^2} \cdots \xrightarrow{d^{r-1}} E_1^{r,p} \xrightarrow{d^r} \cdots \tag{14}$$

is of fundamental importance to the calculus of variations. The space K is the subspace

$$K = \partial(\Omega^{0,p-1}) \subset \Omega^{0,p} \tag{15}$$

and $\kappa : K \to \Omega^{0,p}$ is the inclusion. Spaces $E_1^{r,p}$ are the quotient spaces $\Omega^{r,p}/\partial(\Omega^{r,p-1})$. Operators γ and d^r are defined by the commutative diagram

$$\begin{array}{ccccccccccc}
\Omega^{0,p} & \xrightarrow{\delta} & \Omega^{1,p} & \xrightarrow{\delta} & \Omega^{2,p} & \xrightarrow{\delta} & \cdots & \xrightarrow{\delta} & \Omega^{r,p} & \xrightarrow{\delta} & \cdots \\
\downarrow \varepsilon^0 & \searrow \gamma & \downarrow \varepsilon^1 & & \downarrow \varepsilon^2 & & & & \downarrow \varepsilon^r & & \\
E_1^{0,p} & \xrightarrow{d^0} & E_1^{1,p} & \xrightarrow{d^1} & E_1^{2,p} & \xrightarrow{d^2} & \cdots & \xrightarrow{d^{r-1}} & E_1^{r,p} & \xrightarrow{d^r} & \cdots
\end{array} \qquad (16)$$

where $\varepsilon^r : \Omega^{r,p} \to E_1^{r,p}$ are the canonical projections.

Exactnes of the sequence (13) as well as that of the spectral term (14) follow from the exactess of the augmented bicomplex

$$\begin{array}{ccccccccccccc}
 & & 0 & & 0 & & & & & & & & \\
 & & \downarrow & & \downarrow & & & & & & & & \\
 & & R & = & R & & 0 & & & & 0 & & \\
 & & \downarrow \rho & & \downarrow \sigma & & \downarrow & & & & \downarrow & & \\
0 & \to & \Phi^0 & \xrightarrow{\eta} & \Omega^{0,0} & \xrightarrow{\delta} & \Omega^{1,0} & \xrightarrow{\delta} & \cdots & \xrightarrow{\delta} & \Omega^{r,0} & \xrightarrow{\delta} & \cdots \\
 & & \downarrow d & & \downarrow \partial & & \downarrow \partial & & & & \downarrow \partial & & \\
0 & \to & \Phi^1 & \xrightarrow{\eta} & \Omega^{0,1} & \xrightarrow{\delta} & \Omega^{1,1} & \xrightarrow{\delta} & \cdots & \xrightarrow{\delta} & \Omega^{r,1} & \xrightarrow{\delta} & \cdots \\
 & & \downarrow d & & \downarrow \partial & & \downarrow \partial & & & & \downarrow \partial & & \\
 & & \vdots & & \vdots & & \vdots & & & & \vdots & & \\
 & & \downarrow d & & \downarrow \partial & & \downarrow \partial & & & & \downarrow \partial & & \\
0 & \to & \Phi^p & \xrightarrow{\eta} & \Omega^{0,p} & \xrightarrow{\delta} & \Omega^{1,p} & \xrightarrow{\delta} & \cdots & \xrightarrow{\delta} & \Omega^{r,p} & \xrightarrow{\delta} & \cdots
\end{array} \qquad (17)$$

where

$$0 \longrightarrow R \xrightarrow{\rho} \Phi^0 \xrightarrow{d} \Phi^1 \xrightarrow{d} \cdots \xrightarrow{d} \Phi^p \qquad (18)$$

is the augmented cochain complex of differential forms on R^p. Exactness of rows of the diagram (17) is the consequence of a version of the Poincaré lemma with parameters. The sequence (18) is exact. Exactness of columns for $r > 0$ is proved by constructing a cochain homotopy [10]. Exactness of the remaining column follows by diagram chasing. Exactness of sequences (13) and (14) is also proved by diagram chasing.

Exactness of the complex (13) is closely related to the inverse problem of the calculus of variations. In order to establish the relation we introduce in Ω an operator

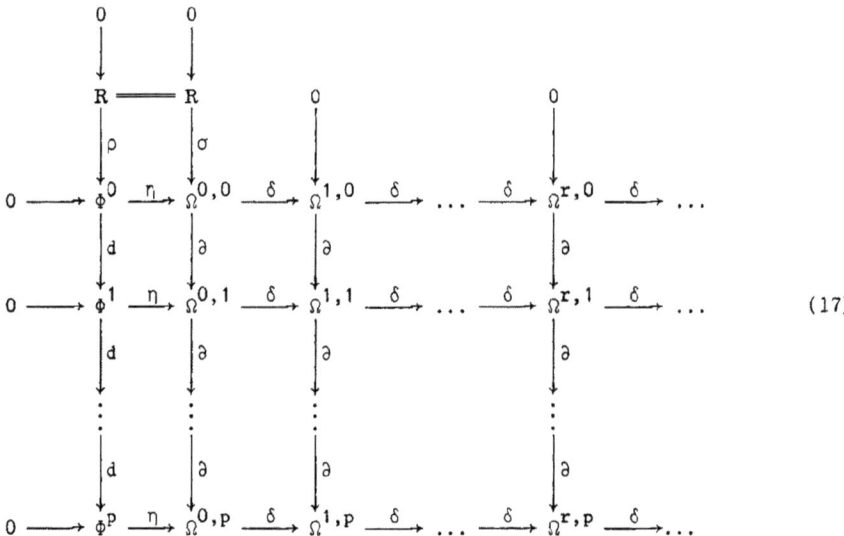

$$\tau = \Sigma_k \Sigma_{\alpha_1,\ldots,\alpha_k} \frac{(-1)^k}{k!} \partial_{\alpha_1} \cdots \partial_{\alpha_k} \theta_k^{\alpha_1\cdots\alpha_k}, \tag{19}$$

where ∂_α are derivations defined by

$$\begin{aligned} \partial_\alpha x^\beta &= \delta_\alpha^\beta, \\ \partial_\alpha x^i{}_{\alpha_1\ldots\alpha_k} &= x^i{}_{\alpha\alpha_1\ldots\alpha_k}, \\ \partial_\alpha dx^\beta &= 0, \\ \partial_\alpha \omega^i{}_{\alpha_1\ldots\alpha_k} &= \omega^i{}_{\alpha\alpha_1\ldots\alpha_k} \end{aligned} \tag{20}$$

and $\theta_k^{\alpha_1\cdots\alpha_k}$ are derivations characterized by

$$\begin{aligned} \theta_k^{\alpha_1\cdots\alpha_k} x^\alpha &= 0, \\ \theta_k^{\alpha_1\cdots\alpha_k} x^i{}_{\beta_1\ldots\beta_k} &= 0, \\ \theta_k^{\alpha_1\cdots\alpha_k} dx^\alpha &= 0, \\ \theta_k^{\alpha_1\cdots\alpha_k} \omega^i &= 0, \\ \theta_0 \omega^i{}_{\alpha_1\ldots\alpha_k} &= \omega^i{}_{\alpha_1\ldots\alpha_k}, \\ [\theta_k^{\alpha_1\cdots\alpha_k}, \partial_\alpha] &= \sum_{i=1}^{k} \delta_\alpha^{\alpha_i} \theta_{k-1}^{\alpha_1\cdots\hat{\alpha}_i\cdots\alpha_k}. \end{aligned} \tag{21}$$

Relations

$$\tau\partial_\alpha = 0, \quad \tau\tau = \tau\theta_0 \tag{22}$$

are easily established by direct computation and

$$\tau\partial = 0 \tag{23}$$

follows from

$$\partial\mu = \Sigma_\alpha dx^\alpha \wedge \partial_\alpha \mu. \tag{24}$$

Let $D:\Omega \to \Omega$ be defined by

$$D = \Sigma_k \Sigma_{\alpha,\alpha_1,\ldots,\alpha_k} \frac{(-1)^k}{(k+1)!} \iota_\alpha \partial_{\alpha_1} \cdots \partial_{\alpha_k} \theta_{k+1}^{\alpha\alpha_1\cdots\alpha_k}, \tag{25}$$

where ι_α are antiderivations defined by

$$\begin{aligned} \iota_\alpha x^\beta &= 0, \quad \iota_\alpha x^i{}_{\alpha_1\ldots\alpha_k} = 0, \\ \iota_\alpha dx^\beta &= \delta_\alpha^\beta, \quad \iota_\alpha \omega^i{}_{\alpha_1\ldots\alpha_k} = 0. \end{aligned} \tag{26}$$

For each $\mu \in \Sigma_r \Omega^{r,p}$ the formula

$$\tau\mu + \partial D\mu = \theta_0\mu \tag{27}$$

is verified by computation.

PROPOSITION. For $r > 0$ the subspace $\Lambda^r = \tau(\Omega^{r,p}) \subset \Omega^{r,p}$ is a complement of the subspace $\partial(\Omega^{r,p-1}) \subset \Omega^{r,p}$.

Proof. Since $\theta_0\mu = r\mu$ for each $\mu \in \Omega^{r,p}$, it follows from (22) that the operator

$$\tau^r : \Omega^{r,p} \to \Omega^{r,p} : \mu \mapsto \frac{1}{r}\tau\mu \tag{28}$$

is a projection operator. From (23) it follows that

$$\begin{aligned} \mu &= \tau^r\mu + (1 - \tau^r)\mu \\ &= \tau^r\mu + \frac{1}{r}\partial D\mu \end{aligned} \tag{29}$$

is the unique decomposition of $\mu \in \Omega^{r,p}$ into elements $\tau^r\mu \in \Lambda^r$ and $(1 - \tau^r)\mu \in \partial(\Omega^{r,p-1})$. ∎

We conclude from the Proposition that the sequence (13) can be replaced by the exact sequence

$$0 \longrightarrow K \xrightarrow{\kappa} \Omega^{0,p} \xrightarrow{\lambda^0} \Lambda^1 \xrightarrow{\lambda^1} \cdots \xrightarrow{\lambda^{r-1}} \Lambda^r \xrightarrow{\lambda^r} \cdots \tag{30}$$

isomorphic to (13). The operators λ^r are defined by the commutative diagram

$$\begin{array}{ccccccccccc} \Omega^{0,p} & \xrightarrow{\delta} & \Omega^{1,p} & \xrightarrow{\delta} & \Omega^{2,p} & \xrightarrow{\delta} & \cdots & \xrightarrow{\delta} & \Omega^{r,p} & \xrightarrow{\delta} & \cdots \\ & \searrow^{\lambda^0} & \downarrow{\scriptstyle\tau^1} & & \downarrow{\scriptstyle\tau^2} & & & & \downarrow{\scriptstyle\tau^r} & & \\ & & \Lambda^1 & \xrightarrow{\lambda^1} & \Lambda^2 & \xrightarrow{\lambda^2} & \cdots & \xrightarrow{\lambda^{r-1}} & \Lambda^r & \xrightarrow{\lambda^r} & \cdots \end{array} \tag{31}$$

Since τ^r are projection operators and $\delta\mu = d\mu$ for $\mu \in \Sigma_r\Omega^{r,p}$, the operators λ^r are given explicitly by

$$\lambda^r = \tau^{r+1}d. \tag{32}$$

It is easy to see that

$$\lambda^0 : \Omega^{0,p} \to \Lambda^1 \tag{33}$$

is the Euler-Lagrange operator. Hence exactness of the sequence (30) provides a local solution of the inverse problem of the calculus of variations:

THEOREM. If $\mu \in \Lambda^1$ then there exists a form $L \in \Omega^{0,p}$ such that $\mu = \lambda^0 L$ if and only if $\lambda^1\mu = 0$. Also if $L \in \Omega^{0,p}$ then $\lambda^0 L = 0$ if and only if $L \in K = \partial(\Omega^{0,p-1})$.

REFERENCES

[1] I.M. Anderson and T. Duchamp, On the existence of global variational principles, preprint, University of Utah.

[2] J.M. Bordman, Singularities of differentiable maps, Publ. I.H.E.S., 33, pp. 21--57, 1967.

[3] P. Dedecker, Sur un problème inverse du calcul des variations, Bull. Acad. R. Belg. Sc., 36, pp. 63-70, 1950.

[4] P. Dedecker, On the generalization of symplectic geometry to multiple integrals in the calculus of variations, Symposium on differential geometric methods in mathematical physics, Lecture Notes in Mathematics, 570, pp. 395-456, Springer--Verlag, 1977.

[5] G.W. Horndeski, Differential operators associated with the Euler-Lagrange operator, Tensor, 28, pp. 303-318, 1974.

[6] B.A. Kuperschmidt, Geometry of jet bundles and the structure of Lagrangian and Hamiltonian formalisms, preprint, M.I.T.

[7] F. Takens, Symmetries, conservation laws and variational principles, Geometry and Topology: Proceedings of III Latin American School of Mathematics,Lecture Notes in Mathematics, 597, pp. 581-604, 1977.

[8] E. Tonti, Variational formulations of nonlinear differential equations I, Acad. R. Belg. Bull. Cl. Sci. 55, pp. 137-165 and pp. 262-276, 1969.

[9] W.M. Tulczyjew, The Lagrange complex, Bull. Soc. math. France, 105, pp. 419--431, 1977.

[10] W.M. Tulczyjew, The Euler-Lagrange resolution, in this volume.

[11] A.M. Vinogradov, A spectral sequence associated with a nonlinear differential equation, and algebro-geometric foundations of Lagrangian field theory with constraints, Soviet Math. Dokl., 19, pp. 144-148, 1978.

[12] P. Dedecker, On applications of homological algebra to Calculus of Variations and Mathematical Physics, Proceedings of the IVth International Colloquium on Differential Geometry, September 1978, Universidad de Santiago de Compostela, 1979.

GEODESIC FIELDS IN THE CALCULUS OF VARIATIONS OF MULTIPLE INTEGRALS DEPENDING ON DERIVATIVES OF HIGHER ORDER.

T.SZAPIRO

Department of Mathematical Mathods of Physics

Warsaw,University

1. Introduction.

When one considers functional of variational calculus

$$J(\varphi) = \int_a^b L(x, \varphi, \frac{d\varphi}{dx})\, dx$$

then the notion of geodesic fields plays very important role for obtaining sufficient conditions for minimum of such a functional.The theory becomes more complicated,if one passes to multiple integrals.It was developed by Caratheodory,Weyl,de Donder,Lepage and others and reached its final shape in the language of modern differential geometry in a paper [1] of Dedecker.These theories deal only with L depending on derivatives of the first order.But important problems of physics and engineering lead to lagrangians depending on derivatives of higher order ; e.g. oscillations of rigid plate leads to the functional with second derivatives.

The aim of this paper is to sketch a theory in a spirit of Dedecker's theory in case of lagrangians depending on derivatives of higher order i.e. to investigate the functionals on sections of some bundles,which in local description are of the form

$$(1) \qquad J(\phi) = \int L\,(x^{\mu}, \varphi^{A}, \frac{\partial\varphi^{A}}{\partial x^{\mu}}, \frac{\partial^2\varphi^{A}}{\partial x^{\mu}\partial x^{\nu}}, \dots, \frac{\partial^N\varphi^{A}}{\partial x^{i_1}\dots\partial x^{i_N}})\, d^n x$$

We shall develope such fundamental notions as Legendre transformation,Noether theorem and present sufficient condition for the minimum of these general functionals.We shall also deal with an important class of geodesic fields - fields of Mayer.

2. Linearization of variational problem with derivatives of order N.

Let M be a m-dimensional manifold and $\pi : W \longrightarrow M$ fibre bundle with typical fibre R^r. By W^N we shall denote the bundle of jets of order N of sections of bundle W. The fibre of W^N in any point is given by classes of local sections of W, which have the same Taylor's expansion of order N in this point. The projection $\pi^N : W^N \longrightarrow W$ assigns to N-jet of section in a given point the value of this section in fibre of W over this point. When we have local coordinates (x^μ) on M $1 \leq \mu \leq m$, and (x^μ, φ^A) - on W $1 \leq A \leq r$, the section $\varphi : M \longrightarrow W$ (we shall use notation $\varphi \in \Gamma^\infty(M,W)$) is given by means of functions $\varphi^A(x)$,and k-jets - by taking derivatives of $\varphi^A(x)$ up to order k in point x.

Let us denote $\varphi_i^A := \dfrac{\partial^{|i|}\varphi^A}{\partial x^i}$ - i being multiindex describing order and succesion of differentiation. If we allow only nonincreasing multindices then (x^μ, φ_i^A) $0 \leq |i| \leq N$ define local coordinates on W^N.

Definition 1. Let $L : W^N \longrightarrow \Lambda^m T^* M$ be a smooth mapping, such that following diagram

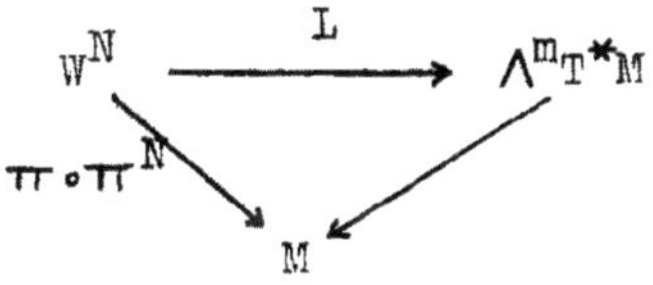

is commutative (therefore L conserves fibres) .

Let G be a compact m-dimensional submanifold of M with boundary ∂G. (W^N, L, G) will be called variational problem on G with derivatives of order N. L is a lagrangian of the problem. It defines functional on sections of W^N, which for N-jets of sections of W is of the form (1) .

Definition 2. Let $\varphi \in \Gamma^\infty(M,W)$. One parmeter family $h_t \in \Gamma^\infty(M,W)$ $-\varepsilon < t < \varepsilon$,such that $h_0 = \varphi$ and jets of N-1 order of h_t are equal on ∂G for any t will be called homotopy of φ with N-1 derivatives fixed on ∂G. φ is extremal section of the problem (W^N,L,G) if for any such homotopy the following condition is fulfilled :

$$\frac{d}{dt}\Big|_{t=0} J_{G,L}(h_t) = \frac{d}{dt}\Big|_{t=0} \int_G L(j_N(h_t)) = 0$$

where $j_N : \Gamma^\infty(M,W) \longrightarrow \Gamma^\infty(M,W^N)$ is an operation of taking jets of order N.

Plainly extremals of (W^N,L,G) satisfy the Euler-Lagrange Equation. Let $V \longrightarrow M$ be a fibre bundle, then any m-form Ω on V defines variational problem on sections $\Psi \in \Gamma^\infty(M,V)$ by means of a following natural lagrangian L_Ω

$$L_\Omega(j^1(\Psi)) = \Psi^*\Omega$$

Definition 3. (V^1,L_Ω,G) is called a linear variational problem of order one.

We will give a procedure of passing from variational problem (W^N,L,G) to a suitable (V^1,L_Ω,G) and we shall describe the correspondance of extremals of both problems. By $\pi^N_{N-1}: W^N \longrightarrow W^{N-1}$ let us denote mapping of jet bundles. Consider the bundle $\tilde{\pi} : \Lambda^m T^*(W^N) \longrightarrow W^N$, and its subbundle $\Lambda^m T^*(\underline{W}^N) \longrightarrow W^N$ of m-covectors on W^N horizontal with respect to π^N_{N-1}. Consider the system of following 1-forms ω^A_i on W^N

$$\omega^A_i := d\varphi^A_i - \varphi^A_{(\mu,i)}\, dx^\mu \qquad 0 \leqslant |i| \leqslant N-1$$

((μ,i) is a multindex given by μ and $i = (i_1,i_2,\ldots,i_{|i|})$).

Let us notice, that if

(2) $$\phi^*(\omega^A_i) = 0$$

for $\phi \in \Gamma^\infty(M,W^N)$, then the component φ^A_i of ϕ is determined : $\varphi^A_{(\mu,i)} = \frac{\partial \varphi^A_i}{\partial x^\mu}$, when (2) is satisfied for all i $0 \leqslant |i| \leqslant N-1$ then

$$\varphi^A_j = \frac{\partial^{|j|}\varphi^A}{\partial x^j} \qquad 0 \leqslant |j| \leqslant N.$$ One can construct a base in $\Lambda^m T^*(\underline{W}^N)$

using ω_i^A, it is composed of m-covectors $\tilde{\omega}^{A_1\ldots A_2}_{(\mu_1,j_1)\ldots,(\mu_k,j_k)}$ formed from $dx^1\wedge\ldots\wedge dx^m$ by putting in places $\mu_1,\ldots,\mu_k$ – $\omega^{A_1}_{j_1},\ldots,\omega^{A_k}_{j_k}$ instead of $dx^{\mu_1},\ldots,dx^{\mu_2}$. Thus $\{dx^1\wedge\ldots\wedge dx^m$,

$\tilde{\omega}^{A_1,A_2,\ldots,A_k}_{(\mu_1,j_1),\ldots,(\mu_k,j_k)}$; $1\leqslant k\leqslant m$, $0\leqslant |j_s|\leqslant N-1$, $1\leqslant s\leqslant k\}$ is a base in fibre $\Lambda^m T^*(\underline{W}^N)$ and therefore induces on $\Lambda^m T^*(\underline{W}^N)$ local coordinates $(x^\mu, \varphi_i^A, \lambda_\alpha^J)$, where μ, A_s, i $1\leqslant s\leqslant k$ are as above, J – passes systems of multindices of required length, $\alpha=(A_1,\ldots,A_k)$ $1\leqslant|\alpha|\leqslant m$.

<u>Definition 4</u>. On $\Lambda^m T^*(\underline{W}^N)$ there exists a <u>canonical m-form $\tilde{\Omega}$</u> given by the following formula

$$\langle \overset{m}{X} \mid \tilde{\Omega}_v\rangle = \langle \tilde{\pi}_* \overset{m}{X} \mid v\rangle$$

– here $v\in\Lambda^m T^*(\underline{W}^N)$ and $\tilde{\pi}_*$ is a tangent mapping of m-covectors induced by $\tilde{\pi}:\Lambda^m T^*(W^N)\longrightarrow W^N$.

Thus in the local description $\tilde{\Omega}_v$ is given with same formula as v.

<u>Definition 5</u>. Let us denote by $V\subset\Lambda^m T^*(\underline{W}^N)$ the set of m-covectors v fulfilling the following requirements :

if $v=\lambda dx^1\wedge\ldots\wedge dx^m+\sum\lambda^{(\mu_1,j_1)}_{A_1}\omega^{A_1}_{(\mu_1,j_1)}+\ldots$ then

1°. $\lambda dx^1\ \ldots\ dx^m = L$

2°. $\lambda^k_A = \dfrac{\partial\lambda}{\partial\varphi^A_k}$ $\quad |k| = N$

3°. $\lambda^{k\ ,l}_{A_k,A_l} - \dfrac{\partial^2\lambda}{\partial\varphi^{A_k}_k\partial\varphi^{A_l}_l}$ is nondegenerate for $|k|=|l|=N$

V is a subbundle of $\Lambda^m T^*(\underline{W}^N)$ and will be called the <u>bundle of Dedecker of the problem of the order N</u> (since it was in case N = 1 introduced by Dedecker in [1]).

The bundle V provided with a canonical m-form $\Omega:=\tilde{\Omega}\big|_V$

defines linear variational problem of the order one. The question arises how the original problem (W^N, L, G) is related to a linearized one (V^1, L_Ω, G). In order to give the answer we have to introduce the following notion :

Definition 6. If components φ_i^A and λ_A^i of section $\phi \in \Gamma^\infty(M,V)$ are given by the following formulas :

$$\varphi_j^A = \frac{\partial^{|j|}\varphi^A}{\partial x^j} \qquad 0 \leqslant |j| \leqslant N$$

$$\lambda_A^i = \sum_{|m|=0}^{N-|i|} \frac{d^{|m|}}{dx^m} \frac{\partial \lambda}{\partial \varphi^A_{(m,i)}} \qquad 1 \leqslant |i| \leqslant N-1$$

then ϕ will be called a lifting of φ (here $(\varphi \in \Gamma^\infty(M,W)$ is W - component of ϕ) and denoted $\varphi^\#$.

Notice, that for given $\varphi \in \Gamma^\infty(M,W)$ there exist in general different possible liftings.

Theorem 1.

If $\phi \in \Gamma^\infty(M,V)$ is extremal section of linearized variational problem of order one (V^1, L_Ω, G) then ϕ is a lifting of $\varphi = \pi^N \circ \tilde{\pi} \circ \phi \in \Gamma^\infty(M,W)$ and ϕ is extremal section of (W^N, L, G).

If $\varphi \in \Gamma^\infty(M,W)$ is extremal section of (W^N, L, G) then any $\varphi^\#$ is extremal section of (V^1, L_Ω, G) .

3. The transformation of Legendre . Noether theorem.

Definition 7. The bundle map $\mathcal{L} : V \longrightarrow \Lambda^m T^*(W^{N-1})$

$$\begin{array}{ccc} \Lambda^m T^*(\underline{W}^N) \supset V & \xrightarrow{\ \mathcal{L}\ } & \Lambda^m T^*(W^{N-1}) \\ \downarrow & & \downarrow \\ W^N & \xrightarrow[\pi^N_{N-1}]{} & W^{N-1} \end{array}$$

given by the formula :

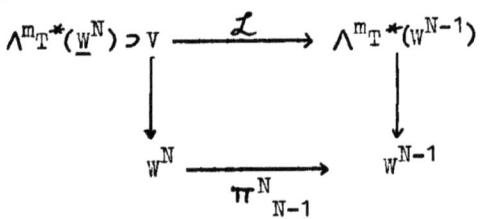

$$\langle \pi^N_{N-1*}(\overset{m}{X}) \mid \mathcal{L}(v)\rangle = \langle \overset{m}{X} | v\rangle$$

where $\overset{m}{X} \in \Lambda^m T(W^N)$, $v \in V$ is called the <u>transformation of Legendre</u>.
We have the following

<u>Theorem 2</u>.

The transformation of Legendre is a local isomorphism.

<u>Remark</u>.The proof of injectivity of the differential of $\mathcal{L}$ follows from the condition 3^0 of definition 5.

We close the present section with Noether theorem.

<u>Definition 8</u>.Let $X \in \Gamma^\infty(TV)$. X <u>preserves dynamics</u> if there exists $\eta \in \Lambda^{m-1}T^*(V)$ such that

$$\mathcal{L}_X\Omega = d\eta + Y \lrcorner d\Omega \qquad Y \in \Gamma^\infty(TV).$$

<u>Theorem 3</u>.

If X preserves dynamics then $\eta_X := \eta - X \lrcorner \Omega$ is conserved (physical) quantity i.e. $\phi^*(d\eta_X) = 0$ for extremal $\phi \in \Gamma^\infty(M,V)$.

The proof :

$$\phi^*(d\eta_X) = \phi^*[d(\eta - X \lrcorner \Omega)] = \phi^*[d\eta - d(X \lrcorner \Omega)] =$$
$$= \phi^*[d\eta - d(X \lrcorner \Omega) - X \lrcorner d\Omega] = \phi^*(d\eta - \mathcal{L}_X\Omega) =$$
$$= \phi^*(Y \lrcorner d\Omega) = 0$$

We have used twice the fact that $\phi^*(Z \lrcorner d\Omega) = 0$ for extremal ϕ and any vector field on V .

4.<u>Geodesic fields</u>.

A general theory of geodesic fields was given for multiple integrals by Lepage and obtained the classical form in papers of Dedecker (N=1).

<u>Definition 9</u>.Let (W^N, L, G) be a variational problem of order N and (V^1, L_Ω, G) the linearized problem of order one. The mapping

$$p : W^k \longrightarrow V \quad \text{such that} \quad d(p^*\Omega) = 0 \qquad 0 \leq k \leq N-1$$

is called <u>geodesic field on W^k</u> for (W^N, L, G).

Definition 10. $\phi \in \Gamma^{\infty}(M,V)$ is embedded in geodesic field p on W^k if 1°. $\phi = p \circ \pi^N_k \circ \tilde{\pi} \circ \phi$

2°. $\phi = (\pi \circ \tilde{\pi} \circ \phi)^{\#}$

Theorem 4.

$\phi \in \Gamma^{\infty}(M,V)$ embedded in geodesic field is extremal section of the problem (V^1, L_{Ω}, G).

Let us consider $\Lambda^m T_W V$ - the space of liftings to $\Lambda^m TV$ of m-covectors tangent to j_N and following bundle $P \longrightarrow W$; $P = \bigcup_{w \in W} P_w$ with the fibre

$$P_w := \{(X,Y) \; ; \; X \in (\Lambda^m T_W V)_{v_1}, \; Y \in (\Lambda^m T_W V)_{v_2}, \pi^N(v_i) = w, \; i = 1,2\}$$

Definition 11. The function $E : P \longrightarrow R^1$ given by the formula

$$E(X,Y) = X \lrcorner \Omega - Y \lrcorner \Omega$$

is called the function of excess.

Theorem 5.

If $\varphi^{\#} \in \Gamma^{\infty}(M,V)$ is embedded in geodesic field p on W^k and $E \geqslant 0$ then φ is a minimal point of problem (W^N, L, G) for simply convex G and φ fixed on ∂G with N-1 derivatives.

Definition 12. The geodesic field p on W is called the field of Mayer if it is locally integrable.

Theorem 6.

Let dim $M > 1$, then every extremal foliation of W determines the field of Mayer.

Concluding remarks.

1°. The whole theory can be written down in an invariant form.

2°. The bundle structure of W is not neccessary to obtain the preceding results . One can deal with the notion of extremal chain in the manifold W. Such approach is more general. The theory is presented here in the bundle language because of its natural physical applications.

Acknowledgments.

I am indebted to dr K.Gawędzki, my thesis advisor, and to Prof. K.Maurin for giving of their time to many stimulating conversations which resulted in substantial improvements in final presentation of my results.I would like to thank to organizers of Conferrence for warm hospitality during stay in Salamanca, and to Prof. S.Sternberg,who drew my attention to refence [4].

References.

[1]. P.Dedecker - Calcul des Variations,Formes Differentielles et Champs Geodesiques,Colloques Internationales du CNRS, Strassbourg 1953.

[2]. P.Dedecker - On the generalization of symplectic geometry to multiple integrals in the calculus of variations, Lecture Notes in Mathematics 570,p.394-456.

[3]. H.Goldschmidt,S.Sternberg - The Hamilton-Cartan formalism in the calculus of variations,Ann.Inst.Fourier 23 1973 p.203-267.

[4]. S.Sternberg - Some preliminary remarks on the formal variational calculus of Gelfand and Dikki,Lecture Notes in Mathematics 670.

[5]. H.Weyl - Geodesic fields,Ann. of Math. 36 1935 ,p.607-629.

SEPARABILITY STRUCTURES ON RIEMANNIAN MANIFOLDS

Sergio Benenti

Institute of Mathematical Physics
Institute of Rational Mechanics
University of Turin, Turin, Italy

1.- Introduction.

Let M be a differentiable manifold of dimension n and let H be a differentiable function on the cotangent bundle T*M. Let $(U,(x^i))$ be a chart on M (i.e. a coordinate system (x^i) defined on the open set $U \subseteq M$) and let $(T^*U,(x^i;p_j))$ be the corresponding canonical chart on T*M $(i,j = 1,\dots,n)$. With this chart we can associate the representative function $H(x^i;p_j)$ of H and the so called reduced Hamilton-Jacobi equation:

$$H(x^i;\frac{\partial S}{\partial x^j}) = e \qquad (1.1)$$

where e is a real parameter.

A complete solution (or a complete integral) of (1.1) is an n-parameters family $S(x^i;c_j)$ of solutions of (1.1) such that:

$$\det\left(\frac{\partial^2 S}{\partial x^i \partial c_j}\right) \neq 0 \qquad (1.2)$$

When a complete solution is known, one can directly find the integral curves of the Hamiltonian vectorfield defined by the Hamiltonian H, by following a well known method.

We are mainly interested in those cases where equation (1.1) is completely separable, i.e. it admits a complete solution which is a sum of functions depending on a single coordinate:

$$S(x^i;c_j) = S_1(x^1;c_j) + S_2(x^2;c_j) + \dots + S_n(x^n;c_j) \qquad (1.3)$$

Although the conditions for the existence of such a kind of complete solution are known (Levi-Civita, 1904 [1]), a detailed treatement of the separable case of the Hamilton-Jacobi equation is so far available only for certain classes of Hamiltonians. Among them, we recall the quadratic Hamiltonians corresponding to the classical mechanical systems, including t-dependent constraints and velocity-dependent potentials. For a hystorical perspective on the separability of the Hamilton-Jacobi equation the reader can refer to the review articles [2] and [3] .

For a better understanding of the separability conditions of equation (1.1), in the present paper we propose an approach based on the concept of separability structure, which seems to allow valuable simplifications. A separability structure is a family of charts at a point $x_0 \in M$ such that the corresponding equations (1.1) have complete integrals of the kind (1.3) representing the same family of functions on a neighborhood of x_0. From this point of view we analyze, in particular, the case of the geodesic Hamiltonian of a Riemannian manifold with definite or indefinite metric. (Throughout this paper we use the term Riemannian manifold in this generalized sense; when we want to distinguish explicitly between definite or indefinite metric we use the terms proper-Riemannian and pseudo-Riemannian manifold respectively). For the sake of simplicity, functions and manifolds are tacitly assumed to be smooth enough to assure the validity of the results.

2.-Separability structures.

A solution of the Hamilton-Jacobi equation associated with the Hamiltonian $H:T^*M \longrightarrow \mathbb{R}$ is a function $S:U \longrightarrow \mathbb{R}$ (where U is an open subset of M) such that:

$$(2.1) \qquad dS(U) \subseteq C_e \quad ,$$

where $dS:M \longrightarrow T^*M$ is the differential of S and $C_e = \{p \in T^*M: H(p) = e\}$. Usually, only regular values of the Hamiltonian are considered, so that C_e is a submanifold of T^*M of codimension 1 (see for instance [4] , p.49 and 204). In fact, inculsion (2.1) is the coordinate free translation of equation (1.1), and we call it (although rather improperly) the reduced Hamilton-Jacobi equation associated with the Hamiltonian H and the energy e.

A complete solution (or a complete integral) of the Hamilton-Jacobi equation associated with the Hamiltonian H is a function $S:\hat{U} \longrightarrow \mathbb{R}$ on an open subset $\hat{U}$ of a product manifold $M \times A \times E$, where E is the energy space ($E \simeq \mathbb{R}$) and A is a parameter space of dimension n-1 (it is not necessary to specify here the nature of such a space: for further information see for instance [5]), such that each function $S_c:U_c \longrightarrow \mathbb{R}$ obtained by fixing an admissible value $c = (a,e) \in A \times E$ ($U_c = \{x \in M: (x,c) \in \hat{U}\}$, $S_c(x) = S(x,c)$) is a solution of the Hamilton-Jacobi equation (2.1):

$$(2.2) \qquad dS_c(U_c) \subseteq C_e \quad ,$$

and the submanifolds $L_c = dS_c(U_c)$, where c covers all its admissible values, form a (local) foliation of T*M. The latter requirement is in fact the geometric counterpart of condition (1.2), where (c_j) must be interpreted as coordinates on $A\times E$ (usually one of the parameters (c_j) coincides with e). We notice that each leaf L_c of the foliation is a section of the canonical projection $\pi_M: T^*M \longrightarrow M$ and a Lagrangian submanifold of T*M with respect to the canonical symplectic 2-form $\omega_M = dx^i \wedge dp_i$. If we consider the natural projection $\sigma_M: M\times A\times E \longrightarrow M$, the open set $U = \sigma_M(\hat{U}) \subseteq M$, which is the union of the open sets (U_c), can be called the <u>base domain</u> of the complete integral $S:\hat{U} \longrightarrow \mathbb{R}$. Then, the foliation (L_c) generated by S is defined in an open set $\tilde{U} \subseteq T^*M$ such that $\pi_M(\tilde{U}) = U$.

(2.3) <u>Definition</u>.- A chart $(U,(x^i))$ on a manifold M is said to be <u>separable</u> with respect to the Hamiltonian $H:T^*M \longrightarrow \mathbb{R}$ if the corresponding Hamilton-Jacobi equation admits a complete solution S whose base domain is U and such that

$$(2.4) \qquad \partial_i \partial_j S = 0 \quad , \quad i \neq j \qquad (^\circ)$$

Let us fix a point $x_0 \in M$ and let us consider the set of the separable charts at x_0.

(2.5) <u>Definition</u>.- Two charts $(U,(x^i))$ and $(U',(x^{i'}))$ are said to be <u>$\mathcal{S}$-compatible</u> with respect to the Hamiltonian H at the point $x_0 \in U \cap U'$, if they are separable and if the corresponding separated complete integrals coincide in an open neighborhood U'' of x_0, i.e. if the two corresponding foliations in T*M coincide in an open set $\hat{U}''$ such that $x_0 \in U'' = \pi_M(\hat{U}'')$.

$\mathcal{S}$-compatibility is clearly an equivalence relation in the family of the separable charts at the point x_0 (which of course could be empty). Hence we are led to the following definition.

(2.6) <u>Definition</u>.- A <u>separability structure</u> (briefly, a <u>$\mathcal{S}$-structure</u>) at the point x_0 is an equivalence class of charts which are $\mathcal{S}$-compatible at x_0 with respect to the Hamiltonian H.

$(^\circ)$ From now on we will use the following abbreviations: $\partial_i = \dfrac{\partial}{\partial x^i}$, $\partial^i = \dfrac{\partial}{\partial p_i}$.

For the cases of major interest to us none of the functions $(\partial^i H)$ vanishes identically. Then we can consider the functions

$$R_i = -\frac{\partial_i H}{\partial^i H} \qquad (2.7)$$

possibly smoothly extended to those points at which the denominators are zero.

(2.8) Definition.- Let $(U,(x^i))$ be a chart at the point $x_o \in M$. We say that a coordinate x^k is of first class at x_o (with respect to the Hamiltonian H) if, in an open set of T*M which projects in a neighborhood of x_o, the corresponding function R_k is linear on the fibers of π_M, i.e.:

$$R_k = B_k^i\, p_k \qquad (2.9)$$

where the functions (B_k^i) depend on the coordinates (x^i) only. In particular, if the functions (B_k^i) vanish , i.e. if $\partial_k H = 0$, we say that the coordinate x^k is ignorable. A coordinate which is not of first class is said to be of second class.

For the sake of simplicity we introduce the following convention.

(2.10) Convention.- Second class coordinates are labeled by indices from the first part of the Latin alphabet $(a,b,c,\ldots)$; those of first class by Greek indices $(\alpha,\beta,\gamma,\ldots)$. Latin indices from the second part of the alphabet $(h,i,j,\ldots)$ are used when the above distinction is not needed. The Einstein summation convention is adapted to the above choice of indices, unless the symbol "n.s." appears together with a distinguished index. The coordinates of any chart at the point x_o are assumed to be ordered in such a way that the first m $(0 \leq m \leq n)$ are of second class, while the others are of first class, so that the indices $(a,b,c,\ldots)$ are assumed to range from 1 to m; $(\alpha,\beta,\gamma,\ldots)$ from m+1 to n and $(h,i,j,\ldots)$ from 1 to n.

The distinction of coordinates into the two classes above (in the particular case of a quadratic Hamiltonian) dates back to Dall'Acqua [6].

When (2.4) holds, by derivation of (1.1) with respect to a coordinate x^i , we obtain $(\partial_i H)_c + (\partial^i H)_c\, \partial_i \partial_i S_c$, where $(\ )_c$ means the substitution $p_j = \partial_j S_c$, i.e. the evaluation on the leaf L_c. Hence, even if $\partial^i H = 0$ at some point of L_c, we can define $(R_i)_c = \partial_i \partial_i S_c$. Moreover, we notice that the separability conditions (2.4) are exactly the integrability conditions of the following system of partial differential equations (see Levi-Civita [1]):

(2.11) $$\partial_i p_j = 0 \quad (i \neq j) \quad , \quad \partial_i p_i = R_i \quad .$$

The integrability conditions are represented by the identities

(2.12) $$\partial_i R_j + \partial^i R_j \, R_i = 0 \qquad (i \neq j \; ; \; i \text{ n.s.}) \, ,$$

which, by (2.7), can be written as follows:

(2.13) $$\partial^i H \, \partial^j H \, \partial_i \partial_j H - \partial^i H \, \partial_j H \, \partial_i \partial^j H - \partial_i H \, \partial^j H \, \partial^i \partial_j H + \partial_i H \, \partial_j H \, \partial^i \partial^j H = 0 \qquad (i \neq j) \; .$$

These are the so called <u>Levi-Civita separability conditions</u>.

Following the above convention, system (2.11) can be written:

(2.14) $$\partial_i p_j = 0 \quad (i \neq j) \; , \; \partial_a p_a = R_a \quad , \; \partial_\alpha p_\alpha = B^i_\alpha \; p_i \quad .$$

However, we have identically

(2.15) $$B^a_\alpha = 0 \quad , \quad \partial_a B^\beta_\alpha = 0 \quad ,$$

so that (2.14) contains an autonomous sub-system in the first class coordinates only:

(2.16) $$\partial_\alpha p_\beta = 0 \quad (\alpha \neq \beta) \quad , \quad \partial_\alpha p_\alpha = B^\gamma_\alpha \; p_\gamma \quad .$$

In fact, from (2.12) it follows in particular: $\partial_a B^i_\alpha \, p_{p_i} + B^a_\alpha \, R_a = 0$ (a n.s.), which implies that R_a is linear in the variables (p_i), against the definition of second class coordinate, unless $B^a_\alpha = 0$; hence also $(2.15)_2$ follows.

Since system (2.16) is linear in the r $(= n - m)$ unknown functions (p_α), in a neighborhood of x_0 there exist r independent solutions, i.e. r^2 functions $(\overset{\beta}{\xi}_\alpha)$ such that:

(2.17) $$\partial_\alpha \overset{\delta}{\xi}_\beta = 0 \quad (\alpha \neq \beta) \; , \quad \partial_\alpha \overset{\delta}{\xi}_\alpha = B^\gamma_\alpha \; \overset{\delta}{\xi}_\gamma$$

and

(2.18) $$\det(\overset{\beta}{\xi}_\alpha) \neq 0 \quad .$$

Then we can consider a new coordinate system (y^i) defined by the equations

(2.19) $$dy^a = dx^a \quad , \quad dy^\alpha = \overset{\alpha}{\xi}_\beta \; dx^\beta \quad .$$

It can be easily checked that the new coordinates (y^i) are $\mathcal{S}$-compatible with (x^i) and that the coordinates (y^α) are ignorable. Hence, since the coordinates are still of second class, we have given a constructive proof of the following theorem.

(2.20) Theorem.- In a $\mathcal{S}$-structure there exist charts where all the first class coordinates are ignorable.

Another important fact is that the number of first (or second) class coordinates is an invariant of a $\mathcal{S}$-structure.

(2.21) Theorem.- Two $\mathcal{S}$-compatible charts have the same number of first (or second) class coordinates.

Proof.- Let (x^i) and $(x^{i'})$ be $\mathcal{S}$-compatible coordinates at x_0. Let (x^a), $(x^{a'})$ be the coordinates of second class ($a=1,\dots,m$; $a'=1,\dots,m'$) and (x^α), $(x^{\alpha'})$ those of first class. They give rise to the same complete integral: $\Sigma_i S_i(x^i;c_j) = \Sigma_{i'} S_{i'}(x^{i'};c_j)$. By derivation with respect to x^α and $x^{a'}$, we obtain:

$$\xi^{\alpha}_{a'}\partial_\alpha\partial_\alpha S_\alpha = \xi^{i}_{a'}\partial_i\,\xi^{i'}_{\alpha}\partial_{i'}S_{i'} + \xi^{a'}_{\alpha}\partial_{a'}\partial_{a'}S_{a'} \qquad (\alpha, a'\ \text{n.s.}),$$

with $\xi^{j'}_i = \partial_i x^{j'}$ and $\xi^i_{j'} = \partial_{j'}x^i$. This means that the following identities hold:

$$\xi^{\alpha}_{a'}R_\alpha = \xi^{i}_{a'}\partial_i\,\xi^{j'}p_{j'} + \xi^{a'}_{\alpha}\,R_{a'} \qquad (\alpha, a'\ \text{n.s.}) \ .$$

If $\xi^{a'}_\alpha \neq 0$, it follows that $R_{a'}$ is linear in the variables $(p_{j'})$(because R_α is also linear in the (p_β) by definition), which is in contrast with the fact that $x^{a'}$ is of second class. Hence, it must be $\xi^{a'}_\alpha = 0$ for each pair of indices (a',α), and (by symmetry) $\xi^a_{\alpha'} = 0$ for each pair (a,α'). It clearly follows that if $m \neq m'$ the matrix $(\xi^{j'}_i)$ cannot be regular, which is absurd. (q.e.d.)

(2.22) Definition.- A separability structure at the point x_0 is said to be of class r (or a $\mathcal{S}_r$-structure) if r is the number of the coordinates of first class at x_0 in any representative chart.

Theorem (2.20) shows that in a $\mathcal{S}_r$-structure there exist charts with r ignorable coordinates. The use of these charts, which we call quasi-normal separable charts, allows of course valuable simplifications in the analysis of separability structures, without any loss of generality. On the other hand, since second class coordinates cannot be ignorable (by definition), and their number is invariant, it follows clearly that:

(2.23) The class of a $\mathcal{S}$-structure is the maximum number of ignorable coordinates that one can find in a representative chart.

This provides in fact an alternative definition of class of a $\mathcal{S}$-structure (see for instance [7]) which is not related with the definition of class of a coordinate.

3.- First integrals associated with a separability structure.

In this section we shall give a short outline of some properties of the first integrals associated with a complete solution of the Hamilton-Jacobi equation in the case of complete separability.

We have already pointed out that a complete integral is represented by a local foliation (L_c) of Lagrangian submanifolds of (T^*M, ω_M) which are sections of the fibration π_M and where the Hamiltonian H is constant (the former requirement could be removed; see for instance [5]). A complete integral can also be represented by a set of real functions (F_i) on an open set of T^*M satisfying the following conditions. i) They are vertically independent, i.e. they are not only functionally independent as functions on T^*M but also, more particularly, functionally independent when restricted to the fibers of T^*M. This fact is represented in natural canonical coordinates by the regularity of the $n\times n$ matrix $(\partial^k F_i)$. ii) The functions (F_i) are in involution, i.e. their Poisson brackets vanish identically: $\{F_i, F_j\} = 0$. iii) The Hamiltonian H depends functionally on the (F_i): $H = \rho(F_i)$ or, equivalently, the Hamiltonian H is in involution with them: $\{H, F_i\} = 0$. In fact, equations of the kind $F_i = c_i$, where $c = (c_i)$ covers a suitable open set of $\mathbb{R}^n$, define a foliation on T^*M of Lagrangian submanifolds which are transversal to the fibers (i.e. sections of π_M) and on which the Hamiltonian H is constant. When such a set of functions is given, then the complete solution S of the Hamilton-Jacobi equation corresponding to H is the integral of the equations $\partial_k S = S_k$, where the functions $S_k(x^i; c_j)$ are obtained by solving the system $F_i(x^j; p_k) = c_i$ with respect to the (p_k): $p_k = S_k(x^i; c_j)$. Conversely, if $S: \hat{U} \longrightarrow \mathbb{R}$ is a complete integral, we can obtain n functions (F_i) satisfying the above conditions, by solving the system $p_k = \partial_k S$ with respect to the coordinates (c_i) of $A\times E$ appearing in the representation of S. We have of course $H = \rho(F_i)$ if $e = \rho(c_i)$ is the representative of the natural projection $\rho_E: A\times E \longrightarrow E$.

Now, let us assume that S is completely separable in the coordinates (x^i). By derivation with respect to the variable x^k of the identity $F_i(x^j; \partial_k S) = c_i$ we have:

$(\partial_k F_i)_c - (\partial^k F_i)_c \partial_k \partial_k S_c = 0$. This implies the identity:

$$\partial_k F_i + \partial^k F_i \; R_k = 0 \qquad (k \text{ n.s.}), \tag{3.1}$$

which can also be written:

$$\partial_k F_i \, \partial^k H - \partial^k F_i \, \partial_k H = 0 \qquad (k \text{ n.s.}). \tag{3.2}$$

We emphasize the fact that in the last formula there is no summation with respect to the index k, so that the Poisson bracket $\{F_i, H\} = \partial_k F_i \partial^k H - \partial^k F_i \partial_k H$ is zero because each term of the sum is zero. Thus we are naturally led to consider something more than the usual involution.

(3.3) <u>Definition</u>. Two functions F and G on T*M are said to be in <u>separable involution</u> (briefly, in <u>$\mathscr{S}$-involution</u>) with respect to the coordinates (x^i) if, for each index k:

$$\partial_k F \, \partial^k G - \partial^k F \, \partial_k G = 0 \qquad (k \text{ n.s.}). \tag{3.4}$$

Hence (3.2) means that each function F_i and H are in $\mathscr{S}$-involution with respect to the separable coordinates (x^i). Moreover, this fact implies that the functions (F_i) are in $\mathscr{S}$-involution themselves: $\partial_k F_i \, \partial^k F_j = -\partial^k F_i \, R_k \partial^k F_j = \partial^k F_i \, \partial_k F_j \quad (k \text{ n.s.})$.

Conversely, let (F_i) be a set of n vertically independent functions in $\mathscr{S}$-involution with respect to a coordinate system (x^i). Since for each index k there exists at least a function F_h such that $\partial^k F_h$ does not vanish identically, we can define the function:

$$R_k = -\frac{\partial_k F_h}{\partial^k F_h} \tag{3.5}$$

Hence, for each function F_i : $\partial_k F_i + \partial^k F_i \; R_k = 0$. Let S be a function generated by the family (F_i) as shown at the beginning of this section, i.e. satisfying the equations $F_i(x^j; \partial_k S(x^h; c_\ell)) = c_i$. By derivation with respect to x^k we obtain: $(\partial_k F_i)_c +$ $+ (\partial^j F_i)_c \partial_j \partial_k S_c = 0$, where, as in section 2, $(\;)_c$ means the evaluation on the leaf L_c generated by the function S_c obtained by fixing the value $c = (c_i)$. Hence: $(k$ n.s.$)$ $0 = -(\partial^k F_i R_k)_c + (\partial^j F_i)_c \partial_j \partial_k S_c = (\partial^j F_i)_c (\partial_j \partial_k S_c - \delta_{jk} (R_k)_c)$. Since $\det(\partial^j F_i) \neq 0$, it follows $\partial_j \partial_k S_c = \delta_{jk} (R_k)_c$. Then, we have in particular $\partial_j \partial_k S_c =$ for $j \neq k$, which means that S is of the kind (1.3). As we know, such a function S is the complete integral of the Hamilton-Jacobi equation corresponding to each Hamiltonian H functionally depending on (F_i) (in this case, as it is easy to check, H is in $\mathscr{S}$-involution with each F_i). Hence we have proved the following theorem.

(3.6) Theorem.- Let (F_i) be a set of n vertically independent functions on T*M. A coordinate system (x^i) on M is separable with respect to a Hamiltonian H functionally dependent on (F_i) if and only if (F_i) are in $\mathcal{S}$-involution with respect to (x^i).

It clearly follows that, if the functions (F_i) are in $\mathcal{S}$-involution with respect to the coordinates (x^i), they are in $\mathcal{S}$-involution with respect to all coordinate systems which are $\mathcal{S}$-compatible, with respect to H, at a point x_0. In fact, these coordinates give rise to the same complete integral (by definition of $\mathcal{S}$-compatibility), that is to say, to the same foliation represented by the functions (F_i). Hence we conclude that to a $\mathcal{S}$-structure there corresponds a group of functions in $\mathcal{S}$-involution of dimension n (i.e. generated by n independent functions). Furthermore we notice that the $\mathcal{S}$-structure is of class r if and only if, among the functions (3.5), there are exactly r which are linear in the variables (p_j).

4.- Separability structures on Riemannian manifolds and normal separable coordinates.

By separability structure on a Riemannian manifold (M,g) we mean a $\mathcal{S}$-structure corresponding to the geodesic Hamiltonian, which is represented, in natural coordinates $(x^i;p_j)$, by a quadratic form:

$$H = \frac{1}{2} g^{ij} p_i p_j \qquad , \tag{4.1}$$

where (g^{ij}) are the contravariant components of the metric tensor g in the coordinates (x^i). Of course, also the definition of first and second class coordinates is understood to be taken with respect to this Hamiltonian.

Let us denote by $\imath : TM \longrightarrow T^*M$ the natural diffeomorphysm defined by the metric tensor g. In natural coordinates $(x^i;\dot{x}^j)$ and $(x^i;p_j)$ of TM and T*M respectively, this diffeomorphysm is represented by the equations $p_j = g_{ji}\dot{x}^i$, where (g_{ji}) are the covariant components of g, or by the inverse relations $\dot{x}^i = g^{ij} p_j$. Let us denote by $\imath^* f$ the pull-back of a function $f: T^*M \longrightarrow \mathbb{R}$, i.e.: $\imath^* f = f \circ \imath$.

In the present case, the functions (2.7) are explicitly given by:

$$R_i = -\frac{1}{2} \frac{\partial_i g^{hk} p_h p_k}{g^{ij} p_j} \qquad . \tag{4.2}$$

Of course R_i is linear in the variables (p_j) (that is to say x^i is a first class coordinate) if and only if its pull-back

$$(4.3) \qquad 2\,{}^*R_i = \frac{1}{2}\,\frac{\partial_i g_{hk}\dot{x}^h\dot{x}^k}{\dot{x}^i}$$

is linear in the variables $(\dot{x}^i)$, i.e. if and only if the quadratic polynomial $\partial_i g_{hk}\dot{x}^h\dot{x}^k$ is divisible by $\dot{x}^i$. Hence:

(4.4) <u>A coordinate x^i is of first class at a point $x_0 \in M$ if and only if, in a neighborhood of x_0, $\partial_i g_{hk} = 0$ for each $h,k \neq i$.</u>

In particular, a coordinate x^i is ignorable when $\partial_i g_{hk} = 0$ for each pair of indices (h,k). As it is well known, an ignorable coordinate corresponds to a <u>Killing vectorfield</u> (briefly, a <u>K-vector</u>) on the Riemannian manifold (M,g), i.e. to a vectofield X satisfying the <u>Killing equation</u> $L_X g = 0$ (where L_X is the Lie derivative symbol).

As a consequence of (4.4), and accordingly to definition (2.22) and convention (2.10), we notice that:

(4.5) <u>A $\mathscr{S}$-structure at a point x_0 of a Riemannian manifold (M,g) is of class r if and only if in any representative chart there exist exactly r coordinates (x^α) such that, in a neighborhood of x_0:</u>

$$(4.6) \qquad \partial_\alpha g_{hk} = 0 \qquad (h,k \neq \alpha) \quad .$$

Therefore, theorem (2.20) has the following corollary ([8]).

(4.7) <u>Theorem.- If in a separable chart $(U,(x^i))$ of a Riemannian manifold (M,g) equations (4.6) hold in a neighborhood of $x_0 \in U$ for r coordinates (x^α), then there exist $\mathscr{S}$-compatible charts at x_0 with r ignorable coordinates. If (M,g) admits at x_0 a $\mathscr{S}_r$-structure, then, in a neighborhood of x_0, there exist r commuting K-vectors, independent at each point, or, equivalently, an Abelian r-parameters group of isometries acts freely. In particular, if (M,g) admits a $\mathscr{S}_n$-structure at x_0 $(n = \dim M)$, then M is flat at x_0.</u>

The quasi-normal separable charts (see the end of section 2) can be found by following the constructive proof of theorem (2.20), where now the coefficients (B^β_α) have, as it is easy to check, the following expressions:

$$(4.8) \qquad B^\beta_\alpha = g^{\beta k}\partial_\alpha g_{\alpha k} - \frac{1}{2}\,g^{\alpha\beta}\partial_\alpha g_{\alpha\alpha} \qquad (\alpha \text{ n.s.}) \quad .$$

Another method of construction, of a pure algebraic character, will be shown below.

When investigating the relations between $\mathscr{S}$-compatible coordinates, we notice, as a general property of the $\mathcal{S}$-structures, that a rescaling of a coordinate x^i (i.e. a reversible transformation of x^i involving no other coordinate) preserves the class of the coordinate and the separability.

(4.9) Theorem.- A chart $(U,(x^i))$ is $\mathcal{S}$-compatible with a given separable chart $(U',(x^{i'}))$ at a point $x_0 \in U \cap U'$ if and only if the second class coordinates are simply related by a rescaling, i.e.:

$$dx^{a'} = \delta^{a'}_a f_a(x^a)\, dx^a \qquad (4.10)$$

(a suitable reordering of the coordinates is understood) and moreover

$$dx^{\alpha'} = \xi^{\alpha'}_i\, dx^i \qquad (4.11)$$

where functions $(\xi^{\alpha'}_i)$ satisfy, in a neighborhood of x_0, the following equations:

$$\partial_i \xi^{\alpha'}_j + \xi^{\beta'}_i \xi^{\beta'}_j B'^{\alpha'}_{\beta'} = 0 \qquad (i \neq j)\ . \qquad (4.12)$$

Proof.- i) If the two charts are $\mathcal{S}$-compatible, they have the same number of first (and second) class coordinates (theorem (2.21)). By using the same notations as in the proof of theorem (2.21), we can see that: $\partial_i \partial_j S_c = \partial_i \xi^{j'}_j \partial_{j'} S_c + \xi^{i'}_i \xi^{i'}_j (R'_{i'})_c = 0$ for $i \neq j$. Hence, the following identities hold:

$$\partial_i \xi^{j'}_j p_{j'} + \xi^{\beta'}_i \xi^{\beta'}_j B'^{\alpha'}_{\beta'} p_{\alpha'} + \xi^{a'}_i \xi^{a'}_j R'_{a'} = 0 \qquad (i \neq j)\ , \qquad (4.13)$$

where, of course, $(B'^{\alpha'}_{\beta'})$ and $(R'_{a'})$ are defined as in (4.8) and (4.2) with respect to coordinates $(x^{i'})$. Since by the definition of second class coordinate we can see that $\xi^{a'}_i \xi^{a'}_j\, z^*(R'_{a'})$ cannot be linear in the variables $(x^{i'})$, it follows that equations (4.12) necessarily hold as well as: $\partial_i \xi^{a'}_j = 0$, $\xi^{a'}_i \xi^{a'}_j = 0$ $(i \neq j$, a'n.s.). The first set of these equations simply means that each $\xi^{a'}_j$ is function of x^j only; the second set implies that, for a fixed index a', only one of the functions $(\xi^{a'}_i)$ does not vanish identically. Since $\xi^{a'}_\alpha = 0$ (see the proof of theorem (2.21), there exists a second class coordinate x^a which is simply related with $x^{a'}$ by a rescaling. ii) Conversely, if (4.10) - (4.12) hold, it is easy to check, by reversing the above reasoning, that the two charts are $\mathcal{S}$-compatible. (q.e.d.)

We notice in particular that if $(x^{i'})$ are quasi normal separable coordinates (i.e. $(x^{\alpha'})$ are ignorable, hence $B'^{\alpha'}_{\beta'} = 0$), for all $\mathcal{S}$-compatible coordinates (x^i) the following relations hold in a neighborhood of x_0 , apart from a reordering and a rescaling:

(4.14) $$dx^{a'} = dx^{a} \quad , \qquad dx^{\alpha'} = \xi^{\alpha'}_{i}\, dx^{i} \quad ,$$

<u>where each</u> $\xi^{\alpha'}_{i}$ <u>is function of</u> x^{i} <u>only</u>.

For a Hamiltonian such as in (4.1), Levi-Civita separability conditions (2.13) are polynomial identities in the variables (p_j). If we write them for second class coordinates $x^a \neq x^b$ as follows:

(4.15) $$\partial^a H\,(\partial^b H\,\partial_a\partial_b H - \partial_b H\,\partial_a\partial^b H) = \partial_a H\,(\partial^b H\,\partial_b\partial^a H - \partial_b H\,\partial^a\partial^b H) \qquad (a\neq b\ ,\text{n.s.})$$

we observe (with Levi-Civita [1]) that, since $\partial^a H$ is not a divisor of $\partial_a H$, it must be a divisor of the polynomial

(4.16) $$\partial^b H\,\partial_b\partial^a H - \partial_b H\,\partial^a\partial^b H \qquad (a\neq b,\ \text{n.s.})$$

Then, by the pull-back $\imath^*$, we see that $\dot{x}^a = \imath^*(\partial^a H)$ must be a divisor of

$$g^{ak}\partial_b g_{hk}\,\dot{x}^h\dot{x}^b - \tfrac{1}{2}\,g^{ab}\,\partial_b g_{hk}\,\dot{x}^h\dot{x}^k \qquad (a\neq b,\ \text{n.s.})$$

This implies,in particular:

(4.17) $$g^{ab}\,\partial_b g_{hk} = 0 \qquad (a\neq b;\ \ h,k\neq a,b\ ;\ \ b\ \text{n.s.})\ .$$

Since x^b is a second class coordinate, there exist at least two indices $i,j\neq b$ such that $\partial_b g_{ij}\neq 0$ at some point of each neighborhood of x_0 . If,moreover, $i,j\neq a$, from (4.17) it follows $g^{ab} = 0$ $(a\neq b)$ in a neighborhood of x_0. On the other hand,the exceptional case

(4.18) $$\partial_b g_{ij} = 0 \qquad (a\neq b;\ \ i,j\neq a,b\ )$$

is irrelevant because, as it is easy to see, we can always pass to $\mathcal{S}$-compatible coordinates for which this no longer holds,while conditions like (4.18) are invariant, since second class coordinates are related by a rescaling. Hence:

(4.19) <u>If</u> $(U,(x^i))$ <u>is a separable chart, for two distinct coordinates</u> x^a, x^b , <u>of second class at</u> x_0, <u>we have, in a neighborhood of</u> x_0:

(4.20) $$g^{ab} = 0 \qquad (a\neq b) \qquad .$$

Now we see that the polynomial (4.16), which is divisible by $\partial^a H$, simply reduces, in a neighborhood of x_0, to $\partial^b H\,\partial_b\partial^a H$. Since $\partial^a H$ is not a divisor of $\partial^b H$, it follows that $\partial_b \partial^a H = f^{ab} \partial^a H$, where f^{ab} is a function of the coordinate (x^i) only, and thus:

(4.21) $$\partial_b g^{ai} = f^{ab} g^{ai} \quad (a \neq b) \quad .$$

Since in the last formula the function f^{ab} does not depend on the choice of the index i , we can see at once that:

(4.22) <u>If $(U,(x^i))$ is a separable chart, for two distinct coordinates x^a, x^b of second class at a point $x_0 \in U$, the following equations hold in a neighborhood of x_0:</u>

(4.23) $$g^{ai} \partial_b g^{aj} = g^{aj} \partial_b g^{ai} \qquad (a \neq b) \ .$$

In Dall'Acqua's paper [6], conditions (4.20) have been proved through cumbersome calculations and under the assumption $g^{aa} \neq 0$ for each second class coordinate (which in a proper-Riemannian manifold is certainly satisfied). We have followed here the very simple proof given by Agostinelli in [9], in which,however,the possibility of the exceptional case (4.18) is not considered. In fact,this obstacle cannot be overcome so easily without the concept of separability structure and without knowing that second class coordinates remain essentially unchanged in $\mathcal{S}$-compatible charts.

From proposition (4.19) and theorem (4.9) it follows in particular:

(4.24) <u>The coordinates of a $\mathcal{S}_0$-structure at a point x_0 are uniquely determined, up to a rescaling, and orthogonal ($g^{ij} = 0$ for $i \neq j$) in a neighborhood of x_0.</u>

(4.25) <u>Remark</u>.- When g is analytic, by (4.5) we notice that a coordinate is of first class at each point of the domain of the chart if it is of first class at least at one point (consequently, a second class coordinate at a point is of second class everywhere). Hence,the above results hold with respect to all points of the domain of a separable chart. In other words, it is not necessary to work at a distinguished point. This remark must be kept in mind also in the sequel.

Together with the first classification given by the class, for $\mathcal{S}$-structures on Riemannian manifolds we can introduce a second classification by another integer invariant: the <u>index</u> of the $\mathcal{S}$-structure. Actually this second classification works non-trivially only for $\mathcal{S}$-structures on pseudo-Riemannian manifolds of class different from the dimension.

(4.26) <u>Definition</u>.-The <u>index</u> of a $\mathcal{S}$-structure at a point x_0 of a Riemannian manifold

(M,g) is the number of the second class coordinates of any representative chart for which, in a neighborhood of x_0, $g^{aa}=0$.

Notice that the definition makes sense because, in virtue of theorem (4.9), the condition $g^{aa}=0$ for a second class coordinate is invariant under $\mathscr{S}$-compatible transformations.

(4.27) Remark.- It could happen that for some second class coordinate x^a the condition $g^{aa}=0$ is satisfied on a closed submanifold of M containing x_0, or at x_0 only. In the present paper we do not consider such a kind of $\mathscr{S}$-structures, which we can call singular.

From now on we adopt, together with (2.10), the following convention concerning second class coordinates.

(4.28) Convention.- A $\mathscr{S}_{r;d}$-structure is a (non singular) $\mathscr{S}_r$-structure of index d. This abbreviation is used only when it is necessary to specify the index. $\mathscr{S}_n$-structures ($n = \dim M$: no second class coordinates) have of course to be considered of index zero. Second class coordinates are labeled by $\tilde{a},\tilde{b},\tilde{c},\ldots$ if $g^{\tilde{a}\tilde{a}}\neq 0,\ldots$, and by $\bar{a},\bar{b},\bar{c},\ldots$ if, on the contrary, $g^{\bar{a}\bar{a}}=0,\ldots$. Second class coordinates are assumed to be ordered in such a way that $\tilde{a},\tilde{b},\tilde{c},\ldots$ range from 1 to $m-d$ and $\bar{a},\bar{b},\bar{c},\ldots$ range from $m-d+1$ to m (m being the total number of second class coordinates). When the distinction into these two sub-classes is not needed, we use the unaffected indices $a,b,c,\ldots$ which range from 1 to m. Summation convention is adapted to this choice unless the symbol "n.s." appears together with a distinguished index.

(4.29) Theorem.- In a $\mathscr{S}_{r;d}$-structure on a Riemannian manifold (M,g) there exist charts $(V,(y^i))$ with r ignorable coordinates (y^α) ($\alpha = m-1,\ldots,n$; $m = n-r$) such that for the metric tensor components ($\overset{ij}{g}$) the following conditions hold ($\tilde{a} = 1,\ldots,m-d$; $\bar{a},\bar{b} = m-d+1,\ldots,n-r$):

$$(4.30) \qquad \overset{\tilde{a}\tilde{a}}{g} \neq 0 \quad , \quad \overset{\tilde{a}i}{g} = 0 \quad (i\neq\tilde{a}) \quad , \quad \overset{\bar{a}\bar{b}}{g} = 0 \quad .$$

Proof.- Let $(U,(x^i))$ be a quasi normal separable chart of a $\mathscr{S}_{r;d}$-structure at the point $x_0\in M$. Let (x^α) be the ignorable coordinates (theorem (2.20)) and g^{ij} the components of g. If $g^{\tilde{a}\tilde{a}}\neq 0$, from (4.23) it follows in particular: $\partial_b(g^{\tilde{a}\alpha}(g^{\tilde{a}\tilde{a}})^{-1}) = 0$ for $b\neq\tilde{a}$. This means that, in a neighborhood of x_0, and for each pair of indices $(\tilde{a},\alpha)$, there exists a function $\theta^\alpha_{\tilde{a}}$ of $x^{\tilde{a}}$ only, such that: $g^{\tilde{a}\alpha} = \theta^\alpha_{\tilde{a}}\, g^{\tilde{a}\tilde{a}}$ ($\tilde{a}$ n.s.). Then we can define new coordinates (y^i) at x_0 by the equations: $dy^a = dx^a$, $dy^\alpha = dx^\alpha - \theta^\alpha_{\tilde a}\, dx^{\tilde a}$. These coordinates are clearly $\mathcal{S}$-compatible (compare with (4.14)) and, in particular, (y^α) are ignorable. The corresponding components $(\overset{ij}{g})$ satisfy conditions (4.30) since: $\overset{ab}{g} = g^{ab}$ and $\overset{\tilde a\alpha}{g} = g^{\tilde a\alpha} - \theta^\alpha_{\tilde a}\, g^{\tilde a\tilde a} = 0$ ($\tilde a$ n.s.). (q.e.d.)

(4.31) <u>Definition</u>.- A chart satisfying the conditions of theorem (4.29) is called <u>normal separable chart</u>; the corresponding coordinates are called <u>normal separable coordinates</u>.

In normal separable coordinates the matrix $(\overset{ij}{g})$ has the following form:

(4.32)
$$
\begin{array}{c}
m-d \\ d \\ r
\end{array}
\left\{
\left\|
\begin{array}{c|c|c}
\begin{matrix} \ddots & & 0 \\ & \overset{\tilde a\tilde a}{g} & \\ 0 & & \ddots \end{matrix} & 0 & 0 \\
\hline
0 & 0 & \overset{\bar a\alpha}{g} \\
\hline
0 & \overset{\alpha\bar a}{g} & \overset{\alpha\beta}{g}
\end{array}
\right\|
\right.
\qquad
\underbrace{\quad}_{m-d}\ \underbrace{\quad}_{d}\ \underbrace{\quad}_{r}
$$

For $\mathcal{S}$-structures of index 0 (as in proper-Riemannian manifolds) the central rows and columns disappear.

(4.33)
$$
\begin{array}{c}
m \\ r
\end{array}
\left\{
\left\|
\begin{array}{c|c}
\begin{matrix} \ddots & & 0 \\ & \overset{aa}{g} & \\ 0 & & \ddots \end{matrix} & 0 \\
\hline
0 & \overset{\alpha\beta}{g}
\end{array}
\right\|
\right.
\qquad
\underbrace{\quad}_{m}\ \underbrace{\quad}_{r}
$$

(4.34) <u>Remark</u>.- We emphasize that separable coordinates (y^i) are normal separable coordinates if and only if: i) all first class coordinates are ignorable, ii) for each second class coordinate y^a such that $g^{aa} \neq 0$, and for each ignorable coordinate y^α, $g^{a\alpha} = 0$.

5.- The general form of the metric tensor components in separable coordinates.

In order to construct a method of integration by separation of variables of the geodesic Hamilton-Jacobi equation of a Riemannian manifold it is necessary to know the general form of the contravariant components of a metric tensor in separable coordinates. This general form can be easily obtained by the above considerations on separability structures.

(5.1) Definition.- Let $(\underset{h}{\varphi}^j)$ be a $n\times n$ matrix of functions of n variables (x^i) $(h,i,j,\dots = 1,\dots,n)$. We say that $(\underset{h}{\varphi}^j)$ is a Stäckel matrix in the variables (x^i) if it is regular everywhere and if each element $\overset{h}{\varphi}_j$ of the inverse matrix $(\overset{h}{\varphi}_j)$ (defined by the equations $\overset{h}{\varphi}_i \underset{k}{\varphi}^i = \delta^h_k \Longleftrightarrow \overset{h}{\varphi}_i \underset{h}{\varphi}^j = \delta^j_i$) is a function of the variable x^j corresponding to the lower index only.

A matrix of this kind appears in the statement of Stäckel's theorem [9] on the separability of orthogonal coordinates (see section 7). Actually, Stäckel matrices are closely related with questions concerning separability structures not only in this particular case.

(5.2) Theorem.- In normal separable coordinates (y^i) of a $\mathcal{S}_{r;d}$-structure on a Riemannian manifold (M,g), the contravariant components $(\overset{ij}{g})$ of g have the following form:

$$
(5.3)\qquad \begin{cases} \overset{\tilde a i}{g} = 0 \quad (i\neq \tilde a) \ , \quad \overset{\bar a \bar b}{g} = 0 \ , \\ \overset{\tilde a\tilde a}{g} = \underset{m}{u}^{\tilde a} \ , \quad \overset{\bar a \alpha}{g} = \theta^{\alpha}_{\bar a}\, \underset{m}{u}^{\bar a} \quad (\bar a \text{ n.s.}) \ , \quad \overset{\alpha\beta}{g} = \zeta^{\alpha\beta}_{a}\, \underset{m}{u}^{a} \end{cases}
$$

$(m = n-r \ ; \tilde a = 1,\dots,m-d \ ; \ \bar a, \bar b = m-d+1,\dots,m \ ; \ \alpha,\beta = m+1,\dots,n)$ where: i) $(\underset{m}{u}^a)$ is the m-th row of a Stäckel matrix $(\underset{b}{u}^a)$ in the variables (y^a), ii) $(\theta^{\alpha}_{\bar a})$ and $(\zeta^{\alpha\beta}_a)$ are functions of the variable corresponding to the lower index only.

Proof.- $(5.3)_{1,2}$ are already known by theorem (4.29)(see (4.30)). Hence, in normal separable coordinates the Hamilton-Jacobi equation becomes:

$$
(5.4)\qquad \overset{\tilde a\tilde a}{g}(\partial_{\tilde a}S)^2 + 2\overset{\bar a\alpha}{g}\partial_{\bar a}S\,\partial_{\alpha}S + \overset{\alpha\beta}{g}\partial_{\alpha}S\,\partial_{\beta}S = 2e \quad .
$$

Since the coordinates (y^α) are ignorable, the complete integral has the form:

$$
(5.5)\qquad S = \sum_{a=1}^{n-r} S_a(y^a;c_i) + c_\alpha y^\alpha \quad .
$$

We can always assume that $(c_i) = (c_a; c_\alpha)$, i.e. that the r constants of motion (c_α) corresponding to the ignorable coordinates coincide with r of the parameters (c_i) and, moreover, that one of the remaining parameters (c_a) coincides with 2e: for instance, $c_m = 2e$. From (4.22) it follows in particular:

$$(5.6) \qquad \overset{\bar{a}\alpha}{g}\, \partial_b \overset{\bar{a}\beta}{g} = \overset{\bar{a}\beta}{g}\, \partial_b \overset{\bar{a}\alpha}{g} \quad (b \neq \bar{a}) .$$

For a fixed index $\bar{a}$ there exists at least one index β, say for instance $\beta = n$, such that $\overset{\bar{a}n}{g} \neq 0$ (otherwise $\det(\overset{ij}{g}) = 0$); hence, by (5.6): $\partial_b (\overset{\bar{a}\alpha}{g} (\overset{\bar{a}n}{g})^{-1}) = 0 \quad (b \neq \bar{a})$. This means that for each index $\bar{a}$ there exist r functions $(\theta^\alpha_{\bar{a}})$ of $y^{\bar{a}}$ only, such that

$$(5.7) \qquad \overset{\bar{a}\alpha}{g} = \theta^\alpha_{\bar{a}}\, \gamma^{\bar{a}} \quad (\bar{a} \text{ n.s.}) ,$$

where $\gamma^{\bar{a}}$ is a suitable function which does not depend on the index α (we can choose in particular $\gamma^{\bar{a}} = g^{\bar{a}n}$; hence $\theta^n_{\bar{a}} = 1$). If we set

$$(5.8) \qquad u_{\tilde{a}} = (\partial_{\tilde{a}} S_{\tilde{a}})^2 , \qquad u_{\bar{a}} = 2c_\alpha\, \theta^\alpha_{\bar{a}}\, \partial_{\bar{a}} S_{\bar{a}} ,$$

from (5.4) it follows:

$$(5.9) \qquad \overset{\tilde{a}\tilde{a}}{g}\, u_{\tilde{a}} + \gamma^{\bar{a}}\, u_{\bar{a}} + \overset{\alpha\beta}{g}\, c_\alpha c_\beta = c_m .$$

By derivation of (5.9), with respect to c_b , we obtain

$$(5.10) \qquad \overset{\tilde{a}\tilde{a}}{g}\, \overset{b}{u}_{\tilde{a}} + \gamma^{\bar{a}}\, \overset{b}{u}_{\bar{a}} = \delta^b_m ,$$

where:

$$(5.11) \qquad \begin{cases} \overset{b}{u}_{\tilde{a}} = \dfrac{\partial u_{\tilde{a}}}{\partial c_b} = 2\, \partial_{\tilde{a}} S_{\tilde{a}}\, \dfrac{\partial}{\partial c_b}\, \partial_{\tilde{a}} S_{\tilde{a}} , \\ \overset{b}{u}_{\bar{a}} = \dfrac{\partial u_{\bar{a}}}{\partial c} = 2\, c_\alpha\, \theta^\alpha_{\bar{a}}\, \dfrac{\partial}{\partial c_b}\, \partial_{\bar{a}} S_{\bar{a}} . \end{cases}$$

Since $\det(\frac{\partial}{\partial c_b} \partial_a S) = \det(\frac{\partial}{\partial c_i} \partial_j S) \neq 0$, the $m \times m$ matrix $(\overset{b}{u}_a) = (\overset{b}{u}_{\tilde{a}}, \overset{b}{u}_{\bar{a}})$ is everywhere regular except on the surfaces $c_\alpha \theta^\alpha_{\bar{a}} = 0$. However, these surfaces do not belong to the domain of definition of the complete integral S, since, as we shall see in the next section, they represent singularities for the integration of (5.4). If $(\underset{b}{u}^a)$ is the inverse matrix of $(\overset{b}{u}_a)$, from (5.10) it follows

$$(5.12) \qquad \overset{\tilde{a}\tilde{a}}{g} = \underset{m}{u}^{\tilde{a}} , \qquad \gamma^{\bar{a}} = \underset{m}{u}^{\bar{a}} .$$

On the other hand, by derivation of (5.9) with respect to c_α and c_β , we obtain:

$$(5.13) \qquad \overset{\alpha\beta}{g} = \zeta^{\alpha\beta}_{\tilde a}\,\overset{\tilde a\tilde a}{g} + \zeta^{\alpha\beta}_{\bar a}\,\gamma^{\bar a}$$

with

$$(5.14) \qquad \zeta^{\alpha\beta}_{a} = -\frac{\partial^2 u_a}{\partial c_\alpha \partial c_\beta} \quad .$$

From (5.7), (5.12) and (5.13) expressions (5.3) follow. By (5.8), (5.11) and (5.14) we see that conditions i) and ii) are satisfied. (q.e.d.)

(5.15) Theorem.- Let $(U,(x^i))$ be a separable chart of a $\mathcal{S}_{r;d}$-structure at a point x_0 of a n-dimensional Riemannian manifold (M,g). In a neighborhood of x_0 the contravariant components (g^{ij}) take the following form:

$$(5.16) \qquad \begin{cases} g^{\tilde a\tilde a} = \underset{m}{u}^{\tilde a} \quad , \quad g^{\bar a\bar a} = 0 \quad , \quad g^{ab} = 0 \quad (a \neq b) \quad , \\ g^{a\alpha} = \underset{\beta}{\xi}^{\alpha}\, \theta^{\beta}_{a}\, \underset{m}{u}^{a} \qquad (a \text{ n.s.}) \quad , \\ g^{\alpha\beta} = \underset{\gamma}{\xi}^{\alpha}\, \underset{\delta}{\xi}^{\beta}\, \eta^{\gamma\delta}_{a}\, \underset{m}{u}^{a} \quad , \end{cases}$$

$(m = n-r\ ; \tilde a = 1,\dots,m-d\ ;\ \bar a = m-d+1,\dots,m\ ;\ a,b = 1,\dots,m\ ;\ \alpha,\beta,\gamma,\delta = m-1,\dots,n)$ where: i) $(\underset{m}{u}^{a})$ is the m-th row of a Stäckel matrix $(\underset{b}{u}^{a})$ in the m variables (x^a); ii) $(\underset{\beta}{\xi}^{\alpha})$ is a Stäckel matrix in the r variables (x^α); iii) (θ^{β}_{a}) and $(\eta^{\gamma\delta}_{a})$ are functions of the variable corresponding to the lower index only.

Proof.- Let $(y^i) = (y^a,y^\alpha)$ be normal separable coordinates of the given $\mathcal{S}$-structure. Since all second class coordinates (y^α) are ignorable, by (4.14) we observe that for any other $\mathcal{S}$-compatible coordinate system (x^i), apart from a suitable reordering and rescaling of the second class coordinates, the following relations hold in a neighborhood of x_0:

$$(5.17) \qquad dy^a = dx^a \quad , \qquad dy^\alpha = \overset{\alpha}{\xi}_i\, dx^i \quad ,$$

where each function $\overset{\alpha}{\xi}_i$ depends on the variable x^i only. The $r \times r$ sub-matrix $(\overset{\alpha}{\xi}_\beta)$ of $(\overset{\alpha}{\xi}_i)$ is regular: $\det(\overset{\alpha}{\xi}_\beta) = \det(\frac{\partial y^i}{\partial x^j}) \neq 0$; hence, the inverse matrix $(\underset{\alpha}{\xi}^{\beta})$ is a Stäckel matrix. Let (g^{ij}) and $(\overset{ij}{g})$ be the contravariant components of g in the coordinates (x^i) and (y^i) respectively. The following relations hold:

$$g^{ab} = \overset{ab}{g} \quad , \quad g^{a\alpha} = \underset{\beta}{\xi}^{\alpha}\,(\overset{a\beta}{g} - \overset{\beta}{\xi}_b\,\overset{ab}{g}) \quad ,$$

$$g^{\alpha\beta} = \underset{\gamma}{\xi}^{\alpha}\,\underset{\delta}{\xi}^{\beta}\,(\overset{\gamma\delta}{g} + \overset{\gamma}{\xi}_a\,\overset{\delta}{\xi}_b\,\overset{ab}{g} - \overset{\gamma}{\xi}_a\,\overset{a\delta}{g} - \overset{\delta}{\xi}_a\,\overset{a\gamma}{g}) \quad .$$

Since (y^i) are normal separable coordinates, by theorem (5.2) we have:

$$g^{\tilde a\tilde a} = \underset{m}{u}^{\tilde a} \ , \quad g^{\bar a\bar a} = 0 \ , \quad g^{ab} = 0 \quad (a \neq b) \ ,$$

$$g^{\tilde a\alpha} = -\underset{\beta}{\xi}^{\alpha}\overset{\beta}{\xi}_{\tilde a}\underset{m}{u}^{\tilde a} \quad (\tilde a \text{ n.s.}) \ ,$$

$$g^{\bar a\alpha} = \underset{\beta}{\xi}^{\alpha}\overset{\beta}{\theta}_{\bar a}\underset{m}{u}^{\bar a} \quad (\bar a \text{ n.s.}) \ ,$$

$$g^{\alpha\beta} = \underset{\gamma}{\xi}^{\alpha}\underset{\delta}{\xi}^{\beta}\Big(\zeta^{\gamma\delta}_{a}\underset{m}{u}^{a} + \overset{\gamma}{\xi}_{\tilde a}\overset{\delta}{\xi}_{\tilde a}\underset{m}{u}^{\tilde a} - \big(\overset{\gamma}{\xi}_{\bar a}\,\theta^{\delta}_{\bar a} + \overset{\delta}{\xi}_{\bar a}\,\theta^{\mu}_{\bar a}\big)\underset{m}{u}^{\bar a}\Big) \ .$$

Furthermore, if we set:

$$(5.18) \qquad \begin{cases} \theta^{\beta}_{\tilde a} = -\overset{\beta}{\xi}_{\tilde a} \ , \quad \eta^{\gamma\delta}_{\tilde a} = \zeta^{\gamma\delta}_{\tilde a} + \overset{\gamma}{\xi}_{\tilde a}\overset{\delta}{\xi}_{\tilde a} \ , \\ \eta^{\gamma\delta}_{\bar a} = \zeta^{\gamma\delta}_{\bar a} - \overset{\gamma}{\xi}_{\bar a}\theta^{\delta}_{\bar a} - \overset{\delta}{\xi}_{\bar a}\theta^{\gamma}_{\bar a} \ , \end{cases}$$

expressions (5.16) follow. (q.e.d.)

The contravariant components of a metric tensor with respect to separable coordinates could be presented in many other forms, appparently more general than the one described in theorem (5.15) (or (5.12)) (for the case $d = 0$, see for instance [8],[11], [12],[13]). It seems that (5.16),and (5.3) for normal separable coordinates, provide the simplest representation. For instance, in [11] we obtained for $\overset{\alpha\beta}{g}$ the expressions $\overset{\alpha\beta}{g} = \zeta^{\alpha\beta}_{a}\underset{m}{u}^{a} + \zeta^{\alpha\beta}_{o}$ where $(\zeta^{\alpha\beta}_{o})$ are constant. However, if we set $\tilde\zeta^{\alpha\beta}_{a} = \zeta^{\alpha\beta}_{o}\overset{m}{u}_{a} + \zeta^{\alpha\beta}_{a}$ we obtain the expression $\overset{\alpha\beta}{g} = \tilde\zeta^{\alpha\beta}_{a}\underset{m}{u}^{a}$, which is of the same kind given in $(5.3)_3$. Again with the reference to the case $d = 0$, in [12] the components (g^{aa}) are given in the form $g^{aa} = \sum_{b=1}^{m}\underset{b}{v}^{a}$, where $(\underset{b}{v}^{a})$ is still a Stäckel matrix in the second class variables (x^a). We can thus consider a more general expression like $g^{aa} = k^b\underset{b}{v}^{a}$ where (k^b) are constant (compare with the t-dependent case considered in [13]). Actually this form can be obtained automatically, following the same proof as in theorem (5.2)without the non restrictive assumption $c_m = 2e$. On the other hand, apart from what concerns the separability structure theory, the equivalence of the two representations follows directly from the following theorem on Stäckel matrices.

(5.19) Theorem.- Let $(\underset{b}{v}^{a})$ be a Stäckel matrix in the m variables (x^a). For any element $(k^b) \in \mathbb{R}^m$ different from zero, there exists a Stäckel matrix $(\underset{b}{u}^{a})$, still in the variables (x^a), such that $\underset{m}{u}^{a} = k^b\underset{b}{v}^{a}$.

Proof.- Let us take a regular $m \times m$ constant matrix $(\overset{b\cdot}{k}_{a})$ such that $\overset{b}{k}_{m} = k^b$. Let us consider the inverse matrix (k^a_b) and let us set $\overset{c}{u}_a = k^c_b \overset{b}{v}_a$. Notice that $\overset{c}{u}_a$ is a function of the corresponding variable x^a only; moreover, the matrix $(\overset{c}{u}_a)$ is obviously regular, since it is a product of two regular matrices. Hence, the inverse matrix (u^a_c) is a Stäckel matrix and moreover $u^a_c = \overset{b}{k}_c v^a_b$, so that $u^a_m = \overset{b}{k}_m v^a_b = k^b v^a_b$. (q.e.d.)

6.- Separation of the variables in the geodesic Hamilton-Jacobi equation.

Let us consider a chart $(U,(x^i))$ on (M,g) and let us assume that it is separable (Levi-Civita conditions (2.13) are satisfied for the geodesic Hamiltonian (4.1)). This implies the existence of a complete integral of the kind (1.3). In order to reduce the Hamilton-Jacobi equation to a system of separated equations, we can proceed as follows. First of all we must recognize the class of each coordinate at a point $x_0 \in U$ (definition (2.8), proposition (4.4)). When this is done, we know the class and the index of the $\mathcal{S}$-structure determined by the chart at x_0 (the class and the index could be independent on the choice of the point x_0). Then, we check if the chart is a normal separable chart (theorem (4.29), remark (4.34)). If the answer is negative, we can get a normal separable chart in two steps: i) by reduction to a quasi normal separable chart (i.e. to a maximal number of ignorable coordinates), following the method described in the proof of theorem (2.20); ii) by passing to normal separable coordinates through a coordinate transformation of the kind described in the proof of theorem (4.29). One could also proceed as follows: i) detecting, by a suitable algebraic process, the functions (ξ^α_β), (θ^β_a), $(\zeta^{\gamma\delta}_{\bar a})$ (and possibly (u^a_b) too) representing the metric tensor components as in (5.16); ii) performing a coordinate transformation of the kind (5.17), where (ξ^β_α) are obtained by inverting the matrix (ξ^α_β) and $(\xi^\alpha_{\bar a})$, $(\xi_{\bar a})$ by reversing the relations (5.18). When we are in normal separable coordinates, we can apply the following theorem.

(6.1) Theorem.- If in a coordinate system (y^i) the contravariant metric tensor components $(\overset{ij}{g})$ are in the form (5.3), then a complete integral of the kind (5.5) is obtainable by the integration of the following separated system of ordinary differential equations:

$$
(6.2)\qquad
\begin{cases}
\left(\dfrac{dS_{\tilde a}}{dy^{\tilde a}}\right)^2 + \zeta^{\alpha\beta}_{\tilde a}\, c_\alpha c_\beta = c_b \overset{b}{u}_{\tilde a}\,, \\[2ex]
2c_\alpha \theta^\alpha_{\bar a} \dfrac{dS_{\bar a}}{dy^{\bar a}} + \zeta^{\alpha\beta}_{\bar a}\, c_\alpha c_\beta = c_b \overset{b}{u}_{\bar a}\,.
\end{cases}
$$

Proof.- By (5.3) we can write equation (5.4) as follows:

(6.3) $$\underset{m}{u}^a \Phi_a = 2e \quad ,$$

where:

(6.4) $$\Phi_{\bar{a}} = (\partial_{\bar{a}} S)^2 + \zeta_{\bar{a}}^{\alpha\beta} c_\alpha c_\beta \quad , \quad \Phi_{a} = 2c_\alpha \theta_{\bar{a}}^{\alpha} \partial_{\bar{a}} S + \zeta_{\bar{a}}^{\alpha\beta} c_\alpha c_\beta \quad , \quad c_\alpha = \partial_\alpha S \ .$$

Equation (6.3) is satisfied by

(6.5) $$\Phi_a = c_b \overset{b}{u}_a$$

where $(c_b) \in R^m$, $c_m = 2e$. Equations (6.5) are nothing but (6.2), since $\partial_a S = \dfrac{dS_a}{dy^a}$. It can be directly verified that, when the real parameters $(c_b ; c_\alpha)$ range in a suitable domain of R^n , the solution of the geodesic Hamilton-Jacobi equation obtained by equations (6.2) is complete. (q.e.d.)

In section 3 we pointed out that a complete integral S can be also represented by n equations of the kind $F_i = c_i$, where (F_i) are n vertically independent functions in involution. From this point of view we can observe, through (6.4) and (6.5) (which gives $c_b = \underset{b}{u}^a \Phi_a$), that:

(6.6) In normal separable coordinates (y^i) the complete integral S is defined by the n functions:

(6.7) $$\begin{cases} F_b = \underset{b}{u}^{\bar{a}} (p_{\bar{a}})^2 + 2\, \underset{b}{u}^{\bar{a}} \theta_{\bar{a}}^{\alpha} p_{\bar{a}} p_\alpha + \underset{b}{u}^{a} \zeta_{a}^{\alpha\beta} p_\alpha p_\beta & , \\ F_\alpha = p_\alpha & . \end{cases}$$

where (p_i) are the corresponding momenta.

By the general theory developed in section 3 we also know that functions (6.7) are vertically independent and in $\mathcal{S}$-involution with respect to the coordinates (y^i) (and to any other $\mathcal{S}$-compatible system). Therefore we realize that the function group associated with a $\mathcal{S}_r$-structure on a Riemannian manifold is generated by r linear first integrals and n − r (homogeneous) quadratic first integrals. These first integrals are of course determined up to reversible transformations of the kind:

(6.8) $$F'_c = k_c^b F_b + k_c^{\alpha\beta} F_\alpha F_\beta \quad , \qquad F'_\rho = k_\rho^\alpha F_\alpha \quad ,$$

where all the coefficients are constant.

The above remarks can be translated in terms of Killing vectors and Killing tensors: a $\mathcal{S}_r$-structure at a point x_0 of a Riemannian manifold gives rise (in a neighborhood of x_0) to r K-vectors (see also theorem (4.7)) and n - r K-tensors of order 2, which commute in the Schouten-Nijenhuis Lie algebra of the contravariant symmetric tensors .

If the contravariant components (g^{ij}) in normal separable coordinates are known, finding a Stäckel matrix $(\underset{b}{u}^a)$ and the functions $(\theta^\alpha_{\tilde{a}})$, $(\zeta^{\alpha\beta}_a)$ entering the representation (5.3) is a pure algebraic problem. When these functions are known, by $(6.7)_1$ one can immediately write the quadratic first integrals associated with the $\mathcal{S}$-structure (from the choice $c_m = 2e$ it follows in particular $F_m = 2H$), or the corresponding K-tensors:

$$(6.9) \qquad K_b = \underset{b}{u}^{\tilde{a}}\, \partial_{\tilde{a}} \otimes \partial_{\tilde{a}} + \underset{b}{u}^{\tilde{a}} \theta^\alpha_{\tilde{a}} (\partial_{\tilde{a}} \otimes \partial_\alpha + \partial_\alpha \otimes \partial_{\tilde{a}}) + \underset{b}{u}^a\, \zeta^{\alpha\beta}_a\, \partial_\alpha \otimes \partial_\beta \qquad \left(\partial_i = \frac{\partial}{\partial y^i}\right) \quad .$$

However, the problem of finding the matrix $(\underset{b}{u}^a)$ can be simplified by knowing how to express, in a simple manner, the elements of a generic Stäckel matrix (of the required order) in terms of functions depending on a single variable, that is to say by knowing what can be called a <u>canonical</u> <u>representation</u> of a Stäckel matrix of a certain order. A canonical representation of a Stäckel matrix is of course obtainable by the algebraic relations between its elements and the elements of the inverse matrix, which are indeed functions of a single variable (definition (5.1)). For instance (see [14]), a 2×2 Stäckel matrix $(\underset{b}{u}^a)$, such that $\underset{2}{u}^1 \neq 0$ and $\underset{2}{u}^2 \neq 0$, can always be represented as follows:

$$\underset{1}{u}^1 = \frac{\psi_1 \varphi_2}{\varphi_1 + \varphi_2} \quad , \quad \underset{1}{u}^2 = \frac{-\psi_2 \varphi_1}{\varphi_1 + \varphi_2} \quad ,$$

$$\underset{2}{u}^1 = \frac{\psi_1}{\varphi_1 + \varphi_2} \quad , \quad \underset{2}{u}^2 = \frac{\psi_2}{\varphi_1 + \varphi_2} \quad ,$$

where (ψ_1, φ_1) and (ψ_2, φ_2) are functions of x^1 and x^2 respectively.

The so called <u>canonical</u> <u>forms</u> of the K-tensors (in particular, of the metric tensor) corresponding to a $\mathcal{S}_{n-m}$-structure correspond to canonical representations of the Stäckel matrices of order m. Canonical forms for $\mathcal{S}$-structures of class n - 2 and n - 3 are discussed in [14] and [15] respectively, in the case of index zero.

7.- Orthogonal separability structures.

The separability of the geodesic Hamilton-Jacobi equation can be considered in the particular case of orthogonal coordinates: $g^{ij}=0$ for $i \neq j$. The first fundamental result in this field was obtained by Stäckel at the end of the last century [9, 16]. (In fact, he considered a more general Hamiltonian including an additional "potential" function, as suggested by mechanics). Bearing in mind definition (5.1), Stäckel's theorem can be stated as follows:

(7.1) Theorem.- An orthogonal coordinate system (x^i) is separable if and only if $g^{ii} = \underset{n}{\varphi}^{i}$, where $(\underset{n}{\varphi}^{i})$ is the row of a Stäckel matrix $(\underset{k}{\varphi}^{i})$ in the variables (x^i). In this case the functions $F_k = \underset{k}{\varphi}^{i}\,(p_i)^2$ are first integrals in involution and vertically independent.

The second part of this theorem can be considered as a corollary of the following proposition concerning Stäckel matrices.

(7.2) A $n \times n$ matrix $(\underset{k}{\varphi}^{i})$ of C^1 real functions of n variables (x^i) is a Stäckel matrix if and only if the functions $F_k = \underset{k}{\varphi}^{i}\,(p_i)^2$ are vertically independent and in involution.

It is clear that functions (F_k) are interpreted as functions on the cotangent bundle of the domain of definition of the matrix, and moreover that they are vertically independent if and only if the matrix is regular (we exclude of course the points of the zero section: $p_i = 0$). If we assume that (F_k) are in involution, we see at once that they are in particular in $\mathscr{S}$-involution, hence: $\partial_i \underset{k}{\varphi}^{j}\, \underset{h}{\varphi}^{i} = \partial_i \underset{h}{\varphi}^{j}\, \underset{k}{\varphi}^{i}$ (i n.s.). If we multiply this relation by $\overset{h}{\varphi}_{\ell}\, \overset{k}{\varphi}_{m}$ (which are the elements of the inverse matrix, see definition (5.1)), by the summation over the repeated indices h and k, we obtain: $\delta^i_\ell\, \underset{k}{\varphi}^{j}\, \partial_i \overset{k}{\varphi}_{m} = \delta^i_m\, \underset{h}{\varphi}^{j}\, \partial_i \overset{h}{\varphi}_{\ell}$; hence, by setting in particular $m=i$, it follows: $\delta^i_\ell\, \partial_i \overset{h}{\varphi}_{i} = \partial_i \overset{h}{\varphi}_{\ell}$, i.e. $\partial_i \overset{h}{\varphi}_{\ell} = 0$ for $i \neq \ell$. This means that $\overset{h}{\varphi}_{\ell}$ is a function of x^ℓ only, hence that $(\underset{k}{\varphi}^{i})$ is a Stäckel matrix. Conversely, if $(\underset{k}{\varphi}^{i})$ is a Stäckel matrix, the following identities hold: $\partial_i \underset{k}{\varphi}^{j} = - \underset{k}{\varphi}^{i}\, \partial_i \overset{\ell}{\varphi}_{i}\, \underset{\ell}{\varphi}^{j}$ (i n.s.). Then: $\partial_i \underset{k}{\varphi}^{j}\, \underset{h}{\varphi}^{i} = - \underset{k}{\varphi}^{i}\, \underset{h}{\varphi}^{i}\, \partial_i \overset{\ell}{\varphi}_{i}\, \underset{\ell}{\varphi}^{j}$ and the Poisson bracket $\{F_h, F_k\}$ vanishes.

From the point of view of the separability structures theory, the subject acquires new aspects.

(7.3) Definition.- A $\mathscr{S}$-structure on a Riemannian manifold is said to be orthogonal if it contains orthogonal separable charts, i.e. if there exist representative charts with orthogonal coordinates.

By theorem (4.7), proposition (4.24), definition (4.26) and theorem (4.29) we see at once that:

(7.4) i) $\mathscr{S}$-structures of class $n = \dim M$ and 0 are orthogonal; ii) $\mathscr{S}$-structures of class 1 and index 0 are orthogonal; iii) orthogonal $\mathscr{S}$-structures have index 0 .

The last statement is a consequence of the fact that among the orthogonal coordinates of an orthogonal $\mathscr{S}$-structure we can find normal separable coordinates. If (x^i) are orthogonal separable coordinates of a $\underset{r}{\mathscr{S}}$-structure, the first class coordinates (x^α) are defined by the conditions: $\partial_\alpha g_{ii} = 0$ (or, equivalently, $\partial_\alpha g^{ii} = 0$) for $i \neq \alpha$. On the other hand, the functions (B_α^β) defined in (4.8) reduce simply to:

$$B_\alpha^\beta = \delta_\alpha^\beta B_\alpha \quad , \quad B_\alpha = \tfrac{1}{2} g^{\alpha\alpha} \partial_\alpha g_{\alpha\alpha} \quad (\alpha \text{ n.s.}). \tag{7.5}$$

We know by the general theory developed in section 2 that these functions do not depend on the second class coordinates (x^a) (see $(2.15)_2$). Now, the integrable system (2.16) becomes: $\partial_\alpha p_\beta = 0$ $(\beta \neq \alpha)$, $\partial_\alpha p_\alpha = B_\alpha p_\alpha$ (α n.s.), hence each B_α is a function of the corresponding variable x^α only. Thus, we conclude that the process shown in the proof of theorem (2.20) for finding ignorable coordinates simply reduces to a rescaling of the first class coordinates:

$$dy^\alpha = (\exp \int B_\alpha dx^\alpha)\, dx^\alpha \quad (\alpha \text{ n.s.}) \tag{7.6}$$

By leaving the second class coordinates unchanged $(y^a = x^a)$, we obtain normal separable coordinates (y^i) which are still orthogonal. Then, accordingly to theorem (5.2), the corresponding contravariant components $(\overset{ii}{g})$ can be put in the form:

$$\overset{aa}{g} = \underset{m}{u}{}^a \quad , \quad \overset{\alpha\alpha}{g} = \zeta_a^\alpha \underset{m}{u}{}^a \tag{7.7}$$

where $(\underset{m}{u}{}^a)$ is a row of a Stäckel matrix $(\underset{b}{u}{}^a)$ and (ζ_a^α) are functions of the coordinate corresponding to the lower index only.

8.- Separability structures with zero index.

The existence of separability structures on a Riemannian manifold is characterized by some local geometrical properties of the manifold itself. In this last section we consider the case of $\mathcal{S}$-structures with zero index.

(8.1) Theorem.- At a point x_0 of a n-dimensional Riemannian manifold (M,g) there exists a $\mathcal{S}_{r;0}$-structure if and only if, in a neighborhood of x_0, the following conditions hold:

i) there exist r K-vectors (X_α) and $n-r$ K-tensors of order 2 (K_a) which commute: $[X_\alpha, X_\beta] = 0$, $[X_\alpha, K_a] = 0$, $[K_a, K_b] = 0$ $(a,b = 1,\dots,n-r;\ \alpha,\beta = n-r+1,\dots,n)$, and such that the corresponding first integrals are functionally independent;

ii) the K-tensors have $n-r$ common eigenvectorfields (X_a) such that the n vectors $(X_i) = (X_a; X_\alpha)$ $(i = 1,\dots,n)$ are linearly independent and, moreover, $g(X_a, X_i) = 0$ for $i \neq a$ and $[X_a, X_i] = 0$;

iii) for each vector X_a, in the set (X_i) there are two vectors different from X_a such that $L_{X_a} g\,(X_i, X_j) \neq 0$.

Proof.- Let us assume that conditions i) and ii) are satisfied. The vector fields (X_i) give rise, in a neighborhood of x_0, to a coordinate system (x^i) such that $\partial_i = X_i$ and $g_{ai} = 0$ for $i \neq a$. Since (∂_a) are eigenvectors of the K-tensors (K_a), their components (K_a^{ij}) satisfy the conditions $K_a^{bi} = 0$ for $b \neq i$. If (p_i) are the momenta corresponding to the coordinates (x^i), then $F_a = K_a^{bb}(p_b)^2 + K_a^{\alpha\beta} p_\alpha p_\beta$ and $F_\alpha = p_\alpha$ are the first integrals corresponding to the K-tensors (K_a) and the K-vectors (X_α) respectively. The commutation relations $\{F_\alpha, F_a\} = 0$ imply $\partial_\alpha F_a = 0$, hence $\partial_\alpha K_a^{bb} = 0$ and $\partial_\alpha K_a^{\beta\gamma} = 0$. Since we have also $\partial_i F_\alpha = 0$, the matrix $(\partial_i F_j, \partial^i F_j)$ has maximal rank if and only if the matrix $(\partial^i F_j)$ has maximal rank. Thus, the hypotesis of independence of (F_i) implies the vertical independence. On the other hand, the commutation relations $\{F_a, F_b\} = 0$ can be written as follows:

$$(\partial_d K_a^{cc} K_b^{dd} - \partial_d K_b^{cc} K_a^{dd})(p_c)^2 p_d - (\partial_d K_a^{\alpha\beta} K_b^{dd} - \partial_d K_b^{\alpha\beta} K_a^{dd})\, p_\alpha p_\beta p_d = 0 .$$

This implies:

$$\partial_d K_a^{cc} K_b^{dd} - \partial_d K_b^{cc} K_a^{dd} = 0 , \quad \partial_d K_a^{\alpha\beta} K_b^{dd} - \partial_d K_b^{\alpha\beta} K_a^{dd} = 0 \qquad (d \text{ n.s.})$$

Hence: $\partial_d F_a \partial^d F_b = \partial_d F_b \partial^d F_a$ (d n.s.). Then we can see that all the functions (F_i) are in $\mathcal{S}$-involution with respect to the coordinates (x^i). Since they are also vertically independent and $\{H,F_i\} = 0$, by theorem (3.6) it is proved that the coordinates (x^i) are separable. In particular the coordinates (x^α) (which correspond to the K-vectors (X_α)) are ignorable, hence of first class. The remaining coordinates (x^a) are of second class because of condition iii) , which can now be written: $\partial_a g_{ij} = 0$ for at least a pair of indices $i, j \neq a$ (see definition (4.4)). Then the coordinates (x^i) define a $\mathscr{S}$-structure of class r and index 0 .

Conversely, if at x_0 there exists a $\mathscr{S}_{r;0}$-structure, we can consider a normal separable coordinate system (x^i). Then the vectors $X_i = \partial_i$ and the $n-r$ K-tensors constructed by the method shown in section 6 (see formula (6.9) with no index of the kind $\bar{a}$)obviously satisfy conditions i), ii) and iii). (q.e.d.)

(8.2) Remark.- Condition iii) guarantees that the class of the $\mathscr{S}$-structure defined by conditions i) and ii) is exactly the number of the K-vectors (X_α). Without this condition the class could be greater. Let us assume, for instance, that for a vector X_a condition iii) is not satisfied. This means of course that the corresponding coordinate x^a is of first class. Since $g^{ai} = 0$ for $i \neq a$ (x^a is "orthogonal" to the remaining coordinates), by following a reasoning analogous to that described in the last part of the preceding section, we can conlcude that there exists a rescaling of the coordinate x^a leading to an ignorable coordinate, or, in other words, that there exists a function f_a , depending on x^a only, such that $f_a X_a$ is a K-vector which still commutes with the remaining vectors $(X_i, i \neq a)$. Hence, condition iii) in theorem (8.1) can be substituted by the following one: iii') for each vector X_a there exists no function f_a such that $L_{X_i} f_a = 0$ for $i \neq a$ and $L_{f_a X_a} g = 0$.

(8.3) Remark.- Conditions i), ii) and iii) in theorem (8.1) can be used to define a $\mathscr{S}_{r;0}$-structure in a global sense.

The K-vectors (X_α) occurring in theorem (8.1) define on M a local foliation of r-dimensional submanifolds which are flat in the induced metric (in fact they are the orbits of the Abelian group of local isometries considered in theorem (4.7)). On the other hand, vectors (X_a) define a complementary orthogonal foliation of $(n-r)$-dimensional submanifolds, which are of course isometric; the induced metric is $g_{aa}\, dx^a \otimes dx^a$, being (x^a) the orthogonal coordinates corresponding to vectors (X_a). The contravariant components of the induced metric coincides with the corresponding components (g^{aa}) of g, and, by theorem (5.2) restricted to the case of index 0, they have the form $g^{aa} = \underset{m}{u}^a$ where $(\underset{m}{u}^a)$ is a line of a Stäckel matrix. Hence, by Stäckel's theorem (7.1), we can see that the coordinates (x^a) are separable in the induced metric. Thus, we have proved the following theorem (see also [17]).

(8.4) Theorem.- A n-dimensional Riemannian manifold which admits a $\mathcal{S}_{r;o}$-structure (at a point x_o) has (locally) two orthogonal transversal foliations, one made of r-dimensional submanifolds which are flat in the induced metric and the other made of (n - r)-dimensional isometric submanifolds admitting an orthogonal $\mathcal{S}$-structure.

* * *

References

[1] .- T.Levi-Civita, Math. Ann. 59, 383 - 397 (1904).

[2] .- S.Benenti & M.Francaviglia, The theory of separability of the Hamilton-Jacobi equation and its applications to general relativity, in "Einstein Memorial Volume", A.Held Ed., Ch. 14 , Plenum, New-York , forthcoming.

[3] .- A.Huaux, Ann. Mat. Pura Appl. 108, 251 - 282 (1976).

[4] .- R.Abraham & J.Marsden, Foundations of Mechanics 2nd ed., Benjamin-Cummings(1978).

[5] .- S.Benenti & W.M.Tulczyjew, The geometrical meaning and globalization of the Hamilton-Jacobi method, Proc. of The Intern. Colloquium "Differential Geometrical Methods in Mathematical Physics", Aix-en-Provence 1979, this volume.

[6] .- A.Dall'Acqua, Rend. Circ. Mat. Palermo 33, 341 - 351 (1912).

[7] .- S.Benenti, Structures de séparabilité sur une variété riemannienne de signature quelconque, C. R. Acad. Sci., forthcoming.

[8] .- S.Benenti, Rend. Sem. Mat. Univ. Pol. Torino 34, 431 - 463 (1975/76).

[9] .- C.Agostinelli, Mem. Reale Acc. Sci. Torino 69 (I), 1 - 54 (1937).

[10] .- P.Stäckel, Math. Ann. 42, 537 - 563 (1893).

[11] .- S.Benenti, Rep. Math. Phys. 12, 311 - 316 (1977).

[12] .- P.Havas, J. Math. Phys. 16, 1461 - 1468 (1975).

[13] .- M.S.Iarov-Iarovoi, J. Appl. Math. Mech. 27, 1499 - 1519 (1964).

[14] .- S.Benenti & M.Francaviglia, Gen. Rel. Grav. 10, 79 - 92 (1979).

[15] .- S.Benenti & M.Francaviglia, Ann. Inst. Henri Poincaré 30, 143 - 157 (1979).

[16] .- P.Stäckel, C. R. Acad. Sci. 116, 485 - 487 (1893).

[17] .- S.Benenti, C. R. Acad. Sci. 283, 215 - 218 (1976).

The present study has been sponsored by Consiglio Nazionale delle Ricerche - Gruppo Nazionale per la Fisica Matematica (Italian National Research Council - National Group for Mathematical Physics).

Vol. 670: Fonctions de Plusieurs Variables Complexes III, Proceedings, 1977. Edité par F. Norguet. XII, 394 pages. 1978.

Vol. 671: R. T. Smythe and J. C. Wierman, First-Passage Perculation on the Square Lattice. VIII, 196 pages. 1978.

Vol. 672: R. L. Taylor, Stochastic Convergence of Weighted Sums of Random Elements in Linear Spaces. VII, 216 pages. 1978.

Vol. 673: Algebraic Topology. Proceedings 1977. Edited by P. Hoffman, R. Piccinini and D. Sjerve. VI, 278 pages. 1978.

Vol. 674: Z. Fiedorowicz and S. Priddy, Homology of Classical Groups Over Finite Fields and Their Associated Infinite Loop Spaces. VI, 434 pages. 1978.

Vol. 675: J. Galambos and S. Kotz, Characterizations of Probability Distributions. VIII, 169 pages. 1978.

Vol. 676: Differential Geometrical Methods in Mathematical Physics II, Proceedings, 1977. Edited by K. Bleuler, H. R. Petry and A. Reetz. VI, 626 pages. 1978.

Vol. 677: Séminaire Bourbaki, vol. 1976/77, Exposés 489–506. IV, 264 pages. 1978.

Vol. 678: D. Dacunha-Castelle, H. Heyer et B. Roynette. Ecole d'Eté de Probabilités de Saint-Flour. VII-1977. Edité par P. L. Hennequin. IX, 379 pages. 1978.

Vol. 679: Numerical Treatment of Differential Equations in Applications, Proceedings, 1977. Edited by R. Ansorge and W. Törnig. IX, 163 pages. 1978.

Vol. 680: Mathematical Control Theory, Proceedings, 1977. Edited by W. A. Coppel. IX, 257 pages. 1978.

Vol. 681: Séminaire de Théorie du Potentiel Paris, No. 3, Directeurs: M. Brelot, G. Choquet et J. Deny. Rédacteurs: F. Hirsch et G. Mokobodzki. VII, 294 pages. 1978.

Vol. 682: G. D. James, The Representation Theory of the Symmetric Groups. V, 156 pages. 1978.

Vol. 683: Variétés Analytiques Compactes, Proceedings, 1977. Edité par Y. Hervier et A. Hirschowitz. V, 248 pages. 1978.

Vol. 684: E. E. Rosinger, Distributions and Nonlinear Partial Differential Equations. XI, 146 pages. 1978.

Vol. 685: Knot Theory, Proceedings, 1977. Edited by J. C. Hausmann. VII, 311 pages. 1978.

Vol. 686: Combinatorial Mathematics, Proceedings, 1977. Edited by D. A. Holton and J. Seberry. IX, 353 pages. 1978.

Vol. 687: Algebraic Geometry, Proceedings, 1977. Edited by L. D. Olson. V, 244 pages. 1978.

Vol. 688: J. Dydak and J. Segal, Shape Theory. VI, 150 pages. 1978.

Vol. 689: Cabal Seminar 76–77, Proceedings, 1976–77. Edited by A.S. Kechris and Y. N. Moschovakis. V, 282 pages. 1978.

Vol. 690: W. J. J. Rey, Robust Statistical Methods. VI, 128 pages. 1978.

Vol. 691: G. Viennot, Algèbres de Lie Libres et Monoïdes Libres. III, 124 pages. 1978.

Vol. 692: T. Husain and S. M. Khaleelulla, Barrelledness in Topological and Ordered Vector Spaces. IX, 258 pages. 1978.

Vol. 693: Hilbert Space Operators, Proceedings, 1977. Edited by J. M. Bachar Jr. and D. W. Hadwin. VIII, 184 pages. 1978.

Vol. 694: Séminaire Pierre Lelong – Henri Skoda (Analyse) Année 1976/77. VII, 334 pages. 1978.

Vol. 695: Measure Theory Applications to Stochastic Analysis, Proceedings, 1977. Edited by G. Kallianpur and D. Kölzow. XII, 261 pages. 1978.

Vol. 696: P. J. Feinsilver, Special Functions, Probability Semigroups, and Hamiltonian Flows. VI, 112 pages. 1978.

Vol. 697: Topics in Algebra, Proceedings, 1978. Edited by M. F. Newman. XI, 229 pages. 1978.

Vol. 698: E. Grosswald, Bessel Polynomials. XIV, 182 pages. 1978.

Vol. 699: R. E. Greene and H.-H. Wu, Function Theory on Manifolds Which Possess a Pole. III, 215 pages. 1979.

Vol. 700: Module Theory, Proceedings, 1977. Edited by C. Faith and S. Wiegand. X, 239 pages. 1979.

Vol. 701: Functional Analysis Methods in Numerical Analysis, Proceedings, 1977. Edited by M. Zuhair Nashed. VII, 333 pages. 1979.

Vol. 702: Yuri N. Bibikov, Local Theory of Nonlinear Analytic Ordinary Differential Equations. IX, 147 pages. 1979.

Vol. 703: Equadiff IV, Proceedings, 1977. Edited by J. Fábera. XIX, 441 pages. 1979.

Vol. 704: Computing Methods in Applied Sciences and Engineering, 1977, I. Proceedings, 1977. Edited by R. Glowinski and J. L. Lions. VI, 391 pages. 1979.

Vol. 705: O. Forster und K. Knorr, Konstruktion verseller Familien kompakter komplexer Räume. VII, 141 Seiten. 1979.

Vol. 706: Probability Measures on Groups, Proceedings, 1978. Edited by H. Heyer. XIII, 348 pages. 1979.

Vol. 707: R. Zielke, Discontinuous Čebyšev Systems. VI, 111 pages. 1979.

Vol. 708: J. P. Jouanolou, Equations de Pfaff algébriques. V, 255 pages. 1979.

Vol. 709: Probability in Banach Spaces II. Proceedings, 1978. Edited by A. Beck. V, 205 pages. 1979.

Vol. 710: Séminaire Bourbaki vol. 1977/78, Exposés 507–524. IV, 328 pages. 1979.

Vol. 711: Asymptotic Analysis. Edited by F. Verhulst. V, 240 pages. 1979.

Vol. 712: Equations Différentielles et Systèmes de Pfaff dans le Champ Complexe. Edité par R. Gérard et J.-P. Ramis. V, 364 pages. 1979.

Vol. 713: Séminaire de Théorie du Potentiel, Paris No. 4. Edité par F. Hirsch et G. Mokobodzki. VII, 281 pages. 1979.

Vol. 714: J. Jacod, Calcul Stochastique et Problèmes de Martingales. X, 539 pages. 1979.

Vol. 715: Inder Bir S. Passi, Group Rings and Their Augmentation Ideals. VI, 137 pages. 1979.

Vol. 716: M. A. Scheunert, The Theory of Lie Superalgebras. X, 271 pages. 1979.

Vol. 717: Grosser, Bidualräume und Vervollständigungen von Banachmoduln. III, 209 pages. 1979.

Vol. 718: J. Ferrante and C. W. Rackoff, The Computational Complexity of Logical Theories. X, 243 pages. 1979.

Vol. 719: Categorial Topology, Proceedings, 1978. Edited by H. Herrlich and G. Preuß. XII, 420 pages. 1979.

Vol. 720: E. Dubinsky, The Structure of Nuclear Fréchet Spaces. V, 187 pages. 1979.

Vol. 721: Séminaire de Probabilités XIII. Proceedings, Strasbourg, 1977/78. Edité par C. Dellacherie, P. A. Meyer et M. Weil. VII, 647 pages. 1979.

Vol. 722: Topology of Low-Dimensional Manifolds. Proceedings, 1977. Edited by R. Fenn. VI, 154 pages. 1979.

Vol. 723: W. Brandal, Commutative Rings whose Finitely Generated Modules Decompose. II, 116 pages. 1979.

Vol. 724: D. Griffeath, Additive and Cancellative Interacting Particle Systems. V, 108 pages. 1979.

Vol. 725: Algèbres d'Opérateurs. Proceedings, 1978. Edité par P. de la Harpe. VII, 309 pages. 1979.

Vol. 726: Y.-C. Wong, Schwartz Spaces, Nuclear Spaces and Tensor Products. VI, 418 pages. 1979.

Vol. 727: Y. Saito, Spectral Representations for Schrödinger Operators With Long-Range Potentials. V, 149 pages. 1979.

Vol. 728: Non-Commutative Harmonic Analysis. Proceedings, 1978. Edited by J. Carmona and M. Vergne. V, 244 pages. 1979.

Vol. 729: Ergodic Theory. Proceedings, 1978. Edited by M. Denker and K. Jacobs. XII, 209 pages. 1979.

Vol. 730: Functional Differential Equations and Approximation of Fixed Points. Proceedings, 1978. Edited by H.-O. Peitgen and H.-O. Walther. XV, 503 pages. 1979.

Vol. 731: Y. Nakagami and M. Takesaki, Duality for Crossed Products of von Neumann Algebras. IX, 139 pages. 1979.

Vol. 732: Algebraic Geometry. Proceedings, 1978. Edited by K. Lønsted. IV, 658 pages. 1979.

Vol. 733: F. Bloom, Modern Differential Geometric Techniques in the Theory of Continuous Distributions of Dislocations. XII, 206 pages. 1979.

Vol. 734: Ring Theory, Waterloo, 1978. Proceedings, 1978. Edited by D. Handelman and J. Lawrence. XI, 352 pages. 1979.

Vol. 735: B. Aupetit, Propriétés Spectrales des Algèbres de Banach. XII, 192 pages. 1979.

Vol. 736: E. Behrends, M-Structure and the Banach-Stone Theorem. X, 217 pages. 1979.

Vol. 737: Volterra Equations. Proceedings 1978. Edited by S.-O. Londen and O. J. Staffans. VIII, 314 pages. 1979.

Vol. 738: P. E. Conner, Differentiable Periodic Maps. 2nd edition, IV, 181 pages. 1979.

Vol. 739: Analyse Harmonique sur les Groupes de Lie II. Proceedings, 1976–78. Edited by P. Eymard et al. VI, 646 pages. 1979.

Vol. 740: Séminaire d'Algèbre Paul Dubreil. Proceedings. 1977–78. Edited by M.-P. Malliavin. V, 456 pages. 1979.

Vol. 741: Algebraic Topology, Waterloo 1978. Proceedings. Edited by P. Hoffman and V. Snaith. XI, 655 pages. 1979.

Vol. 742: K. Clancey, Seminormal Operators. VII, 125 pages. 1979.

Vol. 743: Romanian-Finnish Seminar on Complex Analysis. Proceedings, 1976. Edited by C. Andreian Cazacu et al. XVI, 713 pages. 1979.

Vol. 744: I. Reiner and K. W. Roggenkamp, Integral Representations. VIII, 275 pages. 1979.

Vol. 745: D. K. Haley, Equational Compactness in Rings. III, 167 pages. 1979.

Vol. 746: P. Hoffman, τ-Rings and Wreath Product Representations. V, 148 pages. 1979.

Vol. 747: Complex Analysis, Joensuu 1978. Proceedings, 1978. Edited by I. Laine, O. Lehto and T. Sorvali. XV, 450 pages. 1979.

Vol. 748: Combinatorial Mathematics VI. Proceedings, 1978. Edited by A. F. Horadam and W. D. Wallis. IX, 206 pages. 1979.

Vol. 749: V. Girault and P.-A. Raviart, Finite Element Approximation of the Navier-Stokes Equations. VII, 200 pages. 1979.

Vol. 750: J. C. Jantzen, Moduln mit einem höchsten Gewicht. III, 195 Seiten. 1979.

Vol. 751: Number Theory, Carbondale 1979. Proceedings. Edited by M. B. Nathanson. V, 342 pages. 1979.

Vol. 752: M. Barr, *-Autonomous Categories. VI, 140 pages. 1979.

Vol. 753: Applications of Sheaves. Proceedings, 1977. Edited by M. Fourman, C. Mulvey and D. Scott. XIV, 779 pages. 1979.

Vol. 754: O. A. Laudal, Formal Moduli of Algebraic Structures. III, 161 pages. 1979.

Vol. 755: Global Analysis. Proceedings, 1978. Edited by M. Grmela and J. E. Marsden. VII, 377 pages. 1979.

Vol. 756: H. O. Cordes, Elliptic Pseudo-Differential Operators – An Abstract Theory. IX, 331 pages. 1979.

Vol. 757: Smoothing Techniques for Curve Estimation. Proceedings, 1979. Edited by Th. Gasser and M. Rosenblatt. V, 245 pages. 1979.

Vol. 758: C. Năstăsescu and F. Van Oystaeyen; Graded and Filtered Rings and Modules. X, 148 pages. 1979.

Vol. 759: R. L. Epstein, Degrees of Unsolvability: Structure and Theory. XIV, 216 pages. 1979.

Vol. 760: H.-O. Georgii, Canonical Gibbs Measures. VIII, 190 pages. 1979.

Vol. 761: K. Johannson, Homotopy Equivalences of 3-Manifolds with Boundaries. 2, 303 pages. 1979.

Vol. 762: D. H. Sattinger, Group Theoretic Methods in Bifurcation Theory. V, 241 pages. 1979.

Vol. 763: Algebraic Topology, Aarhus 1978. Proceedings, 1978. Edited by J. L. Dupont and H. Madsen. VI, 695 pages. 1979.

Vol. 764: B. Srinivasan, Representations of Finite Chevalley Groups. XI, 177 pages. 1979.

Vol. 765: Padé Approximation and its Applications. Proceedings, 1979. Edited by L. Wuytack. VI, 392 pages. 1979.

Vol. 766: T. tom Dieck, Transformation Groups and Representation Theory. VIII, 309 pages. 1979.

Vol. 767: M. Namba, Families of Meromorphic Functions on Compact Riemann Surfaces. XII, 284 pages. 1979.

Vol. 768: R. S. Doran and J. Wichmann, Approximate Identities and Factorization in Banach Modules. X, 305 pages. 1979.

Vol. 769: J. Flum, M. Ziegler, Topological Model Theory. X, 151 pages. 1980.

Vol. 770: Séminaire Bourbaki vol. 1978/79 Exposés 525–542. IV, 341 pages. 1980.

Vol. 771: Approximation Methods for Navier-Stokes Problems. Proceedings, 1979. Edited by R. Rautmann. XVI, 581 pages. 1980.

Vol. 772: J. P. Levine, Algebraic Structure of Knot Modules. XI, 104 pages. 1980.

Vol. 773: Numerical Analysis. Proceedings, 1979. Edited by G. A. Watson. X, 184 pages. 1980.

Vol. 774: R. Azencott, Y. Guivarc'h, R. F. Gundy, Ecole d'Eté de Probabilités de Saint-Flour VIII-1978. Edited by P. L. Hennequin. XIII, 334 pages. 1980.

Vol. 775: Geometric Methods in Mathematical Physics. Proceedings, 1979. Edited by G. Kaiser and J. E. Marsden. VII, 257 pages. 1980.

Vol. 776: B. Gross, Arithmetic on Elliptic Curves with Complex Multiplication. V, 95 pages. 1980.

Vol. 777: Séminaire sur les Singularités des Surfaces. Proceedings, 1976-1977. Edited by M. Demazure, H. Pinkham and B. Teissier. IX, 339 pages. 1980.

Vol. 778: SK_1 von Schiefkörpern. Proceedings, 1976. Edited by P. Draxl and M. Kneser. II, 124 pages. 1980.

Vol. 779: Euclidean Harmonic Analysis. Proceedings, 1979. Edited by J. J. Benedetto. III, 177 pages. 1980.

Vol. 780: L. Schwartz, Semi-Martingales sur des Variétés, et Martingales Conformes sur des Variétés Analytiques Complexes. XV, 132 pages. 1980.

Vol. 781: Harmonic Analysis Iraklion 1978. Proceedings 1978. Edited by N. Petridis, S. K. Pichorides and N. Varopoulos. V, 213 pages. 1980.

Vol. 782: Bifurcation and Nonlinear Eigenvalue Problems. Proceedings, 1978. Edited by C. Bardos, J. M. Lasry and M. Schatzman. VIII, 296 pages. 1980.

Vol. 783: A. Dinghas, Wertverteilung meromorpher Funktionen in ein- und mehrfach zusammenhängenden Gebieten. Edited by R. Nevanlinna and C. Andreian Cazacu. XIII, 145 pages. 1980.

Vol. 784: Séminaire de Probabilités XIV. Proceedings, 1978/79. Edited by J. Azéma and M. Yor. VIII, 546 pages. 1980.

Vol. 785: W. M. Schmidt, Diophantine Approximation. X, 299 pages. 1980.

Vol. 786: I. J. Maddox, Infinite Matrices of Operators. V, 122 pages. 1980.

MIX
Papier aus verantwortungsvollen Quellen
Paper from responsible sources
FSC® C105338

If you have any concerns about our products,
you can contact us on
ProductSafety@springernature.com

In case Publisher is established outside the EU,
the EU authorized representative is:
Springer Nature Customer Service Center GmbH
Europaplatz 3, 69115 Heidelberg, Germany

Printed by Libri Plureos GmbH
in Hamburg, Germany